Interdisciplinary Applied Mathematics

Volume 5

Editors
S.S. Antman   **J.E. Marsden**
L. Sirovich

Geophysics and Planetary Sciences

Mathematical Biology
L. Glass, J.D. Murray

Mechanics and Materials
R.V. Kohn

Systems and Control
S.S. Sastry, P.S. Krishnaprasad

Problems in engineering, computational science, and the physical and biological sciences are using increasingly sophisticated mathematical techniques. Thus, the bridge between the mathematical sciences and other disciplines is heavily traveled. The correspondingly increased dialog between the disciplines has led to the establishment of the series: *Interdisciplinary Applied Mathematics*.

The purpose of this series is to meet the current and future needs for the interaction between various science and technology areas on the one hand and mathematics on the other. This is done, firstly, by encouraging the ways that mathematics may be applied in traditional areas, as well as point towards new and innovative areas of applications; and, secondly, by encouraging other scientific disciplines to engage in a dialog with mathematicians outlining their problems to both access new methods and suggest innovative developments within mathematics itself.

The series will consist of monographs and high-level texts from researchers working on the interplay between mathematics and other fields of science and technology.

For other titles published in this series, go to
http://www.springer.com/series/1390

Rüdiger Seydel

Practical Bifurcation and Stability Analysis

Third Edition

Rüdiger Seydel
Universität zu Köln
Mathematisches Institut
Weyertal 86-90
50931 Köln
Germany
seydel@math.uni-koeln.de

Series Editors
S.S. Antman
Department of Mathematics and
    Institute for Physical Science and
    Technology
University of Maryland
College Park, MD 20742
USA
ssa@math.umd.edu

J.E. Marsden
Control and Dynamical Systems
Mail Code 107-81
California Institute of Technology
Pasadena, CA 91125
10122 Torino
USA
marsden@cds.caltech.edu

L. Sirovich
Department of Biomathenatics
Laboratory of Applied Mathematics
Mt. Sinai School of Medicine
Box 1012
NYC 10029
USA
Lawrence.Sirovich@mssm.edu

ISBN 978-1-4614-2530-4   e-ISBN 978-1-4419-1740-9
DOI 10.1007/978-1-4419-1740-9
Springer New York Dordrecht Heidelberg London

MSC 2010: 34C, 65P, 34A, 34B, 34D, 35B32, 37M, 37N, 65L07, 70Kxx, 70-08, 70K, 76E30, 92-08

Springer is part of Springer Science+Business Media (www.springer.com)

Preface

Fifteen years have elapsed after the second edition of *Practical Bifurcation and Stability Analysis* was published. During that time period the field of computational bifurcation has become mature. Today, bifurcation mechanisms are widely accepted as decisive phenomena for explaining and understanding stability and structural change. Along with the high level of sophistication that bifurcation analysis has reached, the research on basic computational bifurcation algorithms is essentially completed, at least in ordinary differential equations. The focus has been shifting from mathematical foundations towards applications.

The evolution *from equilibrium to chaos* has become commonplace and is no longer at the cutting edge of innovation. But the corresponding methods of *practical bifurcation and stability analysis* remain indispensable instruments in all applications of mathematics. This constant need for practical bifurcation and stability analysis has stimulated an effort to maintain this book on a present-day level. The author's endeavor has resulted in this third edition. It is based on more than three decades of practical experience with the subject, and on many courses given at several universities.

Like the previous editions, this third edition consists of three parts. In the first part (Chapters 1 to 3) an introduction into bifurcation and stability phenomena is given, basically restricted to models built of ordinary differential equations. Phenomena such as birth of limit cycles, hysteresis, or period doubling are explained. The second part (Chapters 4 to 7) introduces computational methods for analyzing bifurcation and stability. This includes continuation and branch switching as basic means. The final part (Chapters 8 and 9) gives qualitative insight that may help in understanding and assessing computational results. Such an interpretation of numerical results is based on singularity theory, catastrophe theory, and chaos theory.

This book emphasizes basic principles and shows the reader how the methods result from combining and, on occasion, modifying the underlying principles. The book is written to address the needs of scientists and engineers and to attract mathematicians. Mathematical formalism is kept to a minimum; the style is not technical, and is often motivating rather than proving. Compelling examples and geometrical interpretations are essential ingredients in the style. Exercises and projects complete the text. The book is self-contained, assuming only basic knowledge in calculus. The extensive bibliography includes many references on analytical and numerical methods, applications in science and engineering, and software. The references may serve as first steps in finding additional material for further research.

The book attempts to provide a practical guide for the performance of parameter studies.

New in This Edition

This third edition has been partly reorganized. The main change is a newly written Chapter 3. The third chapter of the second edition was removed, part of its contents was added to the fourth chapter. The new Chapter 3 is devoted to applications and extensions of standard ODE approaches. It includes brief expositions on delay differential equations, on differential-algebraic equations, and on pattern formation. This last aspect is concentrating on reaction-diffusion problems with applications in nerve models. Finally, this new third chapter addresses the aspect of deterministic risk, which can be tied to bifurcation. Applications include production of blood cells, dry friction, a flip-flop circuit, Turing bifurcation, and an electric power generator.

In addition to the new Chapter 3, several new sections have been inserted. In Chapter 5, the new Section 5.5 summarizes the information on second-order derivatives. In Chapter 7, on periodic orbits, the Section 7.6 on numerical aspects of bifurcation was enlarged. In Chapter 9, the section on fractal dimensions has been extended, and a new section has been added on the control of chaos, with focus on the OGY method.

Apart from these expanded sections, the entire book has been thoroughly reworked and revised. There are many new figures, while other figures have been improved. A considerable number of new references guide the reader to some more recent research or applications. The additions of this third edition are substantial; this may be quantified by the increase in the number of pages (+16%), figures (+19%), or references (+22%). The author has attempted to follow the now generally adopted practice to use *branching* and *bifurcation* as synonyms.

How to Use the Book

A path is outlined listing those sections that provide the general introduction into bifurcation and stability. Readers without urgent interest in computational aspects may wish to concentrate on the following:

Sections 1.1 to 1.4;
all of Chapter 2;
part of Chapter 3;
part of Section 5.4.2, and Sections 5.5, 5.6.4, and 5.6.5;
Section 6.1, example in Section 6.4, and Section 6.8;

Sections 7.1 to 7.4, 7.7, and 7.8;
all of Chapter 8; and
all of Chapter 9.

Additional information and less important remarks are set in small print. On first reading, the reader may skip these parts without harm. Readers with little mathematical background are encouraged to read Appendices 1 to 3 first. Solutions to several of the exercises are given later in the text. References are not meant as required reading, but are hints to help those readers interested in further study. The figures framed in boxes are immediate output of numerical software.

I hope that this book inspires readers to perform their own experimental studies. The many examples and figures should provide a basis and motivation to start right away.

Köln, September 2009   *Rüdiger Seydel*

Contents

Notation

Problem-Inherent Variables

λ scalar parameter to be varied (bifurcation parameter)

$\mathbf{y}$ vector of state variables, vector function, solution of an equation

n number of components of vectors $\mathbf{y}$ or $\mathbf{f}$

$\mathbf{f}$ vector function, defines the dynamics of the problem
that is to be solved; typical equation $\mathbf{f}(\mathbf{y}, \lambda) = \mathbf{0}$

t independent variable, often time

$\dot{\mathbf{y}}$ derivative of $\mathbf{y}$ with respect to time, $\dot{\mathbf{y}} = d\mathbf{y}/dt$

$\mathbf{y}'$ derivative of $\mathbf{y}$ with respect to a general independent variable

a, b define an interval in which t varies, $a \leq t \leq b$

x spatial variable, may be scalar or vector with up to three components

$\mathbf{r}$ vector function, often used to define boundary conditions as
in $\mathbf{r}(\mathbf{y}(a), \mathbf{y}(b)) = \mathbf{0}$

T period in case of a periodic oscillation

γ additional scalar parameter

Notations for a General Analysis

In particular examples, several of the following meanings are sometimes superseded by a local meaning

Specific Versions of $\mathbf{y}$ and λ

λ_0 specific parameter value of a bifurcation point

$\mathbf{y}_0$ specific n-vector of a bifurcation point

y_i ith component of vector $\mathbf{y}$

$\mathbf{y}^j$ jth continuation step (j is not an exponent here), specific solution

λ_j specific parameter value, corresponds to $\mathbf{y}^j$

$\mathbf{y}^{(\nu)}$ iterates of a map. For example, Newton iterate;
for $\nu = 1, 2, \ldots$ sequence of vectors converging to a solution $\mathbf{y}$

$\mathbf{y}^s$ stationary solution ($\mathbf{y}_s$ in Sections 6.6 and 6.7)

Integers

k frequently, the kth component has a special meaning
N number of nodes of a discretization
$i,\ j,\quad l,m,\nu$ other integers
      (Note that i denotes the imaginary unit.)

Scalars

$[\mathbf{y}]$ a scalar measure of $\mathbf{y}$ (cf. Section 2.2)
ρ radius
ϑ angle
ω frequency
ϵ accuracy, error tolerance
δ distance between two solutions, or parameter
η value of a particular boundary condition
τ test function indicating bifurcation
Δ increment or decrement, sometimes acting as operator on the
      following variable; for instance, $\Delta\lambda$ means an increment in λ
s arclength
$u,\ v$ functions, often solutions of scalar differential equations
σ step length
p parameterization, or phase condition, or polynomial
c_i constants
μ $=\alpha+i\beta$ complex-conjugate eigenvalue
$\zeta,\ \xi$ further scalars with local meaning

Vectors

$\mathbf{z}$ n-vector (column) in various roles, as tangent, or initial vector of a
      trajectory, or emanating solution, or eigenvector
$\mathbf{z}^{tr}$ row vector (transposed column)
$\mathbf{d}$ difference between two n-vectors
$\mathbf{h}$ n-vector, solution of a linearization; $\mathbf{h}_0$ or $\bar{\mathbf{h}}$ are related n-vectors
$\mathbf{e}_i$ ith unit vector (cf. Appendix A.2)
φ $\varphi(t;\mathbf{z})$ is the trajectory starting at $\mathbf{z}$ (Eq. (7.7))
$\mathbf{w}$ eigenvector, also $\mathbf{w}^k$
μ $=\alpha+i\beta$ vector of eigenvalues
Λ vector of parameters
$\mathbf{Y}$ vector with more than n components, contains $\mathbf{y}$ as subvector
$\mathbf{F}$ vector with more than n components, contains $\mathbf{f}$ as subvector
$\mathbf{R}$ vector with more than n components, contains $\mathbf{r}$ as subvector
$\mathbf{P}$ map, Poincaré map
$\mathbf{q}$ argument of Poincaré map

n^2-Matrices (n rows, n columns)

I   identity matrix
J   Jacobian matrix $\mathbf{f_y}$ of first-order partial derivatives of $\mathbf{f}(\mathbf{y})$ w.r.t. $\mathbf{y}$
M   monodromy matrix
A, B derivatives of boundary conditions (Eq. (6.12))
E, $\mathbf{G}_j$ special matrices of multiple shooting (Eq. (6.21), Eq. (6.22))
$\varPhi$, Z fundamental solution matrices (cf. Section 7.2)
S   element of a group $\mathcal{G}$

Further Notations

Ω hypersurface
$\mathcal{M}$ manifold
$\mathcal{G}$ group, see Appendix A.7
$\in$ "in," element of a set, $\notin$ for "not in"
^{tr} as superscript means "transposed"
ln   natural logarithm
Re   real part
Im   imaginary part
∂ partial derivative
$\bar{\mathbf{y}}, \bar{\lambda}, \ \bar{\mathbf{h}}, \bar{\mathbf{z}}$ overbar characterizes approximations
∇u gradient of u (∇ is the "del" operator)
$\nabla^2 u$ Laplacian operator (summation of second-order derivatives)
$\nabla \cdot u$ divergence of u
$:=$ defining equation; the left side is "new" and is defined by the
         right-hand side; see, for example, Eq. (4.14)
$O(\sigma)$ terms of order of σ
$\| \ \|$ vector norm, see Appendix A.1

Abbreviations

t.h.o. terms of higher order
w.r.t. with respect to
DAE   differential-algebraic equation (cf. Section 3.3)
ODE   ordinary differential equation
OGY   Ott-Grebogi-Yorke method (cf. Section 9.6)
PDE   partial differential equation
UPO   unstable periodic orbit

1 Introduction and Prerequisites

1.1 A Nonmathematical Introduction

Every day of our lives we experience changes that occur either gradually or suddenly. We often characterize these changes as quantitative or qualitative, respectively. For example, consider the following simple experiment (Figure 1.1). Imagine a board supported at both ends, with a load on top. If the load λ is not too large, the board will take a bent shape with a deformation depending on the magnitude of λ and on the board's material properties (such as stiffness, K). This state of the board will remain *stable* in the sense that a small variation in the load λ (or in the stiffness K) leads to a state that is only slightly perturbed. Such a variation (described by Hooke's law) would be referred to as a quantitative change. The board is deformed within its elastic regime and will return to its original shape when the perturbation in λ is removed.

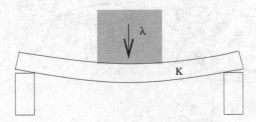

Fig. 1.1. Bending of a board

The situation changes abruptly when the load λ is increased beyond a certain *critical level* λ_0 at which the board breaks (Figure 1.2b). This sudden action is an example of a qualitative change; it will also take place when the material properties are changed beyond a certain limit (see Figure 1.2a). Suppose the shape of the board is modeled by some function (solution of an equation). Loosely speaking, we may say that there is a solution for load values $\lambda < \lambda_0$ and that this solution ceases to exist for $\lambda > \lambda_0$. The load λ and stiffness K are examples of *parameters*. The outcome of any experiment, any event, and any construction is controlled by parameters. The practical problem is to *control the state* of a system—that is, to find parameters such

R. Seydel, *Practical Bifurcation and Stability Analysis*,
Interdisciplinary Applied Mathematics 5, DOI 10.1007/978-1-4419-1740-9_1,
© Springer Science+Business Media, LLC 2010

that the state fulfills our requirements. This role of parameters is occasionally emphasized by terms such as *control parameter*, or *design parameter*. Varying a parameter can result in a transition from a quantitative change to a qualitative change. The following pairs of verbs may serve as illustrations:

$$\text{bend} \rightarrow \text{break}$$
$$\text{incline} \rightarrow \text{tilt over}$$
$$\text{stretch} \rightarrow \text{tear}$$
$$\text{inflate} \rightarrow \text{burst}.$$

The verbs on the left side stand for states that are stable under small perturbations; the response of each system is a quantitative one. This behavior ends abruptly at certain critical values of underlying parameters. The related drastic and irreversible change is reflected by the verbs on the right side. Close to a critical threshold the system becomes most sensitive; tiny perturbations may trigger drastic changes. To control a system may mean to find parameters such that the state of the system is safe from being close to a critical threshold. Since reaching a critical threshold often is considered a failure, the control of parameters is a central part of risk control.

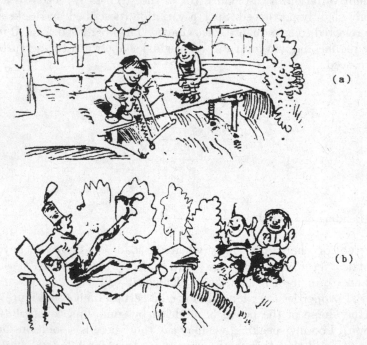

(a)

(b)

Fig. 1.2. From W. Busch [Bus62]. After the original hand drawing in Wilhelm-Busch-Museum, Hannover

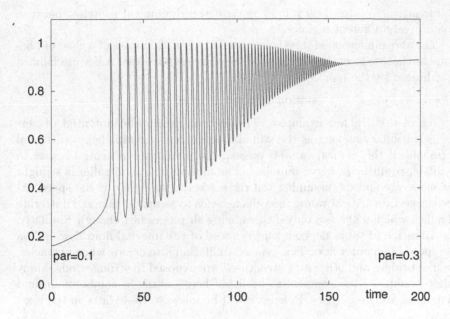

par=0.1   par=0.3

Fig. 1.3. Response of a system to a parameter that gradually grows from 0.1 to 0.3

For example, the response of a system to variation of a parameter might look as the situation in Figure 1.3. We see the temporal variation of a reaction in a chemical or biological system, where the parameter drifts from the value $\lambda = 0.1$ to the value $\lambda = 0.3$ within the time interval $0 \leq t \leq 200$. The vertical axis might represent the temperature of a reaction, and the parameter λ could be the opening of a valve. Initially, for $t \approx 0$, the observed state variable reacts only moderately to the growing parameter. Then, all of a sudden at around time $t \approx 30$, when the parameter λ passes approximately the value 0.13, a large-amplitude oscillation sets in. The regime has changed drastically. With the parameter growing further, the oscillation slowly dies out. Finally, the state becomes again stationary ($t \approx 150$, $\lambda \approx 0.25$), and the state of the system has entered another regime. This third regime differs from the first regime by its significantly higher level. Later in this text we shall understand what has happened when the parameter passed the interval $0.1 \leq \lambda \leq 0.3$: Two critical threshold values were passed, and there was a "hard loss of stability" of the first regime, which goes along with a jump in the state variable. Analyzing the system under consideration closer, reveals the underlying structure, which is the *skeleton* of the dynamical behavior. This is illustrated by Figure 1.4, where the skeleton is built in ("bifurcation diagram" from Chapter 2, Example 2.14). The two horizontal axes of the parameter and of the time match perfectly. Two threshold values (in Figure 1.4 the "bifurcations" of the heavy line) initiate the dynamical switches between qualitatively different regimes.

The above-mentioned threshold values are first examples of a class of phenomena that we later shall denote with the term *bifurcation*. A key mechanism is indicated by the pair

$$\text{stationary state} \leftrightarrow \text{motion}.$$

Let us mention a few examples. The electric membrane potential of nerves is stationary as long as the stimulating current remains below a critical threshold; if this critical value is passed, the membrane potential begins to oscillate, resulting in nerve impulses. The motion of a semitrailer is straight for moderate speeds (assuming the rig is steered straight); if the speed exceeds a certain critical value, the vehicle tends to sway. Or take the fluttering of a flag, which will occur only if the moving air passes fast enough. Similarly, the vibration of tubes depends on the speed of the internal fluid flow and on the speed of an outer flow. This type of oscillation also occurs when obstacles, such as bridges and other high structures, are exposed to strong winds. Many other examples—too complex to be listed here—occur in combustion, fluid dynamics, and geophysics. Reference will be made to these later in the text.

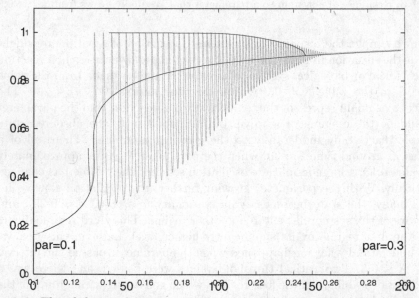

Fig. 1.4. The skeleton of Figure 1.3: The underlying bifurcation diagram from Example 2.14 explains the structure of the dynamics. horizontal axis: both time $0 \le t \le 200$ and parameter $0.1 \le \lambda \le 0.3$

The transition from a stationary state to motion, and vice versa, is also a qualitative change. Here, speaking again in terms of *solutions*—of governing equations—we have a different quality of solution on either "side" of a critical parameter (Figures 1.3, 1.4). Let the parameter in question again be denoted by λ, with critical value λ_0. Thinking, for instance, in terms of the state variable wind speed, the state (e.g., of a flag or bridge) is stationary for $\lambda < \lambda_0$ and oscillatory for $\lambda > \lambda_0$. Qualitative changes may come in several steps, as indicated by the sequence

stationary state

regular motion

irregular motion.

The transition from regular to irregular motion is related to the onset of turbulence, or "chaos." — As a first tentative definition, we will denote a qualitative change caused by the variation of some physical (or chemical or biological, etc.) parameter λ as *bifurcation* or *branching*. We will use the same symbol λ for various kinds of parameters. Some examples of parameters are listed in Table 1.1.

TABLE 1.1.   Examples of parameters.

Phenomenon	Controlled by a typical parameter
Bending of a rod	Load
Vibration of an engine	Frequency or imbalance
Combustion	Temperature
Nerve impulse	Generating potential
Superheating	Strength of external magnetic field
Oscillation of an airfoil	Speed of plane relative to air
Climatic changes	Solar radiation

Some important features that may change at bifurcations have already been mentioned. The following list summarizes various kinds of qualitative changes:

stable $\leftrightarrow$ unstable

symmetric $\leftrightarrow$ asymmetric

stationary $\leftrightarrow$ periodic (regular) motion

regular $\leftrightarrow$ irregular

order $\leftrightarrow$ chaos

Several of these changes may take place simultaneously in complicated ways.

The quality of solutions or states is also distinguished by their geometrical shape—that is, by their *pattern*. For example, the five patterns in Figure 1.5 characterize five possibilities of how a state variable varies with time. The solution profile of Figure 1.5(a) is "flat" or stationary, the state remains at a constant level. Figure 1.5(b) shows a wavy pattern with a simple periodic structure. The patterns of Figure 1.5(c) and (d) are again wavy but less regular, and the pattern of Figure 1.5(e) appears to be irregular (chaotic). The five different patterns of Figure 1.5 arise for different values of a parameter λ;

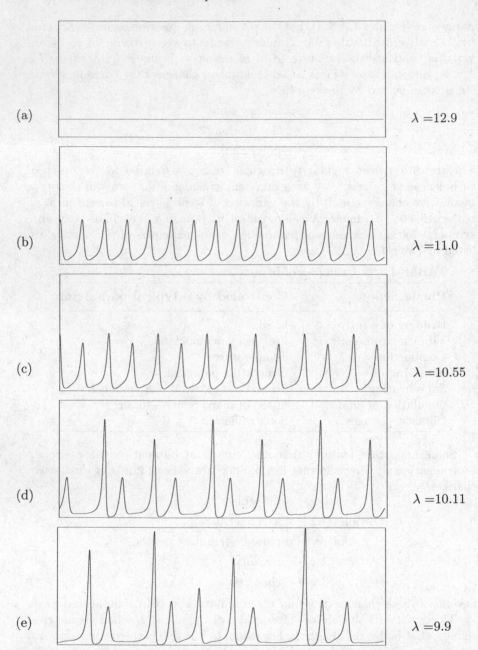

Fig. 1.5. Changing structure or pattern: growing complexity with decreasing parameter λ; (a): stationary state, (b): periodic state, (c): periodic with double period, (d): fourfold period, (e): aperiodic motion ("chaos") — For readers who have already visited Chapter 7: Example 7.8 of an isothermal reaction; parameter values shown on the right; each of the five boxes depicts $y_1(t)$ for $0 \leq t \leq 10$, the vertical axes are scaled such that $0 \leq y_1 \leq 85$

new patterns form when the parameter passes critical values. This example illustrates why such bifurcation phenomena are also called *pattern formation*. — Figure 1.5 shows an example of an isothermal reaction. Such transitions are typical for a wide range of problems. A similar sequence of patterns is, for example, the velocity of the reaction front in [BaM90], where the first profile (a) stands for a uniformly propagating combustion front, and the wavy pattern (b) represents a regularly pulsating front.

So far this introduction has stressed the situation where the state of the system varies with time—that is, the focus has been on *temporal dynamics*. In addition, the state of a system may also vary with space. For example, animal coats may have spots or stripes, which can be explained by variations of morphogens. If the morphogen is non-uniformly distributed (the *heterogeneous* state) a pattern of spots or stripes develops. No pattern develops in case the morphogens are distributed homogeneously. The pair

$$\text{homogeneous} \leftrightarrow \text{heterogeneous}$$

is the spatial analog to the pair "stationary $\leftrightarrow$ motion" that stresses temporal dynamics. Problems in full generality will often display both temporal and spatial dynamics. For example, a chemical reaction may show a concentration with spiral-wave pattern that migrates slowly across the disk.

Transitions among different patterns (as depicted in Figure 1.5) are ubiquitous. For instance, cardiac rhythm is described by similar patterns. One of the possible patterns may be more desirable than others. Hence, one faces the problem of how to switch patterns. By means of a proper external stimulus one can try to give the system a "kick" such that it hopefully changes its pattern to a more favorable state. For example, heart beat can be influenced by electrical stimuli. The difficulties are to decide how small a stimulus to choose, and how to set the best time instant for stimulation. One pattern may be more robust and harder to disturb than another pattern that may be highly sensitive and easy to excite. Before manipulating the transition among patterns, mechanisms of *pattern selection* must be studied. Which *structure* is most *attractive*? Which states are stable? For which values of the parameters is the system most sensitive?

To obtain related knowledge, a thorough discussion of bifurcation phenomena is necessary, which requires the language of mathematics. Before proceeding to the mathematical analysis of stability, bifurcation, and pattern formation, we shall review some important mathematical tools and concepts.

1.2 Stationary Points and Stability (ODEs)

Many kinds of qualitative changes can be described by systems of ordinary differential equations (ODEs). In this section some elementary facts are recalled and notation is introduced; compare also Appendix A.3.

1.2.1 Trajectories and Equilibria

Suppose the state of the system is described by functions $y_1(t)$ and $y_2(t)$. These *state variables* may represent, for example, the position and velocity of a particle, concentrations of two substances, or electric potentials. The independent variable t is often "time." Let the physical (or chemical, etc.) law that governs $y_1(t)$ and $y_2(t)$ be represented by two ordinary differential equations

$$\dot{y}_1 = f_1(y_1, y_2),$$
$$\dot{y}_2 = f_2(y_1, y_2). \tag{1.1}$$

In vector notation, equation (1.1) can be written as $\dot{\mathbf{y}} = \mathbf{f}(\mathbf{y})$. This system is called *autonomous*, because the functions f_1 and f_2 do not depend explicitly on the independent variable t.

The autonomous system $\dot{\mathbf{y}} = \mathbf{f}(\mathbf{y})$ is a *dynamical system*. A dynamical system consists of three ingredients, namely,

time,

state space,

law of evolution.

For $\dot{\mathbf{y}} = \mathbf{f}(\mathbf{y})$, time t varies continuously, $t \in \mathbb{R}$. The state space is $\mathbb{R}^2$, and is also called *phase plane*. The evolution is defined by $\mathbf{f}$. The law $\mathbf{f}$ does not change with time, in contrast to a non-autonomous system described by $\mathbf{f}(t, \mathbf{y})$. The system $\dot{\mathbf{y}} = \mathbf{f}(\mathbf{y})$ represents the classical deterministic scenario.

As illustrated in Figure 1.6, solutions $y_1(t)$, $y_2(t)$ of this *two-dimensional* system form a set of *trajectories* (*flow lines*, *orbits*) densely covering part or all of the (y_1, y_2)-plane. A figure like Figure 1.6 depicting selected trajectories is called a *phase diagram*. A specific trajectory is selected by requesting that it pass through a prescribed point (z_1, z_2) of *initial values*

$$y_1(t_0) = z_1, \quad y_2(t_0) = z_2.$$

Because the system is autonomous, we can take $t_0 = 0$. Imagine a tiny particle floating over the (y_1, y_2)-plane, the tangent of its path being given by the law of evolution $\mathbf{f}$ of the differential equations. After time t_1 has elapsed, the particle will be at the point

$$(y_1(t_1), \ y_2(t_1))$$

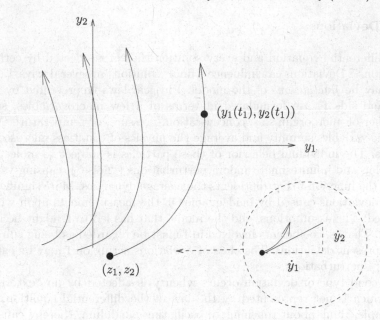

Fig. 1.6. Trajectories in a phase plane

(see Figure 1.6). The set of all trajectories that start from some part of the
y-plane is called the *flow* from that part.

Of special interest are *equilibrium* points (y_1^s, y_2^s), which are defined by

$$\dot{y}_1 = 0, \quad \dot{y}_2 = 0.$$

In equilibrium points, the system is at rest; equilibrium solutions are constant solutions. These points (y_1^s, y_2^s) are also called *stationary* solutions, and
sometimes singular points, critical points, rest points, or fixed points. Stationary points are solutions of the system of equations given by the *right-hand
side* **f** of the differential equation,

$$f_1(y_1^s, y_2^s) = 0,$$
$$f_2(y_1^s, y_2^s) = 0.$$

In vector notation, this system of equations is written as $\mathbf{f}(\mathbf{y}^s) = \mathbf{0}$, which
defines stationary solutions $\mathbf{y}^s$. For a practical evaluation of stationary points
in two dimensions, we observe that each of these two equations defines a curve
named *null cline*. Hence, stationary solutions in the plane are intersections
of null clines. This suggests starting with drawing the null clines in a phase
plane. In this way, not only the equilibria are obtained, but also some information on the global behavior of the trajectories. Recall that trajectories
intersect the null cline defined by $f_1(y_1, y_2) = 0$ vertically and intersect the
null cline of $f_2(y_1, y_2) = 0$ horizontally.

1.2.2 Deviations

Every differential equation and every solution thereof is affected by certain "deviations." Deviations can influence a final "solution" in several ways. First, there may be *fluctuations* of the model. Physical laws (represented by the right-hand side $\mathbf{f}$) are formulated in terms of a few macrovariables, such as number of microorganisms, concentration, pressure, or temperature. Macroscopic variables simplify and average the physics of countless microscopic particles. The individual behavior of these particles is subject to molecular interaction and infinitesimal random perturbations ("noise"), causing variations in the function $\mathbf{f}$ that represents the average dynamics. More significant are the deviations caused by inadequacies in the measurements upon which $\mathbf{f}$ is based. The assumptions, and the model that has led to $\mathbf{f}$ might be inadequate. That is, coefficients involved in $\mathbf{f}$ must be regarded as being subject to ever-present deviations. The class of deviations acting on $\mathbf{f}$ may be called "interior" perturbations.

A second type of deviation occurs when $\mathbf{y}$ is affected by an "external" force, which is not represented by the law of the differential equation. As an example, think about touching an oscillating pendulum, thereby causing momentarily a sudden change in the dynamic behavior. Mathematically speaking, such "outer" perturbations can be described as a jump in the trajectory at some instant t_1,

$$(y_1(t_1),\ y_2(t_1)) \rightarrow (z_1, z_2).$$

When the perturbation ends at $t_2 > t_1$, the system (e.g., pendulum, chemical reaction) is again governed by the differential equation (1.1) starting from the initial values

$$y_1(t_2) = z_1,\ y_2(t_2) = z_2$$

and traveling along a neighboring trajectory (Figure 1.7).

Further deviations occur when the differential equation is integrated numerically. Then discretization errors of a numerical approximation scheme and rounding errors are encountered. These inevitable errors can be seen as a kind of noise affecting the calculated trajectory or interpreted as a great number of small perturbations of the kind illustrated in Figure 1.7.

Now assume that our "traveling particle" $\mathbf{y}$, its itinerary given by differential equation (1.1), is in a stationary state, $(z_1, z_2) = (y_1^{\mathrm{s}}, y_2^{\mathrm{s}})$. Theoretically, the particle would remain there forever, but in practice there is rarely such a thing as "exactness." Due to perturbations in $\mathbf{y}$ or $\mathbf{f}$, either the position (z_1, z_2) of the particle or the location of the equilibrium $(y_1^{\mathrm{s}}, y_2^{\mathrm{s}})$ varies slightly. This results in a deviation at, say, time $t_0 = 0$,

$$(z_1, z_2) \neq (y_1^{\mathrm{s}},\ y_2^{\mathrm{s}}).$$

For simplicity, we assume that for $t > 0$ no further perturbations or fluctuations arise. Then it is the differential equation that decides whether trajectories

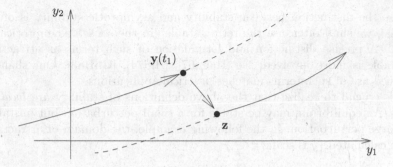

Fig. 1.7. Perturbation of a trajectory in a phase plane

starting from $\mathbf{z}$ remain near the equilibrium or depart from it. The question is how solutions $\mathbf{y}(t)$ of the *initial-value problem* (in vector notation)

$$\dot{\mathbf{y}} = \mathbf{f}(\mathbf{y}), \quad \mathbf{y}(0) = \mathbf{z}$$

behave for $t > 0$.

1.2.3 Stability

Informal Definition 1.1   *A stationary solution* $\mathbf{y}^{\mathrm{s}}$ *is **asymptotically stable** if the response to a small perturbation approaches zero as the time approaches infinity. An asymptotically stable equilibrium is also called **sink**.*

More formally, this means that

$$\mathbf{y}(t) \xrightarrow{\;t\to\infty\;} \mathbf{y}^{\mathrm{s}}$$

for $\mathbf{y}(t)$ starting from initial values that are sufficiently close to the equilibrium. Asymptotically stable equilibria are examples of *attractors*. The set of all initial values $\mathbf{z}$ from which trajectories converge to the attractor is called *domain of attraction*.

Informal Definition 1.2   *The stationary solution* $\mathbf{y}^{\mathrm{s}}$ *is **stable** if the response to a small perturbation remains small as the time approaches infinity. Otherwise the stationary solution is called **unstable** (the deviation grows).*

As we see, asymptotic stability implies stability. The above classical definitions apply to general systems of ODEs, including the simplest "system" of one scalar differential equation. Let us illustrate stability with some scalar examples; the simplest is as follows (we omit the subscript 1):

$\dot{y} = \lambda y$; solution: $y(t) = \exp(\lambda t)$
    equilibrium: $y^{\mathrm{s}}(t) \equiv 0$
    y^{s} is stable for $\lambda \leq 0$, asymptotically stable for $\lambda < 0$, and unstable for $\lambda > 0$.

Because the distinction between stability and asymptotic stability is often artificial, we shall often use the term "stable" in the sense "asymptotically stable." A precise distinction and formulation of such terms as attractive and stable is more involved; see [Hahn67], [HiS74], [GuH83]. Our slightly simplified use of such terms matches practical applications.

It is crucial to realize that the above definitions of stability are *local* in nature. An equilibrium may be stable for a small perturbation but unstable for a large perturbation. In the following example, the domain of attraction can be calculated (y is scalar):

$$\dot{y} = y(y^2 - \alpha^2), \quad y(0) = z.$$

The closed-form solution is

$$y \equiv 0 \qquad\qquad\qquad\qquad\qquad \text{for } z = 0,$$
$$y(t) = (\alpha^{-2} + (z^{-2} - \alpha^{-2})\exp(2\alpha^2 t))^{-1/2} \qquad \text{for } z \neq 0$$

(Exercise 1.1). For $|z| < |\alpha|$ the radicand approaches infinity for $t \to t_\infty$. From this we infer that the equilibrium $y^s = 0$ is stable. The local character of this result is illustrated by the observation that for $|z| > |\alpha|$ solutions diverge. The domain of attraction $|z| < |\alpha|$ of the stable equilibrium $y^s = 0$ is bounded by two unstable equilibria $y^s = \pm\alpha$. Although locally stable, the equilibrium $y^s = 0$ is globally unstable when "large" perturbations occur. In a practical situation where the radius of the domain of attraction is smaller than the magnitude of possible perturbations, the local stability result is meaningless. This situation is made plausible by a model of a hill with a cross section, as depicted in Figure 1.8. Imagine a ball (rock) placed in the dip on top of the hill. Apparently the ball is stable under small vibrations but unstable under the influence of a violent quake. The undulating curve in Figure 1.8 may be seen as representing the potential energy. Figure 1.8 depicts two "weakly" stable equilibria and one "strongly" stable equilibrium.

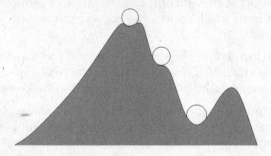

Fig. 1.8. Stability of a ball

The above scalar examples and the global stability behavior of their solutions were analyzed using a closed-form solution. In general, problems are so complicated that no closed-form solution can be calculated. Accordingly, global stability results ("How large is the domain of attraction?") are usually difficult to obtain [DíF02]. The situation is much better with local stability results. Here the principle of *linearized stability* provides insight into what happens "close" to an equilibrium.

1.2.4 Linear Stability; Duffing Equation

A Taylor series expansion of f_1 about (y_1^s, y_2^s) gives

$$\dot{y}_1 = f_1(y_1, y_2) = f_1(y_1^s, y_2^s) + \frac{\partial f_1}{\partial y_1}(y_1^s, y_2^s)(y_1 - y_1^s)$$

$$+ \frac{\partial f_1}{\partial y_2}(y_1^s, y_2^s)(y_2 - y_2^s) + \text{ terms of higher order.}$$

The terms of higher order are at least of order 2; they are denoted $O(|\mathbf{y} - \mathbf{y}^s|^2)$, or t.h.o. Expanding also the f_2 of the second differential equation, observing that $f_1(y_1^s, y_2^s) = f_2(y_1^s, y_2^s) = 0$, and dropping the high-order terms give two differential equations that are *linear* in $y_1 - y_1^s$ and $y_2 - y_2^s$. This system of equations is easily written down using the notation of the *Jacobian matrix* $\mathbf{f_y}$ of the first-order partial derivatives. Evaluated at the equilibrium, the Jacobian is

$$\frac{\partial \mathbf{f}}{\partial \mathbf{y}}(\mathbf{y}^s) = \begin{pmatrix} \frac{\partial f_1}{\partial y_1}(y_1^s, y_2^s) & \frac{\partial f_1}{\partial y_2}(y_1^s, y_2^s) \\ \frac{\partial f_2}{\partial y_1}(y_1^s, y_2^s) & \frac{\partial f_2}{\partial y_2}(y_1^s, y_2^s) \end{pmatrix}.$$

This matrix with constant coefficients, evaluated at the stationary solution $\mathbf{y}^s$, will be denoted $\mathbf{f_y^s}$. Now the linearized system in vector notation reads

$$\dot{\mathbf{h}} = \mathbf{f_y^s}\,\mathbf{h}. \tag{1.2}$$

The vector $\mathbf{h}$ represents first-order approximations

$$h_1(t) \approx y_1(t) - y_1^s,$$
$$h_2(t) \approx y_2(t) - y_2^s$$

of small distances between the points $\mathbf{y}(t)$ on the trajectories and the stationary point $\mathbf{y}^s$. The smaller the distance $\mathbf{y} - \mathbf{y}^s$, the better the approximation $\mathbf{h}$. That is, under certain assumptions to be stated later, $\mathbf{h}(t)$ describes the *local* behavior of the solution. The approximation $\mathbf{h}(t)$ indicates how the process evolves when the initial state deviates slightly from its equilibrium values. Thus, the question of local stability is reduced to a discussion of the linear system equation (1.2). The standard procedure is to insert the *ansatz* (hypothesis) $\mathbf{h}(t) = e^{\mu t}\mathbf{w}$,

$$h_1(t) = e^{\mu t}w_1,$$
$$h_2(t) = e^{\mu t}w_2 \tag{1.3}$$

into equation (1.2). A straightforward calculation shows that the *eigenvalue problem*

$$(\mathbf{f}_{\mathbf{y}}^{\mathrm{s}} - \mu \mathbf{I})\mathbf{w} = \mathbf{0} \tag{1.4}$$

results, with eigenvalue μ, constant eigenvector $\mathbf{w}$, and identity matrix

$$\mathbf{I} = \begin{pmatrix} 1 & 0 \\ 0 & 1 \end{pmatrix}.$$

First, the eigenvalues μ_1 and μ_2 are calculated. The existence of a nontrivial solution $\mathbf{w} \neq 0$ of equation (1.4) reflects the singularity of the matrix, and requires μ_1 and μ_2 to be the roots of the *characteristic equation*

$$0 = \det(\mathbf{f}_{\mathbf{y}}^{\mathrm{s}} - \mu \mathbf{I}).$$

In the two-dimensional situation, this is a quadratic equation; its roots μ_1 and μ_2 are easily obtained. Then eigenvectors $\mathbf{w}$ can be calculated as solutions of the specific system of linear equations that results from inserting μ_1 and μ_2 into equation (1.4). From this point on, the procedure is best illustrated by studying an example.

Example 1.3   Duffing Equation (without external forcing)
Consider the second-order scalar differential equation

$$\ddot{u} + \dot{u} - u + u^3 = 0, \tag{1.5}$$

which is a special case of the Duffing equation

$$\ddot{u} + a\dot{u} + bu + cu^3 = d\cos\omega t.$$

Duffing equations are used to describe oscillations in series-resonance circuits, where a nonlinear inductor or capacitor is modeled by a cubic polynomial [Duf18], [NaM79]. For our purpose of illustrating some calculus, we choose in equation (1.5) simple coefficients and omit the excitation ($d = 0$). Later, more general Duffing equations will be treated.

We introduce a vector $\mathbf{y}$ via $y_1 = u$, $y_2 = \dot{u}$. The equation (1.5) is written as a first-order system of ODEs,

$$\dot{y}_1 = y_2 =: f_1(y_1, y_2),$$
$$\dot{y}_2 = y_1 - y_1^3 - y_2 =: f_2(y_1, y_2).$$

The phase plane associated with the scalar second-order equation (1.5) displays u and $\dot{u}$. Calculating stationary points, one finds that $y_2^{\mathrm{s}} = 0$ and that y_1^{s} is a solution of

$$0 = y_1(1 - y_1^2).$$

This equation has three distinct roots, and the above system thus has three distinct stationary points $(y_1^{\mathrm{s}}, y_2^{\mathrm{s}})$,

$$(0,0), \quad (1,0), \quad (-1,0).$$

The Jacobian matrix of the first-order partial derivatives is

$$\mathbf{f_y} = \begin{pmatrix} 0 & 1 \\ 1 - 3y_1^2 & -1 \end{pmatrix}.$$

The three equilibrium points (y_1^s, y_2^s) produce three eigenvalue problems; see equation (1.4). In this particular example, the analysis simplifies because the Jacobian $\mathbf{f_y}$ and therefore equation (1.4) are identical for the two equilibria $(\pm 1, 0)$. Pursuing the analysis in more detail, we obtain the following results:

$(y_1^s, y_2^s) = (0, 0)$:

$$\mathbf{f_y^s} = \begin{pmatrix} 0 & 1 \\ 1 & -1 \end{pmatrix},$$

characteristic equation: $0 = \mu^2 + \mu - 1$,
roots: $\mu_{1,2} = \frac{1}{2}(-1 \pm \sqrt{5})$.

Two distinct roots imply the existence of two eigenvectors, which are obtained by equation (1.4),

$$\begin{pmatrix} -\mu & 1 \\ 1 & -1 - \mu \end{pmatrix} \begin{pmatrix} w_1 \\ w_2 \end{pmatrix} = \begin{pmatrix} 0 \\ 0 \end{pmatrix}.$$

The first component of this system of linear equations is

$$-\mu w_1 + w_2 = 0,$$

which suggests choosing $w_1 = 1$. This implies that $w_2 = \mu$ and generates the two eigenvectors:

$$\mathbf{w} = \begin{pmatrix} 1 \\ \mu \end{pmatrix}, \quad \mu = \mu_1, \mu_2.$$

$(y_1^s, y_2^s) = (\pm 1, 0)$:

$$\mathbf{f_y^s} = \begin{pmatrix} 0 & 1 \\ -2 & -1 \end{pmatrix},$$

characteristic equation: $0 = \mu^2 + \mu + 2$,
roots: $\mu_{1,2} = \frac{1}{2}(-1 \pm \sqrt{-7})$ (complex conjugates).

We omit calculation of the complex eigenvectors because we do not need them.   □

This example will be analyzed further at the end of this section. In the meantime, we discuss important types of qualitative behavior of trajectories close to an equilibrium. What follows is based on the ansatz $\mathbf{h}(t) = e^{\mu t}\mathbf{w}$ in equation (1.3) with the associated eigenvalue problem equation (1.4).

Nodes: μ_1, μ_2 real, $\mu_1 \cdot \mu_2 > 0$, $\mu_1 \neq \mu_2$

In this case, where both eigenvalues are real and have the same sign, the stationary point $\mathbf{y}^s$ is called a *node*. This case splits into two subcases:

(a) $\mu < 0$ implies $\lim_{t \to \infty} e^{\mu t} = 0$.
  Therefore, $\mathbf{h}(t)$ tends to zero and $\mathbf{y}(t)$ converges to $\mathbf{y}^s$ in a sufficiently small neighborhood of the node. This type of node is called a *stable node*. Small perturbations die out at stable nodes.

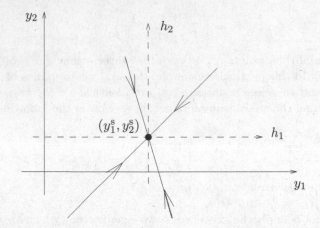

Fig. 1.9. Motion along the eigenspaces $\mathcal{E}$

(b) $\mu > 0$ implies $\lim_{t\to\infty} e^{\mu t} = \infty$.

As a consequence, $\mathbf{h}(t)$ "explodes," which means locally that the trajectories $\mathbf{y}(t)$ leave the neighborhood of the node. This type of node is an *unstable node*.

The two real eigenvectors $\mathbf{w}$ that exist in both these subcases have a geometrical meaning (see Figure 1.9, which depicts a stable node). They define two straight lines passing through the stationary point. Each half-ray is a trajectory of the linear system equation (1.2) (see Exercise 1.3). If $|\mu|$ is large (small), we call the corresponding motion fast (slow). A negative eigenvalue μ means that the trajectory is directed toward the equilibrium point $\mathbf{y}^s$ (stable node), while a positive eigenvalue is associated with motion away from $\mathbf{y}^s$ (unstable node). For the nonlinear problem, the situation may be as in Figure 1.10, where a stable node is illustrated. At the stationary point, the four heavy solid trajectories are tangent to the lines (*eigenspaces*) of Figure 1.9; compare Appendix A.5. For an unstable node, the directions in Figures 1.9 and 1.10 are reversed.

Saddles: μ_1, μ_2 real, $\mu_1 \cdot \mu_2 < 0$

When the real eigenvalues have different signs, the stationary point is called a *saddle point*. Because one of the eigenvalues is positive, two of the four trajectories associated with the eigenspaces leave the equilibrium. A saddle point, therefore, is always unstable (see Figure 1.11).

Foci: μ_1, μ_2 complex conjugates with nonzero real part

Set

$$\mu_1 = \alpha + \mathrm{i}\beta, \quad \mu_2 = \alpha - \mathrm{i}\beta$$

(i: imaginary unit), where α and $\pm\beta$ denote the real and imaginary parts, respectively. Going back to ansatz equation (1.3), the time-dependent part

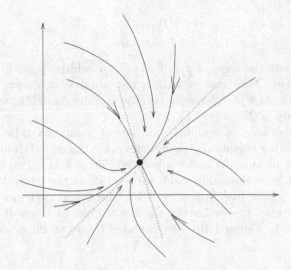

Fig. 1.10. Stable node in a phase plane, nonlinear case (eigenspaces are dotted)

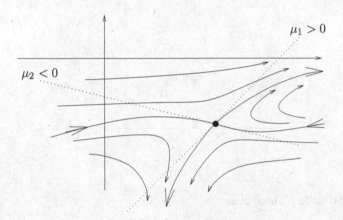

Fig. 1.11. Saddle in a phase plane (eigenspaces indicated by their eigenvalue μ)

of $\mathbf{h}(t)$ is

$$e^{(\alpha + i\beta)t} = e^{\alpha t} e^{i\beta t}.$$

The factor $e^{i\beta t} = \cos \beta t + i \sin \beta t$ represents a rotation counterclockwise in the complex plane through β radians if $\beta > 0$, and clockwise through $-\beta$ radians if $\beta < 0$. The factor $\exp(\alpha t)$ is a radius, which is increasing if $\alpha > 0$ and decreasing if $\alpha < 0$. Any trajectory close to the equilibrium thus resembles a *spiral*. The corresponding equilibrium is called an *unstable focus* ($\alpha > 0$) or a *stable focus* ($\alpha < 0$). Figure 1.12 depicts an unstable focus. In order to determine whether the rotation is clockwise or counterclockwise in the phase plane, one picks a test point $\mathbf{z}$ close to the equilibrium and evaluates

$$\dot{z}_1 = f_1(z_1, z_2),$$
$$\dot{z}_2 = f_2(z_1, z_2).$$

For instance, study the signs of $f_j(\mathbf{y}^s + \epsilon \mathbf{e}_k)$ for arbitrary small $|\epsilon|$ and the kth unit vector $\mathbf{e}_k$. Note that picking a test point is not always required: For an important class of differential equations some straightforward results simplify the analysis (Exercise 1.8).

In the neighborhood of equilibria, subsets of points $\mathbf{z}$ can be designated according to whether emanating trajectories reach the equilibrium for $t \to \infty$ or $t \to -\infty$. For instance, consider a saddle (Figure 1.11). The two specific trajectories that leave the saddle point for $t \to \infty$ are the *unstable manifolds* (*outsets*) of the saddle (see Appendix A.5). The two trajectories that enter the saddle point when $t \to \infty$ are the *stable manifolds* or *insets* of the saddle. The stable node in Figure 1.10 is surrounded by a two-dimensional stable manifold.

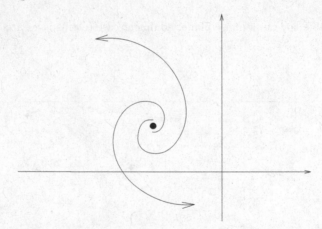

Fig. 1.12. Focus in a phase plane

Example 1.4   Duffing Equation, continued

Having recalled the above three classes of stationary points, we continue with the discussion of equation (1.5). The above analysis reveals that

(0,0) is a saddle point ($\mu_1 \approx 0.62$, $\mu_2 \approx -1.62$),
($\pm 1, 0$) are stable foci.

The test point (1.1, 0) close to the focus (1,0) will drift "downward," since $\dot{y}_1 = 0$, $\dot{y}_2 < 0$. This indicates a clockwise rotation around the focus, as illustrated in Figure 1.13 (compare Exercise 1.8). The two insets of the saddle point (0,0) are examples of *separatrices*. A separatrix divides (parts of) the phase space into attracting basins. Here the phase plane is separated into two basins, the shaded and unshaded regions in Figure 1.13. All particles that start in the shaded region ultimately approach the focus (1, 0); all the other trajectories end up at the focus $(-1,0)$. What is depicted in Figure 1.13 extends the local results of the analysis of the stationary points to a global picture. Thereby, global results that can be obtained by numerical integration of the differential equation (1.1) are anticipated. In particular, the calculation of a separatrix provides global insight. Because only the final limit point (the saddle) of a separatrix is known, the integration must be *backward*, $t \to -\infty$, starting from suitable initial values $\mathbf{y}(0) = \mathbf{z}$. An obvious choice for $\mathbf{z}$ would be to take the coordinates of the saddle $(0,0)$, trusting that rounding and discretization errors cause $\mathbf{f}(\mathbf{y}) \neq \mathbf{0}$, thereby enabling the calculations to leave the saddle point. This initial phase tends to be slow. Also, the resulting trajectory may be quite far away from the separatrix. If rounding and discretization errors happen to be zero, this method will not work at all. This situation suggests starting from $\mathbf{z}$ not identical but close to the saddle point and close to the separatrix. Such a starting point is provided by a point on the separatrix of the linearized problem—that is, by the eigenvector $\mathbf{w}$ associated with the negative eigenvalue: $\mathbf{z} = \mathbf{y}^\mathrm{s} + \epsilon\mathbf{w}$. In our example this eigenvector

$$\mathbf{w} = \begin{pmatrix} 1 \\ -\frac{1}{2}(1 + \sqrt{5}) \end{pmatrix}$$

produces the starting point

$$z_1 = \epsilon, \quad z_2 = -\frac{1}{2}\epsilon(1 + \sqrt{5})$$

for, say, $\epsilon = 0.0001$. The trajectory obtained this way by integrating in reverse time can be expected to be close to the exact separatrix in the neighborhood of the saddle point. In this example, the separatrices spiral around the "core" part of the plane (see Figure 1.13 again), and the two attracting basins spiral around, encircling each other (not shown in the figure). □

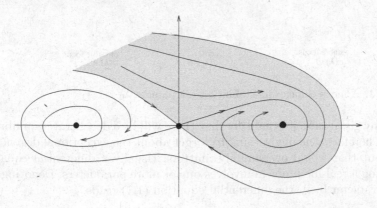

Fig. 1.13. Phase plane of the Duffing equation (1.5), schematically

1.2.5 Degenerate Cases; Parameter Dependence

As seen above, the eigenvalues of the Jacobian matrix evaluated at an equilibrium point determine the dynamic behavior in the neighborhood of the equilibrium. This holds for all three types of equilibrium points (nodal points, saddle points, and focal points). These type of points are *generic* in that the assumptions ($\mu_1 \neq \mu_2$, $\mu_1 \cdot \mu_2 \neq 0$, or $\alpha = \mathrm{Re}(\mu) \neq 0$) almost always apply. For an analysis of the exceptional cases

$$\mu_1 = \mu_2 \qquad \text{(a special node)},$$
$$\mu_1 \cdot \mu_2 = 0 \qquad \text{(require nonlinear terms)},$$
$$\mu_{1,2} = \pm\beta\mathrm{i} \qquad \text{(a \textit{center} of concentric cycles)}.$$

we might refer to some hints in Section 8.7 and to ODE textbooks, such as [Har64], [Arn73], [HiS74]. But in some way this entire book is devoted to the study of these cases.

Definition 1.5 *The equilibrium* $\mathbf{y}^s$ *is called* **hyperbolic** *or* **nondegenerate** *when the Jacobian* $\mathbf{f_y}(\mathbf{y}^s)$ *has no eigenvalue with zero real part.*

The exceptional cases $\mu_1\mu_2 = 0$ and $\mu_{1,2} = \pm\mathrm{i}\beta$ are called *nonhyperbolic* or *degenerate*.

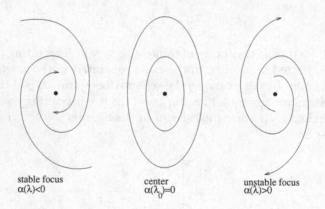

stable focus   center   unstable focus
$\alpha(\lambda)<0$ $\alpha(\lambda_0)=0$ $\alpha(\lambda)>0$

Fig. 1.14. A focus changes its stability for λ passing λ_0

In any given case, it is unlikely that there will be a degenerate equilibrium point. Therefore, one might think to forget about the exceptional degenerate cases, but that would oversimplify matters. Usually a differential equation describing a real-life problem involves one or more parameters. Denoting one such parameter by λ, the differential equation (1.1) reads

$$\dot{y}_1 = f_1(y_1, y_2, \lambda),$$
$$\dot{y}_2 = f_2(y_1, y_2, \lambda).$$

Because the systems $\dot{\mathbf{y}} = \mathbf{f}(\mathbf{y}, \lambda)$ depend on λ, we speak of a *family* of differential equations. Solutions now depend on both the independent variable t and the parameter λ,

$$\mathbf{y}(t; \lambda).$$

Consequently, stationary points, Jacobian matrices, and the eigenvalues μ depend on λ,

$$\mu(\lambda) = \alpha(\lambda) + \mathrm{i}\beta(\lambda).$$

Upon varying the parameter λ, the position and the qualitative features of a stationary point can vary. For example, imagine a stable focus ($\alpha(\lambda) < 0$) for some values of λ. When λ passes some critical value λ_0, the real part $\alpha(\lambda)$ may change sign and the stationary point may turn into an unstable focus. During this transition, a degenerate focus (a center) is encountered for an instant (see Figure 1.14). Other metamorphoses of a stable spiral to an unstable spiral are possible (see Exercise 1.12). Summarizing, we point out that degenerate cases can occur "momentarily." Often, qualitative changes such as a loss of stability are encountered when a degenerate case is passed. This makes the occurrence of "degenerate" cases especially interesting. There will be examples and exercises in the following section (Exercises 1.10 and 1.11). A main concern of this book is to treat phenomena initiated by degenerate solutions.

Throughout this book, the parameter λ is considered constant, although it may be varied. This variation of λ is called *quasi-static* variation; a parameter λ with this characterization $\dot{\lambda} = 0$ is sometimes called "adiabatic constant." In a practical example, a change of a parameter value may mean $\dot{\lambda} \neq 0$ for some time interval. Then *transient* behavior of the trajectories occurs until the trajectory is caught by some attractor. In this book we mostly discuss asymptotic states in that we assume that any given parameter is constant long enough such that transients have died out. For an example of transient behavior, see Figure 1.3. The final asymptotic behavior of the system for $t \to \infty$ is called *steady state*.

1.2.6   Generalizations

So far our analysis has concentrated on a system of two first-order differential equations—that is, the vector functions $\mathbf{y}(t)$ and $\mathbf{h}(t)$, the initial vector $\mathbf{z}$, and the eigenvectors $\mathbf{w}$ have consisted of two components. Most of the above results generalize to n-dimensional vectors. Separatrices in $\mathbb{R}^n$ are $(n-1)$-dimensional surfaces. The Jacobian matrix $\mathbf{f_y}$ consists of n^2 first-order partial derivatives $(\partial f_k / \partial y_j)$, $k, j = 1, ..., n$. Evaluated at an equilibrium $\mathbf{y}^s$, this matrix $\mathbf{f_y^s}$ has constant coefficients. The stability of the equilibria is characterized by the eigenvalues $\mu_1, ..., \mu_n$ of the Jacobian $\mathbf{f_y^s}$. For the case in which the matrix $\mathbf{f_y^s}$ has n linear independent eigenvectors $\mathbf{w}^k$, $k = 1, ..., n$, the stability situation of the linear system equation (1.2)

$$\dot{\mathbf{h}} = \mathbf{f}_{\mathbf{y}}^{\mathrm{s}}\,\mathbf{h}$$

can be decided easily. Then the solution $\mathbf{h}(t)$ is a linear combination of n vector functions $\exp(\mu_k t)\mathbf{w}^k$; see Appendix A.3. The factor $\mathrm{e}^{\mu t}$ satisfies

$$|\mathrm{e}^{\mu t}| = |\mathrm{e}^{\alpha t}||\mathrm{e}^{\mathrm{i}\beta}| = \mathrm{e}^{\alpha t} = \mathrm{e}^{\mathrm{Re}(\mu)t}\,.$$

For $t \to \infty$ it grows to infinity in case $\mathrm{Re}(\mu) > 0$, and decays to zero in case $\mathrm{Re}(\mu) < 0$. Hence, the sign of the real part $\mathrm{Re}(\mu)$ decides on the stability. The vectors $\mathbf{w}^k$ span the eigenspaces. For a pair of simple eigenvalues $\alpha \pm \mathrm{i}\beta$, with eigenvectors $\mathrm{Re}(\mathbf{w}^k) \pm \mathrm{i}\,\mathrm{Im}(\mathbf{w}^k)$, the two vectors $\mathrm{Re}(\mathbf{w}^k)$ and $\mathrm{Im}(\mathbf{w}^k)$ span the corresponding eigenspace. For $\mathbf{f}_{\mathbf{y}}^{\mathrm{s}}$ having multiple eigenvalues, generalized eigenvectors may be needed, $\mathbf{w}^k$ may depend on t, but the stability argument is the same. The following general stability result on the nonlinear system $\dot{\mathbf{y}} = \mathbf{f}(\mathbf{y})$ is attributed to Liapunov (1892) [Lia66]:

Theorem 1.6   *Suppose $\mathbf{f}(\mathbf{y})$ is two times continuously differentiable and $\mathbf{f}(\mathbf{y}^{\mathrm{s}}) = \mathbf{0}$. The real parts of the eigenvalues μ_j $(j = 1,\dots,n)$ of the Jacobian evaluated at the stationary solution $\mathbf{y}^{\mathrm{s}}$ determine stability in the following way:*
(a) $\mathrm{Re}(\mu_j) < 0$ for all j implies asymptotic stability, and
(b) $\mathrm{Re}(\mu_k) > 0$ for one (or more) k implies instability.

Theorem 1.6 establishes the principle of linearized stability. For hyperbolic equilibria, *local stability* is determined by the eigenvalues of the Jacobian. In order to stress the local character of this stability criterion, this type of stability is also called "conditional stability" [IoJ81] or "linear stability."

   The number of cases generated by the various combinations of the signs of the eigenvalues increases dramatically with n. For $n = 3$, there are already 10 different types of hyperbolic stationary points [Arn73]. Classification is not as simple as in the planar situation. For illustrations see the Example 1.7 below, and Figure 8.12.

   In Section 1.2.4 we used fundamental results that relate the nonlinear ODEs with the linearized equations. Specifically, the flow (the set of trajectories) of the nonlinear ODE $\dot{\mathbf{y}} = \mathbf{f}(\mathbf{y}, \lambda)$ in a neighborhood of an equilibrium is compared with the flow of its linearization $\dot{\mathbf{h}} = \mathbf{f}_{\mathbf{y}}^{\mathrm{s}}\mathbf{h}$ (see Appendix A.5). Corresponding theorems hold in n-dimensional space for the neighborhoods of hyperbolic equilibria. The theorem of Grobman and Hartman assures for hyperbolic equilibria that the *local* solution behavior of the nonlinear differential equation is qualitatively the same as that of the linearization. Insets and outsets (stable and unstable manifolds) at equilibria are tangent to the corresponding eigenspaces of the linearization, which are formed by linear combinations of the eigenvectors of the stable eigenvalues or of the unstable eigenvalues, respectively. This is illustrated in the planar case of Figures 1.10 and 1.11; the eigenspaces of the linearization are spanned by the dashed lines. In Figure 1.10 the depicted part of the (y_1, y_2)-plane forms a two-dimensional stable manifold. In Figure 1.11 there is one stable and one unstable manifold;

the opposite directions tangent to the same eigenspace are identified. Generally the dimension of the stable manifold equals the number of eigenvalues with negative real parts; the dimension of the unstable manifold is equal to the number of the eigenvalues with positive real part (Appendix A.5). For related theorems, refer to books on ordinary differential equations, and to [Wig90]; for a generalization to PDEs see [Cos94].

Example 1.7   Lorenz Equation

The possible cases in the planar case ($n = 2$) are easy to comprehend. Higher-dimensional examples are more ambitious. Here we discuss an $n = 3$ example more closely, namely, the Lorenz equation [Lor63]

$$\dot{y}_1 = P(y_2 - y_1)$$
$$\dot{y}_2 = -y_1 y_3 + R y_1 - y_2$$
$$\dot{y}_3 = y_1 y_2 - b y_3$$

with the specific choice of parameters $P = 16$, $R = 40$, $b = 4$. For the background of the equation see Section 2.8, where $\lambda := R$ will serve as our main parameter. Here we discuss equilibria, as well as stable and unstable manifolds.

We first calculate the stationary states. The first equation tells us $y_1 = y_2$. Substituting the expression for y_3 from the third equation into the second yields the equation

$$0 = -y_1^3/b + R y_1 - y_1 \,,$$

which has the three roots

$$y_1 = 0, \quad y_1 = \pm(bR - b)^{1/2} \,.$$

For convenience, we denote the square root by S,

$$S := (bR - b)^{1/2} \,.$$

With the corresponding y_2 and y_3, one finds that the equilibria are

$$(y_1, y_2, y_3) = (0, 0, 0) \quad \text{for all } R,$$
$$(y_1, y_2, y_3) = (\pm S, \pm S, R - 1) \quad \text{for } R \geq 1.$$

(The nontrivial state parameterized by R emerges from the trivial state at $R = 1$.) In order to check for stability, we need the Jacobian matrix

$$\begin{pmatrix} -P & P & 0 \\ -y_3 + R & -1 & -y_1 \\ y_2 & y_1 & -b \end{pmatrix} .$$

For the chosen values for the parameters P, R, b, the spectrum of the matrix that belongs to $\mathbf{y}^s = (0, 0, 0)$ is

eigenvalue	−34.887	−4	17.887
eigenvector	−0.646	0	−0.427
	0.763	0	−0.904
	0	1	0

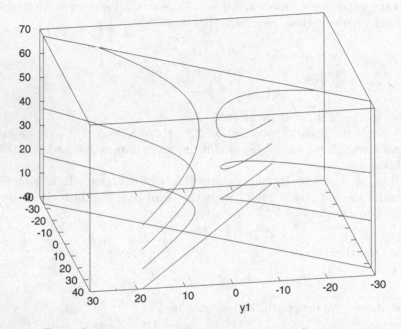

Fig. 1.15. Example 1.7, Lorenz equation linearized about $\mathbf{y}^s = \mathbf{0}$; simulation within the bounding box $-30 \le y_1 \le 30$, $-40 \le y_2 \le 40$, $0 \le y_3 \le 70$; the plane of the stable manifold and the line of the unstable manifold are shown.

Figure 1.15 depicts the dynamics of the linearized equation. The plane in the bounding box represents the stable manifold that corresponds to the eigenvalues -34.887 and -4. Some trajectories are illustrated, starting in the stable manifold, but eventually departing into the direction of the unstable manifold. Both manifolds intersect in the equilibrium $\mathbf{y}^s = \mathbf{0}$, which in Figure 1.15 sits in the middle of the bottom face of the box. Notice that the flow indicated in the figure is that of the equation linearized about $\mathbf{y}^s = \mathbf{0}$, namely, $\dot{\mathbf{y}} = \mathbf{f_y}(\mathbf{0})\mathbf{y}$. The flow of the original nonlinear equation resembles the "nonlinear" flow only in a neighborhood of $\mathbf{0}$.

For the equilibrium $(y_1^s, y_2^s, y_3^s) = (S, S, R - 1)$ and its Jacobian the spectrum is

eigenvalue	-21.422	$0.211 \pm 15.264i$
eigenvector	-0.864	$0.266 \pm 0.309i$
	0.293	$-0.028 \pm 0.566i$
	0.409	0.716

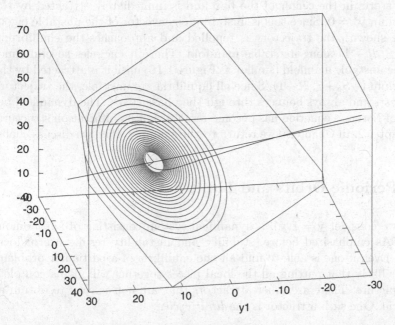

Fig. 1.16. Example 1.7, Lorenz equation linearized about $\mathbf{y}^s = (S,\, S,\, R-1)$, bounding box as in Figure 1.15, the plane of the unstable manifold and the line of the stable manifold are shown.

The stable manifold of the linear system is one-dimensional and defined by $\mathbf{y}^s + \sigma \mathbf{b}$, $\sigma \in \mathbb{R}$, where $\mathbf{b}$ is the eigenvector, $\mathbf{b} = (-0.864, 0.293, 0.409)$. The positive real part of the complex-conjugate pair of eigenvalues describes the dynamics of the unstable manifold. Hence the vectors $(0.266, -0.028, 0.716)$ and $(0.309, 0.566, 0)$ span this two-dimensional plane. The local behavior around $\mathbf{y}^s$ is defined by the linearized equation

$$\dot{\mathbf{y}} = \mathbf{f}_{\mathbf{y}}(\mathbf{y}^s)(\mathbf{y} - \mathbf{y}^s).$$

Figure 1.16 illustrates the dynamical behavior of the linearized equation. The plane of the unstable manifold is indicated by its intersection with the bounding box. The stable manifold is the line that intersects the plane in the equilibrium. A single trajectory of the linearized equation is shown starting at the right-hand edge ($y_1 = -30$), encircling the line of the stable manifold while rapidly approaching the unstable manifold. Once "inside," the depicted unstable focus winds outbound. The dynamical behavior around the third equilibrium $(-S, -S, R-1)$ is analogous.

For the nonlinear system, the above linear behavior is valid only locally. In particular, each of the three equilibria possesses an unstable and a stable manifold, which are conflicting in $\mathbb{R}^3$. So, trajectories squeeze through the various different dynamical regions. This is shown by Figure 1.17. The trajectory starts in the center of the box and is immediately attracted by the equilibrium $\mathbf{y}^s = \mathbf{0}$. Since this is unstable (eigenvector of the unstable eigenvalue is shown), the trajectory is repelled and approaches the equilibrium $(-S, -S, R-1)$ along its stable manifold. There it encircles several times close the unstable manifold (similar as Figure 1.16) until it is attracted by the equilibrium $(+S, +S, R-1)$. Since all equilibria are unstable, the trajectory never rests and always bounces through the phase space. The dynamical behavior of Lorenz's equation has become famous. For its bifurcation structure, see Chapter 2. In Chapter 9 we return to the example when we discuss chaos.

1.3   Periodic Orbits and Limit Sets

Again we consider $\dot{\mathbf{y}} = \mathbf{f}(\mathbf{y})$, a dynamical system consisting of autonomous ODEs. As emphasized before, stability and instability results are of local nature. Even if one is able to find all the equilibria of a particular problem, it is not likely that putting all the local pieces together will give a complete global picture. There are other attractors that are not as easy to obtain as equilibria. One such attractor is the *limit cycle*.

1.3.1   Simulation

One method of studying global behavior of trajectories is a *numerical simulation*. Such a simulation proceeds as follows. Starting from one or more initial points $\mathbf{z}$, a numerical integration method is applied to calculate for each $\mathbf{z}$ the trajectory for a finite time interval $0 \leq t \leq t_f$. This was done for Figures 1.15, 1.16, 1.17.

Algorithm 1.8   Simulation.   *Choose $\mathbf{z}$ and t_f. Integrate equation (1.1), starting from $\mathbf{y}(0) = \mathbf{z}$ and terminating at $t = t_f$.*

The choice of $\mathbf{z}$ and t_f strongly affects the amount of information provided by the numerical simulation. It makes sense to analyze possible stationary points first.

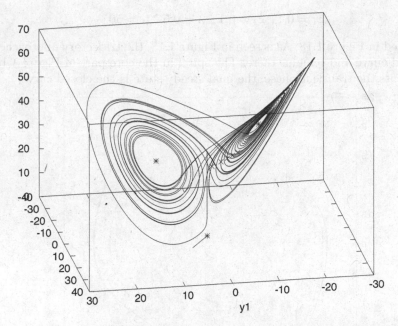

Fig. 1.17. Example 1.7; the "full" nonlinear Lorenz equation, parameters and bounding box as in the previous figures. Simulation for $0 \le t \le 20$ starting from $(0.001, 0.001, 35.001)$. A few hit points where the trajectory cuts the plane $y_3 = 29$ from above are shown by crosses; asterisks: stationary points. The unstable manifold is depicted for $(0, 0, 0)$.

Example 1.9   Van der Pol Equation

Several electrical and biological oscillations can be modeled by the van der Pol equation [Van27]. This equation is a differential equation of second order where the nonlinearity results from damping,

$$\ddot{u} - \lambda(1 - u^2)\dot{u} + u = 0. \tag{1.6}$$

The van der Pol equation (1.6) can be transformed into a system by means of, for instance, $y_1 = u$, $y_2 = \dot{u}$, leading to

$$\dot{y}_1 = y_2,$$
$$\dot{y}_2 = \lambda(1 - y_1^2)y_2 - y_1.$$

This system has one stationary solution at $(y_1^s, y_2^s) = (0, 0)$.

For $0 < \lambda < 2$, this point is an unstable focus (Exercise 1.10). To discuss the dynamics of the system for small positive values of λ, one tests by means of simulation where the trajectories tend as $t \to \infty$. The unstable focus at

(0,0) of the van der Pol equation suggests starting simulations near (0,0). The trajectory for the data

$$z_1 = 0.1\,,\ z_2 = 0.1\,,\ \lambda = 0.5\,,\ t_{\mathrm{f}} = 50$$

is plotted in Figure 1.18. As is seen in Figure 1.18, the trajectory approaches a closed curve and remains there. The spiral in the core part of Figure 1.18 represents the transient phase; the final steady state is the closed curve.

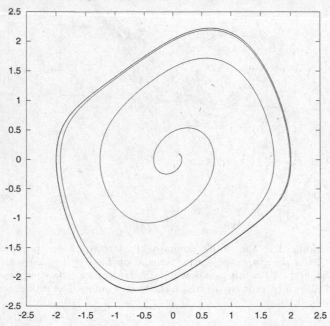

Fig. 1.18. Example 1.9, (y_1, y_2)-phase plane of the van der Pol equation, $\lambda = 0.5$; trajectory that winds on the limit cycle, starting from $(0.1, 0.1)$

1.3.2 Periodic Orbits

Such a closed curve toward which the trajectory winds is termed a *limit cycle*. A limit cycle is a *periodic solution* (periodic orbit). After some time period T is elapsed, the values of the solution $\mathbf{y}$ of a periodic orbit are the same,

$$\mathbf{y}(t + T) = \mathbf{y}(t)\,.$$

The minimum $T > 0$ satisfying this equation is called the *period*. For the above van der Pol example, for the chosen λ, we have a limit cycle with the period $T \approx 6.2$. For the chosen starting point $\mathbf{z}$, the transient phase before the limit cycle is reached is about $0 \le t \le 30$. Trajectories starting outside the limit cycle also wind toward it. Speaking in terms of stability, limit cycles are stable. Limit cycles are approached by nearby trajectories. There are also unstable periodic orbits, here trajectories leave their neighborhoods (see Exercise 1.14). Unstable periodic orbits are (surprisingly) important; we shall see examples of UPOs later in the text.

Limit cycles represent regular motions. Examples are nerve impulses, currents in electrical circuits, vibrations of violin strings, the flutter of panels, and laser light. Additional examples will be mentioned later.

Generating a limit cycle by means of a simulation is fairly easy. But in general, simulation is not necessarily the best way to create limit cycles; other methods will be discussed in Chapter 7.

1.3.3   An Analytical Method

Analytical methods for approximating limit cycles dominated for several decades. Since one object of this book is to make use of modern computers, we confine ourselves to illustrating one of the simplest analytical methods. We briefly illustrate van der Pol's *method of slowly varying amplitude* for Example 1.9 [equation (1.6)]. In this method, one seeks a solution in the form

$$u(t) = A(t) \cos \omega t$$

with frequency ω and amplitude $A(t)$. The amplitude $A(t)$ is assumed to be slowly varying,

$$|\dot{A}| \ll |A|\omega.$$

This assumption justifies approximating $\dot{u}$ and $\ddot{u}$ by the expressions

$$-\omega A \sin \omega t \approx \dot{u},$$

$$-2\omega \dot{A} \sin \omega t - \omega^2 A \cos \omega t \approx \ddot{u}.$$

Inserting the ansatz for u and these approximations to its derivatives into the van der Pol equation yields

$$0 = \sin \omega t [-2\omega \dot{A} + \lambda \omega A - \tfrac{1}{4}\lambda \omega A^3]$$
$$+ \cos \omega t [A - A\omega^2] - \tfrac{1}{4}\lambda \omega A^3 \sin 3\omega t.$$

The method is based on expanding the nonlinear term $u^2 \dot{u}$ into a Fourier series and neglecting subharmonic modes—that is, dropping the term $\sin 3\omega t$. Since the equality holds for all t, the two bracketed terms must vanish. Two equations result for amplitude A and frequency ω,

$$\dot{A} = \tfrac{1}{2}\lambda A - \tfrac{1}{8}\lambda A^3,$$
$$\omega = 1.$$

The equation for A is a Bernoulli differential equation. Its solution describes the initial transient phase, thereby revealing the stability characteristics of the limit cycle (Exercise 1.10). The limit cycle itself is obtained by studying the asymptotic behavior of A. After transient effects have died out, we have

$\dot{A} = 0$, which implies $A^2 = 4$. In summary, the method of slowly varying amplitude yields the circular motion

$$u(t) \approx 2\cos(t)$$

as a rough approximation of the limit cycle.

For an extensive discussion of how to apply analytical methods in the context of limit cycles, refer to [Sto50], [KiS78], [AnVK87], [Moi92], [MoC96].

1.3.4   Trajectories in Phase Space

So far we have discussed two types of attractors—namely, stable equilibria and limit cycles. These are specific trajectories, represented in phase space by points and closed curves. As a consequence of the existence and uniqueness theorems of ODEs (Appendix A.3) there are exactly three kind of trajectories, namely,

(1) stationary solutions: $\mathbf{y}(t) \equiv \mathbf{y}^s$, $\mathbf{f}(\mathbf{y}^s) = \mathbf{0}$;
(2) periodic solutions: $\mathbf{y}(t + T) = \mathbf{y}(t)$; and
(3) one-to-one solutions: $\mathbf{y}(t_1) \neq \mathbf{y}(t_2)$ for $t_1 \neq t_2$.

The last are also called "injective solutions." Type (3) means that trajectories never intersect, apart from the specific trajectories equilibria and periodic orbits. Injective trajectories are represented in phase space by *curves* that do not intersect. Since for $n > 2$ graphical illustrations of phase portraits are difficult, one must resort to *projections* of the curves onto two-dimensional subspaces, mostly to the (y_1, y_2)-plane. The projections of the curves may intersect although the trajectories in n-dimensional space $(n > 2)$ do not.

In the planar situation there are criteria for the existence of limit cycles. Two famous criteria attributed to Poincaré and Bendixson are included in Theorem 1.10.

Theorem 1.10   *Assume an ODE with n=2 on a planar domain $\mathcal{D}$.*

(a) Let $\mathcal{D}$ be a finite domain that contains no stationary point and from which no trajectory departs. Then $\mathcal{D}$ contains a limit cycle.

(b) If the expression

$$\frac{\partial f_1}{\partial y_1} + \frac{\partial f_2}{\partial y_2}$$

does not change sign within a domain $\mathcal{D}$, then no limit cycle can exist in $\mathcal{D}$.

Criterion (a) can be derived by geometrical arguments, whereas criterion (b) follows from the divergence theorem. Theorem 1.10 can be formulated in this two-dimensional situation because trajectories do not intersect; no trajectory can escape the planar domain given by the "interior" of a limit cycle. In more than two dimensions this "escaping" is possible and the dynamics can be much richer; phenomena such as chaos or trajectories on tori are possible.

Accordingly, the criteria of Theorem 1.10 do not hold for $n \geq 3$. Examples of limit cycles in higher-dimensional space will be given later.

1.3.5 Orbits with Infinite Period

In addition to stable stationary trajectories and limit cycles, there are limit sets that consist of trajectories connecting equilibria. When distinct saddles are connected, one encounters a *heteroclinic orbit* (illustrated by the phase diagram of Figure 1.19a); a heteroclinic orbit may also join a saddle to a node, or vice versa. A *homoclinic orbit* connects a saddle point to itself (see Figure 1.19b). For example, the specific Duffing equation $\ddot{w} - w + w^3 = 0$ has a homoclinic solution $w = \sqrt{2}/\cosh(t)$; compare also Exercise 3.7. Heteroclinic orbits and homoclinic orbits have an infinite "period," because the motion gets arbitrarily slow ($\dot{\mathbf{y}} \to \mathbf{0}$) for trajectories approaching the equilibrium. Several heteroclinic orbits may form a closed path, called a *homoclinic cycle* (see Figure 1.19d). In a nonplanar two-dimensional case (say, on a torus or on a cylinder in the case of a pendulum), and for $n > 2$, even further limit sets are possible. For example, Figure 1.19c shows a homoclinic orbit of a saddle-focus equilibrium in three-dimensional space; this orbit is also called *Silnikov connection* [GuH83], [Wig90]. For heteroclinic connections between periodic orbits, see [KoLMR00].

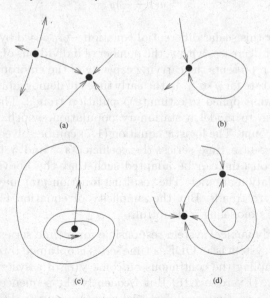

(a)   (b)

(c)   (d)

Fig. 1.19. Heteroclinic and homoclinic orbits

Heteroclinic and homoclinic orbits are of great interest in applications because they form the profiles of *traveling wave* solutions of many reaction-diffusion problems. We come back to this in Section 3.5.

1.4   Maps

In many applications, the state $\mathbf{y}$ of a system is determined not for "all" values of t, but for discrete time instances $t_1, t_2, t_3, \ldots$. Accordingly the sequence of n-vectors $\mathbf{y}$ forms a discrete *time series*

$$\mathbf{y}(t_1), \ \mathbf{y}(t_2), \ \mathbf{y}(t_3), \ldots.$$

To give an example (scalar case, $n=1$) we list the number of the human population in the years 1965, 1975, 1985, ... (in millions)

$$3289, \ 4070, \ 4842, \ldots.$$

Often a law or a differential equation on which the time series is based is unknown. Finding a model equation that matches a time series can be a formidable task. Here we give a simple model related to the growth of populations.

Example 1.11   Logistic Differential Equation
A differential equation that can be related to population time series is the logistic differential equation

$$\dot{y} = \delta(1 - \frac{y}{\gamma})y. \tag{1.7}$$

Characteristic for this scalar differential equation—proposed by Verhulst—is a growth rate that decreases with y, the number of individuals of a population. The parameter γ represents the carrying capacity of the environment and δ is the linear growth rate for $y \ll \gamma$. In the early times of demography, this simple model equation was applied to estimate population trends. The equilibrium $y = \gamma$ allows one to model a stationary population, which takes limited resources into account. The logistic equation (1.7) can be solved analytically. From three values of a time series, the coefficients δ and γ together with the integration constant can be adapted such that the above numbers of the human population are met. The resulting function $y(t)$ may be carefully applied to estimate trends. But the simplicity of equation (1.7) contrasts reality, and so this modeling is doubtful.   □

Sometimes a dynamical system responsible for a time series is known. In case a dynamical system is an ODE, a time series is obtained by discretization —that is, by sampling the continuous outcome $\mathbf{y}(t)$ of a system at discrete time instances t_ν. (Exercise 1.16) But frequently the sequence of vectors is not derived from a continuous differential equation, but is defined by some discrete *map*. A map is given by a vector function $\mathbf{P}(\mathbf{y})$, which maps a vector $\mathbf{y}^{(\nu)}$ onto a vector $\mathbf{y}^{(\nu+1)}$,

$$\mathbf{y}^{(\nu+1)} = \mathbf{P}(\mathbf{y}^{(\nu)}). \tag{1.8}$$

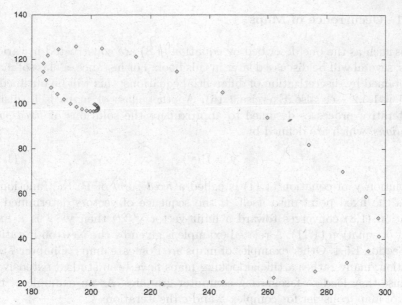

Fig. 1.20. Example 1.12, iteration (1.10), phase-diagram of an orbit of $(y_1(t_\nu), y_2(t_\nu))$-values for $\nu = 0, 1, ..., 40$ and $\lambda = 50$, $b = 0.002$, starting from (200,20).

In case $\mathbf{P}(\mathbf{y})$ reflects dynamical behavior, the vectors $\mathbf{y}^{(\nu)}$ have the interpretation $\mathbf{y}(t_\nu)$. In this case ν can be seen as discrete time, and we speak of a *discrete dynamical system*. The sequence $\mathbf{y}^{(1)}, \mathbf{y}^{(2)}, \mathbf{y}^{(3)}$, or $\mathbf{y}_1, \mathbf{y}_2, \mathbf{y}_3, \ldots$ is iteratively determined by (1.8), and is called *orbit* of the map. As a scalar example of a map, consider the function $P(y) = \lambda y(1 - y)$, which leads to the *logistic map*

$$y^{(\nu+1)} = \lambda y^{(\nu)} \left(1 - y^{(\nu)}\right). \tag{1.9}$$

Example 1.12   Student Migration

In this model of student migration within a mathematics graduate school, $y_2(t_\nu)$ denotes the number of students choosing in year t_ν to major in a specific field (numerical analysis, say), and $y_1(t_\nu)$ is the number of students who prefer any other field. The model [ScS00] (motivated by [Fei92]) is

$$\begin{aligned}
y_1(t_{\nu+1}) &= y_1(t_\nu) + \lambda - b\, y_1(t_\nu)\, y_2(t_\nu) + \beta\, y_2(t_\nu) - d_1\, y_1(t_\nu) \\
y_2(t_{\nu+1}) &= y_2(t_\nu) + b\, y_1(t_\nu)\, y_2(t_\nu) - \beta\, y_2(t_\nu) - d_2\, y_2(t_\nu),
\end{aligned} \tag{1.10}$$

with artificial constants

$$d_1 = 0.1, \ d_2 = 0.3, \ \beta = 0.1.$$

λ is the constant input rate of fresh students per year, b describes the migration, dominated by communication, β is the backflow, and d_1, d_2 are the successful examination rates. Figure 1.20 shows a simulation, where the iteration is attracted by an equilibrium.

1.4.1 Occurrence of Maps

Maps such as the one described by equation (1.8) are constructed in various ways; several will be discussed later in this book. For instance, a class of maps is obtained by discretization of differential equations; this will be outlined in Section 1.5.2 (see also Exercise 1.16). Another class of maps is furnished by iterative processes designed to approximate the solutions of *fixed-point equations*, which are defined by

$$\mathbf{y} = \mathbf{P}(\mathbf{y}). \tag{1.11}$$

A solution $\mathbf{y}$ of equation (1.11) is called a *fixed point* of $\mathbf{P}$; the function $\mathbf{P}$ maps the fixed point onto itself. If the sequence of vectors determined by equation (1.8) converges toward a limit vector $\mathbf{y}^{(\infty)}$, then $\mathbf{y}^{(\infty)}$ is a fixed point of equation (1.11). A related example is given by the Newton iteration; see Section 1.5.1. Other examples of maps are Poincaré maps (Chapter 7). In addition, many rather artificial looking maps have been studied extensively, because they lead to fascinating computer graphics. As an example of this kind of map, consider for complex z and c the iteration

$$z_{\nu+1} = z_\nu^2 + c. \tag{1.12}$$

The complex equation (1.12) is equivalent to two real scalar equations—that is, equation (1.12) maps points of a z_ν-plane into the $z_{\nu+1}$-plane. Starting from $z_0 = 0$ one decides by means of computer experiments for given c whether the sequence of points z_ν diverges. Depending on the outcome of the test a color is assigned to c. Proceeding in this way for all c from a lattice that corresponds to pixels of a screen, one obtains beautiful patterns in the c-plane.

"Divergence" is assumed when for $\nu = \nu_{\max}$ the iterate z_ν exceeds some bound ρ, $|z_\nu| > \rho$. Both $\nu_{\max}$ and ρ have to be chosen large enough, say, in the range between 10 and 1000. Assign to c the color "black" if and only if $|z_\nu| < \rho$ for $\nu = 1, 2, ..., \nu_{\max}$. The black area in the c-plane is called *Mandelbrot set*; illustrations are well-known (see, e.g., [Man83], [PeR86]). The boundary of this set is no smooth curve, but has a *fractal* structure. The complementary set (for which "divergence" occurs) can be colored according to the ν for which the bound ρ is reached. Other examples of sets with fractal boundaries are often provided by basins of attraction. For instance, we may ask for the "escape set" of those z for which the iteration of equation (1.12) converges to infinity. The boundary of such a a basin of attraction is called *Julia set*.

1.4.2 Stability of Fixed Points

We discuss the stability of fixed points $\mathbf{y}$—that is, the behavior of $\mathbf{y}^{(\nu)}$ close to $\mathbf{y}$. The definitions of "stable" and "asymptotically stable" parallel those for equilibria of ODEs; see Definitions 1.1 and 1.2. Consider the dynamics of the distances $\mathbf{y}^{(\nu)} - \mathbf{y}$ between iterates $\mathbf{y}^{(\nu)}$ and a fixed point $\mathbf{y}$ ($\nu = 0, 1, 2, \ldots$). Equations (1.8) and (1.11) imply

$$\mathbf{y}^{(\nu+1)} - \mathbf{y} = \mathbf{P}(\mathbf{y}^{(\nu)}) - \mathbf{P}(\mathbf{y}) = \frac{\partial \mathbf{P}(\mathbf{y})}{\partial \mathbf{y}}(\mathbf{y}^{(\nu)} - \mathbf{y}) + \text{ t.h.o.}$$

Hence, we expect the local stability to be governed by the linearization (up to terms of higher order, t.h.o.),

$$\mathbf{y}^{(\nu+1)} - \mathbf{y} = \mathbf{J} \cdot (\mathbf{y}^{(\nu)} - \mathbf{y}),$$

which implies

$$\mathbf{y}^{(\nu)} - \mathbf{y} = \mathbf{J}^{\nu}(\mathbf{y}^{(0)} - \mathbf{y}). \tag{1.13}$$

Here $\mathbf{J}$ denotes the Jacobian matrix $\mathbf{P_y}$ of the n^2 first-order partial derivatives of $\mathbf{P}$, evaluated at $\mathbf{y}$. To simplify the analysis, we concentrate on diagonalizable matrices $\mathbf{J}$. Then there exists a set of n linear independent eigenvectors $\mathbf{w}^k$ with eigenvalues μ_k such that the initial distance $\mathbf{y}^{(0)} - \mathbf{y}$ can be written as a linear combination, and equation (1.13) implies

$$\mathbf{y}^{(\nu)} - \mathbf{y} = \mathbf{J}^{\nu} \sum_{k=1}^{n} c_k \mathbf{w}^k = \sum_{k=1}^{n} c_k \mathbf{J}^{\nu} \mathbf{w}^k = \sum_{k=1}^{n} c_k \mu_k^{\nu} \mathbf{w}^k,$$

which shows that convergence $\mathbf{y}^{(\nu)} \to \mathbf{y}$ can only take place for $\nu \to \infty$ when $|\mu_k| < 1$ for all $k = 1, \ldots, n$. The general result is expressed by the following theorem:

Theorem 1.13   *Assume that all eigenvalues μ_k of $\mathbf{J}$ lie inside the unit circle, $|\mu_k| < 1$. Then, locally, the iterates $\mathbf{y}^{(\nu)}$ converge toward $\mathbf{y}$, which is a stable fixed point.*

In analogy to Definition 1.5, a fixed point is called *hyperbolic* if none of the eigenvalues μ of $\mathbf{J}$ lies on the unit circle, $|\mu_k| \neq 1$ for $k = 1, \ldots, n$. At first glance, the difference between the stability results for equilibria of differential equations (Theorem 1.6: $\text{Re}(\mu) < 0$) and those for fixed points of maps (Theorem 1.13: $|\mu| < 1$) might be surprising. To clarify differences, we repeat in Table 1.2 corresponding relations. The explanation lies in the properties of the complex function $\exp(\mu)$, which maps the "left" half of the complex plane onto the disk inside the unit circle (Exercise 1.17).

1.4.3   Cellular Automata

As will be explained in Section 3.5, excitable media can be described by PDEs of the reaction-diffusion type. Let the state variable of such a PDE be $y(x, t)$, where the space variable x, time t, and y vary continuously—that is, take values in $\mathbb{R}$. *Cellular automata* represent the counterpart to PDEs in that both space x and time t are discrete, and even the state variable y only takes values from a finite discrete set. In the simplest cases the values

TABLE 1.2.   Ingredients of stability analysis.

$\dot{\mathbf{y}} = \mathbf{f}(\mathbf{y})$	$\mathbf{y}^{(\nu+1)} = \mathbf{P}(\mathbf{y}^{(\nu)})$
$t \to t + \Delta$	$\nu \to \nu + 1$
$\mathbf{f}(\mathbf{y}_0) = \mathbf{0}$	$\mathbf{y}_0 = \mathbf{P}(\mathbf{y}_0)$
$\mathbf{A} = \mathbf{f_y}$	$\mathbf{J} = \mathbf{P_y}$
$\mathbf{A}\mathbf{w_A} = \mu_\mathbf{A}\mathbf{w_A}$	$\mathbf{J}\mathbf{w_J} = \mu_\mathbf{J}\mathbf{w_J}$
$\dot{\mathbf{y}} = \mathbf{A}(\mathbf{y} - \mathbf{y}_0) + \text{t.h.o.}$	$\mathbf{y}^{(\nu+1)} - \mathbf{y}_0 = \mathbf{J}(\mathbf{y}^{(\nu)} - \mathbf{y}_0) + \text{t.h.o.}$
$\dot{\mathbf{h}} = \mathbf{A}\mathbf{h}$	$\mathbf{h}^{(\nu+1)} = \mathbf{J}\mathbf{h}^{(\nu)}$
$\mathbf{h}(t) = \sum c \exp(\mu_\mathbf{A} t)\mathbf{w_A}$	$\mathbf{h}^{(\nu)} = \sum c\mu_\mathbf{J}^\nu \mathbf{w_J}$
$\exp(\mu_\mathbf{A} t) \to 0$	$\mu_\mathbf{J}^\nu \to 0$
$\mathrm{Re}(\mu_\mathbf{A}) < 0$	$|\mu_\mathbf{J}| < 1 .$

are integers. Hence, calculating propagating signals in cellular automata is less costly than solving PDEs. (But PDEs seem to reflect underlying physical laws better than cellular automata do.)

Cellular automata are excitable media consisting of two ingredients: a lattice of *cells* representing discrete locations in space, and a set of *rules* that assigns a state to each cell. To visualize a cell (or *site*) think of a tree in a forest, a nerve cell within a membrane, or an individual of a population. Each cell is surrounded by neighboring cells, and there is some interaction among the cells. The cells of cellular automata form a lattice, either finite or infinite. For a two-dimensional example, think of the cells to be arranged as the sites of a checkerboard. The state of a cell can be "excited," "rested," or "recovering." (Depending on the application, these states are also called "firing," "ready," "refractory.") "Excitation" may mean catching fire for the tree in the forest or catching a disease in case of a population. The spread of the excited state is governed by rules that define the state of a cell at time $\nu + 1$ depending on the situation at time ν. The rule incorporates the interaction among the cells within an appropriate neighborhood. For a finite two-dimensional lattice, each site (cell) is represented by the location of an element in a rectangular matrix $\mathbf{C}$. Hence, a matrix $\mathbf{C}^{(\nu)}$ can be used to describe and store the state of the lattice at time ν. Writing the rules symbolically as $\mathbf{P}(\mathbf{C})$ shows that a cellular automaton can be seen as an application of maps, $\mathbf{C}^{(\nu+1)} = \mathbf{P}(\mathbf{C}^{(\nu)})$. A finite one-dimensional array then is represented by $\mathbf{y}^{(\nu+1)} = \mathbf{P}(\mathbf{y}^{(\nu)})$. The rules constituting $\mathbf{P}$ are discrete, defined with "if ... then ..." statements. For a finite string or array of cells special provisions are required for those cells that lie at the edge. Frequently one assumes a *wrap-around* situation in which the edges of the lattice are glued together. For instance, in a one-dimensional chain of cells, a ring results and the rules are periodic.

Example 1.14   Cells in a Ring

Consider n equal cells arranged in a one-dimensional closed ring [MaS91]. The cells are numbered with index k, $k = 1, \ldots, n$, and the state of the k-th cell at time ν is described by $y_k^{(\nu)}$. Define a neighborhood of cell k by the five cells $k-2$, $k-1$, k, $k+1$, $k+2$. We say, for instance, that cell k is excited when $y_k = 3$, at rest when $y_k = 0$, and in a refractory state when $y_k = 2$ or $y_k = 1$. Let j be the number of cells excited within the neighborhood of cell k. A cell k will get excited at the next time $\nu+1$ when two conditions are satisfied: the cell is at rest $(y_k^{(\nu)} = 0)$, and at least m cells in a neighborhood are excited. The number m serves as a *threshold*, which we set to 1. In summary, the rules for the step $\nu \to \nu + 1$ are as follows:

$$\text{if } y_k^{(\nu)} = 0 \text{ and } j \geq 1 : \text{ set } y_k^{(\nu+1)} = 3,$$
$$\text{if } y_k^{(\nu)} = 0 \text{ and } j < 1 : \text{ set } y_k^{(\nu+1)} = 0,$$
$$\text{if } y_k^{(\nu)} > 0 : \text{ set } y_k^{(\nu+1)} = y_k^{(\nu)} - 1.$$

Choosing $n = 16$ cells and as initial state $\mathbf{y}^{(0)}$ the pattern (closed to a ring),

$$0011223300112233\ldots$$

then this pattern will travel periodically around the ring. The reader is encouraged to check this and test what happens with other initial states.   □

Such a model is generalized in various ways: The size of the neighborhood can be defined via a parameter ρ, and the excited state can be defined to be $N + 1$, which allows for N refractory states. Then the outcome of possible patterns will depend on the choice of ρ, N, m, and n. Cellular automata have been discussed repeatedly, see, for instance, [GrH78], [Wol84], [GeST90], [MaH90].

Between fully continuous PDEs and fully discrete cellular automata there are mixed models possible. Because in the scalar setting three items (space, time, or state) can be chosen to be continuous or discrete, the hierarchy of spatiotemporal dynamical systems with one space variable consists of eight classes. One of the classes consists of systems with a continuous state variable and discrete time and discrete space. Such systems, which have been called *coupled map lattices* or *lattice dynamical systems* [CrK87], are also maps. For instance, assume a chain of cells with logistic dynamics, in which the state of cell k is coupled with the state of the two nearest neighbors $k - 1$ and $k + 1$:

$$y_k^{(\nu+1)} = \lambda y_k^{(\nu)}(1 - y_k^{(\nu)}) + \epsilon(y_{k-1}^{(\nu)} + y_{k+1}^{(\nu)}).$$

The parameter ϵ controls the intensity of the coupling. More generally, a lattice dynamical system with one spatial dimension and "nearest-neighbor coupling" is

$$y_k^{(\nu+1)} = f(y_k^{(\nu)}) + \epsilon_0 g(y_k^{(\nu)}) + \epsilon_R g(y_{k-1}^{(\nu)}) + \epsilon_L g(y_{k+1}^{(\nu)})$$

with $g(y)$, ϵ_0, ϵ_R, ϵ_L describing the coupling dynamics. For the above example of logistic dynamics we have a linear coupling $g(y) = y$ with $\epsilon_0 = 0$,

$\epsilon_R = \epsilon_L = \epsilon$. Clearly, a linear coupling with $-\epsilon_0 = 2\epsilon_R = 2\epsilon_L$ mimics the second-order spatial derivative y_{xx}.

Obviously, this field is tempting for computer freaks. Those who know how to handle graphical output should be inspired to start right away; no exercises or projects need be formulated. Try also asymmetric coupling, and $g(y) = f(y)$.

1.5   Some Fundamental Numerical Methods

In the previous sections we defined stationary solutions and mentioned numerical integration as a method of investigating dynamical behavior near such solutions. In the present section, we briefly review some important numerical methods that help in analyzing nonlinear phenomena. Most of the problems of stability analysis can be reduced or transformed in such a way that their solution requires only a small number of standard tools of numerical analysis. Every student and researcher should put together his own toolbox and not rely solely on "packages." The reader familiar with the Newton method and related iterative procedures for nonlinear equations may want to skip this section.

1.5.1   Newton's Method

Stationary solutions $\mathbf{y}^s$ of systems of ODEs are defined as solutions of the (nonlinear) vector equation $\mathbf{f}(\mathbf{y}) = \mathbf{0}$. Many problems of mathematics can be cast into such an equation. For example, locating fixed points of $\mathbf{P}(\mathbf{y}) = \mathbf{y}$ is equivalent to calculating the roots of

$$\mathbf{f}(\mathbf{y}) := \mathbf{P}(\mathbf{y}) - \mathbf{y} = \mathbf{0}.$$

The classical means to solve $\mathbf{f}(\mathbf{y}) = \mathbf{0}$ is Newton's method. By means of a linearization of $\mathbf{f}$, one obtains

$$\mathbf{0} = \mathbf{f}(\mathbf{y}^s) \approx \mathbf{f}(\mathbf{y}^{(0)}) + \mathbf{f}_{\mathbf{y}}^0 \cdot (\mathbf{y}^s - \mathbf{y}^{(0)}).$$

Here, $\mathbf{y}^{(0)}$ is an approximation of $\mathbf{y}^s$, and $\mathbf{f}_{\mathbf{y}}^0$ is the Jacobian matrix evaluated at $\mathbf{y}^{(0)}$. The above relation establishes a way of calculating a better approximation $\mathbf{y}^{(1)}$ by solving

$$\mathbf{0} = \mathbf{f}(\mathbf{y}^{(0)}) + \mathbf{f}_{\mathbf{y}}^0 \cdot (\mathbf{y}^{(1)} - \mathbf{y}^{(0)}). \tag{1.14}$$

Equation (1.14) is a system of linear equations. Solving it for the quantity $\mathbf{y}^{(1)} - \mathbf{y}^{(0)}$ and adding this quantity to $\mathbf{y}^{(0)}$ gives a better approximation $\mathbf{y}^{(1)}$. Applying equation (1.14) repeatedly produces a series of approximations

$$\mathbf{y}^{(1)}, \ \mathbf{y}^{(2)}, \ \mathbf{y}^{(3)}, \dots .$$

This method is called the Newton method or the Newton–Raphson method. Introducing vectors $\mathbf{d}^{(\nu)}$ for the difference between two iterates, namely, $\mathbf{y}^{(\nu)} - \mathbf{y}^{(\nu+1)}$, we have for $\nu = 0, 1, 2, \ldots$

$$\frac{\partial \mathbf{f}(\mathbf{y}^{(\nu)})}{\partial \mathbf{y}} \mathbf{d}^{(\nu)} = \mathbf{f}(\mathbf{y}^{(\nu)}), \qquad (1.15a)$$

$$\mathbf{y}^{(\nu+1)} = \mathbf{y}^{(\nu)} - \mathbf{d}^{(\nu)}. \qquad (1.15b)$$

Under certain assumptions (the most important of which is nonsingularity of the Jacobian at the solution), one can guarantee *quadratic* convergence in a neighborhood of the solution. That is, both $\mathbf{f}(\mathbf{y}^{(\nu)})$ and $\mathbf{d}^{(\nu)}$ quickly approach the zero vector. Roughly speaking, the number of correct decimal digits doubles on each step. In practice, however, the usual assumptions ensuring this rate of convergence are not always satisfied. Therefore we often encounter limitations on the rate of convergence.

First, locally, we cannot expect quadratic convergence when the Jacobian is replaced by an approximation (see below). The second limitation is by far more severe. For many difficult problems an arbitrary initial guess $\mathbf{y}^{(0)}$ does not lead to convergence of the sequence produced by equation (1.15), because $\mathbf{y}^{(0)}$ may be too far away from a solution. In order to enlarge the domain of convergence, one modifies Newton's method as follows: A minimization technique replaces equation (1.15b) by

$$\mathbf{y}^{(\nu+1)} = \mathbf{y}^{(\nu)} - \delta \mathbf{d}^{(\nu)} \qquad (1.15b')$$

with δ such that $\|\mathbf{f}(\mathbf{y}^{(\nu+1)})\|$ is minimal, or at least closer to zero than $\|\mathbf{f}(\mathbf{y}^{(\nu)})\|$. The resulting *damped* Newton method, equations (1.15a) and (1.15b'), has reasonable and often global convergence properties. But one must be aware that a fast quadratic convergence is still restricted to a (small) neighborhood of the solution; the global initial phase may be slow (linear convergence).

The evaluation of the Jacobian in equation (1.15) mostly is expensive. Because analytic expressions for the partial derivatives of $\mathbf{f}$ are often unavailable, implementations of the Newton method make use of *numerical differentiation*. Let $\mathbf{z}^k$ be the kth column of the Jacobian $\mathbf{f}_y$. The following algorithm can be used to calculate an approximation of $\mathbf{f}_y(\mathbf{y})$ ($\mathbf{U}$ and $\mathbf{V}$ are two auxiliary n-vectors):

Algorithm 1.15   Numerical Differentiation
Evaluate $\mathbf{U} = \mathbf{f}(\mathbf{y})$
Choose a small $\epsilon > 0$
Loop for $k = 1, 2, \ldots, n$:
    $\epsilon_1 = \epsilon \max\{1, y_k\} y_k / |y_k|$
    $y_k = y_k + \epsilon_1$
    evaluate the vector $\mathbf{V} = \mathbf{f}(\mathbf{y})$
    $y_k = y_k - \epsilon_1$

$$\mathbf{z}^k = (\mathbf{V} - \mathbf{U})/\epsilon_1 \ .$$

The choice of ϵ depends on the number of mantissa digits l available on a particular computer. Usually,

$$\epsilon = 10^{-l/2}$$

leads to a reliable approximation of the Jacobian. The cost of approximating the n^2 matrix $\mathbf{f_y(y)}$ by means of Algorithm 1.15 is that of evaluating the vector function $\mathbf{f}$ n times (the evaluation of $\mathbf{U}$ was usually carried out earlier).

Evaluating the Jacobian or approximating it by Algorithm 1.15 is so expensive that remedies are needed. The simplest strategy is to replace equation (1.15a) by

$$\mathbf{f_y^0} \cdot \mathbf{d}^{(\nu)} = \mathbf{f}(\mathbf{y}^{(\nu)}) \ .$$

This *chord method* involves the same approximate Jacobian for all iterations or at least for a few iterations. The chord method is a slow procedure. There are other variants of the Newton method with convergence rates ranging between the linear convergence of the chord method and the quadratic convergence of the original Newton method. The Jacobian can be approximated by algorithms that are cheaper (and coarser) than the procedure in Algorithm 1.15. Procedures, called update methods (e.g., Broyden's method [Bro65]), are in wide use.

In such methods, a matrix of rank one, which requires only one function evaluation, is added to the approximate Jacobian of the previous step. For details we refer to textbooks on numerical analysis, for instance, [StB80], [DaB08]. The use of rank-one updates reduces the convergence rate per iteration step to *superlinear* convergence. The saving in function evaluations is, however, significant, and the overall computing time is usually reduced. This strategy of combining numerical differentiation with rank-one updates is excellent. One takes advantage of rank-one updates especially when the convergence is satisfactory. This kind of *quasi-Newton method* is a compromise between the locally fast but expensive Newton method and the cheap but more slowly converging chord method [DeM77], which is highly recommended.

A general treatment of methods for solving nonlinear equations is given by Ortega and Rheinboldt [OrR70]. The linear equation (1.15a) can be solved by Gaussian elimination or by **LU** decomposition. Recall that the latter means the calculation of a lower triangular matrix **L** and an upper triangular matrix **U** such that

$$\mathbf{f_y} = \mathbf{LU} \ .$$

The diagonal elements of **L** are normalized to unity. Because this **LU** decomposition does not always exist, a permutation of the rows of $\mathbf{f_y}$, which reflects the pivoting of the Gaussian elimination, may be necessary. A sophisticated algorithm for solving linear equations is given in [WiR71]; see also [GoL96].

Because equation (1.15) can be written in the form $\mathbf{y}^{(\nu+1)} = \mathbf{P}(\mathbf{y}^{(\nu)})$, Newton iterations are examples of maps. A solution to $\mathbf{f(y)} = \mathbf{0}$ is the fixed point of the mapping defined in equation (1.15). Another way to derive a fixed-point equation from $\mathbf{f}(\mathbf{y}) = \mathbf{0}$ (and vice versa) is via $\mathbf{P}(\mathbf{y}) = \mathbf{f}(\mathbf{y}) + \mathbf{y}$. Each root of $\mathbf{f}(\mathbf{y}) = \mathbf{0}$ defines a basin of attraction: the set of all initial vectors $\mathbf{y}^{(0)}$ such that the series defined by the Newton method converges to the root. Basins of attraction of zeros of $\mathbf{f}(\mathbf{y}) = \mathbf{0}$ can have a complicated structure. For instance, consider the Newton iteration for approximating the three complex roots of the map $f(z) = z^3 - 1$ for complex z. This map is given by

$$z_{\nu+1} = \frac{1}{3}\left(2z_\nu - \frac{1}{z_\nu^2}\right);$$

the three roots are 1, $e^{2\pi i/3}$, and $e^{4\pi i/3}$. The three basins of attraction are highly interlaced, with a fractal boundary (Julia set).

1.5.2 Integration of ODEs

Another basic tool of numerical analysis is numerical methods for integration of the initial-value problem

$$\dot{\mathbf{y}} = \mathbf{f}(t, \mathbf{y}) , \quad \mathbf{y}(t_0) = \mathbf{z} .$$

The solution $\mathbf{y}(t)$ is approximated at discrete values t_k of the independent variable t,

$$\mathbf{y}(t_1), \ \mathbf{y}(t_2), \ \mathbf{y}(t_3), \ldots .$$

Here, for notational convenience, we also use the letter $\mathbf{y}$ for the numerical approximation of the exact solution. The difference between two consecutive t_ks is the *step size*, $\Delta_k = t_{k+1} - t_k$. Typically one connects the chain of calculated vectors $\mathbf{y}(t_k)$ by an interpolating curve, or by pieces of straight lines, as in Figure 1.21. There are numerous algorithms for numerical integration. We discuss some basic principles for a generic *step* from t to $t + \Delta$. That is, $\mathbf{y}(t)$ is assumed to be calculated, and $\mathbf{y}(t + \Delta)$ is the target to be calculated next as approximation of the true solution at $t + \Delta$.

Among the most reliable integrators are the Runge–Kutta methods with automatic step control, such as the Runge–Kutta–Fehlberg methods (RKF). Methods of Runge–Kutta type are one-step methods in that calculation of the approximation to $\mathbf{y}(t_{k+1})$ involves only data of one previous step $\mathbf{y}(t_k)$. One Runge–Kutta step has the form

$$\mathbf{y}(t + \Delta) = \mathbf{y}(t) + \Delta \cdot \sum_{l=1}^{m} c_l \mathbf{f}(\tilde{t}_l, \tilde{\mathbf{y}}^l) . \tag{1.16}$$

In this m-*stage* formula, the right-hand side $\mathbf{f}$ is evaluated at certain values of $\tilde{t}_l$ and $\tilde{\mathbf{y}}^l$. That is, in order to proceed from $\mathbf{y}(t)$ to $\mathbf{y}(t + \Delta)$, equation (1.16) averages over m values of the slope $\dot{\mathbf{y}}$, evaluated at suitable points in the flow between t and $t + \Delta$ (see Appendix A.6). The step size Δ is varied from step to step so that certain accuracy requirements are met [Feh69].

The most elementary integration scheme is the *explicit Euler method* or "forward Euler"

$$\mathbf{y}(t + \Delta) = \mathbf{y}(t) + \Delta \mathbf{f}(t, \mathbf{y}), \tag{1.17}$$

which is of first order. For second-order accuracy, two-stage methods are required ($m \geq 2$). A class of two-stage solvers is given by

$$\mathbf{y}(t + \Delta) = \mathbf{y}(t) + \Delta \left\{ (1 - c)\, \mathbf{f}(t, \mathbf{y}) + c\, \mathbf{f}(t + \frac{\Delta}{2c}, \mathbf{y} + \frac{\Delta}{2c}\, \mathbf{f}(t, \mathbf{y})) \right\}.$$

For instance, for $c = 1/2$ and autonomous systems, this Runge–Kutta map simplifies to Runge's method

$$\mathbf{y}(t + \Delta) = \mathbf{y}(t) + \frac{\Delta}{2}\, \mathbf{f}(\mathbf{y}) + \frac{\Delta}{2}\, \mathbf{f}(\mathbf{y} + \Delta \mathbf{f}(\mathbf{y})). \tag{1.18}$$

All the above numerical integration schemes provide explicit expressions for $\mathbf{y}(t + \Delta)$, and can be written in the form

$$\mathbf{y}(t + \Delta) = \mathbf{P}(\mathbf{y}(t)) , \quad \text{or} \quad \mathbf{y}^{(\nu+1)} = \mathbf{P}(y^{(\nu)}) .$$

Many ODE problems can be integrated with the above class of explicit methods of Runge–Kutta type. But there are ODE applications where these methods hardly work. This holds in particular for *stiff* differential equations. Stiffness can be defined in the easiest way for linear systems of ODEs in the form $\dot{\mathbf{y}} = \mathbf{A}\mathbf{y}$, where the matrix $\mathbf{A}$ has only eigenvalues with negative real parts, which differ by orders of magnitude. For such stiff systems, *implicit* methods are required. Typically, for implicit integrators, the right-hand side $\mathbf{f}$ is evaluated for the unknown next step, $\mathbf{f}(t + \Delta, \mathbf{y}(t + \Delta))$. The prototype example is the *implicit Euler method*

$$\mathbf{y}(t + \Delta) = \mathbf{y}(t) + \Delta \mathbf{f}(t + \Delta, \mathbf{y}(t + \Delta)), \tag{1.19}$$

which is of first order. Implicit methods are generated, for example, by multistep methods. For other implicit methods see the literature, for example, [HaNW87], [HaW91], [Lam91], [HaLW02]. For each step $t \to t + \Delta$, an implicit method such as (1.19) calculates a solution of the nonlinear implicit equation for $\mathbf{y}(t + \Delta)$ iteratively. In simple cases, for small enough step sizes $|h|$, one step of a fixed-point iteration may suffice. But for stiff problems the nonlinear systems must be solved by a Newton-like method, as described in Section 1.5.1.

Another highly successful class of integrators are the *collocation methods*. Collocation methods define at each step a local polynomial $\mathbf{u}$ by requiring that it solves the differential equation at specified points $t + c_j \Delta$. Then $\mathbf{y}(t + \Delta)$ is taken as $\mathbf{u}(t + \Delta)$. That is, the polynomial $\mathbf{u}$ in the step from t to $t + \Delta$ is defined by

$$\mathbf{u}(t) = \mathbf{y}(t) ; \quad \dot{\mathbf{u}}(t + c_j \Delta) = \mathbf{f}(t + c_j \Delta, \mathbf{u}(t + c_j \Delta)),$$

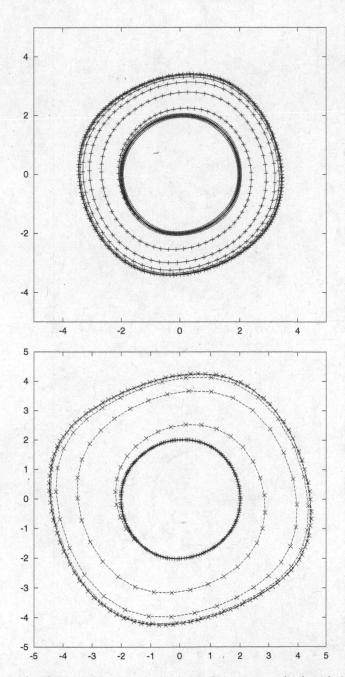

Fig. 1.21. $(u, \dot{u})$-phase plane of the van der Pol equation (1.6) with $\lambda = 0.05$. In each figure the trajectories are calculated by two integrators: Runge's scheme (1.18) and explicit Euler scheme (1.17). starting point for both: $(u, \dot{u}) = (-1.244, -1.6045)$. top: step length $\Delta = 0.1$, bottom: $\Delta = 0.2$

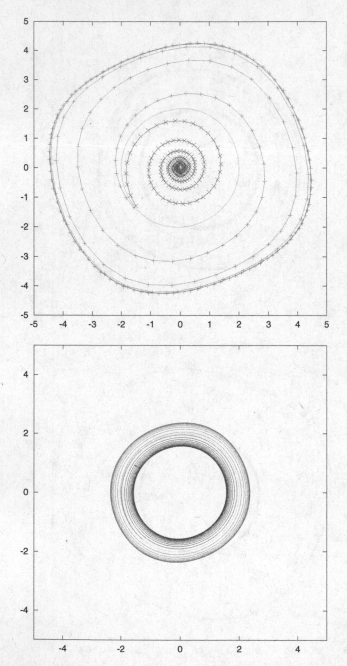

Fig. 1.22. As in Figure 1.21: van der Pol equation (1.6) with $\lambda = 0.05$; trajectories obtained by two integrators: explicit Euler scheme (1.17) and implicit Euler (1.19). top: step length $\Delta = 0.2$, bottom: $\Delta = 0.02$. The true limit cycle is shown separating the trajectories of the Euler schemes.

for $j = 1, \ldots, m$. By choosing suitable weights c_j different collocation methods are defined. Simple examples for $m = 1$ are known already from the above: $c_1 = 0$ leads to the explicit Euler, and $c_1 = 1$ recovers the implicit Euler (the reader may check). For $c_1 = 0.5$ the *implicit midpoint rule* results, which is of second order. Powerful methods for $m = 2$ include the *implicit trapezoidal rule* ($c_1 = 0$, $c_2 = 1$), and the *Gauss method of fourth order* ($c_{1,2} = \frac{1}{2} \mp \frac{\sqrt{3}}{6}$). The latter also has favorable properties in preserving invariants of the differential equation [HaLW02]. A practical advantage of the collocation approach is that via **u** a continuous approximation is given right away.

Of course, the quality of the approximations depends on the step size Δ, and is characterized by its order. But the situation is more subtle because even the quality of the attractors may change by the discretization. Let us illustrate this by means of an experiment.

Figure 1.21 illustrates how the location of a discrete-analog limit cycle depends on the step size Δ. We compare for a van der Pol equation (1.6) with parameter $\lambda = 0.05$ results obtained by explicit Euler for the two crude constant step sizes $\Delta = 0.1$ and $\Delta = 0.2$. Each trajectory starts from the same point on the limit cycle of the ODE. The steps $\mathbf{y}(t_k)$ are marked by crosses. The limit cycle obtained by the map based on Euler's method (1.17) (outer loops in Figure 1.21) differ significantly from the correct continuous-time limit cycle. The latter is reflected by the inner loop, which is obtained by Runge's method (1.18). In comparison to Euler's method, the second-order Runge method essentially shows reasonable accuracy for both choices of Δ. — The experiment (following [HaLW02]) is extended in Figure 1.22 with the application of the implicit Euler method (1.19). If the step size Δ is too large, the implicit Euler does not see a limit cycle, and apparently converges to a different type of attractor, namely, to the stationary solution (inner curve in Figure 1.22, top). A discrete integration scheme can stabilize an unstable state. If the step size is small enough ($\Delta = 0.02$ in Figure 1.22, bottom), then both Euler methods have a limit cycle, but still far away from its correct location. This experiment should warn the reader that a numerical discretization may have other attractors than the ODE. This rarely happens when the step lengths are really small; for the above experiment Δ was chosen intentionally large.

We also learn from the above experiment that the order of the integrator is not the whole story. Integrators differ also in their ability to satisfy certain conservation rules the original ODE may fulfill. For example, the classical predator-prey model of Exercise 1.9 has a function which is invariant along the true solutions, but is not invariant for the results produced by many integrators. For a discussion of which integrators preserve which structure, see [HaLW02].

The relation between the attractors of the ODE and those of their discrete counterparts has been investigated repeatedly. For Runge–Kutta methods, in case $\mathbf{y}(t)$ is equilibrium of an autonomous differential equation, $\mathbf{f}(\mathbf{y}) = 0$, the recursive construction of the $\mathbf{f}(\tilde{t}_l, \tilde{\mathbf{y}}^l)$ implies that they all vanish (see Appendix A,6). Then $\mathbf{y}(t + \Delta) = \mathbf{y}(t)$ holds, which guarantees that equilibria of the differential equation are also equilibria (fixed points) of the Runge–Kutta maps. It is interesting to note that in general the reverse does not hold. Runge–Kutta maps can have additional fixed points, which are not equilibria of the differential equation. Such artificial "solutions" are called *ghost solutions*, or *spurious* solutions. In order to disclose ghost solutions, it is recommended to recalculate results using another step length Δ. Typically, ghost solutions sensitively depend on Δ, which should make possible a distinction between true solutions and spurious solutions (Exercise 1.21). Note that for the explicit Euler method of equation (1.17) the fixed points of the map are the same as the stationary points of the differential equation. Generally, it is recommendable to double check results by applying a different method.

For the dynamics of numerical schemes for ODEs see, for instance, [BrUF84], [Beyn87a], [Ise90], [SaY91], [CaP92], [YeeS94], [StH98]. In the context of bifurcation, we refer to [Bohl79], [Beyn80], [BeD81], [PeSS81]. Stability of fixed points is related to domains of attraction, which can have a fractal structure (Exercise 1.22). In general, multistep methods suffer less from ghost solutions [Ise90].

1.5.3 Calculating Eigenvalues

Let $\mathbf{A}$ be some $n \times n$ matrix with eigenvalue problem $\mathbf{A}\mathbf{w} = \mu\mathbf{w}$. In case all the n eigenvalues need to be calculated, the QR algorithm is the standard routine. For references on how the QR method works see standard texts on numerical methods or [GoL96]. The costs of calculating the eigenvalues by means of the QR method is of the order $O(n^3)$. Hence, applying this algorithm may be prohibitive for larger n.

There are other methods that approximate a single or a few eigenvalues and their eigenvectors. A simple class of methods for calculating the largest eigenvalue in magnitude are the *power methods*, which start from some vector $\mathbf{z}^{(0)}$, which may be randomly chosen, and iterate for $\nu = 1, 2, \ldots$

$$\mathbf{z}^{(\nu+1)} = \mathbf{A}\mathbf{z}^{(\nu)} . \tag{1.20}$$

If the matrix $\mathbf{A}$ is diagonalizable, and if there is a dominant eigenvalue μ_1 in absolute value,

$$|\mu_1| > |\mu_2| \geq \ldots \geq |\mu_n| , \tag{1.21}$$

then the power method equation (1.20) converges. The convergence is faster the smaller $|\mu_2|/|\mu_1|$ is. The iteration is normalized—that is, the iteration of equation (1.20) splits into

$$\mathbf{p} = \mathbf{A}\mathbf{z}^{(\nu)}, \quad \mathbf{z}^{(\nu+1)} = \frac{1}{\sigma}\mathbf{p} , \tag{1.22}$$

with a scaling factor σ taken, for instance, as the maximum component of $\mathbf{p}$. The convergence of the normalized iterates $\mathbf{z}^{(\nu)}$ is toward the eigenvector $\mathbf{w}$ that belongs to μ_1. As a by-product, a sequence can be set up that converges toward the eigenvalue (Exercise 1.23).

Let us denote by $\bar{\mu}$ a guess or an approximation of the eigenvalue μ. In case such an approximation $\bar{\mu}$ is known, the power method can be applied to the matrix

$$\tilde{\mathbf{A}} := (\mathbf{A} - \bar{\mu}\mathbf{I})^{-1},$$

which has the eigenvalues

$$\frac{1}{\mu_1 - \bar{\mu}}, \ldots, \frac{1}{\mu_n - \bar{\mu}}.$$

Frequently, there is a unique simple eigenvalue μ_j closest to $\bar{\mu}$. Then the eigenvalue

$$\tilde{\mu}_1 := \frac{1}{\mu_j - \bar{\mu}}$$

of $\tilde{\mathbf{A}}$ dominates all the others, and the power method converges to $\tilde{\mu}_1$. The power method applied to $\tilde{\mathbf{A}}$ is called *inverse power method*. It amounts to solve a system of linear equations,

$$(\mathbf{A} - \bar{\mu}\mathbf{I})\, \mathbf{z}^{(\nu+1)} = \mathbf{z}^{(\nu)}, \tag{1.23}$$

and is easily set up for complex eigenvalues $\bar{\mu}$. From the resulting limit $\tilde{\mu}_1$ the corresponding eigenvalue of $\mathbf{A}$ is obtained,

$$\mu = \bar{\mu} + \frac{1}{\tilde{\mu}_1}. \tag{1.24}$$

There is another class of iterative methods for approximating several of the eigenvalues largest in magnitude, namely, the Arnoldi iteration [GoL96].

1.5.4 ODE Boundary-Value Problems

ODE boundary-value problems can be solved by shooting methods, finite-difference methods, finite-element methods, or collocation methods; consult, for instance, [StB80], [AsMR88].

Shooting methods integrate initial-value problems in a systematic way in order to find the initial values of the particular trajectory that satisfies the boundary conditions. The initial and final values of an arbitrary trajectory of the differential equation do not in general satisfy the boundary conditions. To see how the shooting idea works, suppose that the boundary conditions are given by $\mathbf{r}(\mathbf{y}(a), \mathbf{y}(b)) = \mathbf{0}$. We denote the trajectory that starts from an initial vector $\mathbf{z}$ by $\varphi(t; \mathbf{z})$, $a \leq t \leq b$. One wants to find $\mathbf{z}$ such that the *residual* $\mathbf{r}(\mathbf{z}, \varphi(b; \mathbf{z}))$ vanishes. This establishes an equation of the type $\mathbf{f}(\mathbf{z}) = \mathbf{0}$, with $\mathbf{f}(\mathbf{z}) := \mathbf{r}(\mathbf{z}, \varphi(b; \mathbf{z}))$. A combination of Newton's method with an integrator serves as the simplest shooting device. Note that each evaluation of **f** means the integration of a trajectory (Exercises 1.24, 1.25); **f** is not given explicitly. Shooting codes also use the *multiple shooting* approach, in which the integration interval is suitably subdivided. The resulting shortening of the integration intervals enlarges the domain of attraction of the Newton method and dampens the influence of "bad" initial guesses [StB80], [Lory80].

1.5.5 Further Tools

The above list of numerical methods is far from complete. One requires interpolation methods and algorithms for graphics. As examples of the more complex methods we mention methods for numerically solving partial differential equations. In some cases suitable discretizations enable to transform the PDEs to large systems of ODEs. Apart from such techniques there are numerous methods specifically developed for the PDE case. We shall briefly describe the method of *inertial manifolds* in Chapter 6. We conclude this section with a list of the main tools required in the numerical analysis of nonlinear phenomena. These tools are:

> linear equation solver,
> Newton method (with modifications),
> ODE integrator, nonstiff and stiff,
> methods for calculating eigenvalues (QR method and power
> methods),
> solver for ODE boundary-value problems (such as shooting), and
> plotting devices.

General reference and texts on numerical methods and software include [Ral65], [StB80], [KaMN89], [PrFTV89], [Sch89], [GoL96], [QuSS00], [DaB08].

Exercises

Exercise 1.1

(a) Go through the procedure of calculating the closed-form solutions of the scalar differential equation $\dot{y} = y(y^2 - \alpha^2)$.
(b) Show that solutions of the initial-value problem with $y(0) = z$ possess a singularity for $|z| > |\alpha|$, that is, $|y(t)| \to \infty$ for $t \to t_\infty$.
(c) Calculate t_∞ and note that it is a *movable singularity*, $t_\infty = t_\infty(z)$.
(d) For $\alpha = 1$, sketch some solutions ($|z| < 2$).

Exercise 1.2
Solve the system of two differential equations

$$\dot{y}_1 = y_2 \,,$$
$$\dot{y}_2 = -y_1 - \lambda y_2$$

for $\lambda = -1$, $\lambda = 0$, $\lambda = 1$. When is the equilibrium stable, asymptotically stable, unstable? (Hint: Appendix A.3 summarizes how to solve a linear system.)

Exercise 1.3
Consider a node or saddle point in the plane with two intersecting straight lines determined by the eigenvectors (see Figure 1.9). Show that a trajectory of equation (1.2) starting on such a line remains there.

Exercise 1.4
(Cf. Exercise 1.2.) Discuss the stability of the equilibria of

$$\dot{y}_1 = y_2 \,,$$
$$\dot{y}_2 = -y_1 - \lambda y_2$$

for all values of the parameter λ.

Exercise 1.5
Perform the stability analysis of the three equilibria of equation (1.5) in detail.

Exercise 1.6   Model of Heartbeat
A simple mathematical model of heartbeat [Zee77] is given by

$$\dot{y}_1 = -a \left(y_1^3 - by_1 + y_2\right),$$
$$\dot{y}_2 = y_1 - c \,.$$

($y_1(t)$: length of a muscle fiber (plus a constant); $y_2(t)$: electrochemical control.)
Discuss the equilibrium (diastole) for $a = 100$, $b = 1$, $c = 1.1$. Sketch some trajectories.

Exercise 1.7
Consider the differential equations

$$\dot{y}_1 = y_2 + \tfrac{1}{2} \exp(y_1^2 - 1) \,,$$
$$\dot{y}_2 = y_1^2 + y_1 \,.$$

Discuss equilibria and stability and sketch trajectories for $-1.5 \leq y_1 \leq 0.5$, $-1 \leq y_2 \leq 0$.

Exercise 1.8
Assume a scalar differential equation of second order for $y_1(t)$, which is transformed to a system of ODEs in the usual way, with $\dot{y}_1 = y_2$. Show that

(a) all equilibria are located on the y_1-axis;
(b) the rotation about a focus is clockwise; and
(c) if μ is an eigenvalue of the Jacobian, then $\begin{pmatrix} 1 \\ \mu \end{pmatrix}$ is eigenvector.

Exercise 1.9   Predator-Prey Model
The Lotka–Volterra model of the interaction of two species u and v is

$$\dot{u} = (a - bv)\, u$$
$$\dot{v} = (-c + du)\, v$$

for positive constants a, b, c, d. The terms in parentheses represent growth rates of the two species.

(a) Argue why $u(t)$ can be seen as the number of predators, and $v(t)$ as the number of prey.
(b) Discuss the equilibria, and sketch some trajectories.
(c) For $F(u, v) := c \log u - d\, u + a \log v - b\, v$ show that F is constant along each trajectory. (The function F is an invariant of the flow; every solution lies on a level curve $F(u, v) = constant$.)
(d) Use the transformations $p := \log u$ and $q := \log v$ to show that the system of differential equations for $\dot{p}$, $\dot{q}$ is *Hamiltonian* in the sense: There exists a function $H(p, q)$ such that

$$\dot{p} = H_q(p, q)$$
$$\dot{q} = -H_p(p, q)\,.$$

Exercise 1.10
Consider the van der Pol equation (1.6). Discuss the stability properties of the equilibria as they vary with the parameter λ. Solve the Bernoulli differential equation for $A(t)$ that results from the method of slowly varying amplitude.

Exercise 1.11   FitzHugh's Nerve Model
The system
$$\dot{y}_1 = 3(y_1 + y_2 - y_1^3/3 + \lambda)\,,$$
$$\dot{y}_2 = -(y_1 - 0.7 + 0.8y_2)/3\,,$$

describing impulses of nerve potentials, has been discussed by FitzHugh [Fit61] (see also [HaHS76]). The variable y_1 represents the membrane potential, while y_2 is a measure of accommodation and refractoriness. The parameter λ is the stimulating current.

(a) For $y_1 = 0.4$, calculate the values of λ and y_2 of the equilibrium.
(b) Prove that there is a stable limit cycle for the λ of part (a). (Hint: Construct a domain $\mathcal{D}$ from which no trajectory departs. Choose a rectangle.)

Exercise 1.12
Figure 1.14 shows a metamorphosis of a stable spiral into an unstable spiral. Find two more ways in which a spiral can lose its stability. Draw "snapshots" of these phase planes illustrating the metamorphoses. (Hint: Encircle the spiral with a limit cycle of varying radius.)

Exercise 1.13
Consider the system of ODEs

$$\dot{y}_1 = -y_2 + y_1(y_1^2 + y_2^2)^{-1/2}(1 - y_1^2 - y_2^2),$$
$$\dot{y}_2 = y_1 + y_2(y_1^2 + y_2^2)^{-1/2}(1 - y_1^2 - y_2^2).$$

Use polar coordinates ρ, ϑ,

$$y_1 = \rho\cos\vartheta, \quad y_2 = \rho\sin\vartheta,$$

to transform the above system into

$$\dot{\rho} = 1 - \rho^2, \quad \dot{\vartheta} = 1.$$

Solve the transformed system to show that the unit circle in the (y_1, y_2)-plane is a stable limit cycle.

Exercise 1.14
Consider the system of ODEs

$$\dot{y}_1 = -y_2 + y_1(y_1^2 + y_2^2 - 1),$$
$$\dot{y}_2 = y_1 + y_2(y_1^2 + y_2^2 - 1).$$

Show that the periodic orbit is unstable.

Exercise 1.15
In the (y_1, y_2)-plane ($y_1 = \rho\cos\vartheta$, $y_2 = \rho\sin\vartheta$), two differential equations are given by

$$\dot{\rho} = \rho - \rho^2,$$
$$\dot{\vartheta} = \lambda - \cos 2\vartheta.$$

The values of $\lambda = 1$ and $\lambda = -1$ separate regions of different dynamic behavior. Calculate all stationary states and sketch five (y_1, y_2)-phase portraits illustrating the dynamics.

Exercise 1.16   Logistic Equation
Use the finite-difference approximation $\frac{1}{\Delta}(\mathbf{y}^{(\nu+1)} - \mathbf{y}^{(\nu)})$ in equation (1.7) for $\dot{y}$ and derive equation (1.9).

Exercise 1.17
Explain what is summarized in Table 1.2.

Exercise 1.18   Student Migration
Equation (1.10) can be written in fixed-point notation as $\mathbf{y}^{(\nu+1)} = \mathbf{P}(\mathbf{y}^{(\nu)})$.
Establish the fixed points $\mathbf{y} = \mathbf{P}(\mathbf{y})$ for $y_2 = 0$ and for $y_2 > 0$, and investigate
their stability.

Exercise 1.19   Duffing Equation
Consider the example of the Duffing equation (1.5). Integrate the separatrix
numerically as described in Section 1.2.

Exercise 1.20   Van der Pol Equation
Consider the van der Pol equation (1.6). For $\lambda = 1$, carry out a numerical
simulation (Algorithm 1.8), starting from several points $\mathbf{z}$ inside and outside
the limit cycle (see Figure 1.18).

Exercise 1.21
Study the fixed points of Runge's method equation (1.18) for the example
$\dot{y} = f(y) = y(1-y)$ (logistic equation equation (1.7) with $\delta = \gamma = 1$):
Calculate the two ghost solutions, and show that they approach infinity for
$\Delta \to 0$.

Exercise 1.22
Next study the map obtained when the Euler method is applied to equation
(1.7) (cf. Exercise 1.21). Show that there is a linear coordinate transformation
$z = \delta y + \gamma$ that transforms the map into the form $z_{\nu+1} = z_\nu^2 + c$ of equation
(1.12).

Exercise 1.23
Set up a sequence of numbers as a by-product of the power method equation
(1.22) that converges toward the dominant eigenvalue.

Exercise 1.24
Suppose some second-order differential equation is given together with boundary conditions $u(0) = 0$, $u(1) = 0$. Give a geometrical interpretation of
the shooting method. To this end, consider the initial vector $\mathbf{z} = (u(0),$
$u'(0)) = (0, \delta)$ for various δ. Draw a sketch and indicate $\mathbf{f}$.

Exercise 1.25
(Project.) Formulate a driving algorithm for a simple shooting method. To
this end, use the Newton method and an ODE integrator as black boxes.

2 Basic Nonlinear Phenomena

Beginning with this chapter, nonlinearity and parameter dependence will play a crucial role. We shall assume throughout that λ is a real parameter, and we shall study solutions of a system of ODEs,

$$\dot{\mathbf{y}} = \mathbf{f}(\mathbf{y}, \lambda), \tag{2.1}$$

or solutions of a system of "algebraic" equations,

$$\mathbf{0} = \mathbf{f}(\mathbf{y}, \lambda). \tag{2.2}$$

Sometimes boundary conditions must be attached to equation (2.1). As in Chapter 1, the vectors $\mathbf{y}$ and $\mathbf{f}$ have n components. If a particular example involves more than one parameter, we assume for the time being that all except λ are kept fixed. Clearly, solutions $\mathbf{y}$ of equation (2.1) or (2.2) in general vary with λ. We shall assume throughout that $\mathbf{f}$ depends smoothly on $\mathbf{y}$ and λ—that is, $\mathbf{f}$ is to be sufficiently often continuously differentiable. This hypothesis is usually met by practical examples.

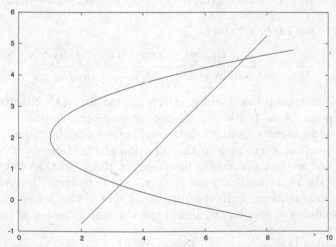

Fig. 2.1. Solutions of equation (2.4), y over λ

R. Seydel, *Practical Bifurcation and Stability Analysis*,
Interdisciplinary Applied Mathematics 5, DOI 10.1007/978-1-4419-1740-9_2,
© Springer Science+Business Media, LLC 2010

2.1 A Preparatory Example

We first study a simple scalar example (illustrated by Figure 2.1). We have drawn a parabola and a straight line. Both curves are solutions of the scalar equation

$$0 = f(y,\lambda) = [1 - \lambda + (y-2)^2](y - \lambda + 2.75)\,. \tag{2.3}$$

Assume now that we want to investigate the solutions of equation (2.3) without knowing the result depicted in Figure 2.1. That is, we start from

$$0 = y^3 - y^2(\lambda + 1.25) + y(3\lambda - 6) + \lambda^2 - 7.75\lambda + 13.75\,. \tag{2.4}$$

We shall not try to calculate a solution $y(\lambda)$ of equation (2.4) analytically. The structure of equation (2.4) enables us to calculate easily λ-values for given values of y. For example, $y = 1$ implies $0 = (\lambda - 3.75)(\lambda - 2)$. This produces two points on the solution curve(s); we choose the point $\lambda = 2, y = 1$. Starting from this point, we can generate a solution curve by calculating the derivative $dy/d\lambda$. Differentiating the identity

$$0 \equiv f(y(\lambda), \lambda)$$

with respect to λ by using the chain rule yields

$$0 = \frac{\partial f(y,\lambda)}{\partial y}\frac{dy}{d\lambda} + \frac{\partial f(y,\lambda)}{\partial \lambda}\,. \tag{2.5}$$

In our equation (2.4), differentiation gives

$$0 = \{3y^2 - y(2\lambda + 2.5) + (3\lambda - 6)\}\frac{dy}{d\lambda} - (y^2 - 3y - 2\lambda + 7.75)\,,$$

which leads to the explicit formula

$$\frac{dy}{d\lambda} = \frac{y^2 - 3y - 2\lambda + 7.75}{3y^2 - y(2\lambda + 2.5) + 3\lambda - 6} = -\frac{\partial f/\partial \lambda}{\partial f/\partial y}\,. \tag{2.6}$$

This expression establishes a differential equation for $y(\lambda)$. Starting from the chosen point ($\lambda = 2, y = 1$), we can integrate this initial-value problem in order to generate nearby solutions. The slope of the "first" tangent is $dy/d\lambda = -0.5$. A small step in the direction of this tangent produces a new point, and we can evaluate a new tangent there. In this way we can approximate the curve until a point with $f_y = 0$ is reached (Exercise 2.1). A vanishing denominator f_y in the right-hand side of the differential equation (2.6) produces a singularity hindering a successful *tracing* of the entire solution curve.

The situation is as follows (see Figure 2.2): Let (y^*, λ^*) be a solution of $f(y,\lambda) = 0$, the scalar version of equation (2.2). We encounter three different cases along a solution curve:

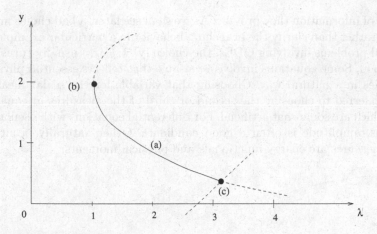

Fig. 2.2. Detail of Figure 2.1

(a) $f_y(y^*, \lambda^*) \neq 0$: The curve can be uniquely continued in a λ-neighbor-hood—that is, there are nearby values $\lambda \neq \lambda^*$ such that $f(y, \lambda) = 0$ has a solution y close to y^*.
(b) $f_y(y^*, \lambda^*) = 0$ and $f_\lambda(y^*, \lambda^*) \neq 0$: The tangent is not defined.
(c) $f_y(y^*, \lambda^*) = 0$ and $f_\lambda(y^*, \lambda^*) = 0$: The tangent is not unique; with additional effort, tangents can be calculated.

Apparently something "happens" in cases (b) and (c), where f_y becomes zero. A significant part of our attention will be focused on cases (b) and (c). The standard situation (a) is characterized by the implicit function theorem (Appendix A.4).

2.2 Elementary Definitions

We now return to the general vector equations (2.1) and (2.2). In order to illustrate graphically the dependence of $\mathbf{y}$ on λ in a two-dimensional diagram, we require a scalar measure of the n-vector $\mathbf{y}$. We shall use the notation $[\mathbf{y}]$ for such a scalar measure of $\mathbf{y}$. Examples are

$$[\mathbf{y}] = y_k \text{ for any of the } n \text{ indices, } 1 \leq k \leq n;$$
$$[\mathbf{y}] = \|\mathbf{y}\|_2 := (y_1^2 + y_2^2 + \ldots + y_n^2)^{1/2}; \tag{2.7}$$
$$[\mathbf{y}] = \|\mathbf{y}\|_\infty := \max\{|y_1|, |y_2|, \ldots, |y_n|\}.$$

The two latter choices are norms; see Appendix A.1. In Exercise 2.2, $[\mathbf{y}]$ was taken to be yet another scalar measure. The three examples in equation (2.7) are appropriate for stationary solutions $\mathbf{y}$. The formulas are easily adapted to solutions $\mathbf{y}(t)$ of ODEs for $\mathbf{y}(t)$ evaluated at some specific value $t = t_0$. The first of the scalar measures in equation (2.7) then reads $[\mathbf{y}] = y_k(t_0)$. Note that the values of $[\mathbf{y}]$ vary with λ. The various choices of $[\mathbf{y}]$ are not equal in the amount of information they provide. As we shall see later, a bad choice might conceal rather than clarify the branching behavior in a particular example. In practical problems involving ODEs, the choice $[\mathbf{y}] = y_k(t_0)$ usually turns out to be good. Some equations involve a variable that reflects essential physical properties in a natural way. Choosing that variable for the scalar measure $[\mathbf{y}]$ is preferred to choosing the second or third of the measures in equation (2.7), which are somewhat artificial. For differential equations with oscillating solutions, amplitude is often a good candidate. Other naturally occurring scalar measures are energy functionals and physical moments.

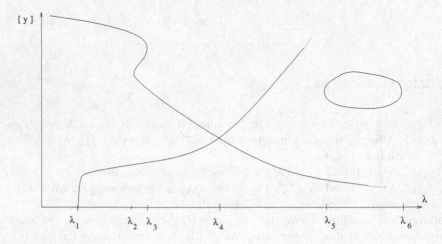

Fig. 2.3. A fictive bifurcation diagram

Equipped with a scalar measure $[\mathbf{y}]$ of the vector $\mathbf{y}$, we are able to depict the solutions of equation (2.1) or (2.2) in a diagram like Figure 2.1. For a particular example, such a diagram may look like Figure 2.3.

Definition 2.1 *A diagram depicting* $[\mathbf{y}]$ *versus* λ, *where* $(\mathbf{y}, \lambda)$ *solves equation (2.1) or (2.2), will be called a* **bifurcation diagram** *(or* **branching diagram** *or* **response diagram***).*

Bifurcation diagrams will turn out to be extremely useful. As Figures 2.1 and 2.3 illustrate, solutions may form continua, reflected by continuous curves in bifurcation diagrams. The continua of solutions are called *branches*. Equation (2.2) consists of n scalar equations with $n + 1$ scalar unknowns $(\mathbf{y}, \lambda)$; the branches are therefore one-dimensional curves in $\mathbb{R}^{n+1}$. Upon varying λ, the number of solutions may change. New branches may emerge, branches may end, or branches may intersect. The list in Table 2.1 reports on the *multiplicity* of solutions as shown in the fictitious bifurcation diagram of Figure 2.3. The significance of the values $\lambda_1, \lambda_2, \ldots$ gives rise to the following informal definition of a branch point:

Definition 2.2   *A bifurcation point or branch point (with respect to λ) is a solution $(\mathbf{y}_0, \lambda_0)$ of equation (2.1) or (2.2), where the number of solutions changes when λ passes λ_0.*

TABLE 2.1.   Solutions in Figure 2.3.

λ-interval	Number of solutions y
$\lambda < \lambda_1$	1
$\lambda_1 \leq \lambda < \lambda_2$	2
λ_2	3
$\lambda_2 < \lambda < \lambda_3$	4
λ_3	3
$\lambda_3 < \lambda < \lambda_4$	2
λ_4	1
Etc.	

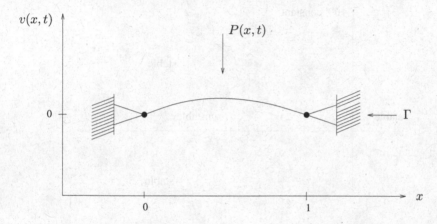

Fig. 2.4. Deflection v of a beam

2.3   Buckling and Oscillation of a Beam

Consider the following situation (see Figure 2.4). A beam is subjected at its ends to a compressive force Γ along its axis. The beam is excited by a harmonic excitation $P(x, t)$ that depends on the spatial variable x (with $0 \leq x \leq 1$) and on the time t. This experiment may represent a beam supporting a machine with a rotating imbalance. We denote the viscous damping by δ and the membrane stiffness by K. Small deflections $v(x, t)$ of the beam are described by solutions of the partial differential equation

$$v_{xxxx} + \left(\Gamma - \mathrm{K} \int_0^1 (v_x(\xi,t))^2 \, d\xi \right) v_{xx} + \delta v_t + v_{tt} = P$$

[Hua68], [Hol79]. Reasonable boundary conditions at $x = 0$ and $x = 1$ are $v = v_{xx} = 0$. We simplify the analysis by assuming perfect symmetry of $P(x,t)$ around $x = 0.5$, in which case the response of the beam is likely to be in the "first mode." That is, we take both P and v to be sinusoidal in x,

$$P(x,t) = \gamma \cos \omega t \sin \pi x \,,$$
$$v(x,t) = u(t) \sin \pi x \,.$$

The ansatz for P reflects the additional assumption of a harmonic excitation with frequency ω. Inserting the expressions for P and the single-mode approximation of v into the PDE leads to a Duffing equation describing the temporal behavior of the displacement of the beam,

$$\ddot{u} + \delta \dot{u} - \pi^2(\Gamma - \pi^2)u + \tfrac{1}{2}K\pi^4 u^3 = \gamma \cos \omega t \,. \tag{2.8}$$

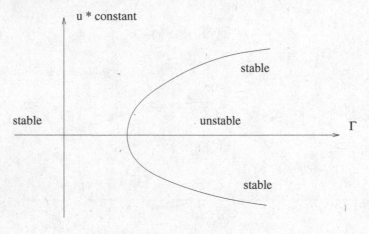

Fig. 2.5. Solutions to equation (2.8), $\gamma = 0$

We first discuss the stability of deflections when the driving force is zero ($\gamma = 0$). It turns out (Exercise 2.4) that for $\Gamma > \pi^2$ there are two stable equilibria with $u \neq 0$, whereas for $\Gamma < \pi^2$ there is only one equilibrium ($u = 0$, stable). Here we interpret the force Γ as our bifurcation parameter λ and depict the results in a bifurcation diagram (Figure 2.5). The stationary solution $(u, \Gamma) = (0, \pi^2)$ of equation (2.8) with $\gamma = 0$ is an example of a bifurcation point. As Figure 2.5 indicates, this bifurcation point separates domains of different qualitative behavior. In particular, the "trivial" solution $u = 0$ loses its stability at $\Gamma = \pi^2$. The value $\Gamma = \pi^2$ is called *Euler's first buckling load*, because Euler calculated the critical load where a beam buckles (this will be discussed in some detail in Section 6.5). The symmetry of the

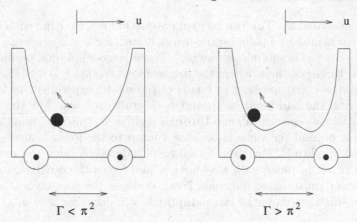

Fig. 2.6. A model experiment

bifurcation diagram with respect to the Γ-axis reflects the basic assumption of perfect symmetry.

So far no external energy has entered the system ($\gamma = 0$). Now we study equation (2.8) with excitation ($\gamma \neq 0$). Note that for $\gamma \neq 0$ the ODE is no longer autonomous. The possible responses of the beam can be explained most easily by discussing the experiment illustrated in Figure 2.6. Imagine a vehicle with a ball rolling inside on a cross section with one minimum ($\Gamma < \pi^2$) or two minima ($\Gamma > \pi^2$). Moving the vehicle back and forth in a harmonic fashion corresponds to the term $\gamma \cos \omega t$. We expect interesting effects in the case $\Gamma > \pi^2$. Choose (artificially)

$$\Gamma = \pi^2 + 0.2\pi^{-2}, \quad K = 16\pi^{-4}/15, \quad \gamma = 0.4, \quad \delta = 0.04\,.$$

This choice leads to the specific Duffing equation

$$\ddot{u} + \tfrac{1}{25}\dot{u} - \tfrac{1}{5}u + \tfrac{8}{15}u^3 = \tfrac{2}{5}\cos \omega t\,. \tag{2.9}$$

The harmonic responses can be calculated numerically; we defer a discussion of the specific methods to Section 6.1 and focus our attention on the results. The parameter Γ is now kept fixed, and we choose the frequency ω as bifurcation parameter λ. This parameter is considered to be constant; it is varied in a quasi-static way. The main response of the oscillator to the sinusoidal forcing is shown in Figure 2.7. As a scalar measure of $u(t)$, amplitude A is chosen. The bifurcation diagram Figure 2.7 shows a typical response of an oscillator with a *hardening spring*. We find two bifurcation points with parameter values

$$\omega_1 = 0.748, \quad \omega_2 = 2.502\,.$$

Hysteresis effects, such as that depicted in Figure 2.7, are ubiquitous in science. Characteristic for hysteresis effects are jump phenomena, which here take place at ω_1 and ω_2. This can be explained as follows. Starting with an oscillation represented by a point on the upper branch for $\omega < \omega_2$, $\omega \approx \omega_2$ (that is, close to B), the amplitude is "large." These large-amplitude oscillations pass all the three positions, where the autonomous version $\gamma = 0$ of (2.8) has stationary states. Or, speaking in terms of the model experiment in Figure 2.6 ($\Gamma > \pi^2$), the ball bounces around both stable minima. For the beam these oscillations represent a snap-through regime. In this situation, slowly increasing ω beyond the value ω_2 causes a jump to the lower branch—that is, to small amplitude. For these small-amplitude oscillations the external excitation is not in phase with the motion, and keeps the oscillation closer to the "former" unstable equilibrium. Next, decrease the parameter ω, and a jump from small amplitude to large amplitude will occur for $\omega \approx \omega_1$ (Figure 2.7C).

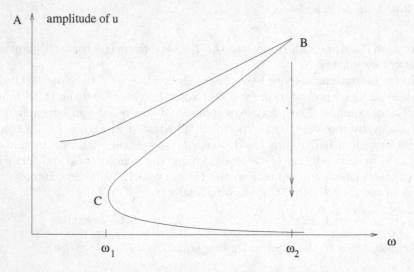

Fig. 2.7. Duffing equation (2.9), response diagram

Transient phenomena and asymmetric solutions are not shown in Figure 2.7. The response of the middle part of the bifurcation diagrams (Figure 2.7 between B and C) is unstable and thus cannot be realized in any experiment. Jump phenomena of oscillations caused by varying frequencies occur in centrifuges and in many other commonly used instruments.

The most important results of this section are summarized in the two bifurcation diagrams Figures 2.5 and 2.7. Comparing these two figures, we notice the different qualities of the bifurcation phenomena. At this point, a classification of such phenomena is needed.

2.4   Turning Points and Bifurcation Points: The Geometric View

In the previous section, we encountered bifurcations, without defining them precisely. It is easy to define these bifurcation points geometrically. We introduce these phenomena by simple examples, point out differences between them, and continue with an algebraic definition in the next section. Terms such as "turning point" and "bifurcation point" are used in mathematics. In applications, other names—which place more emphasis on practical features—are often used. In structural engineering, such phenomena are called *snapping* and *buckling* [Riks72]. We shall first concentrate on bifurcations among stationary solutions, deferring bifurcation into periodic orbits to Section 2.6.

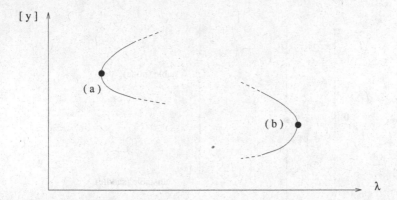

Fig. 2.8. Turning points

A commonly encountered class of bifurcations are the turning points. They can be introduced using the scalar equations

$$0 = y^2 - \lambda \quad \text{or} \quad 0 = y^2 + \lambda.$$

The solutions $y(\lambda)$ of $0 = y^2 - \lambda$ form a parabola that is defined only for $\lambda \geq 0$. At $\lambda = 0$ there is only one solution ($y = 0$), whereas for $\lambda > 0$ there are two solutions, $y = +\sqrt{\lambda}$ and $y = -\sqrt{\lambda}$. The point where solutions begin to exist ($\lambda = 0$, $y = 0$ in this example) is a turning point. The bifurcation diagram in Figure 2.8 depicts a typical situation and prompts the name "turning point." The branch comes from one side ("side" refers to the λ-direction) and turns back at the turning point (emphasized by the dots in Figure 2.8). In (a), the branch opens to the right. The case of a turning point on a branch opening to the left is depicted in Figure 2.8(b), the simplest example of which is $0 = y^2 + \lambda$. Terms "right" or "left" in connection with bifurcation diagrams refer to larger or smaller values of the branching parameter λ. Turning points frequently arise in pairs, resulting in hysteresis effects, as in equation (2.9);

see Figure 2.7. Other names for turning points are *fold bifurcations, saddle nodes*, or *limit points*.

The name "saddle node" is motivated by the stability behavior of the solutions when they are interpreted as equilibria of differential equations. We study this via the scalar differential equation

$$\dot{y} = \lambda - y^2, \tag{2.10}$$

which is related to the above scalar examples. Clearly, the equilibrium $+\sqrt{\lambda}$ is stable, whereas $-\sqrt{\lambda}$ is unstable. The standard stability analysis recalled in Chapter 1 reduces in the scalar case to the investigation of the sign of f_y. The stability results of this example can be presented graphically, as in Figure 2.9. This figure follows the well-established convention of distinguishing between stable and unstable equilibria by using a heavy continuous curve for the former and a dashed curve for the latter.

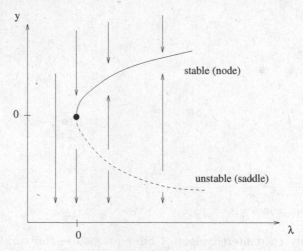

Fig. 2.9. "Saddle node"

The question arises as to whether a turning point always separates stable equilibria from unstable equilibria. The answer is no. It is easy to construct a counterexample in higher dimensions:

$$\dot{y}_1 = \lambda - y_1^2, \quad \dot{y}_2 = y_1 + cy_2. \tag{2.11}$$

A stability analysis of the equilibria of equation (2.11) reveals that one equilibrium is a saddle and the other is a node. The stability of the node is determined by the sign of c; the result is shown in Figure 2.10. We learn from this example that both the half-branches meeting at a turning point can be unstable. Later we shall see examples of this type that are important in practical applications (for example, the problem of a continuous stirred tank reactor in Sections 2.9 and 8.6). The following terminology will turn out to be useful.

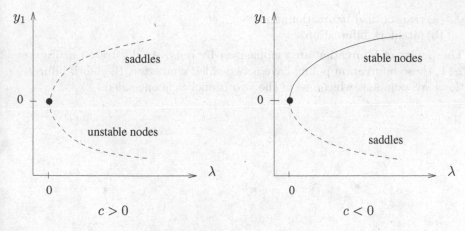

Fig. 2.10. Equilibria of equation (2.11)

We call a branch or a part of a branch stable (unstable) if all its solutions are stable (unstable). A branch will be called periodic, symmetric, or stationary if its solutions are periodic, symmetric, or stationary.

The stability analysis discussed in this book is valid under small perturbations that conform with the underlying model or assumptions. It can easily happen that a branch that was claimed to be stable becomes unstable when the model is enhanced. For example, suppose that, based on some similarity assumption, a three-dimensional flow has been modeled by a two-dimensional PDE. In order to obtain valid stability results for two-dimensional solutions, three-dimensional disturbances should be analyzed. Unfortunately this stability analysis can be more expensive than the calculation of the solutions themselves. Various types of instabilities can be defined for various three-dimensional patterns (see [ClB74], [Bus82]). Other examples of how changes in models or assumptions affect bifurcation and stability results will come up later in this book. The reader should bear in mind that subsequent stability claims will generally be of a limited nature because of restrictions of particular models or assumptions.

Let us summarize the geometrical features of a turning point. Locally there are no solutions on one side of a turning point and two solutions on the other side. At a turning point two solutions are born or two solutions annihilate each other. Viewing the two half-branches that meet in a turning point as parts of one smooth branch, we realize that only one branch is involved in a turning point.

In the early days of bifurcation, turning points were not regarded as member of the family of bifurcation points. Rather, the name "bifurcation point" was reserved to situations where two (ore more) branches with distinct tangents intersect. Two such types of bifurcation in a narrower sense are shown in Figure 2.11. These are

(a) transcritical bifurcation, and
(b) pitchfork bifurcation.

The points of bifurcation are emphasized by dots. As illustrated in Figure 2.11, these bifurcation points have a two-sided character. Pitchfork bifurcations are points at which one of the two branches is one-sided.

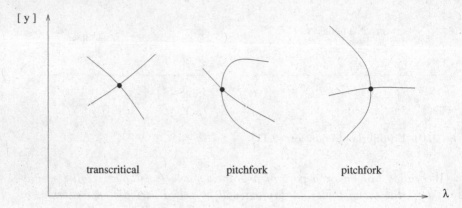

Fig. 2.11. Bifurcation points where two branches intersect

Model examples of bifurcations where two branches intersect, are furnished by the following scalar differential equations:

Example 2.3   Transcritical Bifurcation
The simplest example of a transcritical bifurcation is

$$\dot{y} = \lambda y - y^2. \tag{2.12}$$

The equilibria are $y = 0$ and $y = \lambda$. Figure 2.12 indicates the stable branches by heavy lines. The *trivial branch* $y = 0$ loses stability at the bifurcation point $(y, \lambda) = (0, 0)$. ("lose" in the sense of a growing parameter; otherwise it would gain stability.) Simultaneously, there is an *exchange of stability* to the other branch.   □

Example 2.4   Supercritical Pitchfork Bifurcation
The simplest example of a supercritical pitchfork bifurcation is

$$\dot{y} = \lambda y - y^3. \tag{2.13}$$

For $\lambda > 0$ there are two nontrivial equilibria, $y = \pm\sqrt{\lambda}$. The transition of stability is illustrated by Figure 2.13. We recall the example of the buckling of a beam, which exhibits this kind of loss of stability, from Section 2.3 (Figure 2.5). In equation (2.13) there is a symmetry: Replacing $y \to -y$ yields the same differential equation—that is, the equation is invariant with respect to the transformation $y \to -y$. This symmetry of the differential equation is

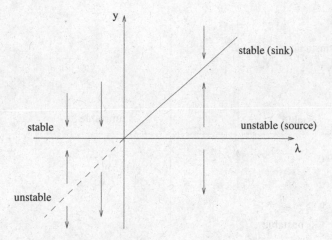

Fig. 2.12. Transcritical bifurcation

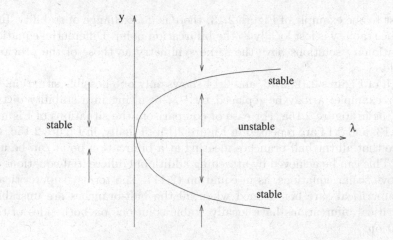

Fig. 2.13. Supercritical pitchfork

also found in the stationary solutions. Apart from the transformation $y \rightarrow -y$ the two half-branches emanating from $(y_0, \lambda_0) = (0,0)$ are identical and thus can be identified. Symmetry plays a crucial role for several bifurcation phenomena; we shall return to this topic later.   •   □

Example 2.5   Subcritical Pitchfork Bifurcation

The simplest example of a subcritical pitchfork bifurcation is

$$\dot{y} = \lambda y + y^3 . \tag{2.14}$$

Equilibria and stability behavior for this equation are shown in Figure 2.14. There is again a loss of stability at the bifurcation point $(y_0, \lambda_0) = (0,0)$. In

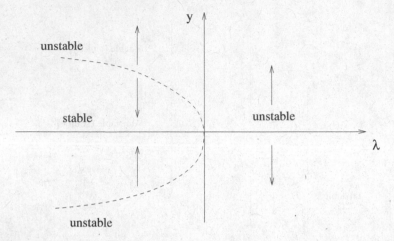

Fig. 2.14. Subcritical pitchfork

contrast to the example of Figure 2.13, there is no exchange of stability. Instead, the stability is lost locally at the bifurcation point. Differential equation and stationary solutions have the same symmetry as those of the previous example.   □

Each of Figures 2.12, 2.13, and 2.14 shows only one possible situation. In another example, λ may be replaced by $-\lambda$, resulting in a stability distribution, as in Figure 2.15a. For ease of comparison, the situations of Figures 2.12, 2.13, and 2.14 are repeated in Figure 2.15b. Finally, in Figure 2.15c, we indicate that all the half-branches meeting at a bifurcation point can be unstable. This can be achieved by attaching additional differential equations to the above scalar equations, as in equation (2.11). The terms "supercritical" and "subcritical" are not defined when all the half-branches are unstable. Supercritical bifurcations have locally stable solutions on both sides of the bifurcation.

> In the literature there is some variance in the use of the words "supercritical" and "subcritical." These terms are occasionally used in a fashion solely oriented toward decreasing or increasing values of the parameter λ [GoS79] and not related to the concept of stability. In this vein a pitchfork bifurcation at $\lambda = \lambda_0$ is said to be supercritical if there is locally only one solution for $\lambda < \lambda_0$—that is, "supercritical" bifurcates to the right and "subcritical" bifurcates to the left (the λ-axis points to the right). This book will follow the more physical convention indicated in Figure 2.15.

So far, we have introduced the concepts of turning point and *simple* bifurcation point, the latter denoting a point where precisely two branches intersect. (Note that this interpretation of the term "simple bifurcation point" does not rule out the possibility that globally the two branches might be connected.) We mention in passing that there are additional branching phenomena. For example, a hysteresis phenomenon may collapse into a situation

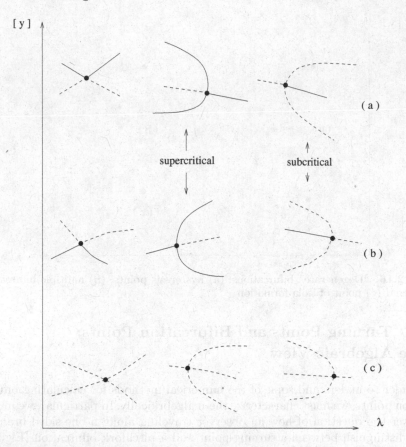

Fig. 2.15. Stability at bifurcations (solid line for stable solutions)

like that shown in Figure 2.16(a). This kind of branch point is called a *hysteresis point* or *cusp point*. A bifurcation point where more than two branches intersect as in Figure 2.16(b) is called a *multiple bifurcation point*. Finally, in Figure 2.16(c) an *isola center* is depicted (also called *point of isola formation*). A neighboring solution cannot be found to the right or to the left of an isola center. Phenomena as illustrated in Figure 2.16 are in some sense rare or degenerate. The occurrence of hysteresis points or isola centers will be discussed in Section 2.10.

The simplest equations that represent a certain type of bifurcation are called *normal forms*. We conclude this section by summarizing the normal forms of the bifurcations discussed so far:

$$0 = \lambda \pm y^2 \qquad \text{turning point, fold bifurcation;}$$
$$0 = \lambda y \pm y^2 \qquad \text{transcritical bifurcation;}$$
$$0 = \lambda y \pm y^3 \qquad \text{pitchfork bifurcation.}$$

Further normal forms can be found in Section 8.4.

[y]

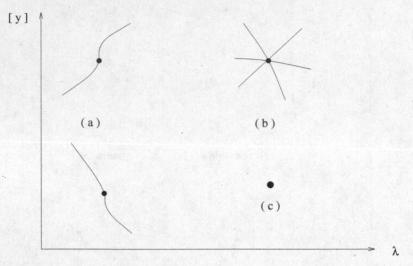

(a)

(b)

(c)

λ

Fig. 2.16. "Degenerate" bifurcations: (a) hysteresis points, (b) multiple bifurcations, and (c) point of isola formation

2.5 Turning Points and Bifurcation Points: The Algebraic View

In order to understand some of the numerical methods for calculating bifurcation points, we must characterize them algebraically. In particular, we must answer the question of how an observer traveling along a one-sided branch can distinguish between a turning point and a pitchfork bifurcation (Figure 2.17). In this section we restrict our attention purely to the stationary case $\mathbf{f}(\mathbf{y}, \lambda) = \mathbf{0}$. This Section 2.5 gives an introduction; for more theory we refer to Chapter 5, in particular to Section 5.5. We first study a simple example that exhibits both phenomena.

Example 2.6
We consider the two scalar equations ($n = 2$)

$$0 = 1 + \lambda(y_1^2 + y_2^2 - 1) =: f_1(y_1, y_2, \lambda)\,,$$
$$0 = 10y_2 - \lambda y_2(1 + 2y_1^2 + y_2^2) =: f_2(y_1, y_2, \lambda)\,. \tag{2.15}$$

The second equation is satisfied for $y_2 = 0$, which implies by means of the first equation

$$y_1^2 = \frac{\lambda - 1}{\lambda}\,.$$

This establishes two branches

$$(y_1, y_2) = \left(\pm\left(\frac{\lambda - 1}{\lambda}\right)^{1/2}, \, 0\right),$$

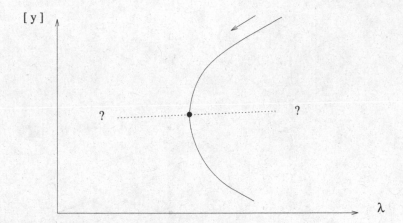

Fig. 2.17. Bifurcation or turning point?

which are defined for $\lambda \geq 1$ and $\lambda < 0$. It remains to discuss the case $y_2 \neq 0$. Substituting the expression

$$y_1^2 + y_2^2 = 1 - \frac{1}{\lambda}$$

from the first equation into the second equation yields

$$0 = 10 - \lambda \left(1 + y_1^2 + \left(1 - \frac{1}{\lambda} \right) \right) = 11 - 2\lambda - \lambda y_1^2 \,.$$

This gives

$$(y_1, y_2) = \left(\pm \left(\frac{11 - 2\lambda}{\lambda} \right)^{1/2}, \pm \left(\frac{3\lambda - 12}{\lambda} \right)^{1/2} \right)$$

as additional branches, defined for $4 \leq \lambda \leq 5.5$. These branches form a smooth loop in the three-dimensional (y_1, y_2, λ)-space. Figure 2.18 shows the branches for $\lambda > 0$ and $y_2 \geq 0$ as solid lines and those for $y_2 < 0$ as dotted lines. The branches are symmetric with respect to the planes defined by $y_1 = 0$ and $y_2 = 0$; the latter plane is indicated by shading. There are three turning points and two pitchfork bifurcation points—namely, those with (y_1, y_2, λ) values

$(0, 0, 1)$	turning point,
$(\pm\frac{1}{2}\sqrt{3}, 0, 4)$	pitchfork bifurcation points,
$(0, \pm 3(11)^{-1/2}, 5.5)$	turning points.

□

We postpone further analysis of bifurcation points to illustrate problems that may arise when the scalar measure [ ] is not chosen properly. As the scalar measure for the bifurcation diagram, we choose

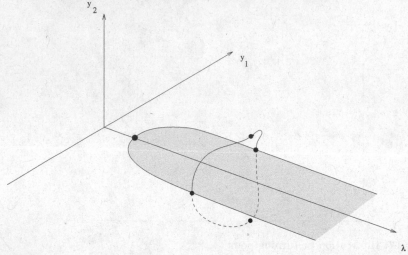

Fig. 2.18. Example 2.6., solutions to equation (2.15)

$$[\mathbf{y}] = y_1^2 + 2y_2^2\,,$$

which is depicted in Figure 2.19 for the example under consideration. We see an effect that is typical for bifurcation diagrams of equations where symmetries are involved. The two half-branches with $1 \le \lambda \le 4$ have the same values of $[\mathbf{y}]$, and Figure 2.19 does not distinguish between them. Part (a) is *covered twice*. This phenomenon is even more significant with the four different parts of the loop branch with $4 \le \lambda \le 5.5$. All four parts have the same values of $[y]$, and part (b) in Figure 2.19 is covered four times. Thus the bifurcation diagram Figure 2.19 conceals the fact that the bifurcations at $\lambda = 4$ are of the pitchfork type and that the dots at $\lambda = 1$ and $\lambda = 5.5$ represent turning points. For a better understanding of this effect, compare Figure 2.19 with Figure 2.18. Certain choices of $[\mathbf{y}]$ lead to better insight into the branching behavior of a particular equation; other choices provide much less information (Exercise 2.7). Figure 2.19 illustrates that in bifurcation diagrams there need not be a curve that passes pitchfork bifurcations or turning points perpendicularly to the λ-axis.

We now continue with equation (2.15), concentrating on characteristic algebraic properties of turning points and pitchfork bifurcation points. Motivated by the example in Section 2.1 and the implicit function theorem, we study the derivatives of $\mathbf{f}(\mathbf{y},\lambda)$. The Jacobian matrix is

$$\mathbf{f_y}(\mathbf{y},\lambda) = \begin{pmatrix} 2\lambda y_1 & 2\lambda y_2 \\ -4\lambda y_1 y_2 & 10 - \lambda - 2\lambda y_1^2 - 3\lambda y_2^2 \end{pmatrix},$$

and the vector of the partial derivatives with respect to the parameter λ is

$$\mathbf{f}_\lambda(\mathbf{y},\lambda) = \begin{pmatrix} y_1^2 + y_2^2 - 1 \\ -y_2(1 + 2y_1^2 + y_2^2) \end{pmatrix}.$$

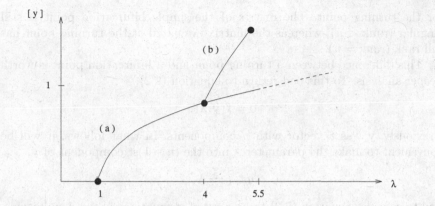

Fig. 2.19. Example 2.6., equation (2.15), bifurcation diagram

Substituting the $(\mathbf{y}, \lambda)$ values of one of the pitchfork bifurcation points and one of the turning points into these derivatives, we obtain the following:

for the pitchfork point $(\frac{1}{2}\sqrt{3},\ 0,\ 4)$:

$$\mathbf{f_y} = \begin{pmatrix} 4\sqrt{3} & 0 \\ 0 & 0 \end{pmatrix}, \quad \mathbf{f_\lambda} = \begin{pmatrix} -\frac{1}{4} \\ 0 \end{pmatrix};$$

for the turning point $(0,\ 3(11)^{-1/2},\ 5.5)$:

$$\mathbf{f_y} = \begin{pmatrix} 0 & 3(11)^{1/2} \\ 0 & -9 \end{pmatrix}, \quad \mathbf{f_\lambda} = \begin{pmatrix} -2/11 \\ -60(11)^{-3/2} \end{pmatrix}.$$

The Jacobian $\mathbf{f_y}$ is *singular* in both cases—that is, the rank is less than $n = 2$, or equivalently (Appendix A.2),

$$\det \mathbf{f_y}(\mathbf{y}_0, \lambda_0) = 0. \tag{2.16}$$

Here and in the remainder of the book, bifurcation points will be denoted by $(\mathbf{y}_0, \lambda_0)$; $\mathbf{y}_0$ being an n-vector.

To distinguish between a turning point and a simple bifurcation point where two branches intersect, we need more information. Upon attaching the vector $\partial\mathbf{f}/\partial\lambda$ to the Jacobian matrix, one obtains the augmented matrix

$$\begin{pmatrix} 4\sqrt{3} & 0 & -\frac{1}{4} \\ 0 & 0 & 0 \end{pmatrix}$$

for the pitchfork bifurcation point, and

$$\begin{pmatrix} 0 & 3\sqrt{11} & -2/11 \\ 0 & -9 & -60(11)^{-3/2} \end{pmatrix}$$

for the turning point. The matrix of the simple bifurcation point is still singular (rank $< n$), whereas the matrix evaluated at the turning point has full rank (rank $= n$).

This difference between a turning point and a bifurcation point is worth deeper analysis. To this end return to equation (2.2),

$$0 = \mathbf{f}(\mathbf{y}, \lambda).$$

Previously, $\mathbf{y}$ was a vector with n components. In what follows, it will be convenient to make the parameter λ into the $(n+1)$st component of $\mathbf{y}$,

$$y_{n+1} := \lambda,$$

and to consider equation (2.2) to be a set of n equations in $n+1$ unknowns,

$$0 = f_i(y_1, y_2, \ldots, y_n, y_{n+1}) \qquad (i = 1, 2, \ldots, n).$$

The rectangular matrix of the partial derivatives consists of $n+1$ columns $\mathbf{z}^i$,

$$(\mathbf{f_y} \mid \mathbf{f}_\lambda) = (\mathbf{z}^1 \mid \mathbf{z}^2 \mid \ldots \mid \mathbf{z}^k \mid \ldots \mid \mathbf{z}^n \mid \mathbf{z}^{n+1}). \tag{2.17}$$

We are free to interpret any of the $n+1$ components (say, the kth component) as a parameter. Call this parameter γ:

$$\gamma = y_k.$$

The dependence of the remaining n components

$$y_1, y_2, \ldots, y_{k-1}, y_{k+1}, \ldots, y_n, y_{n+1}$$

on γ is characterized by the "new" Jacobian that results from equation (2.17) by removing the kth column. As long as the matrix in equation (2.17) has full rank,

$$\text{rank}(\mathbf{f_y}, \mathbf{f}_\lambda) = n, \tag{2.18}$$

it is guaranteed that an index k exists such that after removing column $\mathbf{z}^k$, a nonsingular square matrix remains. Consequently, by the implicit function theorem, the branch of stationary solutions can be continued in a unique way. This is the situation in which turning points are possible but bifurcation to another branch of stationary solutions is ruled out. For a simple bifurcation point between stationary branches ("stationary bifurcation") no such k exists (rank $< n$ for all choices of γ).

This exchange of columns of the matrix equation (2.17) has a striking geometrical interpretation. The transition from parameter λ to parameter γ means that the bifurcation diagram $[\mathbf{y}]$ versus γ is obtained from the diagram $[\mathbf{y}]$ versus λ by a rotation of $90°$. As illustrated in Figure 2.20, the turning point disappears when the branch is parameterized by $\gamma = y_k$. This removal of the singularity by change of parameter is reflected algebraically in the

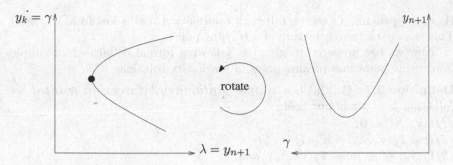

Fig. 2.20. Singularity removed

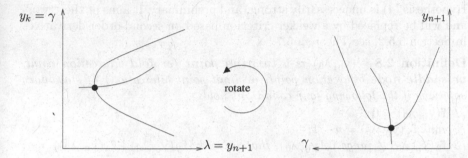

Fig. 2.21. Singularity not removed

nonsingularity of the "new" Jacobian. After this new choice of parameter and rearrangement of components, the implicit function theorem works again.

Trying the same exchange with the matrix associated with a simple bifurcation point, we obtain Figure 2.21. Here the singularity is not removed by changing the parameter; the bifurcation point remains.

Preserving or increasing the rank by attaching the column $\mathbf{f}_\lambda$ to the Jacobian $\mathbf{f}_\mathbf{y}$ can be described in terms of the *range* of a matrix. Assume that the rank of the Jacobian is $n-1$. We first consider the situation of a simple bifurcation point. Here there are constants c_i such that

$$\mathbf{f}_\lambda = \mathbf{z}^{n+1} = \sum_{i=1}^{n} c_i \mathbf{z}^i. \tag{2.19}$$

This can be written

$$\mathbf{f}_\lambda \in \text{range } \mathbf{f}_\mathbf{y}, \tag{2.20}$$

because the vectors $\mathbf{z}^i$ span the range of $\mathbf{f}_\mathbf{y}$. In contrast, for a turning point there are no constants c_i such that $\mathbf{f}_\lambda$ can be expressed as a linear combination of the n column vectors $\mathbf{z}^i$ of the singular matrix $\mathbf{f}_\mathbf{y}$, that is,

$$\mathbf{f}_\lambda \notin \text{range } \mathbf{f}_\mathbf{y}. \tag{2.21}$$

(Loosely speaking, $\mathbf{f}_\lambda$ carries full-rank information that is lost in $\mathbf{z}^1, \ldots, \mathbf{z}^n$.) This is a characteristic feature of a turning point.

Now we are prepared to give the following formal definitions of simple bifurcation point and turning point of stationary solutions:

Definition 2.7 $(\mathbf{y}_0, \lambda_0)$ *is a **simple stationary bifurcation point** if the following four conditions hold:*

(1) $\mathbf{f}(\mathbf{y}_0, \lambda_0) = \mathbf{0}$;

(2) rank $\mathbf{f}_\mathbf{y}(\mathbf{y}_0, \lambda_0) = n - 1$;

(3) $\mathbf{f}_\lambda(\mathbf{y}_0, \lambda_0) \in$ *range* $\mathbf{f}_\mathbf{y}(\mathbf{y}_0, \lambda_0)$; and

(4) exactly two branches of stationary solutions intersect with two distinct tangents.

Hypothesis (4) is unnecessarily strong, and preliminary. It aims at the "two," and will be replaced by a weaker criterion based on second-order derivatives in Section 5.5.2, see Theorem 5.7.

Definition 2.8 $(\mathbf{y}_0, \lambda_0)$ *is a **turning point** (or **fold** bifurcation point, or **saddle-node** bifurcation point, or limit point bifurcation) of stationary solutions if the following four conditions hold:*

(1) $\mathbf{f}(\mathbf{y}_0, \lambda_0) = \mathbf{0}$;

(2) rank $\mathbf{f}_\mathbf{y}(\mathbf{y}_0, \lambda_0) = n - 1$;

(3) $\mathbf{f}_\lambda(\mathbf{y}_0, \lambda_0) \notin$ *range* $\mathbf{f}_\mathbf{y}(\mathbf{y}_0, \lambda_0)$, *that is, rank* $(\mathbf{f}_\mathbf{y}(\mathbf{y}_0, \lambda_0) \mid \mathbf{f}_\lambda(\mathbf{y}_0, \lambda_0)) = n$; and

(4) there is a parameterization $\mathbf{y}(\sigma)$, $\lambda(\sigma)$ *with* $\mathbf{y}(\sigma_0) = \mathbf{y}_0$, $\lambda(\sigma_0) = \lambda_0$, *and* $\mathrm{d}^2\lambda(\sigma_0)/\mathrm{d}\sigma^2 \neq 0$.

Hypotheses (1), (2), and (3) of Definition 2.8 guarantee that the tangent to the branch at $(\mathbf{y}_0, \lambda_0)$ is perpendicular to the λ-axis in the $(n+1)$-dimensional $(\mathbf{y}, \lambda)$-space. Hypothesis (4) implies the "turning property" and prevents $(\mathbf{y}_0, \lambda_0)$ from being a hysteresis point (see Figure 2.16a). In the literature, condition (4) is sometimes replaced by other equivalent conditions.

In classical bifurcation theory, the terms *branching* and *bifurcation* designated exclusively those bifurcation points in the narrower sense where two or more branches intersect [Sta71], [VaT74]. The terms *primary bifurcation* (for a bifurcation from a trivial solution $y = 0$) and *secondary bifurcation* (for a bifurcation from a nontrivial branch) reflected the order in which these phenomena were analyzed. Limitations of the classical terminology became apparent with the rise of such subjects as *catastrophe theory* and by the discovery of new nonlinear effects. The omnipresence of turning points has suggested the use of classical terms in broader senses. The synonym *fold bifurcation* reflects this trend. The great variety of nonlinear phenomena makes it difficult to find convincing names. The terms *regular singular point* [GrR84b] or *isolated nonisolated solution* [Kel81] may serve as examples of the difficulties in finding evident names. The standard turning point (Definition 2.8) was sometimes called *simple quadratic turning point*, and our hysteresis point was also called *simple cubic turning point* [SpJ84]. There is no consensus about which name is to be used for a particular phenomenon. The lack of striking names might explain why several accomplishments of bifurcation theory and singularity theory became popular only through catastrophe theory, chaos theory, or synergetics. Note that the name *turning point* is also used for a completely different phenomenon in perturbation theory. This book will not use this other meaning of turning points.

Today the names *bifurcation point* and *branch point* are used in a most general meaning, including turning points as *saddle-node bifurcations* or *fold bifurcations*. This might cause some confusion because a turning point is no bifurcation in the proper meaning of the word. The respective meaning of the term bifurcation will be clear by the context, or by additional attributes. Our tendency will be to use simple labels for widespread phenomena, restricting complex names to less important effects.

2.6 Hopf Bifurcation

The bifurcation points we have handled so far have one common feature. All the branches have represented equilibria—that is, both branches intersecting in a bifurcation point have consisted of (stationary) solutions of the equation

$$\dot{\mathbf{y}} = \mathbf{0} = \mathbf{f}(\mathbf{y}, \lambda).$$

The term "equilibrium" characterizes a physical situation. Mathematically speaking, we say that the solutions on the emanating branch remain in the same "space"—namely, in the space of n-dimensional vectors. In the previous section, we have called a bifurcation that is characterized by intersecting branches of equilibrium points *stationary bifurcation* (also called *steady-state bifurcation*; see Figure 2.22).

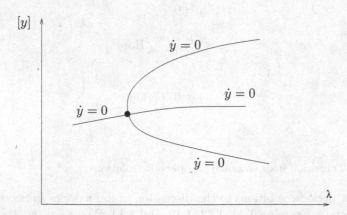

Fig. 2.22. Stationary bifurcation

Physically, an equilibrium represents a situation without "life." It may mean no motion of a pendulum, no reaction in a reactor, no nerve activity, no flutter of an airfoil, no laser operation, or no circadian rhythms of biological clocks [Win75]. Such exciting phenomena are not described by stationary bifurcation, where branching solutions show no more life than the previously known solutions. The full richness of the nonlinear world is not found at equilibria.

The analysis for $\dot{\mathbf{y}} = \mathbf{0}$ is far simpler than the analysis of the situation with $\dot{\mathbf{y}} \neq \mathbf{0}$. The assumption that $\dot{\mathbf{y}} = \mathbf{0}$ (when in reality, e.g., $\dot{\mathbf{y}} = \epsilon \mathbf{d}$ for some vector $\mathbf{d}$ and small $|\epsilon|$) may be justified because equilibria do occur in nature. But nature does not care about simplifying assumptions, and the equilibrium may suddenly be superseded by motion. When the motion can be represented by limit cycles, the oscillations are regular and hence manifestations of an order. In case $\mathbf{y}$ is a macroscopic variable describing, e.g., the dynamic behavior of a flow, its oscillation is seen as self-organization of microscopic parts toward a coherent structure without a spatial gradient; fluctuations around an equilibrium are no longer damped [GlP71], [Hak77].

The mathematical vehicles that describe this kind of "life" are the functions $\mathbf{y}(t)$. The corresponding function spaces include equilibria as the special-case constant solutions (Figure 2.23). The type of bifurcation that connects equilibria with periodic motion is *Hopf bifurcation*. Hopf bifurcation is the door that opens from the small room of equilibria to the large hall of periodic solutions, which in turn is just a small part of the realm of functions.

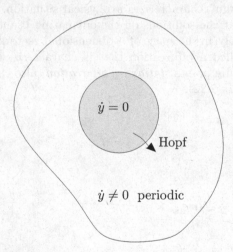

Fig. 2.23. Transition from stationary to periodic solutions

We met periodic solutions in the discussion of limit cycles in Section 1.3; recall especially Exercises 1.11, 1.12, 1.13, and 1.14. We now discuss a simple example of Hopf bifurcation (similar to Exercise 1.14).

Example 2.9

$$\begin{aligned}
\dot{y}_1 &= -y_2 + y_1(\lambda - y_1^2 - y_2^2), \\
\dot{y}_2 &= y_1 + y_2(\lambda - y_1^2 - y_2^2).
\end{aligned} \qquad (2.22)$$

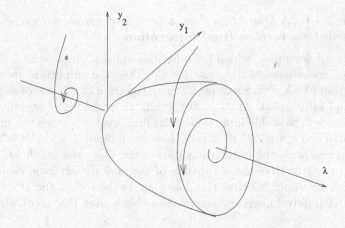

Fig. 2.24. Generation of limit cycles at $\lambda = 0$ (parameter-dependent phase plane)

A straightforward investigation shows that $y_1 = y_2 = 0$ is the only equilibrium for all λ; there is no stationary bifurcation. The Jacobian matrix

$$\begin{pmatrix} \lambda & -1 \\ 1 & \lambda \end{pmatrix}$$

has eigenvalues $\lambda \pm i$. We conclude that the equilibrium is stable for $\lambda < 0$ and unstable for $\lambda > 0$—that is, there is a loss of stability at $\lambda = 0$ but no exchange of stability inside the space of equilibria.

As in Exercises 1.13 and 1.14, a limit cycle can be constructed for equation (2.22). Using polar coordinates ρ, ϑ with

$$y_1 = \rho \cos \vartheta, \; y_2 = \rho \sin \vartheta,$$

we obtain from equation (2.22) the two equations

$$\dot{\rho} \cos \vartheta - \rho \dot{\vartheta} \sin \vartheta = -\rho \sin \vartheta + \rho \cos \vartheta (\lambda - \rho^2),$$
$$\dot{\rho} \sin \vartheta + \rho \dot{\vartheta} \cos \vartheta = \rho \cos \vartheta + \rho \sin \vartheta (\lambda - \rho^2).$$

Multiplying the first equation by $\cos \vartheta$ and the second equation by $\sin \vartheta$ and adding yields the differential equation

$$\dot{\rho} = \rho(\lambda - \rho^2). \tag{2.23}$$

Similarly, one obtains

$$\dot{\vartheta} = 1.$$

For $\rho = \sqrt{\lambda}$ we have $\dot{\rho} = 0$. Hence, there is a periodic orbit $\rho(t) \equiv \sqrt{\lambda}$, $\vartheta(t) = t$ for $\lambda > 0$, and the amplitude of the orbit grows with $\sqrt{\lambda}$. Because

$$\dot{\rho} < 0 \quad \text{for } \rho > \sqrt{\lambda},$$
$$\dot{\rho} > 0 \quad \text{for } 0 < \rho < \sqrt{\lambda},$$

the orbit is stable. For varying λ the orbits form a branch, which merges at $\lambda_0 = 0$ with the branch of equilibria. Figure 2.24 summarizes the results. □

Definition 2.10 *A bifurcation from a branch of equilibria to a branch of periodic oscillations is called* **Hopf bifurcation.**

This is a broad definition, which includes also unusual Hopf points. Usually, one assumes smoothness in the sense $f \in C^2$. Then the amplitude grows with the square root $(\lambda - \lambda_0)^{1/2}$. Example 2.9 with equation (2.22) serves as normal form of Hopf bifurcation. At $\lambda_0 = 0$ there is an exchange of stability from stable equilibrium to stable limit cycle. The limit cycle encircles the unstable equilibrium. The Figure 2.25 illustrates how the temporal behavior of a state variable changes with the bifurcation parameter. Only the stable states are shown, and the growth of the amplitude of the periodic orbits is visible. We observe from Example 2.9 that the Jacobian, evaluated at the Hopf point, has a pair of purely imaginary eigenvalues. Note that the Jacobian is *not* singular.

If $f \in C^1$ but $f \notin C^2$ then the Hopf bifurcation can be unusual: As Hassard points out in [Has00] for the railway bogie [Kaas86], the tangent to the solution branch in the Hopf point is not vertical!

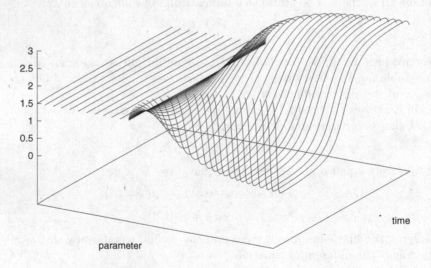

Fig. 2.25. Generation of limit cycles at a Hopf point; dependence on time and parameter (Example 7.15)

We are now prepared to give a general characterization of Hopf bifurcation. The basic results were known to Poincaré; the planar case was handled by Andronov in 1929 [AnVK87]. Because of these early results, bifurcation from equilibria to limit cycles is also called Poincaré–Andronov–Hopf bifurcation. The commonly used name "Hopf bifurcation" may be seen as an abbreviation. It was Hopf who proved the following theorem for the n-dimensional case in 1942 [Hopf42]:

Theorem 2.11 *Assume for* $\mathbf{f} \in C^2$
(1) $\mathbf{f}(\mathbf{y}_0, \lambda_0) = \mathbf{0}$;
(2) $\mathbf{f_y}(\mathbf{y}_0, \lambda_0)$ *has a simple pair of purely imaginary eigenvalues* $\mu(\lambda_0) = \pm \mathrm{i}\beta$
and no other eigenvalue with zero real part; and
(3) $\mathrm{d}(\mathrm{Re}\mu(\lambda_0))/\mathrm{d}\lambda \neq 0$.
Then there is a birth of limit cycles at $(\mathbf{y}_0, \lambda_0)$. *The initial period (of the zero-amplitude oscillation) is*

$$T_0 = \frac{2\pi}{\beta}.$$

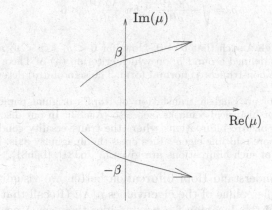

Fig. 2.26. Complex plane of eigenvalues; path of eigenvalues $\mu(\lambda)$ related to Hopf bifurcation

Hypotheses (1), (2), and (3) can be viewed as a formal definition of Hopf bifurcation. Hypothesis (2) can be relaxed in that no eigenvalue is allowed that is an integer multiple of $\pm \mathrm{i}\beta$. Condition (3) is the *transversality hypothesis*; it is "usually" satisfied.

For a proof of Theorem 2.11 we refer to the literature, for example, to [HaKW81], or [Wig90]. Here we only sketch basic lines of a proof. The dynamical behavior close to the Hopf point can be considered on a two-dimensional center manifold. (For an introduction to center manifolds, see Section 8.7.) After transformation to normal form (Section 8.2), one can start the bifurcation analysis from $\mathbf{y}^{\mathrm{s}} = \mathbf{0}$, $\lambda_0 = 0$, restricted to a two-dimensional system

$$\begin{pmatrix} \dot{y}_1 \\ \dot{y}_2 \end{pmatrix} = \begin{pmatrix} \alpha(\lambda) & -\beta(\lambda) \\ \beta(\lambda) & \alpha(\lambda) \end{pmatrix} \begin{pmatrix} y_1 \\ y_2 \end{pmatrix} + \begin{pmatrix} g_1(y_1, y_2, \lambda) \\ g_2(y_1, y_2, \lambda) \end{pmatrix}.$$

This includes (2.22) as special case. The functions g_1, g_2 are nonlinear in y_1, y_2. We assume $\alpha(0) = 0$ and $\beta(0) \neq 0$ to meet the assumptions of Theorem 2.11. Substituting polar coordinates as in Example 2.9, leads to

$$\dot{\rho} = \alpha(\lambda)\rho + a(\lambda)\rho^3 + O(\rho^5)$$
$$\dot{\vartheta} = \beta(\lambda) + b(\lambda)\rho^2 + O(\rho^4)$$

The functions a and b abbreviate larger expressions that occur by transformation to normal form. Being interested in the dynamical behavior near $\lambda = 0$, we Taylor expand α, a, β, b about 0. This yields the normal form in polar coordinates

$$\dot{\rho} = \alpha'(0)\lambda\rho + a(0)\rho^3 + t.h.o.$$
$$\dot{\vartheta} = \beta(0) + \beta'(0)\lambda + b(0)\rho^2 + t.h.o.$$

where $'$ denotes differentiation with respect to λ. In a first step, the terms of higher order are neglected. The resulting truncated system has a periodic orbit with radius

$$\bar{\rho} := \sqrt{-\frac{\lambda\alpha'(0)}{a(0)}} \quad \text{for} \quad \frac{\lambda\alpha'(0)}{a(0)} < 0,$$

for small enough λ such that $\dot{\vartheta} \neq 0$. Now, for $0 < \rho_1 < \bar{\rho} < \rho_2$ an annulus $\rho_1 < \rho < \rho_2$ is defined around $\bar{\rho}$, on which criterion (a) of Theorem 1.10 is applied for the non-truncated normal form. The sign of $a(0)$ determines the stability.

In [MaM76] an English translation of Hopf's original paper [Hopf42] is included. For historical remarks, see also [Arn83]. In our discussion, we have excluded bizarre bifurcations where the transversality condition does not hold or where multiple eigenvalues cross the imaginary axis. References to treatments of such bifurcations are given in [IoJ81], [Lan81].

In order to understand Hopf bifurcation better, we visualize what happens in the complex plane of the eigenvalues $\mu(\lambda)$. (Recall that the $\mu(\lambda)$ are the eigenvalues of the Jacobian $\mathbf{f_y}$; we consider them for λ near λ_0.) Figure 2.26 shows a possible path of the particular pair of eigenvalues that satisfy $\mu(\lambda_0) = \pm i\beta$. Assumption (3) of Theorem 2.11 is responsible for the transversal crossing of the imaginary axis. If all the other eigenvalues have strictly negative real parts, Figure 2.26 illustrates a loss of stability; for a gain of stability, just reverse the arrows. The situation of Hopf bifurcation depicted in Figure 2.26 contrasts with stationary bifurcation, where there is an eigenvalue 0 (see Figure 2.27). For stationary bifurcation, the crossing of imaginary axis is at the real axis.

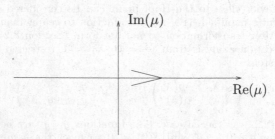

Fig. 2.27. Eigenvalue at a stationary bifurcation

Near Hopf bifurcations there is locally only one periodic solution for each λ (apart from phase shifts). That is, only one half-branch of periodic solutions comes out from the stationary branch (Figure 2.28, left). Recall that in the case of stationary bifurcation two half-branches come out. This difference between stationary bifurcation and Hopf bifurcation may or may not be reflected by bifurcation diagrams. Bifurcation diagrams can be constructed where half-branches of a pitchfork are identified (see, e.g., Figure 2.19). On the other hand, there exist bifurcation diagrams where two curves intersect in a Hopf bifurcation point. In Figure 2.28 (right) two scalar measures $[\mathbf{y}]$ are depicted, namely, the extremal values $y_{\max}$ and $y_{\min}$ of one of the components of $\mathbf{y}$, say of $y_1(t)$. For the Hopf solution at λ_0 the amplitude is zero, hence $y_{\max} = y_{\min}$. In this way the bifurcation diagram of Hopf bifurcation resembles that of pitchfork bifurcation. This is no coincidence, as a closer look at the Hopf normal form equation (2.22) reveals. The corresponding differential equation (2.23) of the radius ρ is equivalent to the normal form equation (2.13) of pitchfork bifurcation; see Example 2.4. and Figure 2.13. The amplitude indicated in Figure 2.28 is a generalization of the radius.

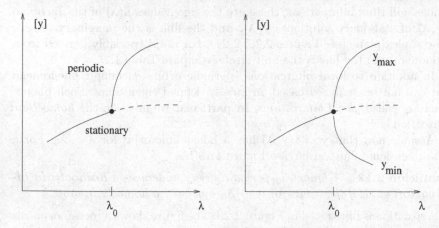

Fig. 2.28. Hopf bifurcation illustrated in bifurcation diagrams

Numerical discretization of the ODEs as described in Section 1.5.2 generally affects the bifurcation. The bifurcation point is replaced by a discretized version with parameter $\overline{\lambda}_0$ which depends on h. Usually, for higher-order integrators and small step lengths h, the discrepancy $\lambda_0 - \overline{\lambda}_0(h)$ is not seen in practice. But one should be aware of the situation. It is easy to artificially construct examples where the difference is noticeable ($\longrightarrow$ Exercise 2.20).

2.7 Bifurcation of Periodic Orbits

As discussed in the previous section, a Hopf bifurcation is the birth place of a branch of periodic orbits. What kind of bifurcations are possible for periodic orbits? We briefly answer this question in this section, leaving details to Chapter 7, which will introduce the necessary instruments.

We have discussed stationary bifurcations, namely, turning points, pitchfork bifurcations, and transcritical bifurcations. These phenomena occur in analogous ways for branches of periodic solutions; we defer an explanation of related mechanisms to Chapter 7. The stability discussion of periodic orbits will be reduced to a stability discussion of maps. Here we only mention the bifurcation phenomenon of *period doubling*. Period doubling is a bifurcation to a branch of periodic orbits that have the double period. The mechanism of period doubling bifurcation will be explained in Section 7.4.2, but examples will occur earlier, for example, in the following Section 2.8.

All these bifurcations are *local bifurcations* in that they can be characterized by locally defined eigenvalues crossing some line. For stationary bifurcations and Hopf bifurcations, these are the eigenvalues $\mu(\lambda)$ of the Jacobian $\mathbf{f}(\mathbf{y}, \lambda)$ of stationary solutions $(\mathbf{y}, \lambda)$, and the line is the imaginary axis in the complex plane (see Figures 2.26, 2.27). For maps (possibly derived from periodic orbits) the line is the unit circle, compare Table 1.2.

In addition to local bifurcations, periodic orbits encounter phenomena that can not be analyzed based on locally defined eigenvalues. Such phenomena are called *global bifurcations*. In particular we mention the *homoclinic bifurcation*.

Assume now that $\dot{\mathbf{y}} = \mathbf{f}(\mathbf{y}, \lambda)$ has a homoclinic orbit for $\lambda = \lambda_0$. For a two-dimensional illustration, see Figure 1.19(b).

Definition 2.12 *A branch of periodic orbits undergoes a **homoclinic bifurcation** at λ_0 if the orbits for $\lambda \to \lambda_0$ approach a homoclinic orbit.*

This situation is illustrated in Figure 2.29. The figure shows a metamorphosis of how a saddle and a periodic orbit approach each other as $\lambda \to \lambda_0$. The periodic branch ends at λ_0 (or is created at λ_0), where a homoclinic orbit is formed, and is broken for λ_{-1} (lower figure). For $\lambda \to \lambda_0$ the periods of the periodic orbits tend to ∞. For a homoclinic bifurcation, a periodic orbit collides with a saddle point. In case a periodic orbit collides with two or more saddle points, we speak of a "heteroclinic bifurcation." A standard reference is [Kuz98].

Example 2.13
The second-order differential equation for $u(t)$

$$\ddot{u} = u - u^2 + \lambda \dot{u} + \frac{1}{2} u \dot{u} \tag{2.24}$$

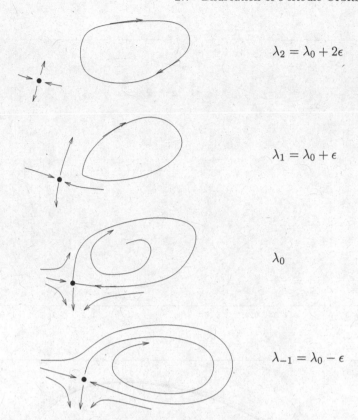

$$\lambda_2 = \lambda_0 + 2\epsilon$$

$$\lambda_1 = \lambda_0 + \epsilon$$

$$\lambda_0$$

$$\lambda_{-1} = \lambda_0 - \epsilon$$

Fig. 2.29. Phase portraits of a homoclinic bifurcation at λ_0

has equilibria $(u, \dot{u}) = (0, 0)$, and $(u, \dot{u}) = (1, 0)$ independent of λ. For $\lambda_0 = -0.5$ there is a Hopf bifurcation. The emanating branch of periodic orbits ends in a homoclinic orbit at $\lambda_0 = -0.429505$ [GuH83], [Beyn87b]. Two phase planes in Figure 2.30 show results of a simulation for $\lambda = -0.45$; this value is in the "middle" of the periodic branch, and not far from the homoclinic bifurcation. The upper figure includes a tangent field, and the trajectory that starts at $(u, \dot{u}) = (0, 0.3)$ with $t = 0$ and terminates at $t_f = 12$. Choosing the initial value of $\dot{u}$ slightly smaller than 0.3 leads to a trajectory that is attracted by the limit cycle. For the trajectory of the lower figure a starting point $(u, \dot{u}) = (0.001, 0.0008)$ is chosen (why?) to simulate where the outset of the saddle is attracted. The limiting cycle is clearly visible. The trajectory of the upper figure closely follows an inset of the saddle, approaching the equilibrium until it drifts away. Knowing the location of the inset and outset of the saddle it is easy to visualize how the periodic orbit squeezes for $\lambda \to -0.4295$ into the cone formed by inset and outset, until the orbit becomes the homoclinic orbit. Comparing Figures 2.30 and 2.29 we see that the scenario of Figure 2.30 is that labeled by λ_1. □

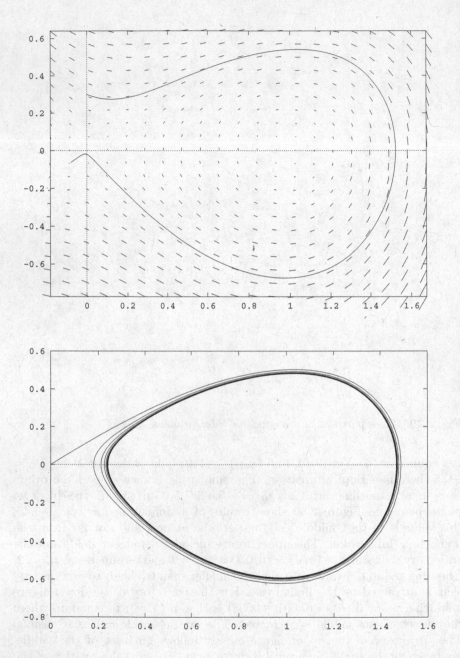

Fig. 2.30. Phase plane of Example 2.13., $\lambda = -0.45$. Top: trajectory close to the inset of the saddle; bottom: outset of the saddle

2.8   Convection Described by Lorenz's Equation

In this section we carry out an evaluation of Hopf bifurcation points for the famous Lorenz equations, which we have encountered already in Example 1.7. In 1963, Lorenz published a system of ODEs [Lor63] that has been given much attention in the literature. The physical background is a flow in a layer of fluid of uniform depth heated from below, with the temperature difference between the upper and lower surfaces maintained at a constant value. This problem is related to the Rayleigh–Bénard problem [Bén1901], [Ray16], [Bus81]. The outcome of Rayleigh–Bénard-type experiments is governed by the values of the Rayleigh number Ra. For values of Ra below a critical value R_c, the system has a stable steady-state solution in which the fluid is motionless and the temperature varies linearly with depth. When the Rayleigh number is increased past R_c, the purely conductive state becomes unstable and convection develops. Driven by buoyancy, the fluid motion organizes in cells or rolls. The onset of certain regular cloud patterns in the atmosphere, the drift of continents, and the granulation observed on the surface of the sun have all been attributed to this phenomenon.

If the convective currents are organized in cylindrical rolls, the governing equations can be written as a set of PDEs in two space variables [Lor63]. Expanding the dependent variables in a Fourier series, truncating these series to one term, and substituting into the PDE yields the three ordinary differential equations we have seen already in Example 1.7

$$\dot{y}_1 = P(y_2 - y_1),$$
$$\dot{y}_2 = -y_1 y_3 + R y_1 - y_2,$$
$$\dot{y}_3 = y_1 y_2 - b y_3$$

(2.25)

for the variables y_1, y_2, y_3. The quantities in equation (2.25) have physical relevance:

y_1 is proportional to the intensity of convection;

y_2 is proportional to the temperature difference between ascending and descending currents;

y_3 is proportional to the distortion of the vertical temperature profile from linearity;

$R = Ra/R_c$ is the relative Rayleigh number;

$P = Pr$ is the Prandtl number (with, e.g., $P = 16$); and

b is a constant (with, e.g., $b = 4$)

The relative Rayleigh number R is our branching parameter λ. For the stationary states, we refer to Example 1.7. The trivial state $\mathbf{y} = \mathbf{0}$ represents the conduction, whereas the nontrivial state $(y_1, y_2, y_3) = (\pm S, \pm S, R - 1)$ stands for convection. Again, the variable S denotes the square root

$$S := (bR - b)^{1/2}.$$

Using Exercise 2.10 it is easy to check that the trivial solution is unstable for $R > 1$.

In order to search for Hopf bifurcation, we substitute the (positive) non-trivial branch into the Jacobian. This gives

$$\begin{pmatrix} -P & P & 0 \\ 1 & -1 & -S \\ S & S & -b \end{pmatrix}.$$

The bifurcation parameter $\lambda = R$ is hidden in S. The eigenvalues μ are calculated via the determinant (compare Section 1.2.4)

$$0 = \begin{vmatrix} -P-\mu & P & 0 \\ 1 & -1-\mu & -S \\ S & S & -b-\mu \end{vmatrix}$$

$$= -(P+\mu)(1+\mu)(b+\mu) - S^2 P - (P+\mu)S^2 + P(b+\mu)$$

$$= -\mu^3 - \mu^2(P+1+b) - \mu(b+bP+S^2) - 2S^2 P.$$

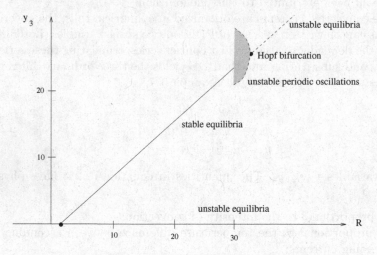

Fig. 2.31. Schematic bifurcation diagram of the Lorenz equation (2.25)

At the Hopf bifurcation point, the characteristic polynomial has a pair of purely imaginary roots $\pm i\beta$. Hence, the polynomial can be written as

$$\pm(\mu^2 + \beta^2)(\mu + \alpha),$$

with $-\alpha$ being the third eigenvalue. The values of β and α are still unknown. They can be calculated conveniently by equating the coefficients of both versions of the characteristic polynomial because they are identical at the Hopf bifurcation. To this end we multiply out, taking the negative sign for convenience,

$$0 = -\mu^3 - \mu^2\alpha - \mu\beta^2 - \beta^2\alpha.$$

Equating the coefficients gives three equations for the three unknowns β, α, and R (via S):

$$\alpha = P + 1 + b,$$
$$\beta^2 = b + bP + S^2,$$
$$\beta^2\alpha = 2S^2P.$$

Substituting the first and second of these expressions into the third gives

$$(b + bP + S^2)(P + 1 + b) = 2S^2P.$$

It follows readily that

$$S^2 = b(R - 1) = \frac{(1 + b + P)b(1 + P)}{P - b - 1},$$

which gives the parameter value $\cdot$

$$\lambda_0 = R_0 = 1 + \frac{(1 + b + P)(1 + P)}{P - b - 1} = \frac{P(P + 3 + b)}{P - b - 1}$$

of Hopf bifurcation. Let us take a numerical example: $b = 4, P = 16$. This choice of constants leads to

$$\lambda_0 = R_0 = \frac{368}{11} = 33.4545\ldots.$$

The initial period of the "first" periodic oscillation is given by

$$T_0 = \frac{2\pi}{\beta} = 2\pi(b + bP + S^2)^{-1/2} = 0.4467\ldots.$$

The third eigenvalue is $-\alpha = -21$. Provided the transversality condition holds (it does), the negative sign of this eigenvalue shows that there is a loss of stability of the equilibrium when R is increased past R_0.

The bifurcation diagram Figure 2.31 anticipates results obtained by calculating periodic oscillations. The equilibria are drawn with a solid line indicating stable equilibria, and a dashed line indicating unstable equilibria. Oscillations are represented in the way indicated by Figure 2.28. A dashed curve represents the range of the $y_3(t)$ values between y_{min} and y_{max}. The oscillations are unstable; this stability result is derived by methods that will be introduced in Chapter 7. As mentioned earlier, only one branch of periodic solutions branches off; the two half-curves represent two different scalar measures for the same periodic solution; compare Figure 2.28.

The graph of the nontrivial branch of equilibria in Figure 2.31 is covered twice. That is, we have two Hopf bifurcations with the same value R_0. The "other" branch bears no new information because there is a symmetry involved. Inspecting the Lorenz equations, one realizes that $(-y_1(t), -y_2(t), y_3(t))$

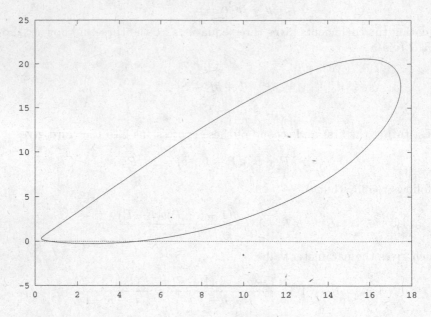

Fig. 2.32. Equation (2.25), (y_1, y_2)-plane (projection) for $R = 22.3$

is a solution of equation (2.25) whenever $(y_1(t), y_2(t), y_3(t))$ is a solution. Geometrically, this is symmetry with respect to the y_3-axis.

Figure 2.32 depicts one of the periodic orbits for $R = 22.3$ in a projection of the three-dimensional phase space onto the (y_1, y_2)-plane. Imagine the other periodic solution for the same value of R located symmetrically with respect to $(0, 0)$.

The question is, what happens with the periodic orbits for decreasing values of R? To understand the dynamics we discuss the projections of the solution with $R = 19.5$ onto the (y_1, y_2)-plane (Figure 2.33, top), and onto the (y_1, y_3)-plane (Figure 2.33, bottom). From the analysis of Example 1.7 we realize that the periodic orbit squeezes to two of the three eigenvectors. Compare Figure 1.17. In particular, the trajectory of this periodic orbit approaches the three-dimensional saddle $(0, 0, 0)$ along the y_3-axis and leaves its neighborhood close to its unstable manifold, which is within the (y_1, y_2)-plane. (Figure 1.17) This indicates that $R = 19.5$ is close to a value R_∞ of a homoclinic orbit. In fact, for $R_\infty \approx 19$ the unstable manifold of the saddle is bent back and inserted into the inset y_3-axis. This is in three dimensions a situation analogous to the planar homoclinic orbit illustrated in Figure 1.19(b). The branch of periodic orbits ends at R_∞.

Interesting things happen "on the other side" of the homoclinic orbit for $R > R_\infty$. As is indicated in Figure 2.33, the periodic orbit $(y_1(t), y_2(t), y_3(t))$ and its counterpart $(-y_1(t), -y_2(t), y_3(t))$ touch for $R = R_\infty$ giving the

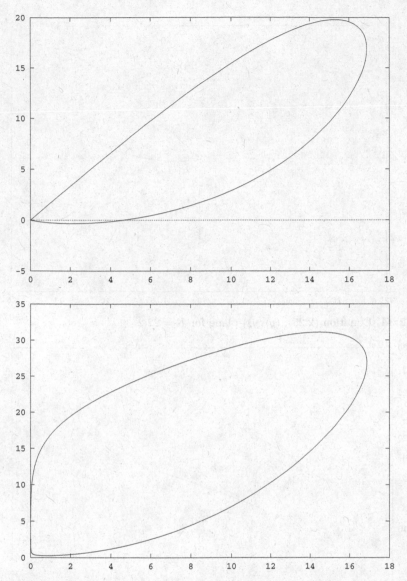

Fig. 2.33. Equation (2.25), (y_1, y_2)-plane (top), and (y_1, y_3)-plane (bottom) for $R = 19.5$

symbol ∞ of R_∞ a graphical meaning. The resulting combined curve, which encircles both nontrivial steady states, is the limiting solution of an infinite number of branches: Take any infinite sequence of $+$ and $-$ signs, with $+$ symbolizing the half orbit with $y_1 \geq 0$, and $-$ symbolizing the half orbit with $y_1 \leq 0$. For example $+ + - + + - \ldots$ symbolizes two times encircling the

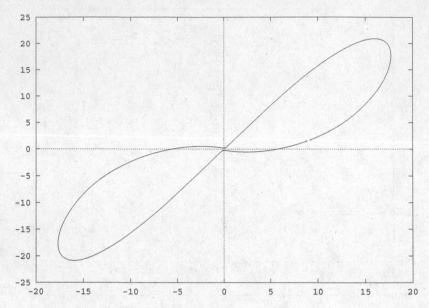

Fig. 2.34. Equation (2.25), (y_1, y_2)-plane for $R = 22.2$

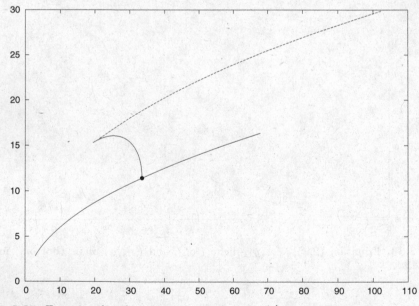

Fig. 2.35. Equation (2.25), bifurcation diagram, $y_1(0)$ versus R

positive steady state, then once encircling the negative steady state, and so
on. For $R > R_\infty$ there is an infinite number of branches of periodic orbits,
which no longer touch $(0, 0, 0)$. This is illustrated in Figure 2.34 for the $+-$

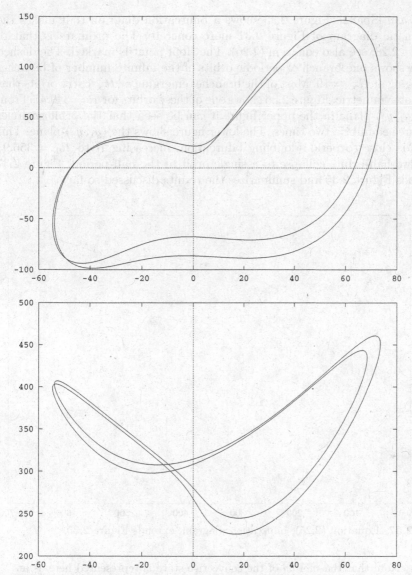

Fig. 2.36. Equation (2.25), periodic orbit "shortly after" a period doubling. (y_1, y_2)-plane (top) and (y_1, y_3)-plane (bottom) for $R = 340.47$

orbit with $R = 22.2$. For increasing R the gap close to the saddle widens; this branch of unstable periodic orbits gains stability at $R_0 = 491.79$. To each sequence of $+$ and $-$ signs there corresponds one branch, the solutions of which have related oscillation behavior. The homoclinic double-orbit at R_∞, final solution of infinitely many branches, has been called *homoclinic explosion* [Spa82]. Figure 2.35 shows a bifurcation diagram that makes the schematic diagram of Figure 2.31 more concrete. The ordinate is that of Figure 2.28; see also equation (2.26). The Hopf point is marked. The dashed curve shows one branch of periodic orbits of the infinite number of branches emerging at $R_\infty \approx 19$. Most of the branches emerging at R_∞ carry orbits that are not symmetric. Figure 2.36 shows one of these orbits for $R = 340.47$. From the (y_1, y_2)-plane in the upper figure it can be seen that the orbit encircles all three equilibria two times. The lower figure shows the (y_1, y_3)-plane. This orbit is close to period-doubling bifurcation; increasing R to $R_0 = 356.93$ the two half-orbits collapse to a single orbit of the half-period. Figure 2.37 extends Figure 2.35 and summarizes the results discussed so far.

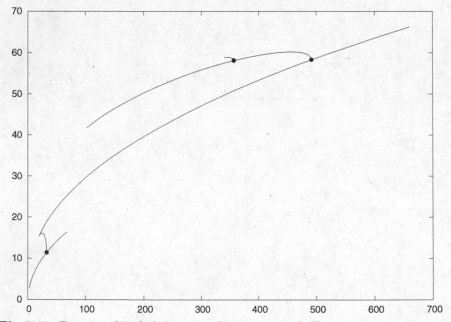

Fig. 2.37. Equation (2.25), bifurcation diagram, extends Figure 2.35.

Note that the *motion* of the convective state is represented here as an equilibrium of an ODE. This seeming contrast between physical motion and mathematical equilibrium occurs because the PDE was established for a *stream function*. The nontrivial equilibrium of equation (2.25) represents a regular convection with steady velocity and temperature profiles. The periodic oscillations of equation (2.25) represent less regular convection.

Calculation of Hopf points, as above, is limited to simple problems with not more than $n = 3$ equations. Note that this analysis did not reveal any information about stability and the "direction" of the emerging branch. As [HaKW81] shows, it is possible to obtain such local information by hand calculation, but the effort required is immense. In general, one must resort to numerical methods (Chapter 7). For an extensive analysis of the Lorenz

equation, see [Spa82]. Because this ODE system stems from a severe truncation of a PDE system, the question of how to obtain more realistic models
has been considered. More modes are included in the analysis of [Cur78].
As shown in [YoYM87], the Lorenz equations also model a geometry where
the flow is constrained in a torus-like tube. In [HaZ94] the homoclinic orbit
was calculated very precisely.

2.9   Hopf Bifurcation and Stability

We have learned that a Hopf bifurcation point connects stationary solutions
with periodic solutions. In the first example of Hopf bifurcation (see Figure
2.24), a stable orbit encircles an unstable equilibrium. In Exercise 2.9 and in
the Lorenz system (2.25), unstable cycles coexist with stable equilibria. In
this section we shall discuss loss (or gain) of stability in some detail.

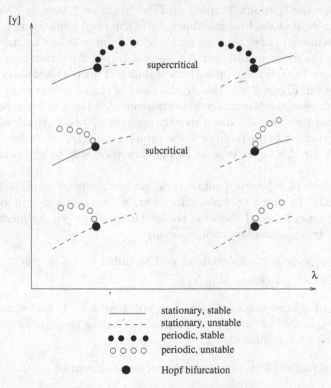

Fig. 2.38. Local stability behavior at Hopf bifurcations

To illustrate bifurcation behavior graphically, we use symbols that distinguish between stable and unstable periodic oscillations. These symbols (dots
and circles) are introduced in Figure 2.38. The branches in Figure 2.38 are
continuous; the gaps between the small circles or dots do not indicate gaps

between solutions. For every λ in the corresponding range there is a periodic orbit. Stability and instability of periodic oscillations will be formally defined in Chapter 7. For the time being, it is sufficient to have the intuitive understanding that a stable periodic orbit is approached by nearby trajectories, whereas trajectories leave a neighborhood of an unstable periodic orbit.

Figure 2.38 illustrates various cases. A heavy continuous line represents stable stationary solutions, and unstable stationary solutions are indicated by dashed lines. A thick dot marks a branch point—here a Hopf bifurcation point. The rule is as follows: Locally—that is, close to the Hopf bifurcation, stable periodic orbits encircle unstable equilibria (first row in Figure 2.38), and unstable periodic orbits encircle stable equilibria (second row). This rule relates the direction of the emanating branch of periodic solutions to the stability properties of these solutions. Following our earlier definition, cases of the first row are supercritical bifurcations, whereas the cases of the second row are subcritical. In the supercritical cases an exchange of stability takes place between the "periodic branch" and the "stationary branch." Note again that these rules are valid in a neighborhood of the Hopf bifurcation point. The stability (instability) of emanating periodic orbits may be lost by mechanisms that will be discussed in Chapter 7. Classical Hopf bifurcation assumes that the stationary branch loses (gains) its stability. This is the situation of the first two rows in Figure 2.38. Also, in the cases of the third row of figures, one eigenvalue crosses the imaginary axis transversally. Here at least one of the eigenvalues of the Jacobian has a strictly positive real part. Hopf bifurcation is then a branching to unstable periodic orbits, no matter what the direction. An example for this latter type of Hopf bifurcation will be given in Section 7.4.1.

In the cases of subcritical bifurcation, an exchange of stability may still occur globally. In order to make this clear, we digress for a moment and discuss a practical way of choosing the scalar measure $[\mathbf{y}]$. As mentioned in Section 2.2, two reasonable candidates are

$$[\mathbf{y}] = y_k(t_0) \quad \text{for some } t_0 \text{ and an index } k, \ 1 \le k \le n \,;$$
$$[\mathbf{y}] = \text{ amplitude of } y_k(t) \,.$$

The choice of k does not cause trouble, just take $k = 1$, for example. The remaining questions of how to pick t_0 and calculate the amplitude can be solved by the compromise

$$[\mathbf{y}] = y_k(t_0) \text{ with } y_k(t_0) = \text{ maximum or minimum of } y_k(t) \,. \tag{2.26}$$

(This scalar measure was anticipated in Figure 2.28.) As we shall see in Section 7.1, there is an easy way to obtain $[\mathbf{y}]$ of equation (2.26) as a by-product of the numerical calculation of an orbit. The choice of equation (2.26) is not unique because there is at least one maximum and one minimum of

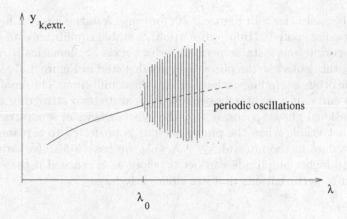

Fig. 2.39. Soft generation of limit cycles

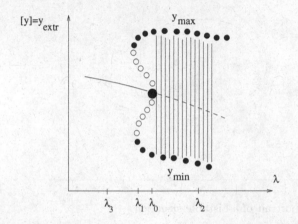

Fig. 2.40. Hard generation of limit cycles

$y_k(t)$. Even when only the maximum or only the minimum is depicted, one can still get a feeling for the evolution of amplitude.

We first illustrate a supercritical Hopf bifurcation with a bifurcation diagram depicting the $[\mathbf{y}]$ of equation (2.26); see Figure 2.39. As we can see, amplitude grows continuously for increasing values of λ. This kind of Hopf bifurcation is called *soft loss of stability* or *soft generation of limit cycles*. Accordingly, supercritical bifurcation has also been called *subtle bifurcation*. An interesting effect occurs when, in the case of a subcritical bifurcation, the periodic branch turns back and gains stability (Figure 2.40). Then, when λ is increased past λ_0, the amplitude suddenly undergoes a jump and takes "large" values. This kind of jump from a stable equilibrium to large-amplitude periodic orbits is called *hard loss of stability* or *hard generation* of limit cycles. This is the phenomenon behind Figure 1.3, and now the skeleton in

Figure 1.4 becomes clear. In Figure 2.40, for every λ in the interval between λ_1 (turning point) and λ_0 (Hopf bifurcation), a stable equilibrium, an unstable periodic orbit, and a stable periodic orbit coexist. Schematically, in two dimensions, this looks like the phase diagram depicted in Figure 2.41. We see two periodic orbits encircling each other and the equilibrium. The smaller orbit (dashed curve) is a separatrix because it separates two attracting basins. (In n-dimensional phase spaces, $n > 2$, the hypersurfaces of separatrices are in general not visible when the phase portrait is projected to a plane.) For parameter values in the interval $\lambda_1 \leq \lambda \leq \lambda_0$ we have a *bistable situation*. The jump to higher amplitude can occur before λ_0 is reached if the system is subjected to perturbations that are sufficiently large.

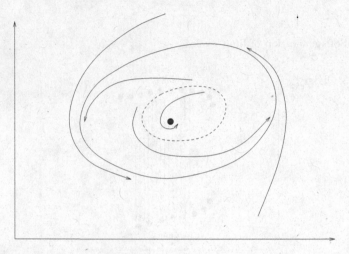

Fig. 2.41. Phase portrait of a bistable situation

When the parameter is decreased from λ_2 (Figure 2.40), a discontinuity occurs again. This time the jump from large values of amplitude to zero amplitude (= equilibrium) takes place at the turning point ($\lambda = \lambda_1$). The bistable situation of hard generation of limit cycles is a hysteresis phenomenon similar to the situation that often occurs with pairs of turning points (see, e.g., Figure 2.7). In a real experiment, unstable solutions are not directly observable. Knowledge of the unstable branches comes from solving the equations. In order to get a feeling for what can be represented physically, disregard all dashed lines and open dots in the above figures—that is, omit all unstable solutions. But the existence of unstable solutions often manifests itself by the role they may play as indicator of how far basins of attraction extend.

A bistable situation with a subcritical Hopf bifurcation enables the onset of *bursts*. A burst consists of a short oscillation of several *spikes* followed by a period of less activity (here stationary solutions). This interval of quiescence may be followed by a burst of several pulses that, after extinction, again gives way to a nearly constant value. Such repetitive burst behavior can be explained using Figure 2.40. Imagine that the control parameter λ depends on another variable or on an "outer" influence in such a way that λ sweeps slowly back and forth through the interval $\lambda_3 \leq \lambda \leq \lambda_2$. This interval contains the narrower interval of bistability where the stable stationary branch (resting mode) and the stable periodic branch (active mode) overlap. Each sweep of the larger interval causes a jump—from burst to quiescence or from quiescence to burst. Such alternation between active and inactive modes has been investigated in examples from chemistry [RiT82b] and nerve physiology [HoMS85]. In the latter paper, the slow sweeping of the parameter interval was modeled by attaching to the equations a "slow oscillator" that controls λ. In a specific system of equations, a bursting phenomenon may be hidden because the role of the slowly and periodically varying parameter λ may be played by some variable that is controlled by a subsystem. For a classification of bursting, see [IzR01]. An interesting application of bistability has been observed in the analysis of railway bogies [XuST92]. For an electronic burster, see [WaS06], and for bursting in nerve models consult [ChFL95]. An example from chemical engineering concludes the section.

Example 2.14   CSTR

This example is a system of ODEs that models a *continuous stirred tank reactor*, commonly abbreviated CSTR. "Continuous" refers to a continuous flow entering (and leaving) the reactor—that is, a CSTR is an *open system*. Human beings and other living organisms that have input of reactants (nutrients) and output of products (wastes) are complex examples of CSTRs.

The specific CSTR model we will discuss is

$$
\begin{aligned}
\dot{y}_1 &= -y_1 + \mathrm{Da}(1 - y_1)\exp(y_2)\,, \\
\dot{y}_2 &= -y_2 + B \cdot \mathrm{Da}(1 - y_1)\exp(y_2) - \beta y_2\,;
\end{aligned}
\tag{2.27}
$$

[UpRP74], [VaSA78]. The exponential term reflects an infinite activation energy. The variables occurring in equation (2.27) are

the Damköhler number $\mathrm{Da} = \lambda$;
y_1, y_2, which describe the material and energy balances;
β, the heat transfer coefficient ($\beta = 3$); and
B, the rise in adiabatic temperature ($B = 16.2$).

Figure 2.42 depicts the result of two simulations. A bifurcation diagram is shown in Figure 2.43. We see a branch of stationary solutions with two Hopf bifurcation points, for which the parameter values and initial periods are

$$
\begin{aligned}
\lambda_0 &= 0.1300175, \quad T_0 = 8.479958, \text{ and} \\
\lambda_0 &= 0.2464796, \quad T_0 = 1.186330\,.
\end{aligned}
$$

Again the two Hopf points are connected by a branch of periodic orbits (not shown in the figure). This example exhibits both subcritical and supercritical

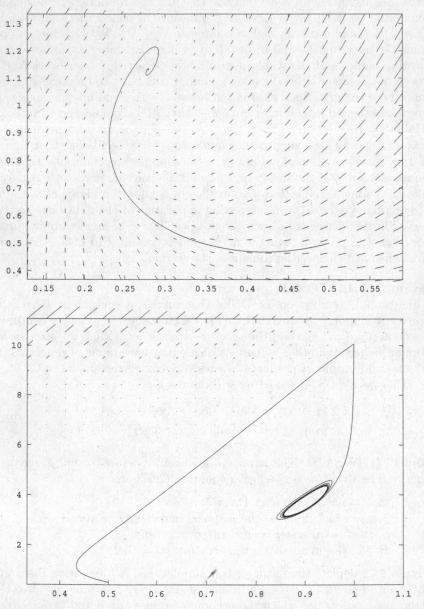

Fig. 2.42. Two phase diagrams of the CSTR Example 2.14 with direction field. top: $\lambda = 0.125$, stable stationary; bottom: $\lambda = 0.24$, limit cycle

bifurcation. At the right Hopf point there is soft generation of limit cycles (for decreasing λ), whereas at the left Hopf point the loss of stability is hard. The branch of stationary solutions has two turning points close to each other with a difference in λ of $\varDelta\lambda = 0.0005$. As is hardly seen in Figure 2.43, the lower turning point is above the left Hopf point. The hysteresis part of the bifurcation diagram thus lies fully in the unstable range between the Hopf points. The papers [UpRP74], [VaSA78] also present results for other values of β and B. $\qquad$ $\square$

The CSTR Example 2.14 is the origin of Figures 1.3, 1.4. The role a bifurcation diagram plays in understanding the dynamics and the structure of a problem is highlighted, for example, by Figure 2.43, or by Figure 2.37. It is justified to speak of the bifurcation diagram as the skeleton or the *signature* of a problem.

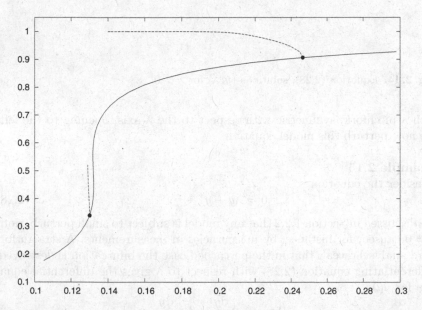

Fig. 2.43. Bifurcation diagram of Example 2.14, $y_1(0)$ versus Da; solid line: stationary, dashed line: periodic orbits. Hopf points are marked.

2.10 Generic Bifurcation

When speaking of bifurcation, we will take the word *generic* as a synonym for "most typical." In this section we will discuss what kind of bifurcation is generic and in what senses other kinds of bifurcation are not generic. Let us first ask which type of bifurcation arises in "nature."

We begin with a motivating example. In Section 2.4, we discussed

$$0 = \lambda y + y^3$$

as a simple scalar example exhibiting pitchfork bifurcation. The resulting bifurcation diagram (Figure 2.14) shows a bifurcation point at $(y, \lambda) = (0, 0)$

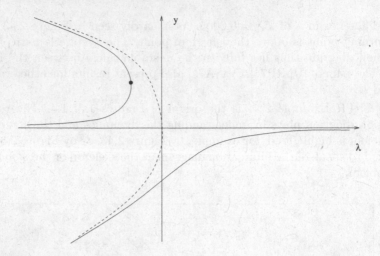

Fig. 2.44. Equation (2.28), solutions (y, λ) for $\gamma > 0$

with a pitchfork, symmetric with respect to the λ-axis, opening to the left. We now perturb this model equation.

Example 2.15
Consider the equation

$$0 = \lambda y + y^3 + \gamma. \tag{2.28}$$

We discussed in Section 1.2.2 that any model is subject to small perturbations $\gamma \neq 0$ caused, for instance, by inaccuracies in measurements. A straightforward analysis reveals that in the perturbed case the bifurcation is destroyed. Differentiating equation (2.28) with respect to λ gives the differential equation for $y(\lambda)$

$$0 = y + \lambda \frac{dy}{d\lambda} + 3y^2 \frac{dy}{d\lambda}.$$

Because $y = 0$ is not a solution of equation (2.28), there is no horizontal tangent $dy/d\lambda = 0$ to the solution curve $y(\lambda)$. Checking for a vertical tangent means considering the dependence of λ on y and differentiating equation (2.28) with respect to y. This yields

$$\frac{d\lambda}{dy} = 0 \quad \text{for} \quad \lambda = -3y^2,$$

which, substituted into equation (2.28), in turn yields the loci

$$(y, \lambda) = \left(\left(\frac{\gamma}{2} \right)^{1/3}, \; -3 \left(\frac{\gamma}{2} \right)^{2/3} \right)$$

of the vertical tangents. There are three solutions of equation (2.28) for

$$\lambda < -3\left(\frac{\gamma}{2}\right)^{2/3},$$

and the diagram looks like Figure 2.44 (solid lines). The point with a vertical tangent is a turning point. For $\gamma < 0$, the curves in Figure 2.44 are reflected in the λ-axis. As $\gamma \to 0$, the curves approach the pitchfork (dashed curve). For an arbitrary perturbation γ, the occurrence of a turning point is typical. Bifurcation is the rare exception because it occurs only for $\gamma = 0$.

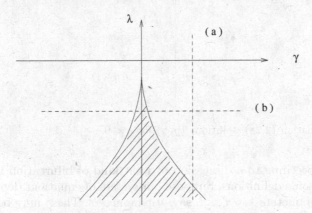

Fig. 2.45. Equation (2.28), parameter chart

Before stating a general rule, let us reconsider this example, focusing on the multiplicity of the solutions. The λ-value at the vertical tangent depends on γ:

$$0 = \left(\frac{\lambda}{3}\right)^3 + \left(\frac{\gamma}{2}\right)^2. \qquad (2.29)$$

Now we consider equation (2.28) as a *two-parameter model*, with γ and λ having equal rights. Equation (2.29) establishes a curve in the (λ, γ)-plane (Figure 2.45) with the form of a *cusp*. For all combinations of (λ, γ)-values inside the cusp (hatched region), equation (2.28) has three solutions; for values outside the cusp, there is only one solution. This becomes clear when bifurcation diagrams are drawn. Figure 2.44 for fixed $\gamma > 0$ depicts the situation for parameter combinations on the dashed line (a) in Figure 2.45. Choosing $\lambda < 0$ leads to the bifurcation diagram of Figure 2.46. The corresponding line in Figure 2.45 (dashed line (b)) crosses the cusp boundary twice. Accordingly, there are two turning points with hysteresis behavior. Clearly, when a line and its parameter combinations cross the cusp curve in the (λ, γ)-plane, a bifurcation takes place. Here it is a jump that occurs when the solution is on an appropriate part of the branch. Because equation (2.28) defines a two-dimensional surface in the three-dimensional (y, λ, γ)-space, each bifurcation diagram is a cross section (Exercise 2.13). Equation (2.28) will be analyzed in a more general context in Section 8.4.   □

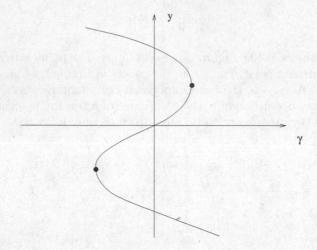

Fig. 2.46. Equation (2.28), solutions (y, γ) for $\lambda < 0$

Before proceeding to a discussion of what kind of bifurcation is generic, we pause for some definitions. Suppose our system of equations depends on a number of parameters $\lambda, \gamma, \zeta, \ldots$, say, ν parameters. These may be physical parameters such as Rayleigh number and Damköhler number, may stand for constants that can be varied, and so on. The parameters form a ν-dimensional vector

$$\Lambda = (\lambda, \gamma, \zeta, \ldots).$$

The *parameter space* of all feasible parameter values is also called the *control space* because the values of the parameters control the solutions of the problem. In our previous example, the control space is the (λ, γ)-parameter plane of Figure 2.45. The loci in a parameter space where a bifurcation phenomenon occurs are called *bifurcation curves*, or *critical boundaries*. In equation (2.28), the bifurcation curves are formed by the solutions of equation (2.29)—that is, the critical boundaries are the curves that form the cusp in Figure 2.45. A bifurcation curve formed by the loci of turning points is also called *fold line*. The subset of the parameter space formed by the critical boundaries is called the *bifurcation set*, or *branching set*, or *catastrophe set*. A graphical representation of a part of the parameter space that displays critical boundaries or multiplicities of solutions will be called a *parameter chart*. Figure 2.45 is an example of a parameter chart. A parameter chart that displays regions of stable and unstable behavior will be called a *stability chart*. It sometimes makes sense to separate a certain bifurcation parameter, say λ, from the remaining $(\nu - 1)$ control parameters (Exercise 2.14). This will be discussed in more detail in Section 8.4.

We mention in passing that so far we have developed four graphical tools for illustrating the behavior of solutions of

$$\dot{\mathbf{y}} = \mathbf{f}(\mathbf{y}, \Lambda),$$

namely,

(a) graphs of $(t, y_k(t))$ for various k ("time history"),
(b) phase diagrams,
(c) bifurcation diagrams, and
(d) parameter charts (stability charts).

It is important that these different tools not be confused (see Exercise 2.15).

After these preliminaries, we address the question of what kind of bifurcation is generic. As we shall see, the answer depends on the number of parameters involved. We start with the principal case of varying only one parameter λ. Any other parameters are kept fixed for the time being.

In one-parameter problems, only turning points and Hopf-type bifurcations (including period doubling) are generic. If the underlying model undergoes small changes, turning points and Hopf points may shift, but they persist. To explain this situation for turning points, let us look again at Definition 2.8. Assumptions (3) and (4) involve the relations "$\not\subseteq$" and "$\neq 0$," which represent a "normal state." If in condition (4), $d^2\lambda/d\sigma^2$ is continuous and $d^2\lambda(\sigma_0)/d\sigma^2 \neq 0$ for some σ_0, then $d^2\lambda/d\sigma^2 \neq 0$ in a neighborhood of σ_0 too. By a similar argument, it is clear that condition (3) is not affected by small perturbations. Small enough perturbations do not cause the assumptions of a turning point (Definition 2.8) to be violated, nor do they cause the conditions that create Hopf bifurcation to be violated. Consequently, turning points and Hopf bifurcations are abundant. Stability such as that of turning points and Hopf bifurcation with respect to small perturbations of the underlying problem is called *structural stability*. Note that perturbations need not be just constants added to the equations. One can study, for instance, periodic perturbations [NaA87].

Hysteresis points and stationary bifurcations are not generic; they are destroyed by arbitrary perturbations. This *unfolding* of bifurcations and hysteresis points is illustrated in Figure 2.47. In the left column of Figure 2.47, we have the nongeneric situation, which is unfolded by a perturbation in one of two ways. The situations in the two right columns are generic because they involve only turning points.

Although pitchfork bifurcations and transcritical bifurcations are not generic, we occasionally see them. As Arnol'd [Arn83] states, these bifurcations are due to some peculiar symmetry or to the inadequacy of the idealization, in which some small effects are neglected. Simplifying assumptions often used in modeling can create artificial bifurcations in equations (see Section 2.11). The onset of a pitchfork bifurcation in a one-parameter model can be seen as an indication of limitations in our modeling capabilities and computers. For example, in considering Euler's buckling beam problem, it is convenient to assume perfect symmetry. It is this simplification that establishes a bifurcation (Exercise 2.14). If material imperfections are taken into consideration, the models are more complex and more difficult to solve. Because it will never be possible to avoid all simplifying assumptions, we cannot rule

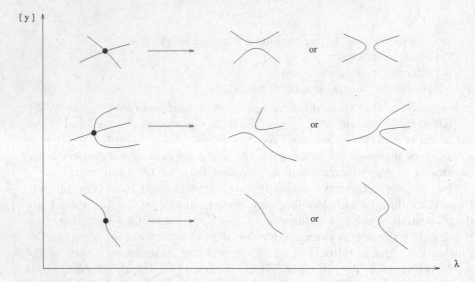

Fig. 2.47. Perturbation of bifurcation phenomena

out the occurrence of pitchfork or transcritical bifurcations in one-parameter problems. Simplifying assumptions such as symmetry play no role in hysteresis. For this reason, hysteresis points hardly ever occur for a given realistic one-parameter problem. The same holds true for multiple bifurcation points and isola centers.

The above discussion might suggest that one can forget about such non-generic cases as hysteresis points. This may be justified for one-parameter problems, but not for multiparameter problems, where such points occur naturally. We illustrate this by studying the situation for a fictitious two-parameter model. Let us again denote the second parameter by γ. Consider a sequence of bifurcation diagrams with respect to λ, say, for $\gamma_1 < \gamma_2 < \gamma_3 < \ldots$. Such a sequence may look like Figure 2.48. For γ_1 we find two branches without connection; the upper branch is an isolated branch. Starting from point A, we come smoothly to point B when λ is increased. For γ_2 the situation is the same although the isolated branch has come closer. Now consider the situation for γ_3, where we no longer encounter an isolated branch; the branches resemble a mushroom in shape. The consequences for the path from A to B are severe. There are two jumps (right figure, double arrows). Apparently there is a situation with a transcritical bifurcation for a value γ_0^B with $\gamma_2 < \gamma_0^B < \gamma_3$. One must go through this bifurcation when γ is increased from γ_2 to γ_3. The value γ_0^B is the instant when the isolated branch touches the lower branch (seen from γ_2), or when two turning points coalesce (seen from γ_3). This corresponds to the first line in Figure 2.47. Hence the bifurcation separates a continuous path A $\to$ B (Figure 2.48) from a path with jumps. This transcritical bifurcation for γ_0^B can be seen as an *organizing*

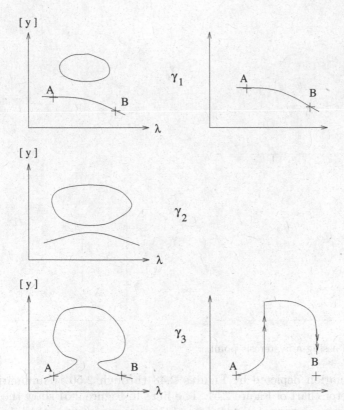

Fig. 2.48. Passing a transcritical bifurcation

center separating a no-jump situation from a two-jump situation. In view of its structure in a two-parameter setting, a bifurcation point may be called *gap center* (compare Figure 2.47).

Increasing γ further, one may obtain Figure 2.49. Here it is assumed that, for increasing γ, the right hysteresis phenomenon disappears and only one jump remains (Figure 2.49, γ_5, right figure). Clearly, between γ_4 and γ_5 there is a value γ_0^{HY} where a hysteresis point occurs. The parameter combination $(\lambda_0^{HY}, \gamma_0^{HY})$ of a hysteresis point is another example of an organizing center and is therefore a *hysteresis center*.

Consider again the first picture of Figure 2.48 ($\gamma = \gamma_1$). We ask what might happen when smaller values of γ are taken, say, γ_{-1}, γ_{-2}, and γ_{-3} with

$$\gamma_1 > \gamma_{-1} > \gamma_{-2} > \gamma_{-3}.$$

Possible bifurcation behavior is depicted in Figure 2.50. Here, for decreasing γ, the isolated branch gets smaller. For some value γ_0^I between γ_{-2} and γ_{-3}, the two turning points coalesce in an isola center. (The unfolding of isola centers can be illustrated as in Figure 2.47.)

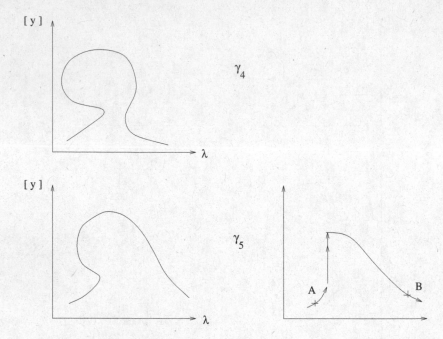

Fig. 2.49. Passing a hysteresis point

The situation depicted by Figures 2.48 through 2.50 is summarized in the parameter chart of Figure 2.51. The lines in Figure 2.51 show the curves where turning points are located. Two of these curves meet in the bifurcation point with parameter pair $(\lambda_0^B, \gamma_0^B)$, two curves meet in $(\lambda_0^{HY}, \gamma_0^{HY})$ (hysteresis point), and two curves merge at $(\lambda_0^I, \gamma_0^I)$, which is the location of an isola center. At these three points a new dynamic behavior is organized (no multiplicity, then no jump, then two jumps, then one jump). This leads us to call these points "organizing centers." A similar interpretation can be found for the onset of a pitchfork bifurcation (Exercise 2.16). Note that examples with such phenomena as those depicted in Figures 2.48 through 2.50 occur in reality. For example, the CSTR problem discussed in [HlKJ70] exhibits these phenomena. The jumps to the upper level (Figure 2.48) are interpreted as ignition of the CSTR because [$\mathbf{y}$] can be related to a temperature. Correspondingly, the jumps to the lower level are extinctions of the reactor. In this example, a turning point is either a point of ignition or a point of extinction. Phenomena such as those depicted in Figures 2.48 through 2.50 can be described by simple equations (Exercise 2.17).

In the above discussion and Figures 2.48 through 2.50, the emphasis is on stationary bifurcations. Hopf bifurcations also vary with parameters and can coalesce or interact with stationary bifurcations. Interactions between Hopf points and stationary bifurcations can trigger interesting effects [Lan79], [Lan81], [Kuz98].

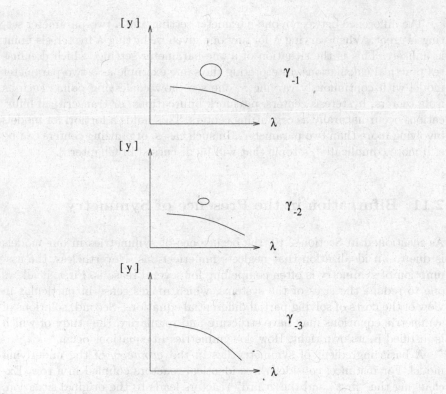

Fig. 2.50. Isola formation

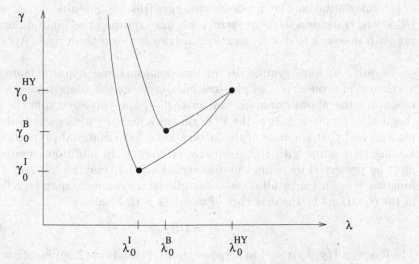

Fig. 2.51. Parameter chart summarizing Figures 2.48 through 2.50

The difference between a one-parameter setting and a two-parameter setting is great. When varying λ for any one given γ, hitting a hysteresis point is unlikely. This is the situation of a one-parameter setting, which dominates practical calculations. Interpreting the same example as a two-parameter model with continuously varying γ, one sees how coalescing points such as isola centers, hysteresis centers, pitchfork bifurcations, or transcritical bifurcations occur naturally as organizing centers. This holds a fortiori for models involving more than two parameters. In such cases, organizing centers can be still more complicated, a topic that will be discussed in Chapter 8.

2.11 Bifurcation in the Presence of Symmetry

As mentioned in Section 2.10, the occurrence of symmetries in our models is due to an idealization that neglects imperfections. Nevertheless, the assumption of symmetry is often compelling for several reasons. First, it allows one to reduce the sizes of the systems, which makes sense in particular in view of the costs of solving partial differential equations. Second, solutions of symmetric equations may have structure and regularity, the study of which is justified in its own right. How do symmetries in equations occur?

A basic ingredient of symmetry lies in the *geometry* of the underlying model. For instance, consider three identical reactors coupled in a row. Exchanging the "first" and the "third" reactors leads to the original situation. Other examples of symmetries of geometry are a square domain of a PDE or a regular hexagonal lattice-dome structure [Hea88]. Suitable rotations and reflections transform the above structures into equivalent configurations. The proper framework to discuss such symmetries is group theory; see Appendix A.7.

In order to have symmetries in the *equations* one requires more than a symmetric geometry. The physical laws must match the symmetry. This means for the above examples that in each of the three reactors the same chemical experiment is run, the PDE on the square is, for instance, the Laplacian, and that the rods of the lattice dome have material properties and loading that agree with the geometric symmetry. In addition, symmetries must be preserved by numerical discretizations that may be inherent in the function $\mathbf{f}(\mathbf{y}, \lambda)$. Under all these assumptions, a symmetry may be reflected in the equations by the existence of matrices $\mathbf{S}$ that satisfy

$$\mathbf{S}\mathbf{f}(\mathbf{y}, \lambda) = \mathbf{f}(\mathbf{S}\mathbf{y}, \lambda). \tag{2.30}$$

The function $\mathbf{f}(\mathbf{y}, \lambda)$ is said to be *equivariant* if equation (2.30) is satisfied for all elements of a group $\mathcal{G}$ of matrices $\mathbf{S}$. $\mathcal{G}$ is the group of symmetries of the *equation*.

The simplest example of a group is the *cyclic group* of order two $\mathbf{Z}_2$, which can be symbolized by the two elements $+1$ and -1 and the multiplication as group operation. For the function $\mathbf{f}(\mathbf{y}, \lambda)$ the element $+1$ is represented by the identity matrix, $\mathbf{S} = \mathbf{I}$. The -1 is represented by any matrix $\mathbf{S}$ that satisfies $\mathbf{S}^2 = \mathbf{I}$. For a scalar equation such as that of the normal form $f(y, \lambda) = \lambda y - y^3$ of pitchfork bifurcation the equivariance condition is clearly satisfied with $Sy = -y$. For the Lorenz system in equation (2.25) the equivariance condition $\mathbf{Sf}(\mathbf{y}, \lambda) = \mathbf{f}(\mathbf{Sy}, \lambda)$ is satisfied for

$$
\mathbf{S} = \begin{pmatrix} -1 & 0 & 0 \\ 0 & -1 & 0 \\ 0 & 0 & 1 \end{pmatrix},
$$

which also satisfies $\mathbf{S}^2 = \mathbf{I}$. Hence the Lorenz equation is $\mathbf{Z}_2$ symmetric. The $\mathbf{Z}_2$ symmetry is the most important symmetry group for bifurcation.

As we have seen for the Lorenz equation, solutions need not be symmetric even though the equation may satisfy the equivariance equation (2.30). Hence, the symmetry group of the equation must be distinguished from the symmetry group of a *solution*. For any solution $\mathbf{y}$ to $\dot{\mathbf{y}} = \mathbf{f}(\mathbf{y}, \lambda)$, the latter is the *isotropy group*

$$
\mathcal{G}_{\mathbf{y}} := \{\mathbf{S} \in \mathcal{G} : \mathbf{Sy} = \mathbf{y}\} .
$$

$\mathcal{G}_{\mathbf{y}}$ is the subgroup of $\mathcal{G}$ consisting of those $\mathbf{S}$ that leave $\mathbf{y}$ invariant. The symmetry of each branch can be described by the isotropy subgroup of its solutions. If the branch carries no symmetry then its solutions are characterized by the trivial subgroup $\{\mathbf{I}\}$. The structures of the group $\mathcal{G}$ and of its subgroups can be used to classify bifurcation behavior. Frequently, at a bifurcation point one of the branches is characterized by a smaller isotropy group than the other branch, which means that one branch is less symmetric than the other. Then one calls the bifurcation a *symmetry-breaking bifurcation*. For example, for the Lorenz equation (see Figure 2.37), the bifurcation at $R_0 = 491.79$ is symmetry breaking. One can label the branches according to their symmetry as indicated in Figure 2.52. This bifurcation is of pitchfork type. The other half-branch is not shown; it is obtained by $\mathbf{Sy}$, when $\mathbf{y}$ is the solution from the shown half-branch and $\mathbf{S}$ describes the $\mathbf{Z}_2$ symmetry.

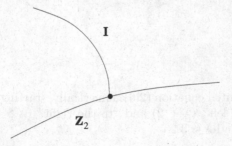

Fig. 2.52. $\mathbf{Z}_2$ symmetry breaking. For each solution on the lower branch $\mathbf{Sy} = \mathbf{y}$ holds, for those on the upper branch $\mathbf{Sy} \neq \mathbf{y}$

Symmetry sheds additional light on genericity and structural stability. Assume, for instance, that $\mathbf{f}(\mathbf{y}, \lambda)$ satisfies a $\mathbf{Z}_2$ equivariance. Then pitchfork bifurcations occur naturally; they are generic *for this restricted class* of functions $\mathbf{f}$. Allowing for arbitrary perturbations of equations the assumption of equivariance is lost. Hence, the symmetry of a problem is structurally unstable unless—by some reason or mechanism—only specific perturbations are assumed or allowed that conserve the equivariance. Symmetry complicates bifurcation analysis but enables the study of suitably reduced equations, the bifurcations of which may be simple again.

Several properties follow for equivariant equations. Clearly, if $\mathbf{y}$ is a solution to $\dot{\mathbf{y}} = \mathbf{f}(\mathbf{y}, \lambda)$, then $\mathbf{Sy}$ is also a solution. The solution $\mathbf{Sy}$ is different from $\mathbf{y}$ unless $\mathbf{S}$ is a symmetry of the solution. A set of solutions connected by group operations forms a *group orbit*. Only one member of a group orbit needs to be calculated; the others are obtained by multiplying $\mathbf{S}$. This allows calculating only one of the emanating half-branches of a $\mathbf{Z}_2$ pitchfork (Figure 2.52).

The $\mathcal{G}$-*symmetric subspace* of all vectors $\mathbf{y}$ with $\mathbf{Sy} = \mathbf{y}$ for all $\mathbf{S} \in \mathcal{G}$ will be denoted by $\mathcal{V}_{\mathcal{G}}$ or $\mathcal{V}$. Differentiating equation (2.30) yields $\mathbf{Sf}_{\mathbf{y}}(\mathbf{y}, \lambda) = \mathbf{f}_{\mathbf{y}}(\mathbf{Sy}, \lambda)\mathbf{S}$. Recall that a stationary bifurcation point $(\mathbf{y}_0, \lambda_0)$ is characterized by $\mathbf{f}_{\mathbf{y}}(\mathbf{y}_0, \lambda_0)\mathbf{h} = \mathbf{0}$ for some eigenvector $\mathbf{h}$. This implies for $\mathbf{y}_0 \in \mathcal{V}$ the validity of $\mathbf{f}_{\mathbf{y}}(\mathbf{y}_0, \lambda_0)\mathbf{Sh} = \mathbf{0}$. Hence, $\mathbf{Sh}$ is also an eigenvector. For many symmetric problems the vectors $\mathbf{h}$ and $\mathbf{Sh}$ are linearly independent. That is, symmetries can force eigenvalues to be multiple. This explains why multiple bifurcations are commonplace for equations with certain symmetries.

For $\mathbf{y} \in \mathcal{V}$, the property, $\mathbf{Sf}(\mathbf{y}, \lambda) = \mathbf{f}(\mathbf{Sy}, \lambda) = \mathbf{f}(\mathbf{y}, \lambda)$ holds, viz., $\mathbf{Sf} = \mathbf{f}$, or $\mathbf{f} \in \mathcal{V}$. This allows the calculation of symmetric solutions of $\mathbf{f}(\mathbf{y}, \lambda) = \mathbf{0}$ by solving a reduced problem $\mathbf{f}_{\mathcal{G}}(\mathbf{y}, \lambda) = \mathbf{0}$. Here $\mathbf{f}_{\mathcal{G}}$ is the restriction of $\mathbf{f}$ to $\mathcal{V}$.

In addition to the groups listed in Appendix A.7, there is a group tailored to periodic solutions, namely, the *circle group* $\mathcal{S}^1$ consisting of all phase shifts $\mathbf{y}(t) \rightarrow \mathbf{y}(t + \theta)$. For periodic solutions, actions under a group $\mathcal{G}$ might be compensated by a phase shift. For literature on symmetry in bifurcation, see, for example, [AlBG92], [BrRR81], [ChLM90], [DeW89], [FäS92], [Fie88], [GoS86], [GoSS88], [Hea88], [Sat79], [Sat80], [TrS91], [WeS84].

Exercises

Exercise 2.1
Integrate the differential equation (2.6) numerically, starting from $\lambda = 2$, $y = 1$. Integrate "to the left" ($\lambda < 2$) and "to the right" ($\lambda > 2$) and see what happens for $\lambda \approx 1$ and $\lambda \approx 3.25$.

Exercise 2.2
Calculate all solutions $y_1(\lambda)$, $y_2(\lambda)$ of the two equations

$$y_1 + \lambda y_1(y_1^2 - 1 + y_2^2) = 0\,,$$
$$10y_2 - \lambda y_2(1 + 2y_1^2 + y_2^2) = 0\,.$$

Sketch $N(\lambda) = y_1^2 + 2y_2^2$ versus λ for all solutions.

Exercise 2.3
Consider equation (2.3) with Figure 2.1. Establish a table of the multiplicity of solutions for various ranges of λ.

Exercise 2.4
Carry out the analysis of equation (2.8) in detail. First derive equation (2.8) using the ansatz for P and v. Then discuss the type of equilibria using Γ as the parameter ($\gamma = 0$).

Exercise 2.5
Consider the Hammerstein equation

$$\lambda y(t) = \frac{2}{\pi} \int_0^\pi (3 \sin \tau \sin t + 2 \sin 2\tau \sin 2t)(y(\tau) + y^3(\tau))d\tau\,.$$

Solutions $y(t; \lambda)$ of this integral equation depend on the independent variable t and the parameter λ.

(a) Show that every solution is of the form

$$y = a(\lambda) \sin t + b(\lambda) \sin 2t\,.$$

(b) Are there nontrivial solutions for $\lambda = 0$?
(c) Calculate all solutions for $\lambda > 0$. (Hint: For $\lambda = 10$ there are nine different solutions.)
(d) For each solution, calculate

$$[y] = \int_0^\pi y^2(t; \lambda)dt$$

and sketch a bifurcation diagram.

Exercise 2.6
(From [IoJ81].) Check the stability of all equilibrium solutions of the scalar differential equation

$$\dot{y} = y(9 - \lambda y)(\lambda + 2y - y^2)([\lambda - 10]^2 + [y - 3]^2 - 1)\,,$$

and draw a bifurcation diagram using dashed lines for unstable equilibria.

Exercise 2.7
Draw bifurcation diagrams depicting the solutions of equation (2.15) using all the scalar measures in equation (2.7).

Exercise 2.8
Consider again the FitzHugh model of the behavior of nerve membrane potential (Exercise 1.11).
(a) Calculate the equilibria. (Hint: Express λ in terms of y_1.)
(b) Calculate the Hopf bifurcation points.
(c) Draw a bifurcation diagram of y_1 versus λ. Distinguish between stable and unstable equilibria. Use the results of Exercise 1.11 and try to formulate a conjecture on how far the periodic branches extend.

Exercise 2.9
[See equation (2.22).] Consider the system of differential equations

$$\dot{y}_1 = -y_2 + y_1(\lambda + y_1^2 + y_2^2),$$
$$\dot{y}_2 = y_1 + y_2(\lambda + y_1^2 + y_2^2).$$

Show that there is a Hopf bifurcation to unstable periodic orbits. Draw a picture similar to Figure 2.24.

Exercise 2.10
By a theorem of Vieta, the roots μ_1, μ_2, μ_3 of the cubic equation $0 = -\mu^3 + a\mu^2 - b\mu + c$ satisfy the relations

$$a = \mu_1 + \mu_2 + \mu_3,$$
$$c = \mu_1\mu_2\mu_3.$$

(a) Show that the eigenvalues of the Jacobian ($n = 3$) satisfy the relation

$$\mu_1\mu_2\mu_3 = \det(\mathbf{f_y}(\mathbf{y}, \lambda)).$$

(b) What do you obtain for the sum $\mu_1 + \mu_2 + \mu_3$?
(c) Show that $\det(\mathbf{f_y}(\mathbf{y}, \lambda)) > 0$ implies local instability for the equilibrium $(\mathbf{y}, \lambda)$.
(d) Investigate which aspects of (a) and (c) might be generalized.

Exercise 2.11
Consider the differential equation

$$\dot{y}_1 = y_2,$$
$$\dot{y}_2 = \delta + \lambda y_2 + y_1^2 + y_1 y_2$$

for the specific constant $\delta = -1$ and for varying parameter λ. Discuss stationary solutions depending on λ and their stability. Sketch phase portraits for characteristic values of λ, and find an argument for the onset of a homoclinic orbit.

Exercise 2.12

Illustrate Figure 2.40 by drawing five snapshots of phase diagrams (for $\lambda = \lambda_3$, $\lambda = \lambda_1$, $\lambda_1 < \lambda < \lambda_0$, $\lambda = \lambda_0$, $\lambda = \lambda_2$) for

(a) increasing λ and
(b) decreasing λ.
(c) Adapt Figure 2.40 to case (b).
(d) Imagine λ slowly sweeping back and forth over the interval $\lambda_3 \leq \lambda \leq \lambda_2$ five times. Sketch qualitatively the bursting phenomenon.

Exercise 2.13

Draw two bifurcation diagrams for equation (2.28):

(a) for a fixed $\lambda > 0$, and
(b) for $\lambda = \gamma + 1$.

Equation (2.28) defines a two-dimensional surface in the three-dimensional (y, λ, γ)-space. Sketch this surface.

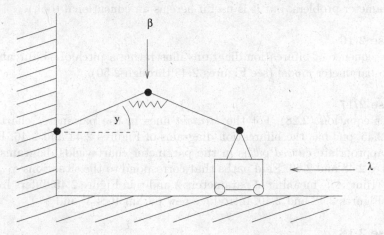

Fig. 2.53. Euler's beam in Exercise 2.14

Exercise 2.14   Euler's Beam

(From [GoS79].) In a finite-element analog of Euler's beam problem, the scalar equation

$$0 = y - \alpha - 2\lambda \sin(y) + \beta \cos(y)$$

occurs (see Figure 2.53). The dependent variable y and the control parameters are

   y, the angle of displacement;
   λ, the compressive force;
   β, the central load; and
   α, the angle at which spring exerts no torque.

(a) Assume $\alpha = 0$ (perfect symmetry, no initial curvature) and $\beta = 0$ (weight-less rods). What solutions exist for $0 < \lambda < 5$? Sketch a bifurcation diagram.

(b) What solutions exist for $0 < \lambda < 5$ when $\alpha = \beta \neq 0$?

(c) Which perturbations must be expected in the general case $\alpha \neq 0$, $\beta \neq 0$? (Hint: Calculate a relation between α and β that establishes a hysteresis point.)

(d) In the (α, β)-plane, draw the critical boundaries of the bifurcation points and the hysteresis points—that is, draw a parameter chart. Shade the region for which every bifurcation diagram with respect to λ has exactly three turning points.

Exercise 2.15
Illustrate the behavior of the solutions of equation (2.22) by drawing four different diagrams: time dependence of y_1, phase diagram(s), a bifurcation diagram, and a parameter chart. (One rarely draws a parameter chart for a one-parameter problem, but it is useful here as an educational tool.)

Exercise 2.16
Draw a sequence of bifurcation diagrams illustrating a pitchfork bifurcation of a two-parameter model (see Figures 2.48 through 2.50).

Exercise 2.17
Consider equation (2.28). For the *straight* lines in the parameter chart of Figure 2.45, one has the bifurcation diagrams of Figures 2.44 and 2.46. Following appropriate *curved* paths in the parameter chart yields diagrams as in Figures 2.48 and 2.49. Find paths that correspond to the situations γ_1, γ_3, and γ_5. (Hint: The meaning of parameters λ and γ in Figure 2.45 differs from that in Figures 2.48 and 2.49. Introduce new parameters λ' and γ'.)

Exercise 2.18
For the stationary solutions of the CSTR model (2.27) calculate the turning points λ_0 with respect to $\mathrm{Da}(= \lambda)$. This establishes a relation of the type $\lambda_0 = \zeta(B, \beta)$ for some function ζ. Use this result to find a formula for the hysteresis points in the (B, β)-parameter plane.

Exercise 2.19
For equation (2.15) in Section 2.5 find a matrix $\mathbf{S}$ such that the equivariance condition equation (2.30) is satisfied.

Exercise 2.20
The van der Pol equation (1.6) has a Hopf bifurcation for $\lambda = 0$. When an explicit Euler scheme (1.17) with step length h is applied, a discretized version is given by the map

$$\mathbf{y}^{(\nu+1)} = \mathbf{P}(\mathbf{y}^{(\nu)}) \quad \text{with} \quad \mathbf{P}(\mathbf{y}) = \begin{pmatrix} y_1 & + & hy_2 \\ y_2 & + & h(\lambda y_2 - \lambda y_1^2 y_2 - y_1) \end{pmatrix}$$

The Hopf bifurcation $(\mathbf{y}_0, \lambda_0) = (\mathbf{0}, 0)$ is characterized by a real part zero of the eigenvalue of the Jacobian $\mathbf{f}_\mathbf{y}$ which corresponds to an eigenvalue $|\mu_J| = 1$, compare Table 1.2.

a) Show $\mu_J = 1 + \frac{h\lambda}{2} \pm \frac{h}{2}\sqrt{\lambda^2 - 4}$,

   and $|\mu_J| = 1$ holds for $\bar{\lambda}_0 = \frac{1}{h}(-1 \pm \sqrt{1 - 2h^2})$.

b) Interpret the results.

3 Applications and Extensions

Ordinary differential equations are the backbone of this book. Symbolically this class of problems can be represented by the ODE prototype equation $\dot{\mathbf{y}} = \mathbf{f}(\mathbf{y})$, which is a short way for

$$\frac{d\mathbf{y}(t)}{dt} = \mathbf{f}(\mathbf{y}(t)), \tag{3.1}$$

with $\mathbf{f}$ sufficiently smooth. For this type of equation we discuss parameter dependence, bifurcation, and stability in detail. Many other bifurcation problems are not of this ODE type. For example, a delay may be involved, or the dynamics fails to be smooth. But even then the ODE background is helpful. On the one hand, methods can be applied that are similar as the ODE approaches. On the other hand, the ODE system (3.1) can be used to approximate non-ODE situations, or to characterize certain special cases.

This chapter outlines how ODEs play an important role also in non-ODE problems. Several applications and extensions offer insight into other areas. Section 3.1 deals with delay differential equations, Section 3.2 discusses important cases of nonsmooth dynamics, and Section 3.3 introduces some aspects of differential-algebraic equations. Partial differential equations enter with the aspect of pattern formation. This is explained in Section 3.4 with nerve models, and more generally in Section 3.5 for reaction-diffusion problems. Finally, in Section 3.6, we point out the importance of the bifurcation machinery for deterministic risk analysis.

3.1 Delay Differential Equations

In (3.1) the derivative $\dot{\mathbf{y}} = \mathbf{f}(\mathbf{y})$ is given as a function of $\mathbf{y}(t)$ —that is, the derivative and $\mathbf{f}(\mathbf{y})$ are both evaluated at time t. For numerous applications this dependence of $\mathbf{f}$ is not adequate. Rather, the derivative $\dot{\mathbf{y}}$ may depend on the state $\mathbf{y}(t-\tau)$ at an earlier time instant $t-\tau$. The past time is specified by the *delay* τ. The *delay differential equation* is then $\dot{\mathbf{y}}(t) = \mathbf{f}(\mathbf{y}(t-\tau))$, or more general,

$$\dot{\mathbf{y}}(t) = \mathbf{f}(t, \mathbf{y}(t), \mathbf{y}(t-\tau)). \tag{3.2}$$

R. Seydel, *Practical Bifurcation and Stability Analysis*,
Interdisciplinary Applied Mathematics 5, DOI 10.1007/978-1-4419-1740-9_3,
© Springer Science+Business Media, LLC 2010

Whereas initial-value problems of ODEs just need one initial vector at t_0, delay equations require an initial *function* ϕ on an interval,

$$\mathbf{y}(t) = \phi(t) \quad \text{for} \quad t_0 - \tau \leq t \leq t_0 \,.$$

Delay equations are used to model many systems, in particular in engineering, biology, and economy.

A nice example of a system where delay occurs naturally is the production of red blood cells in the bone marrow. It takes about four days for a new blood cell to mature. When $b(t)$ represents the number of red blood cells, then their death rate is proportional to $-b(t)$, whereas the reproduction rate depends on $b(t - \tau)$ with a delay τ of about four days. Two pioneering models are given in [WaL76], [MaG77]. Here we analyze the model of [WaL76].

Example 3.1   Production of Blood Cells

For positive δ, p, ν, τ the scalar delay differential equation

$$\dot{b}(t) = -\delta \, b(t) + p \, e^{-\nu b(t - \tau)} \tag{3.3}$$

serves as model for production of blood cells in the bone marrow. The first term on the right-hand side is the decay, with $0 < \delta < 1$ because δ is the probability of death. The second term describes the production of new cells with a nonlinear rate depending on the earlier state $b(t - \tau)$. The equation has an equilibrium b^s, given by $\delta b^s = p \exp(-\nu b^s)$. A Newton iteration for

$$\delta = 0.5\,, \ p = 2\,, \ \nu = 3.694$$

gives the equilibrium value $b^s = 0.5414$. The model (3.3) exhibits a Hopf bifurcation off the stationary solution b^s. We postpone a discussion of the stability of b^s, and start with a transformation to a framework familiar to us from Section 1.4: For an experimental investigation we create an approximating *map*, based on a discretization by forward Euler (1.17).

3.1.1   A Simple Discretization

We stay with Example 3.1. A transformation of (3.3) to normal delay by means of the normalized time $\tilde{t} := t/\tau$ and $u(\tilde{t}) := b(\tau\tilde{t})$ leads to $b(t - \tau) = b(\tau(\tilde{t} - 1)) = u(\tilde{t} - 1)$ and $\frac{du}{d\tilde{t}} = \dot{b} \cdot \tau$, so

$$\frac{du}{d\tilde{t}} = -\delta\tau u + p\tau e^{-\nu u(\tilde{t} - 1)} \,.$$

Introduce a $\tilde{t}$-grid with step size $h = \frac{1}{m}$ for an integer m, and denote the approximation to $u(jh)$ by u_j. Then the forward Euler approximation is

$$u_{j+1} = u_j - \delta h\tau u_j + ph\tau \exp(-\nu u_{j-m}) \,. \tag{3.4}$$

The equation has the equilibrium $u^s (= b^s)$, given by $\delta u^s = p \exp(-\nu u^s)$. With $\mathbf{y}^{(j)} := (u_j, u_{j-1}, \ldots, u_{j-m})^{tr}$ the iteration is written in vector form

$$\mathbf{y}^{(j+1)} := \mathbf{P}(\mathbf{y}^{(j)}, \tau) \,, \tag{3.5}$$

with $\mathbf{P}$ defined by (3.4). This discretization approach has replaced the scalar delay equation (3.3) by an m-dimensional vector iteration (3.5). The stationary solution b^s corresponds to a fixed point $\mathbf{y}^s$ of (3.5).

We consider the delay parameter τ in the role of the bifurcation parameter λ. A simulation for $\tau = 2$ and $\tau = 3$ shows different qualitative behavior, see Figure 3.1. For $\tau = 2$ the equilibrium is stable, whereas for $\tau = 3$ an oscillation sets in. In fact, there is a Hopf bifurcation of the approximating iteration (3.5). An analysis below will show that also the delay differential equation (3.3) has a Hopf bifurcation, with the parameter value $\tau_H = 2.4184$.

The discussion of the stability of the stationary state and of the fixed point follow different lines, compare Section 1.4 and Table 1.2. We still postpone the analysis of the scalar delay equation, and first discuss the stability of the fixed point of the discretized problem. With the matrix of size $(m+1) \times (m+1)$

$$\mathbf{A} := \begin{pmatrix} (1 - \delta h \tau) & 0 & & 0 & 0 \\ 1 & 0 & & & \vdots \\ 0 & 1 & \ddots & & \vdots \\ \vdots & & \ddots & \ddots & 0 & 0 \\ 0 & & 0 & 0 & 1 & 0 \end{pmatrix}$$

the iteration (3.5) is written

$$\mathbf{y}^{(j+1)} = \mathbf{A}\mathbf{y}^{(j)} + \begin{pmatrix} ph\tau \exp(-\nu u_{j-m}) \\ 0 \\ \vdots \\ 0 \end{pmatrix}.$$

The linearization about the equilibrium u^s leads to a Jacobian matrix $\mathbf{P_y}$ which essentially consists of $\mathbf{A}$ with the additional top-right element

$$\gamma := -\nu ph\tau \exp(-\nu u^s) = -\nu h \tau \delta u^s \,.$$

By inspecting the structure of the Jacobian matrix we realize that its eigenvalues z are the zeroes of

$$(z - 1 + \delta h \tau) z^m + \nu h \delta \mu u^s = 0 \,.$$

It is interesting to trace the paths of the eigenvalues of the Jacobian in the complex plane, as they vary with the delay parameter τ, see Figure 3.2. For $\tau = 0$ the eigenvalue equation simplifies to $(z - 1) z^m = 0$. Then $z = 0$ is an m-fold root and $z = 1$ a simple root. For increasing τ, the eigenvalues are initially all inside the unit circle, representing stability of the

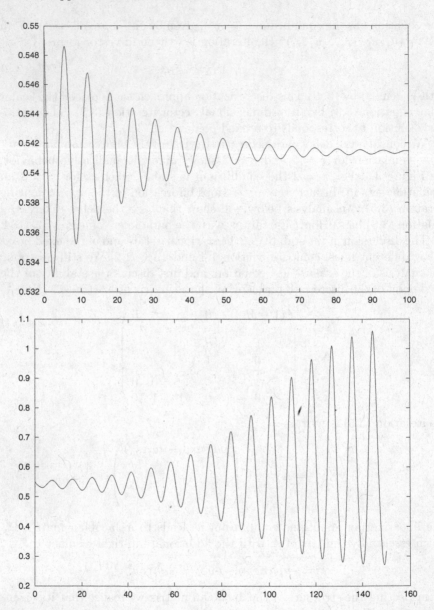

Fig. 3.1. Blood production model of Example 3.1, approximation to $b(t)$; $m = 100$, initial state $\phi \equiv 0.55$; upper figure: delay $\tau = 2$, lower figure: $\tau = 3$

equilibrium u^s for $0 < \tau < \tau_H$, see also [ZhZZ06]. For example, for $m = 6$, two real eigenvalues merge at $\tau = 0.338$ and become a complex conjugate pair. Eventually, for growing τ, all eigenvalues approach the unit circle. For

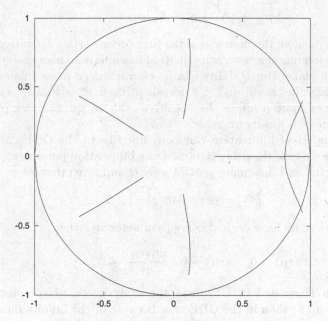

Fig. 3.2. Blood production of Example 3.1, $m = 6$; paths of the m eigenvalues in the complex plane, for $0.01 \leq \tau \leq 3$

the Hopf value τ_H one pair passes the circle, indicating instability of the equilibrium.

3.1.2 Stability Analysis of a Stationary Solution

For an analysis of delay differential equations we study the autonomous system

$$\dot{\mathbf{y}}(t) = \mathbf{f}(\mathbf{y}(t), \mathbf{y}(t - \tau))$$

and assume a stationary solution $\mathbf{y}^s$,

$$\mathbf{f}(\mathbf{y}^s, \mathbf{y}^s) = \mathbf{0}. \tag{3.6}$$

This equation is obviously independent of the delay. We stay with the general vector case although—as seen with Example 3.1—Hopf bifurcation for delay equations is already possible in the scalar case. The analysis of the stability follows the lines of the ODE situation. Small deviations from $\mathbf{y}^s$ are approximated by the solution $\mathbf{h}$ of the linearized system

$$\dot{\mathbf{h}}(t) = \frac{\partial \mathbf{f}(\mathbf{y}^s, \mathbf{y}^s)}{\partial \mathbf{y}(t)} \, \mathbf{h}(t) + \frac{\partial \mathbf{f}(\mathbf{y}^s, \mathbf{y}^s)}{\partial \mathbf{y}(t - \tau)} \, \mathbf{h}(t - \tau), \tag{3.7}$$

compare Section 1.2.4. To find solutions, we start from an ansatz $\mathbf{h}(t) = e^{\mu t}\mathbf{w}$, which yields the nonlinear eigenvalue problem

$$\mu\mathbf{w} = (\mathbf{A}_1 + e^{-\mu\tau}\mathbf{A}_2)\mathbf{w}\,, \tag{3.8}$$

where $\mathbf{A}_1$ and $\mathbf{A}_2$ denote the matrices of the first-order partial derivatives in (3.7). The corresponding characteristic equation has infinitely many solutions μ in the complex plane. But stability can be characterized in an analogous way as with ODEs: In case $\text{Re}(\mu) < 0$ for all μ, then $\mathbf{y}^s$ is locally stable. And in case one or more μ are in the "positive" side of the complex plane ($\text{Re}(\mu) > 0$), then $\mathbf{y}^s$ is locally unstable.

In this setting, Hopf bifurcation can be defined as in the ODE case of Section 2.6. Here we take the delay parameter as bifurcation parameter, and denote the real part and imaginary part of μ by α and β, so that

$$\mu(\tau) = \alpha(\tau) + \mathrm{i}\beta(\tau)\,.$$

Hopf bifurcation occurs for a critical delay parameter τ_H, when

$$\alpha(\tau_\mathrm{H}) = 0\,,\ \beta(\tau_\mathrm{H}) \neq 0\,,\ \frac{\mathrm{d}\alpha(\tau_\mathrm{H})}{\mathrm{d}\tau} \neq 0\,, \tag{3.9}$$

analogously as in Theorem 2.11. Of course, the analysis is more complicated in the delay case (3.8) than in the ODE case because of the infinite number of eigenvalues (Exercise 3.1).

Example 3.2   Business Cycles
After a decision to invest money in a business there is a time delay ("gestation period") until new capital is installed and investment goods are delivered. Following Kalecki (already 1935, see [Kal35], [Szy02]), a net level of investment J can be modeled by the scalar delay equation

$$\dot{J}(t) = aJ(t) - bJ(t - \tau) \tag{3.10}$$

for nonnegative constants a, b, which are lumped parameters [Kal35]. Although this equation is linear, it can be used to study the essential mechanism of Hopf bifurcation; just see J in the role of $\mathbf{h}$, a for $\mathbf{A}_1$ and $-b$ for $\mathbf{A}_2$. The eigenvalue equation (3.8) for Kalecki's scalar model reduces to

$$\mu = a - be^{-\mu\tau}\,.$$

Separating this equation for the complex variable $\mu = \alpha + \mathrm{i}\beta$ into real and imaginary parts results in the pair of real equations

$$\alpha = a - be^{-\alpha\tau}\cos\beta\tau$$
$$\beta = be^{-\alpha\tau}\sin\beta\tau\,.$$

To check for Hopf bifurcation, we ask whether there is a solution $\alpha = 0$ and $\beta \neq 0$ for some τ ($\beta > 0$ without loss of generality). This amounts to the system

$$\frac{a}{b} = \cos \beta \tau$$

$$\frac{\beta}{b} = \sin \beta \tau \,,$$

(3.11)

the solution of which requires $a \leq b$ and $\beta \leq b$. Squaring the equations and adding gives

$$a^2 + \beta^2 = b^2$$

or $\beta = \sqrt{b^2 - a^2}$. For $a < b$ and this value of β the real part α vanishes. For a sufficiently small, a solution $\tau = \tau_H$ to (3.11) exists. By inspecting the second equation in (3.11) more closely (e.g. graphically) we realize that

$$\tau_H > 1/b \,.$$

For a further discussion of the Hopf bifurcation in this example, see [Szy02]. With such a delay model Kalecki explained the occurrence of business cycles. Using his numbers $a = 1.5833$ and $b = 1.7043$, we obtain $\beta = 0.6307$, and hence a time period $\frac{2\pi}{\beta} = 9.962$. This matches the duration of a business cycle of 10 years, as calculated and predicted by Kalecki. For more general nonlinear versions and two-dimensional economic growth models with delay see [Szy03], [ZhW04].

> The delay differential equations are also called *differential difference equations*, or *functional differential equations*, or *retarded equations*. In this subsection we have discussed the common case of a constant delay τ. For the general theory of such equations, see [HaVL93]. Further applications of delay equations in biology include pupil light reflex [BeGMT03], population models [Mur89], nerve models [ChCX06], and predator-prey models [KrC03]. For a delay problem in engineering see [Fof03].
>
> Some "unusual" behavior (as seen from the ODE viewpoint) of solutions of delay equations makes their numerical solution tricky. Specifically designed methods are required. A popular approach is to approximate and replace the delayed term $\mathbf{y}(t - \tau)$ by an interpolating function [ObP81]. In this way the delay equation can be reduced to an ODE. Another approach is Bellman's method of steps, which subdivides the t-axis into intervals of length τ. For each subinterval $[t_0 + (k - 1)\tau, t_0 + k\tau]$ $(k = 2, 3, \ldots)$ a new system is added, and $\mathbf{y}(t - \tau)$ is always given by an ODE subsystem. For a general exposition see [BeZ03].

3.2 Nonsmooth Dynamics

So far we have assumed that both the right-hand side $\mathbf{f}$ and the solutions $\mathbf{y}$ are sufficiently smooth. This is the classical scenario of the ODE (3.1) with well-established theory. But there are many important applications with *nonsmooth* systems. In this section we briefly introduce such problems, and motivate the need for further bifurcation mechanisms.

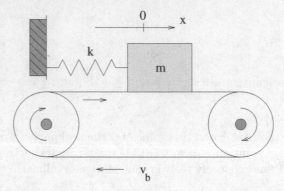

Fig. 3.3. Dry-friction experiment

3.2.1 Dry Friction

A large class of nonsmooth systems involves dry friction. The classical experiment is shown in Figure 3.3. A body with mass m rides on top of a driving belt, which moves with speed v_b. A spring with stiffness k is attached to the body; this completes a one-degree-of-freedom oscillator. The variable x is the coordinate of the center of the mass, with $x = 0$ denoting the point where the outer force supplied by the spring is zero. The dry friction between the mass and the belt causes a friction force $F(v_{rel})$, which depends on the relative velocity $v_{rel} := \dot{x} - v_b$. The simplest ODE-setting of this mechanical model is

$$m\ddot{x} + kx = F(\dot{x} - v_b).\tag{3.12}$$

The left-hand side is the standard linear oscillator with restoring force kx, and the force $F(v_{rel})$ establishes a nonlinear relationship.

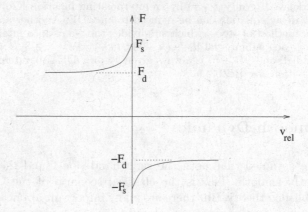

Fig. 3.4. Dry-friction force

Coulomb's theory distinguishes between the static friction for $v_{rel} = 0$, with

$$|F(0)| \leq F_s,$$

and the sliding friction for $v_{rel} \neq 0$ with an average level $|F| \approx F_d$, and $F_d < F_s$. Qualitatively, $F(v_{rel})$ obeys a law like the one illustrated in Figure 3.4. For $x \approx 0$ such that $|kx| < F_s$ the static friction dominates, and the mass sticks to the belt and is transported to the right with speed $\dot{x} = v_b$. During this *sticking phase* with $v_{rel} = 0$, the dry friction force builds up until the restoring force gets large enough, $|kx| > F_s$. Then the sticking ends, and the *slip mode* takes over, with $\dot{x} < v_b$. The mass slips back until again the sticking mode dominates. This qualitative behavior is sketched in Figure 3.5. The sticking mode is along part of the straight line (dashed) $\dot{x} = v_b$, from (b) to (a). Each of the two half planes above and below that line is filled with a family of trajectories. The transition at $\dot{x} = v_b$ is usually nonsmooth. To understand the mechanism, it helps to analyze the linear case for the simplified situation of a constant sliding friction force $F = -\beta$ for $\dot{x} > v_b$ and $F = \beta$ for $\dot{x} < v_b$ (Exercise 3.2).

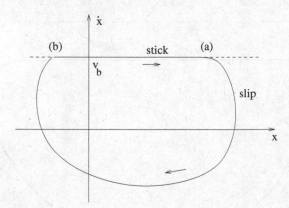

Fig. 3.5. Phase diagram of a stick-slip motion, schematically

3.2.2 Applications

The dry-friction experiment described above represents a large class of mechanical phenomena and devices. Among them are the vibrations of brakes, machine tool chattering, squealing noise of rail wheels, torsional vibrations in oil well drill strings, turbine blade dampers, or peg-in-hole problems. For references on stick-slip phenomena, see [Kaas86], [Ste90], [Pfe92], [GaBB95], [PoHO95], [KuK97], [Lei00], [AwL03], [PuN04].

Part of the research is devoted to seismic phenomena in civil engineering. In that field, an archetypal model experiment is a slender rigid block

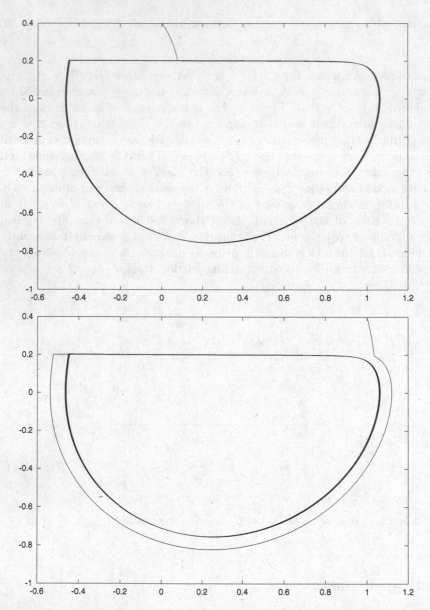

Fig. 3.6. Dry-friction experiment (3.12)/(3.14), approximation with $\eta = 1000$ smoothing, two $(x, \dot{x})$-phase plots with different starting points, limit cycle with two approaching trajectories, here with $\dot{x}(0) > v_b$ initially

on moving foundation [AgS05], [LeR05]. The dynamics involves rocking and bouncing in addition to stick-slip mechanisms.

All of these problems are characterized by discontinuities. In the above classical stick-slip problem, the force $F(\dot{x} - v_{\mathrm{b}})$ is not continuous, compare Figure 3.4. For mechanical problems with elastic support, the restoring force may be continuous but nonsmooth: the derivative is discontinuous. Another discontinuity arises when the system is exposed to outer impacts. Different kind of discontinuities lead to a classification of nonsmooth systems [Lei00].

Nonsmooth systems also occur in electrical engineering, see for example the static var compensator or energy-power system protection, in [DoH06].

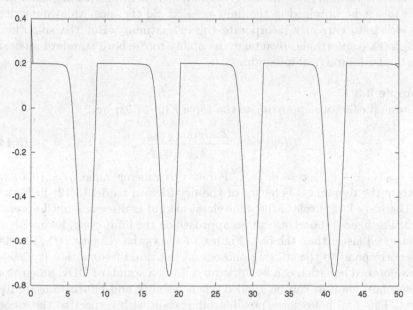

Fig. 3.7. Dry-friction experiment $(3.12)/(3.14)$, $\eta = 1000$; approximation $\dot{x}(t)$, initial point $(x(0),\ \dot{x}(0)) = (0,\ 0.4)$

3.2.3 Numerical Computation

The "nonsmooth" systems are piecewise smooth. For example, the stick-slip problem of Figure 3.3 is smooth for $\dot{x} > v_{\mathrm{b}}$ and smooth for $\dot{x} < v_{\mathrm{b}}$. Within these subdomains no specific numerical method is needed. The difficulty arises at the discontinuity boundary, $\dot{x} = v_{\mathrm{b}}$ in the above example. The build-up of the dry friction for $v_{\mathrm{rel}} = 0$ in Figure 3.4 is a *set-valued function*, which basically consists of the modification $\mathrm{Sign}(x)$ of the $\mathrm{sign}(x)$ function,

$$\mathrm{Sign}(x) := \begin{cases} -1 & \text{for } x < 0 \\ [-1, 1] & \text{for } x = 0 \\ +1 & \text{for } x > 0 \ . \end{cases}$$

Obviously, Sign is hidden in F.

One possible approach to analyze the discontinuous problem is *artificial smoothing*. For instance, the discontinuous $\text{sign}(x)$, $\text{Sign}(x)$ can be approximated by the continuous function

$$\frac{2}{\pi}\arctan(\eta x) \quad \text{for large } \eta. \tag{3.13}$$

The parameter η is the steepness parameter. This approach of artificial smoothing has two drawbacks. First, the systems become stiff (see Section 1.5.2), and their integration gets expensive. Second, there are bifurcation phenomena that can not be obtained in the limit $\eta \to \infty$. So the most valid numerical approach is to correctly incorporate the discontinuity. But the smoothing (3.13) has a comfortable advantage: the ability to use both standard analysis and standard numerical integrators.

Example 3.3
Following [Lei00], we approximate the force F in (3.12) by

$$F(v_{\text{rel}}) = -\frac{2}{\pi}\frac{\arctan(\eta\,v_{\text{rel}})}{1+3|v_{\text{rel}}|}, \tag{3.14}$$

with $v_{\text{rel}} = \dot{x} - 0.2$, $k = m = 1$. For the parameter value $\eta = 1000$, we illustrate the dynamical behavior of the dry-friction model (3.12) in Figure 3.6. There is a limit cycle of the same classical kind as discussed in Chapters 1 and 2. The figures show trajectories approaching the limit cycle, here with an initial speed faster than the belt. Figure 3.7 shows the velocity $\dot{x}(t)$. The flat roofs correspond to the sticking phase, and the rapid free motion is visible. This smoothed equation can be integrated using a standard ODE integrator.

Standard analysis can be applied when stability and bifurcation are discussed. This also holds for a possible bifurcation with respect to the smoothing parameter η. The smoothed equation (3.12)/(3.14) possesses an equilibrium depending on η. Its stability varies with η, and there is a Hopf bifurcation for $\eta \approx 9.6$. See Figure 3.8 for a trajectory winding from the equilibrium to the limit cycle. Here the low value $\eta = 20$ is not relevant for the real nonsmooth case, and the limit cycle does not touch the boundary. (Exercise 3.3)

3.2.4 Switching Conditions

Let us now discuss how to handle the problem without smoothing artificially. Standard procedures can be used until the discontinuity boundary is reached. After an accurate computation of the entry point (e.g., (b) in Figure 3.5), the system is *switched*. That is, the right-hand sides of the ODE are exchanged appropriately. To give a simple example, assume a discontinuity of the following kind: At a specific time instant t_{jump} a relation

$$\mathbf{y}(t_{\text{jump}}^+) = \mathbf{g}(\mathbf{y}(t_{\text{jump}}^-)) \tag{3.15}$$

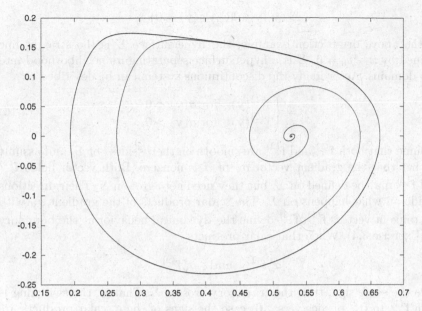

Fig. 3.8. Dry-friction experiment (3.12)/3.14), $(x, \dot{x})$-phase plot, $\eta = 20$ approximation, limit cycle and approaching trajectory starting at the stationary solution

holds for some given function **g**, which may describe an outer impact. (Superscripts $+$ and $-$ stand for the right-hand and left-hand limit.) Then the integration is halted at t_{jump}, and a new initial-value problem for $t \geq t_{\text{jump}}$ starts with the vector obtained from (3.15).

For discontinuities of the stick-slip type of equation (3.12) the switch is from (3.12) to the equation of the belt, $\dot{x} = v_{\text{b}}$ [Lei00]. The general approach is due to Filippov [Fil88], who introduced differential *inclusions* $\mathbf{y}' \in \mathbf{f}(t, \mathbf{y})$ rather than differential equations $\mathbf{y}' = \mathbf{f}(t, \mathbf{y})$. The need for an inclusion becomes clear when the dry-friction force sketched in Figure 3.4 is observed. This is no function in the usual meaning, because it is multivalued for its argument $v_{\text{rel}} = 0$. As mentioned above, the actual value of the force $F(0)$ builds up depending on the current value of the force kx. $F(0)$ varies in the interval $|F(0)| \leq F_{\text{s}}$; the right-hand side of (3.12) is included in this interval. This is the natural setting for the set-valued $\text{Sign}(x)$; F is a set-valued mapping. For $v_{\text{rel}} \neq 0$, the force is

$$F(v_{\text{rel}}) := \begin{cases} -\mu(v_{\text{rel}}) F_{\text{N}} & \text{for } v_{\text{b}} - \dot{x} < 0 \\ +\mu(v_{\text{rel}}) F_{\text{N}} & \text{for } v_{\text{b}} - \dot{x} > 0 \end{cases}$$

where $F_{\text{N}} \geq 0$ is the normal force between body and belt, and μ the coefficient of the sliding friction.

We now turn to a general autonomous situation with one discontinuity boundary Σ defined by a smooth scalar function $\sigma(\mathbf{y})$,

$$\Sigma := \{ \mathbf{y} \in \mathbb{R}^n \mid \sigma(\mathbf{y}) = 0 \} \ .$$

In the above dry-friction example, the hypersurface Σ is the straight line defined by $\dot{x} - v_{\mathrm{b}} = 0$. Σ is a hypersurface separating its neighborhood into two domains. Accordingly the discontinuous system can be described as

$$\dot{\mathbf{y}} = \begin{cases} \mathbf{f}^{(1)}(\mathbf{y}) & \text{for } \sigma(\mathbf{y}) < 0 \\ \mathbf{f}^{(2)}(\mathbf{y}) & \text{for } \sigma(\mathbf{y}) > 0 \ . \end{cases} \tag{3.16}$$

Assume that both $\mathbf{f}^{(1)}$ and $\mathbf{f}^{(2)}$ are smooth on their "sides" of Σ, and assume further that the gradient vector $\sigma_{\mathbf{y}}$ of σ is nonzero. Both vector fields $\mathbf{f}^{(1)}$ and $\mathbf{f}^{(2)}$ may be defined on Σ, but they need not agree on Σ. Their directions decide on what happens on Σ. The scalar products of the gradient $\sigma_{\mathbf{y}}$ with the tangent vectors $\mathbf{f}^{(1)}$, $\mathbf{f}^{(2)}$ define the dynamical behavior at the boundary Σ (Exercise 3.4). When the scalar products

$$\sigma_{\mathbf{y}}^{tr} \mathbf{f}^{(1)} \quad \text{and} \quad \sigma_{\mathbf{y}}^{tr} \mathbf{f}^{(2)}$$

have the same sign, then the trajectory crosses Σ. That is, the switching is from $\mathbf{f}^{(1)}$ to $\mathbf{f}^{(2)}$ or vice versa. In case the signs of these scalar products are different or zero,

$$\sigma_{\mathbf{y}}^{tr} \mathbf{f}^{(1)} \, \sigma_{\mathbf{y}}^{tr} \mathbf{f}^{(2)} \leq 0 \, ,$$

then generically a motion along Σ takes place. When both vector fields push towards Σ, then a point $\mathbf{y}$ once arrived in Σ will stay in Σ. The motion along Σ is often called *sliding mode*. (The usage of "sliding" in this context is not in agreement with the dry-friction experiment, where sliding rather refers to the motion relative to the belt.) The differential equation must be switched to realize the motion on Σ. The switching is to the linear convex combination

$$\dot{\mathbf{y}} = \mathbf{f}^{\Sigma}(\mathbf{y}) := \gamma(\mathbf{y}) \, \mathbf{f}^{(1)}(\mathbf{y}) + (1 - \gamma(\mathbf{y})) \, \mathbf{f}^{(2)}(\mathbf{y})$$

$$\text{with } \gamma(\mathbf{y}) := \frac{\sigma_{\mathbf{y}}^{tr} \mathbf{f}^{(2)}(\mathbf{y})}{\sigma_{\mathbf{y}}^{tr}(\mathbf{f}^{(2)}(\mathbf{y}) - \mathbf{f}^{(1)}(\mathbf{y}))} \tag{3.17}$$

The reader may check that $\sigma_{\mathbf{y}}^{tr} \mathbf{f}^{\Sigma} = 0$ — that is, this motion is tangent to Σ. The motion along Σ ends when one of the vectors $\mathbf{f}^{(1)}$ or $\mathbf{f}^{(2)}$ becomes tangent to Σ. Then the right-hand side must be switched appropriately. For singular situations, when the denominator is zero, or $\dot{\mathbf{y}} = \mathbf{f}^{\Sigma}(\mathbf{y}) = \mathbf{0}$, see special literature such as [KuRG03].

For nonautonomous systems, Filippov's method works similarly. For the above special example of the dry-friction experiment, the force F must be set up in a proper way to be consistent with both Filippov's theory and the mechanical concept of friction [Lei00]. The simplified force of Exercise 3.2 does not fulfill the requirements. In the nonlinear case with a properly defined F (as indicated in Figure 3.4), there is a limit cycle, see Figures 3.5 through 3.8.

3.2.5 Bifurcation

Bifurcations of discontinuous systems where $\mathbf{f}^{(1)}$ and $\mathbf{f}^{(2)}$ in (3.16) depend also on a parameter λ are called *discontinuous bifurcations*. It is not obvious how to define a bifurcation point in this situation. The Definition 2.2 is suitable also for discontinuous bifurcations. For $\mathbf{y} \in \Sigma$ the Jacobian matrix $\mathbf{f_y}$ is typically discontinuous. Accordingly its eigenvalues jump. To see *how* they jump, the jump of the Jacobian is modeled similar as in (3.17) by a convex combination. This defines the "path" of the jump. Depending on how the eigenvalues jump across the imaginary axis, several bifurcations can be classified. [Lei00] gives simple prototype equations for each of the phenomena, similar as done in Chapter 2 for the smooth case. For example, the simplest prototype of a discontinuous saddle-node bifurcation is provided by the equation

$$\dot{y} = \lambda - |y| \,.$$

Since the branching diagram is nonsmooth with a corner (the reader may check), the name "turning point" does not seem appropriate in this situation. Also a discontinuous Hopf bifurcation was found. For each of the classical continuous bifurcations a discontinuous counterpart was found. Periodic orbits can bifurcate too; there are discontinuous phenomena that have no counterpart in the smooth case [Lei00]. A catalog of normal forms is given in [KuRG03]. For further references see, for example, [diBFHH99], [Kun00], [diBKN03], [DoH06], [ZouKB06], [diB08], [Bar09].

3.3 Differential-Algebraic Equations (DAEs)

Our ODEs so far have been explicit—that is, the derivative $\dot{\mathbf{y}}$ was explictly defined by a function $\mathbf{f}$, for example,

$$\dot{\mathbf{y}} = \mathbf{f}(t, \mathbf{y}) \,.$$

In many important applications, the derivative is defined implicitly, as in

$$\mathbf{A}\dot{\mathbf{y}} = \mathbf{f}(t, \mathbf{y}) \,, \tag{3.18}$$

or more general

$$0 = \mathbf{F}(\dot{\mathbf{y}}, \mathbf{y}, t)$$

for some matrix $\mathbf{A}$ or some function $\mathbf{F}$. In case $\mathbf{F}_{\dot{\mathbf{y}}} = \partial\mathbf{F}/\partial\dot{\mathbf{y}}$ is nonsingular, then the implicit function theorem guarantees an explicit version $\dot{\mathbf{y}} = \mathbf{f}(t, \mathbf{y})$. In case of (3.18) this amounts to a nonsingular matrix $\mathbf{A}$ and

$$\dot{\mathbf{y}} = \mathbf{A}^{-1}\mathbf{f}(t, \mathbf{y}) \,.$$

In such a situation we are back in our familiar framework of ODEs. The matrix $\mathbf{A}$ may depend on $\mathbf{y}$.

Frequently, $\mathbf{F}_{\dot{\mathbf{y}}}$ or $\mathbf{A}$ are singular. Then the above simple strategy does not work. This happens, for example, when the problem at hand comes in the form

$$
\begin{aligned}
\dot{\mathbf{y}} &= \mathbf{f}(t, \mathbf{y}, \mathbf{z}) \\
\mathbf{0} &= \mathbf{g}(t, \mathbf{y}, \mathbf{z}) \, .
\end{aligned}
\tag{3.19}
$$

This semi-explicit system consists of two coupled subsystems, namely, an ODE system and an algebraic system. With $\mathbf{Y} := \begin{pmatrix} \mathbf{y} \\ \mathbf{z} \end{pmatrix}$ this coupled system can be written in the form $\mathbf{0} = \mathbf{F}(\dot{\mathbf{Y}}, \mathbf{Y}, t)$, and $\mathbf{F}_{\dot{\mathbf{Y}}}$ is singular. Such systems $\mathbf{0} = \mathbf{F}(\dot{\mathbf{Y}}, \mathbf{Y}, t)$ where $\mathbf{Y}$ and $\mathbf{F}$ have the same dimension and $\mathbf{F}_{\dot{\mathbf{Y}}}$ is singular are called *differential-algebraic equations*, in short DAEs. There is a hierarchy of DAEs, depending on their degree of singularity. The level of difficulty is measured by an "index," see [HaLR89], [HaW91], [HiMT03], [RiT07]. The implicit differential equation (3.18) is included in the DAE hierarchy, and can be seen as an entry point into the DAE world. ($\longrightarrow$ Exercise 3.5) We shall discuss below an example of an implicit differential equation with nonsingular matrix $\mathbf{A}$.

Standard ODE solvers do not work for DAEs. For example, for (3.19) special care must be taken that each integration step stays on the manifold defined by $\mathbf{0} = \mathbf{g}(t, \mathbf{y}, \mathbf{z})$. In this text, we do not attempt to introduce theory and methods for DAEs. Rather our emphasis is on bifurcation—and there are plenty of bifurcation problems of DAEs.

DAEs naturally occur in mechanical and in electrical engineering. In mechanical engineering, classical applications are constrained multibody systems, which include the motion of railway bogies on tracks, and the motion of vehicles on streets. Hopf bifurcation of a bogie model was found, for example, in [Kaas86], based on the model [Coo72]. It turned out that the system under consideration was not smooth enough to apply Hopf's theorem, and the branch of periodic orbits branches off the Hopf point with a tangent *not* vertical to the parameter axis [Has00]. This violates the standard scenario illustrated in Figure 2.28. Wheelset models were studied with collocation methods in [Fra99], [FrF01].

The other natural source of DAE problems are electric circuits. Here Kirchhoff's laws lead to implicit systems of type (3.18). Usually the matrix $\mathbf{A}$ is sparse. In case of a nonsingular $\mathbf{A}$, inversion is not recommendable because sparsity is not preserved in $\mathbf{A}^{-1}$. Inversion is not needed because (3.18) defines a system of linear equations for the vector $\dot{\mathbf{y}}$. But still there are difficulties because the coefficients in $\mathbf{A}$ tend to make the system stiff. So special integrators are required, see for instance [KaRS92]. For applications in power systems, see for instance [MiCR00].

Example 3.4   Flip-Flop
A flip-flop is a bistable electric circuit, as in Figure 3.9, where only one side has a nonzero current, either U_1 or U_2. Figure 3.9 shows a flip-flop circuit consisting of two inverters coupled to a ring. The parameter λ denotes the

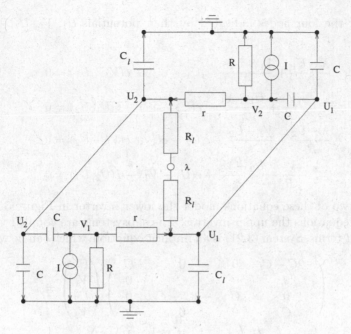

Fig. 3.9. Flip-flop circuit

positive supply voltage that controls the flip-flop. Central elements of each inverter are resistances R_l, R, r and capacitances C_l, C. Take, for example, $\lambda = 5$ V, and ground voltage 0 V (source). The current I depends on U_g (gate) and U_d (drain), $I = I(U_d, U_g)$. For the lower inverter in Figure 3.9, U_2 is the input U_g, and U_1 is the output U_d; for the upper inverter it is the other way round.

Following [ZhD90], the current I is given by

$$I(U_d, U_g) := \begin{cases} 0 & \text{for } U_g \leq V_0 \\ \beta(U_g - V_0)^2(1 + \delta U_d) & \text{for } V_0 < U_g < V_0 + U_d \\ \beta U_d[2(U_g - V_0) - U_d](1 + \delta U_d) & \text{elsewhere} \end{cases}$$

(3.20a)

V_0 is the threshold voltage [ChDK87], and

$$\beta = 8.5 \cdot 10^{-6} \text{ AV}^{-2}, \quad V_0 = 1 \text{ V}, \quad \delta = 0.02 \text{ V}^{-1},$$
$$C = 0.6 \cdot 10^{-15} \text{ F}, \quad R = 10^{15} \text{ } \Omega, \quad r = 60 \text{ } \Omega,$$
$$\lambda = 5 \text{ V}, \quad C_l = 20 \cdot 10^{-15} \text{ F}, \quad R_l = 4.5 \cdot 10^5 \text{ } \Omega.$$

(3.20b)

This defines a specific flip-flop.

To derive differential equations for currents and voltages, Ohm's law $I = U/R$ and the model $I = C\dot{U}$ of the capacitor are applied. Since by Kirchhoff's laws the sum of currents traversing each node of the network equals zero, we get for the four nodes indicated by their potentials U_1, V_1, U_2, V_2 the equations

$$\frac{\lambda - U_1}{R_l} + \frac{V_1 - U_1}{r} - c_l\dot{U}_1 - c\dot{U}_1 + c(\dot{V}_2 - \dot{U}_1) = 0$$

$$-\frac{V_1}{R} + \frac{U_1 - V_1}{r} + c(\dot{U}_2 - \dot{V}_1) - I(U_1, U_2) = 0$$

$$\frac{\lambda - U_2}{R_l} + \frac{V_2 - U_2}{r} - c_l\dot{U}_2 - c\dot{U}_2 + c(\dot{V}_1 - \dot{U}_2) = 0 \tag{3.21}$$

$$-\frac{V_2}{R} + \frac{U_2 - V_2}{r} + c(\dot{U}_1 - \dot{V}_2) - I(U_2, U_1) = 0$$

The first two of these equations model the lower inverter in Figure 3.9, the other two equations the upper inverter. The subsystems are coupled via the nonlinear I terms. System (3.21) is an implicit equation, which can be written

$$\begin{pmatrix} -2C - C_l & 0 & 0 & C \\ 0 & -C & C & 0 \\ 0 & C & -2C - C_l & 0 \\ C & 0 & 0 & -C \end{pmatrix} \begin{pmatrix} \dot{U}_1 \\ \dot{V}_1 \\ \dot{U}_2 \\ \dot{V}_2 \end{pmatrix} =$$

$$= \begin{pmatrix} U_1\left(\frac{1}{R_l} + \frac{1}{r}\right) - V_1\frac{1}{r} - \lambda\frac{1}{R_l} \\ I(U_1, U_2) - U_1\frac{1}{r} + V_1\left(\frac{1}{R} + \frac{1}{r}\right) \\ U_2\left(\frac{1}{R_l} + \frac{1}{r}\right) - V_2\frac{1}{r} - \lambda\frac{1}{R_l} \\ I(U_2, U_1) - U_2\frac{1}{r} + V_2\left(\frac{1}{R} + \frac{1}{r}\right) \end{pmatrix} \tag{3.22}$$

This system is of the type (3.18), with $\mathbf{y} := (U_1, V_1, U_2, V_2)^{tr}$. The small size of system (3.22) allows to make an exception and calculate and multiply the inverse matrix analytically. With the notations

$$\gamma_0 := \frac{1}{C + C_l}, \quad \gamma_1 := \frac{1}{C + C_l} + \frac{1}{C},$$

$$\gamma_2 := \frac{1}{R} + \frac{1}{r}, \quad \gamma_3 := \frac{1}{R_l} + \frac{1}{r}, \quad \gamma_4 := \frac{1}{r}$$

this effort leads to the explicit system

$$\dot{y}_1 = \gamma_0\left\{\gamma_4 y_2 + \lambda\frac{1}{R_l} - \gamma_3 y_1 - I(y_3, y_1) + \gamma_4 y_3 - \gamma_2 y_4\right\}$$

$$\dot{y}_2 = \gamma_1\left\{\gamma_4 y_1 - \gamma_2 y_2 - I(y_1, y_3)\right\} + \gamma_0\left\{\gamma_4 y_4 + \lambda\frac{1}{R_l} - \gamma_3 y_3\right\}$$

$$\dot{y}_3 = \gamma_0\left\{-I(y_1, y_3) + \gamma_4 y_1 - \gamma_2 y_2 + \gamma_4 y_4 + \lambda\frac{1}{R_l} - \gamma_3 y_3\right\} \tag{3.23}$$

$$\dot{y}_4 = \gamma_0\left\{\gamma_4 y_2 + \lambda\frac{1}{R_l} - \gamma_3 y_1\right\} + \gamma_1\left\{-I(y_3, y_1) + \gamma_4 y_3 - \gamma_2 y_4\right\}$$

The ODE system that represents the flip-flop is now fully defined, and we can search for solutions.

After the above inverting and multiplication with $\mathbf{A}^{-1}$ the differential equation (3.23) is scaled with the constants γ_0 and γ_1. Their order of magnitude is 10^{15} —a price we have to pay for setting up an explicit ODE system. These large numbers reflect the ability of the electronic device to toggle quickly; the response is within nanoseconds. Differences in the potentials U_1 and V_1 (or U_2 and V_2, see Figure 3.9) balance almost immediately to $y_1 \approx y_2$ and $y_3 \approx y_4$. This allows to present the dynamics in the two-dimensional (y_1, y_3)-plane.

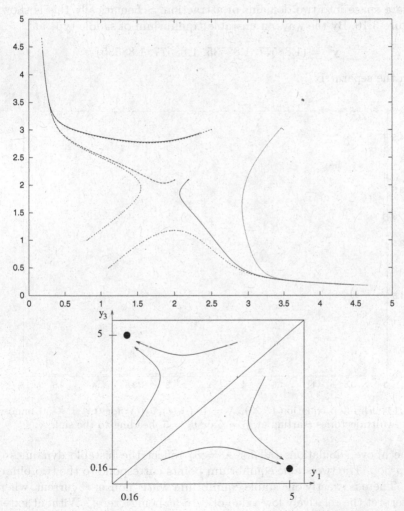

Fig. 3.10. Flip-flop equation (3.23), $\lambda = 5$, top: $(y_1(t), y_3(t))$ for $0 \le t \le 20$ nanoseconds, six trajectories tending to $y_1 = 5$ or $y_3 = 5$; bottom: domains of attraction of the two stable equilibria schematically

For a first set of simulations assume the value $\lambda = 5$. The phase diagram of Figure 3.10 shows six trajectories starting from six arbitrarily chosen starting vectors. Within about 20 nanoseconds trajectories approach one of the two stable equilibria. For starting points below the straight line $y_3 = y_1$ in Figure 3.10, the convergence is to

$$\mathbf{y}^{tr} = (5,\ 5,\ 0.16087,\ 0.1600224)\,,$$

and for starting points above that straight line the other equilibrium is approached. Both are symmetric to each other, which can be seen by exchanging $y_1 \leftrightarrow y_3$, $y_2 \leftrightarrow y_4$. The straight line represents the separatrix that separates the state space into two domains of attraction. Schematically this is shown in Figure 3.10. By the way, an unstable equilibrium of saddle type at

$$\mathbf{y}^{tr} = (1.88577,\ 1.88535,\ 1.88577,\ 1.88535)\,,$$

lies on the separatrix.

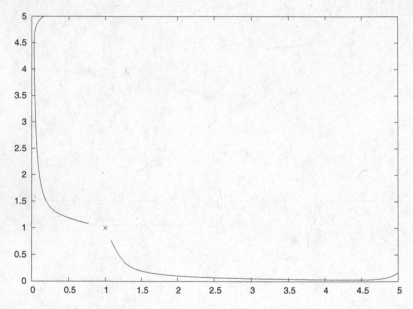

Fig. 3.11. Flip-flop equation (3.23), $\lambda = 1$, $(y_1(t), y_3(t))$ for $0 \le t \le 50$ nanoseconds, two trajectories starting at $y_1 = 5$ or $y_3 = 5$, heading to the sink $\times$.

The above simulations (all for $\lambda = 5$) reflect the bistable dynamics of the flip-flop. The two stable equilibrium points correspond to the two binary states. There is exactly one stable equilibrium with "nonzero" current, where we interpret the relatively low value of 0.16 as being "zero." Without external stimulus the flip-flop can not switch to the other state, the separatrix represents a hill too high to jump over. But by a quick change of the supply voltage $\lambda = 5$, the dynamical behavior of the flip-flop is changed. Apply a

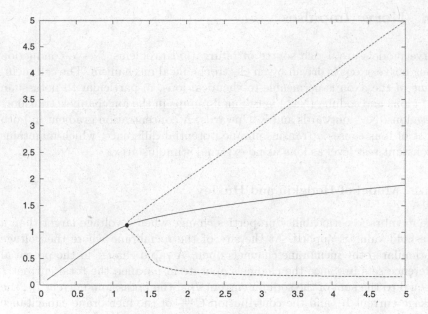

Fig. 3.12. Flip-flop equation (3.23), bifurcation diagram, y_1 (or y_3) over λ

suitable short pulse on λ and the separatrix vanishes. To simulate this, assume for a moment $\lambda = 1$. Figure 3.11 depicts two trajectories which leave the "old" equilibrium position and quickly approach the hypersurface $y_3 = y_1$. A bifurcation study shows that the two large domains of attractions that exist for larger values of λ disappear when λ is lowered below a threshold value $\lambda_0 \approx 1.2$. For example, for $\lambda = 1$ V there is a sink at

$$\mathbf{y}^{tr} = (1, 1, 1, 1).$$

As a simple and crude explanation of the flip-flop, think of the brief lowering of the supply voltage λ to end just in the moment when the trajectory has managed to cross the hypersurface $y_1 = y_3$, $y_2 = y_4$. The equilibria immediately regain stability, and the flip-flop has toggled. The equilibria depending on the supply voltage λ are shown in the bifurcation diagram of Figure 3.12. The solid curve represents the symmetric state $y_1 = y_3$, $y_2 = y_4$. There is a bifurcation at λ_0 where the symmetric state loses stability. For $\lambda > \lambda_0$ the dashed curve is y_1, or y_3 respectively; one state is represented by the upper branch and the other state by the lower branch. As is seen by Figure 3.12, the supply voltage is retained by one of the two inverters. Toggling the flip-flop means to exchange the two states by a quick lowering of λ.

3.4 Nerve Impulses

Nerve models are a rich source of bifurcation problems. Nerve conduction along nerve axons is driven by an electrochemical mechanism. The cell membrane of the axon is permeable to chemical ions, in particular to potassium (K^+) ions and sodium (Na^+) ions. Small pumps in the ion channels transport these ions, Na^+ outwards and K^+ inwards. A concentration gradient in both types of ions causes a transmembrane potential difference, which maintains a constant rest level as long as no external stimulus arises.

3.4.1 Model of Hodgkin and Huxley

The membrane permeability properties change when a voltage larger than a threshold value is applied. At the site of the membrane where the voltage is stimulated the membrane channels open. A rapid change in the potential difference sets in. Since the channels operate in parallel, the total current I is equal to the sum of the sodium current I_{Na}, the potassium current I_K, the leakage current I_L, and the contribution $C\frac{dU}{dt}$ of the membrane capacitance C. (U denotes the potential.) This results in

$$I = C\dot{U} + I_K + I_{Na} + I_L\,. \tag{3.24}$$

The currents due to ion transport across the membrane are modeled as being proportional to the differences between the potential U and the constant equilibrium potentials U_K, U_{Na}, and U_L. For example,

$$I_K = p_K \cdot (U - U_K)\,,$$

where U_K is the stationary K^+ potential. The factors p_K (and, correspondingly, p_{Na}, p_L) are permeabilities.

Hodgkin and Huxley [HoH52] model the permeabilities using three artificial auxiliary variables $V_1(t)$, $V_2(t)$, $V_3(t)$, each assumed to satisfy a differential equation of the type

$$\dot{V} = \alpha(U)\,(1 - V) - \beta(U)\,V\,.$$

The transfer rates $\alpha(U)$ and $\beta(U)$ can be seen as switches that turn the permeability on or off. These rates α, β were calibrated based on measurements. In summary, the famous model of Hodgkin and Huxley consists of the four coupled ODEs

$$
\begin{aligned}
I &= C\dot{U} + \bar{g}_K V_1^4 (U - U_K) + \bar{g}_{Na} V_2^3 V_3 (U - U_{Na}) + \bar{g}_L (U - U_L)\,,\\
\dot{V}_1 &= \alpha_1(U)\,(1 - V_1) - \beta_1(U)\,V_1\,,\\
\dot{V}_2 &= \alpha_2(U)\,(1 - V_2) - \beta_2(U)\,V_2\,,\\
\dot{V}_3 &= \alpha_3(U)\,(1 - V_3) - \beta_3(U)\,V_3\,.
\end{aligned}
\tag{3.25}
$$

The transfer rates are

$$\alpha_1 = 0.01\,(U+10)/\left(\exp\tfrac{U+10}{10}-1\right),\quad \beta_1 = 0.125\,\exp(U/80),$$
$$\alpha_2 = 0.1\,(U+25)/\left(\exp\tfrac{U+25}{10}-1\right),\quad \beta_2 = 4\,\exp(U/18),$$
$$\alpha_3 = 0.07\,\exp(U/20),\qquad\qquad\qquad \beta_3 = 1/\left(\exp\tfrac{U+30}{10}+1\right).$$

The constants and their units are

$$
\begin{array}{lll}
C & 1.0 & (\mu\text{F/cm}^2)\\
U_{\text{Na}} & -115 & (\text{mV})\\
U_{\text{K}} & +12 & (\text{mV})\\
U_{\text{L}} & -10.613 & (\text{mV})\\
\bar{g}_{\text{Na}} & 120 & (\text{m.mho/cm}^2)\\
\bar{g}_{\text{K}} & 36 & (\text{m.mho/cm}^2)\\
\bar{g}_{\text{L}} & 0.3 & (\text{m.mho/cm}^2)
\end{array}
$$

$$(\text{mho} := \text{ohm}^{-1},\ t\text{ in msec})$$

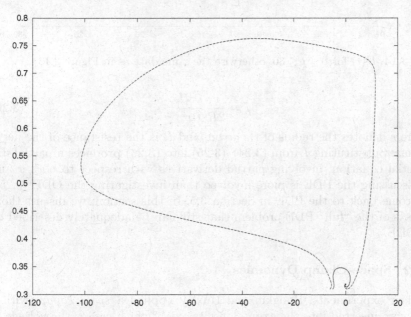

Fig. 3.13. (U, V_1)-phase plane of (3.25); two trajectories with starting values $U(0) = -5$ or -6, and $V_1(0) = 0.31$, $V_2(0) = 0.05$, $V_3(0) = 0.6$, and $I = 0$.

The above equation (3.25) has modeled the local behavior at some site of the excitable nerve membrane. The local potential influences the neighborhood, so that a local variation of U at one site serves as stimulus for adjacent sites. The process repeats, forming a signal which propagates along the axon. As a result, U depends on both t and the space variable x along the axon, $U(x, t)$. The underlying law satisfies the *cable equation*

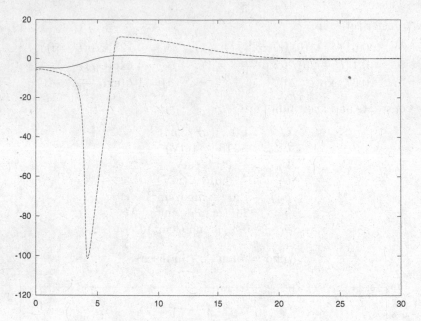

Fig. 3.14. $U(t)$ for $0 \leq t \leq 30$, otherwise the same data as in Figure 3.13

$$I = \frac{r}{2R} \frac{\partial^2 U}{\partial x^2} \tag{3.26}$$

where r denotes the radius of the axon, and R is the resistance of the nerve plasma. Substituting I from (3.24)/(3.25) into (3.26) produces a partial differential equation, involving partial derivatives with respect to both x and t. Discussing the PDE is more involved than investigating the ODE (3.25). We come back to the PDE in Section 3.5. In this section, we discuss those aspects of the "full" PDE problem that still can be adequately described by an ODE.

3.4.2 Space-Clamp Dynamics

In their experiments, Hodgkin and Huxley applied a *space clamp* (with a thin wire inserted into the axon along its axis), which causes the voltage U to be spatially constant, and the current to equal a constant applied value $I = I_a$. For the following discussion we set in (3.25) $I = I_a$, and consider I_a to serve as bifurcation parameter. That is, we discuss the behavior of solutions $U(t), V_1(t), V_2(t), V_3(t)$ depending on the parameter I_a.

We start with $I_a = 0$; no current is applied. By simulation [integrating (3.25)] we easily find the location of an attracting stationary state U_s, V_1, V_2, V_3. To investigate the dynamical behavior in a neighborhood, we perform simulations starting from perturbed values, similarly as illustrated in Figure 1.7. The approximate components V_1, V_2, V_3 of the stationary state serve as initial values $V_1(0)$, $V_2(0)$, $V_3(0)$ for a subsequent integration of (3.25) for $0 \leq t \leq 30$. Only the value $U(0)$ is set artificially to some value $U_0 \neq U_s$, in order to simulate an external stimulus. The Figure 3.13 depicts a phase portrait of two trajectories obtained in his way, for $U_0 = -5$ and $U_0 = -6$. As seen in the figure, the resulting dynamic behavior differs dramatically. For $U_0 = -5$ there is only a small movement back to the stable state U_s. In contrast, for $U_0 = -6$, a significant outbreak of the voltage occurs, which in this nerve context is called "firing." Between $U_0 = -5$ and $U_0 = -6$ there is a threshold value S, which separates the small movement from the spike-like firing. The two different modes are illustrated also in Figure 3.14, where $U(t)$ is shown. The threshold value S is straddled more closely, when $U_0 = -5.6$ and $U_0 = -5.8$ are compared. The results are similar: If we approach the threshold value S, then we find close initial values with drastically different responses. The reader may be inspired to perform related experiments.

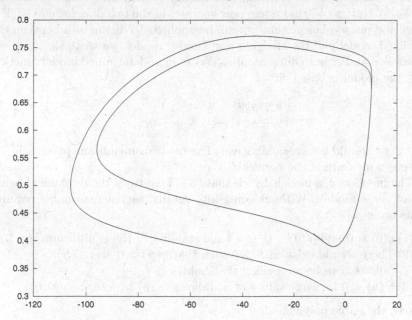

Fig. 3.15. (U, V_1)-phase plane as in Figure 3.13, with $U(0) = -5$ and $I = I_a = -10$.

The above simulation experiment has revealed a large area in phase space, where rather "few" trajectories are found by the experiment described above (Figure 3.13). What happens in that seemingly almost "empty" space? There are nearby periodic solutions in this range. This becomes clear when we study responses of the system to smaller values of the parameter I_a. For instance, a simulation for $I_a = -10$ reveals a periodic orbit, shown in Figure 3.15. This specific limit cycle lies on a branch of periodic solutions, which emerges out of the branch of stationary solutions via a Hopf bifurcation at $I_{\text{Hopf}} = -9.78$. At the Hopf point there is a hard loss of stability, and the branch of unstable periodic orbits extends to a cyclic fold bifurcation with parameter value of $I_a \approx -5.5$, where it gains stability and turns back. (For the turning-point like cyclic fold see Section 7.4.) The periodic orbits represent repetitive firing of the neuron. The resulting bifurcation diagram corresponds to the subcritical case illustrated schematically in Figures 2.40 and 2.41. That is, the model has a range of bistability. The solution data of the Hopf bifurcation at I_{Hopf} are $U = 5.345$, $V_1 = 0.402$, $V_2 = 0.097$, $V_3 = 0.406$, with initial period 10.72. Note that this example exhibits two kind of threshold values, namely, the one with respect to the initial values (S), and the one with respect to the parameter (here the Hopf bifurcation).

3.4.3 FitzHugh's Model

FitzHugh simplified the Hodgkin and Huxley model, assuming constant V_1 and V_3. With $V := V_2$ the system now operates in the two-dimensional (U, V)-plane, and phase-plane arguments can be applied. To distinguish between the simplified model and the Hodgkin and Huxley model, we write the state variables as u and v, both dimensionless. With an undetermined model function $g(u)$, the model is based on

$$\dot{u} = g(u) - v$$
$$\dot{v} = bu - cv, \quad b, \ c > 0. \tag{3.27}$$

In (3.27) u should behave qualitatively like the transmembrane potential $-U$, and v is a measure of the permeability.

The function g is modeled such that (3.27) captures the dynamical behavior of nerve models. Without going into details, we list reasonable requirements on g:

(1) $g(0) = 0$ and $g'(0) < 0$ (for local stability of the equilibrium $u = 0$),
(2) There should exist an $S > 0$ such that $g(S) = 0$ and $g'(S) > 0$
   (allows S to be a repelling threshold)
(3) $g'(u) < 0$ for large values of u (allows (u, v) to return to $(0, 0)$).

Clearly, the cubic polynomial

$$g(u) := u(S - u)(u - 1) \quad \text{for } 0 < S < 1$$

satisfies the requirements. Exercise 3.6 provides more insight into the modeling. The phase plane and the signs of $\dot{u}$ and $\dot{v}$ are crucial for the analysis (Figure 3.17). Here we present a specific example of the type (3.27).

Example 3.5   FitzHugh's Nerve Equation
FitzHugh's model results in the pair of ODEs

$$\dot{y}_1 = 3(y_1 + y_2 - y_1^3/3 + \lambda),$$
$$\dot{y}_2 = -(y_1 - 0.7 + 0.8y_2)/3. \tag{3.28}$$

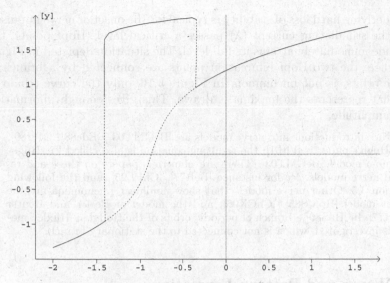

Fig. 3.16. Bifurcation diagram of Example 3.5, $y_1(0)$ versus λ (dotted for unstable)

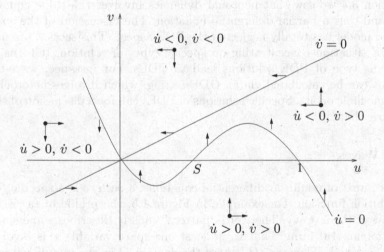

Fig. 3.17. Sketch of the phase plane for the analysis of Exercise 3.6

The analysis of this example has already been done in Exercises 1.11 and 2.8.
The branching diagram including the periodic orbits is shown in Figure 3.16.
This figure depicts the branch of stationary solutions including the unstable
part, and two Hopf bifurcation points. At both Hopf points the bifurcation
is subcritical with hard loss of stability. The bistable regions are narrow:

$$-1.41314 \le \lambda \le -1.40352,$$
$$-0.34647 \le \lambda \le -0.33685.$$

The underlying hard loss of stability is typical for the onset of nerve impulses. When the stimulating current (λ) passes a critical level (Hopf point), the amplitude immediately attains its full level. The situation depicted in Figure 3.16, where the two Hopf bifurcation points are connected by a branch of periodic orbits, is not uncommon. In Figure 3.16, only the curve (chain of dots) that represents the maxima is drawn. This gives enough information on the amplitude.

> Nice introductions into nerve models are [BeGMT03], [Ede88], [Mur89]. FitzHugh's paper is [Fit61]; the resulting equation is also called FitzHugh-Nagumo model [BeGMT03]. There are numerous papers on these and related nerve models, see for instance [BeR77], [ChFL95], and the following Section 3.5. Other nerve models that show nonlinear phenomena are the Plant model [HoMS85], [ChCX06], and the model of Beeler and Reuter [BeR77]. In [HaS89] a branch of periodic orbits of the Hodgkin–Huxley model is investigated, which is not connected to the stationary branch.

3.5 Waves and Pattern Formation

In Section 3.4 we saw that neuronal dynamics involves the cable equation (3.26) and thus a partial differential equation. The discussion of the space-clamped model has totally neglected the PDE aspect. This section discusses the PDE situation concentrating on specific types of solution. It turns out that some type of PDE solutions lead to ODEs. For instance, wave-type solutions can be calculated via an ODE setting, which involves heteroclinic and homoclinic orbits. Specific solutions of PDEs call for a discussion of their geometric shape.

3.5.1 Waves

In our context of nonlinear differential equations, a *pattern* is a specific *form* of a solution function. For example, in Figure 1.5, the profile of (a) is flat, and that of (b) is wavy. The word "pattern" mostly describes some spatial or spatiotemporal form. That is, at least one space variable x is involved, as in $y = u(x,t)$. Then $x \in \Omega$, where the domain Ω in the one-dimensional case often is an interval. The set Ω represents the excitable medium in which patterns may arise. In case the independent variables x and t are separated,

$$u(x,t) = \varphi(x) \cdot \psi(t)$$

for functions φ and ψ, nodes of u (zeros of φ) do not vary with time. Then the pattern represented by φ is pinned and does not propagate across Ω. Frequently this situation is called *standing wave*, in particular when $\varphi(x)$ represents a wave-type profile as in $u(x,t) = a\cos x \sin t$. In this example, the function u is a solution of the wave equation $u_{tt} - u_{xx} = 0$.

"Flat" solutions exist for many PDEs. If the spatial gradient vanishes for all t and all x, $u_x \equiv 0$, we say that u is homogeneous in space, and the stationary case $u_t \equiv 0$ in this context can be characterized as homogeneous in time. For example, consider the scalar PDE with scalar space variable x

$$u_t = u_{xx} + f(u) \tag{3.29}$$

for some function f. Every zero of f constitutes a stationary solution of the PDE. For instance, for FitzHugh's model describing the membrane potential of nerves, f is given by

$$f(u) = u(u-1)(S-u).$$

Here the PDE has three stationary solutions, namely, $u = 0$, $u = 1$, and $u = S$, for all x and all t. A more involved example of fourth order is the Kuramoto–Sivashinsky equation

$$u_t + 4u_{xxxx} + \gamma(u_{xx} + uu_x) = 0, \tag{3.30}$$

which is solved by each constant, in particular, by the trivial solution $u = 0$. The question is, are there nontrivial solutions, and do they emerge from the trivial one?

Next we study waves traveling through their medium. We stay with a PDE with one space variable, as the cable equation (3.26), or (3.29). For motivation, assume a profile U like that of Figure 3.18, which may stand for a nerve impulse. A point that travels along the x-axis with constant speed λ has a position given by $x - \lambda t = \xi$. Hence, in case a function u represents a profile U that travels along the x-axis, it must be of the form

$$u(x,t) = U(x - \lambda t). \tag{3.31}$$

A solution of the PDE of type (3.31) is called *traveling wave* solution. The profile $U(\xi)$ travels along the x-axis with constant *propagation speed* λ.

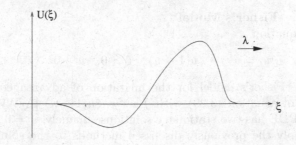

Fig. 3.18. Snapshot of a possible wave; wave profile

The type of PDE with traveling-wave solutions is quite general and includes the equations (3.29) and (3.30). Neither the shape of the wave nor the speed of propagation change. U is the waveform of constant shape. The independent variable $\xi = x - \lambda t$ of U means to introduce the wave-frame (ξ, U)-coordinate system, which travels along the x-axis with the same speed λ as the wave. Another explanation: Along each straight line $\xi = x - \lambda t$ in the (x, t)-plane, $u(x, t)$ has the same value $U(\xi)$.

For this exposition, we stick to the specific class of PDEs (3.29). Substituting the expression for u from (3.31) into (3.29) reduces it to the second-order ODE

$$0 = U'' + \lambda U' + f(U). \tag{3.32}$$

Here $'$ denotes differentiation with respect to the wave-frame coordinate ξ. For the Kuramoto–Sivashinsky equation (3.30), the resulting ODE is of fourth order.

By solving the ODE equation (3.32), one can calculate asymptotic states for the PDE. Let u_1 and u_2 be zeros of f and hence stationary solutions for both the PDE and the ODE. The asymptotic behavior of solutions U for $\xi \to \pm\infty$ determines the type of traveling wave. Every solution with limits

$$\lim_{\xi \to \infty} U(\xi) = u_1 , \quad \lim_{\xi \to -\infty} U(\xi) = u_2 , \quad u_1 \neq u_2$$

is a *front wave* of the PDE. The reacting-medium dynamics described by u switches from one stationary state to the other one when the front wave passes. This corresponds to a heteroclinic orbit of the ODE, connecting the two stationary points $(U, U') = (u_1, 0)$ and $(u_2, 0)$. In case $u_1 = u_2$, for nonconstant U, one encounters a *pulse wave*; the reacting medium returns to its original state after a pulse traverses it (see Figure 3.18). Pulses are characterized by homoclinic orbits in the (U, U')-phase plane. A *wave train* occurs if the medium fires pulse-type spikes at regular intervals; related solutions U of the ODE are periodic. Waves for which the ODE solution U forms a limit cycle in phase space are labeled *rotating waves* [KeNS90].

Example 3.6   Fisher's Model
Consider the equation

$$u_t = u_{xx} + \zeta u(1 - u), \quad \zeta > 0, \quad u \geq 0.$$

This equation—Fisher's model for the migration of advantageous genes—is of the type of equation (3.29) with $f(u) = \zeta u(1 - u)$ [ArW75], [Fis37], [HaR75]. The PDE has two stationary solutions—namely, $u = 0$ and $u = 1$. In order to apply the previously discussed methods to the solution of the ODE of equation (3.32), we transform it via $y_1 = U$, $y_2 = U'$ to a system of first-order differential equations. Its Jacobian matrix is

$$\begin{pmatrix} 0 & 1 \\ -f'(y_1) & -\lambda \end{pmatrix},$$

where the entry $-f' = \zeta(1 - 2y_1)$ takes the values ζ and $-\zeta$ at the equilibria $(y_1, y_2) = (0,0)$ and $(1,0)$. The eigenvalues μ and eigenvectors $\mathbf{w}$ are

$$\mu_{1,2} = \frac{1}{2}(\lambda \pm \sqrt{\lambda^2 - 4f'}), \quad \mathbf{w} = \begin{pmatrix} 1 \\ \mu \end{pmatrix}.$$

Observing that the eigenvalues are real for $\lambda^2 \geq 4f'$, we find

for $(y_1, y_2) = (1, 0)$: μ_i real, $\mu_1 \mu_2 < 0$, that is, saddle;
for $(y_1, y_2) = (0, 0)$: μ_i real for $\lambda \geq 2\sqrt{\zeta}$, $\mu_1 \mu_2 > 0$, hence, stable node.

Consequently, one of the outsets of the saddle $(1,0)$ is attracted by the stable node $(0,0)$ and forms a heteroclinic orbit (Exercise 3.8(a)). The general result is that this wave is unique for $\lambda \geq 2\sqrt{\zeta}$ [Hop75], [Mur89]. The case $\lambda < 2\sqrt{\zeta}$ can be excluded because then the stationary point $(0,0)$ would be a focus with U becoming negative. □

The wave profile $U(\xi)$ is defined for the infinite interval $-\infty < \xi < \infty$. For calculating purposes, this infinite interval is split into three parts, with a finite interval $a < \xi < b$ in the middle and two end intervals $-\infty < \xi < a$ and $b < \xi < \infty$. The above analysis yields the asymptotic behavior of U —that is, analytic expressions for U in the two end parts. Because U is at least continuous, one glues the three parts of U together at the matching points $\xi = a$ and $\xi = b$. This delivers boundary conditions for U in the middle finite interval. In this way, the wave U can be approximated numerically (Exercise 3.8).

3.5.2 Nerve Conduction

We return to the cable equation (3.26) in order to illustrate that the above approach also works for systems. For $C = 1$, with a constant D we have $DU_{xx} = \dot{U} + I_K + I_{Na} + I_L$ or

$$\dot{U} = -I_K - I_{Na} - I_L + DU_{xx}.$$

With FitzHugh's mechanism (Section 3.4.3) this leads to the FitzHugh-Nagumo nerve-conduction model

$$\begin{aligned} \frac{\partial u}{\partial t} &= g(u) - v + D\frac{\partial^2 u}{\partial x^2} \\ \frac{\partial v}{\partial t} &= bu - cv \\ g(u) &= u(S - u)(u - 1) \end{aligned} \qquad (3.33)$$

for $0 < S < 1$. Even though the second equation is of ODE type both solution components vary with x and t, $u(x,t)$, $v(x,t)$. The traveling-wave approach must be applied to both,

$$u(x,t) = U(x - \lambda t)$$
$$v(x,t) = V(x - \lambda t)$$

As a result, we arrive at the system of ODEs

$$-\lambda U' = g(U) - V + DU''$$
$$-\lambda V' = bU - cV.$$

The first of the equations is of second order. The transformations

$$y_1 := U, \; y_2 := U', \; y_3 := V$$

lead to the equivalent first-order system

$$y_1' = y_2$$
$$y_2' = \frac{1}{D}(-\lambda y_2 - g(y_1) + y_3)$$
$$y_3' = \frac{1}{\lambda}(-by_1 + cy_3)$$

for the unknown profiles $U(\xi)$ and $V(\xi)$. This system my be the starting point for computational experiments.

Traveling waves of (3.33) have been reported for $S < \frac{1}{2}$. In a mixed initial boundary-value problem, solutions are investigated on the quarter plane $x > 0$, $t > 0$. Nerves that are initially at rest ($u = v = 0$ for $t = 0$) may be stimulated at an endpoint (boundary conditions for $x = 0$). In case the stimulus is stronger than a critical level, a pulse travels along the axon. For stimuli lower than the threshold, the pulse decays to zero. Apart from a single traveling pulse, wave trains have been observed and calculated. For a discussion of traveling waves in nerves, and particularly in the FitzHugh–Nagumo model, see [Has75], [Has82], [Kee80], [McK70], [RaS78], [ShJ80], [RiK83], [RiT82a], [KhBR92] and the references therein.

Traveling waves occur for a wide range of equations. The Kuramoto–Sivashinsky equation (3.30) is one example, as are beam equations used to model oscillations of suspension bridges [McW90]. More phenomena occur for two- or three-dimensional space variables, represented by $\mathbf{x}$. For a scalar function $u(\mathbf{x},t)$ cases of interest include

⋆   plane waves: $u(\mathbf{x},t) = U(\mathbf{x}^{tr}\mathbf{e_V} - |\mathbf{V}|t)$,
      $|\mathbf{V}|$ is the length of the velocity vector $\mathbf{V}$, $\mathbf{e_V}$ the unit vector in direction of $\mathbf{V}$.
⋆   target patterns: $u(\mathbf{x},t) = U(|\mathbf{x}|,t)$.
⋆   rotating spiral in a plane: $u(\mathbf{x},t) = U(\rho, \vartheta - \lambda t)$,
      $\mathbf{x} = (\rho\cos\vartheta, \rho\sin\vartheta)$, U periodic in the second argument.

In practice, differences between front waves and pulses need not be significant. For instance, when a wave with $u_1 \approx u_2$ is not monotonous it may look like a pulse. A discussion of general patterns carries over to reaction-diffusion problems.

3.5.3 Reaction-Diffusion Problems

Waves are nonequilibrium modes of excitable media. Let the medium be represented by an m-dimensional set $\Omega \subset \mathbb{R}^m$, with $m \in \{1, 2, 3\}$. The set Ω is defined by the application, and $\mathbf{x}$ is the space variable, $\mathbf{x} \in \Omega$. We discuss patterns of vector functions $\mathbf{y}(\mathbf{x}, t) \in \mathbb{R}^n$, which are observed for solutions of PDEs of the reaction-diffusion type

$$\frac{\partial \mathbf{y}}{\partial t} = \mathbf{D} \, \nabla^2 \mathbf{y} + \mathbf{f}(\mathbf{y}, \Lambda), \tag{3.34}$$

for $t > 0$. The Laplacian operator ∇^2 is the sum of the second-order spatial derivatives. In this PDE, $\mathbf{D}$ is the matrix of the diffusion coefficients, which is mostly diagonal with positive entries. ($\mathbf{D}$ is singular in the above FitzHugh-Nagumo model.) The function $\mathbf{f}$ is the reaction term, and Λ represents all parameters involved in $\mathbf{f}$.

Many biological and chemical systems are of the reaction-diffusion type (3.34). These models display a wide variety of pattern formation processes, which range from spots or stripes on animal coats, to oscillations of concentrations of some chemicals, to cardiac arrhythmias [Mur89], [Ede88], [Win90], [Kap95]. In a chemical system, a state that is homogeneous in space represents a well-mixed experiment, no spatial gradient sets in. On the other hand, when there is no mixing, a spatial gradient may lead to some pattern.

The mechanism of *pattern formation* is closely related to stability, in particular to stability of homogeneous states. Again, stability may change with the parameters. When a parameter passes a certain threshold value (bifurcation), a pattern may gain stability. In 1952, Turing laid a chemical basis of pattern formation, or *morphogenesis* [Tur52]. The celebrated paper suggests a reaction-diffusion mechanism that explains how patterns may be formed. In an attempt to describe basic ideas we start from a state that is homogeneous in both space and time. This is the state "no pattern." For some condition of parameters this state may be destabilized, and a state is activated that is still stationary with respect to time, but nonhomogeneous in space. This phenomenon of activating a spatial pattern is called the *Turing bifurcation*, or Turing instability, or *diffusive instability*.

3.5.4 Linear Stability Analysis

Assume a solution $\mathbf{y}^s$ of (3.34) is homogeneous in both time and space,

$$\frac{\partial \mathbf{y}^s}{\partial t} = 0, \quad \nabla \mathbf{y}^s = \mathbf{0}.$$

Consequently, $\mathbf{y}^s$ satisfies $\mathbf{f}(\mathbf{y}^s, \Lambda) = \mathbf{0}$. The question is whether $\mathbf{y}^s$ is stable. To answer this question we set $\mathbf{y}(\mathbf{x}, t) = \mathbf{y}^s + \mathbf{d}(\mathbf{x}, t)$; this yields

$$\frac{\partial \mathbf{d}}{\partial t} = \mathbf{D} \nabla^2 \mathbf{d} + \mathbf{f}(\mathbf{y}^{\mathrm{s}} + \mathbf{d}, \Lambda).$$

For $\mathbf{y}$ close to $\mathbf{y}^{\mathrm{s}}$ ($\|\mathbf{d}\|$ small) truncate

$$\mathbf{f}(\mathbf{y}^{\mathrm{s}} + \mathbf{d}, \Lambda) = \mathbf{0} + \mathbf{f}_{\mathbf{y}}(\mathbf{y}^{\mathrm{s}}, \Lambda)\mathbf{d} + O(\|\mathbf{d}\|^2)$$

after the linear term to obtain the linearized version of (3.34),

$$\frac{\partial \mathbf{h}}{\partial t} = \mathbf{D} \nabla^2 \mathbf{h} + \mathbf{f}_{\mathbf{y}}(\mathbf{y}^{\mathrm{s}}, \Lambda)\mathbf{h}. \tag{3.35}$$

Substituting a separation ansatz $\mathbf{h}(\mathbf{x}, t) = e^{\mu t}\mathbf{w}(\mathbf{x})$ into (3.35) leads to the eigenvalue problem

$$\mu \mathbf{w} = \mathbf{D} \nabla^2 \mathbf{w} + \mathbf{f}_{\mathbf{y}}^{\mathrm{s}}(\mathbf{y}^{\mathrm{s}}, \Lambda)\mathbf{w},$$

which is written

$$\mathbf{0} = \mathbf{D} \nabla^2 \mathbf{w} + (\mathbf{f}_{\mathbf{y}}^{\mathrm{s}} - \mu \mathbf{I})\mathbf{w}. \tag{3.36}$$

In the absence of diffusion, $\mathbf{D} = \mathbf{0}$, stability of $\mathbf{y}^{\mathrm{s}}$ is indicated by the eigenvalues μ^{s} of the Jacobian matrix $\mathbf{f}_{\mathbf{y}}^{\mathrm{s}} := \mathbf{f}_{\mathbf{y}}(\mathbf{y}^{\mathrm{s}}, \Lambda)$. To study how spatial stationary patterns are driven by diffusion, we assume temporal stability of the homogeneous state $\mathbf{y}^{\mathrm{s}}$ —that is, all eigenvalues of the Jacobian $\mathbf{f}_{\mathbf{y}}^{\mathrm{s}}$ have negative real parts. We concentrate on zero-flux boundary conditions

$$(\mathbf{n} \cdot \nabla)\mathbf{w} = \mathbf{0}, \tag{3.37}$$

where $\mathbf{n}$ is the unit-normal vector along the boundary of Ω. (In problems with pattern formation, also periodic boundary conditions make sense.) For zero-flux boundary conditions (3.37), solutions of (3.36) consist of eigenfunctions

$$\mathbf{w}(\mathbf{x}) = \mathbf{a} \cos k_1 x_1 \cos k_2 x_2 \cos k_3 x_3. \tag{3.38}$$

Such a function $\mathbf{w}$ is called a "mode," and each linear combination of modes solves (3.36).

In the case of one space variable $x \in \mathbb{R}$, the resulting ODE-eigenvalue problem for a domain $0 \leq x \leq L$ is

$$\mathbf{D}\mathbf{w}'' + (\mathbf{f}_{\mathbf{y}}^{\mathrm{s}} - \mu \mathbf{I})\mathbf{w} = \mathbf{0}, \quad \mathbf{w}_x(0) = \mathbf{w}_x(L) = \mathbf{0},$$

with eigenfunctions of the type $\mathbf{w}(x) = \mathbf{w}_k(x) = \mathbf{a} \cos kx$, and $k = l\frac{\pi}{L}$, for $l = 0, 1, 2, 3, \ldots$ In this context, k is called the *wave number*. The index l is the *mode number*.

The following analysis is largely identical for 1D, 2D, or 3D problems. For example, for a rectangle $0 \leq x_1 \leq L_1$, $0 \leq x_2 \leq L_2$ we have wave numbers k_1, k_2, and mode numbers l_1, l_2, with $k_i = l_i \frac{\pi}{L_i}$. (For this exposition, we stick to the 2D case of a rectangle Ω.) Substituting (3.38) into (3.36), and using the notation

$$K^2 := \pi^2 \left(\frac{l_1^2}{L_1^2} + \frac{l_2^2}{L_2^2} \right),$$

we find that eigenfunctions $\mathbf{w}(\mathbf{x})$ and eigenvalues μ satisfy the *dispersion equation*

$$\det(-K^2\mathbf{D} + \mathbf{f_y^s} - \mu\mathbf{I}) = 0. \tag{3.39}$$

Note that in this form with the lumped parameter K the dispersion equation is invariant of the space dimension of $\mathbf{x}$. The scalar dispersion equation (3.39) is of the type $\mathcal{F}(\mu, 1, \mathbf{L}, \mathbf{D}, \Lambda) = 0$, where $\mathbf{L} := (L_1, L_2)$ stands for the geometry of the domain Ω, $1 := (l_1, l_2)$ is the number of the mode, $\mathbf{D}$ the diffusion, and Λ represents all the parameters in $\mathbf{f}$.

3.5.5 Turing Bifurcation

The dispersion equation (3.39) in the form $\mathcal{F}(\mu, 1, \mathbf{L}, \mathbf{D}, \Lambda) = 0$ defines implicitly up to n eigenvalues μ for each mode 1. The eigenvalues μ of (3.39) are perturbations of the eigenvalues μ^s, which satisfy $\det(\mathbf{f_y^s} - \mu^s\mathbf{I}) = 0$. The K^2 in (3.39) can be seen as a perturbation parameter that adds diffusion to the system. Increasing K^2 will shift the eigenvalues μ across the complex plane. Thereby it may happen that one or more of the real parts of μ become positive. A loss of stability of $\mathbf{y}^s$ indicated by a positive real part of one of the eigenvalues μ is due to the diffusion term $K^2\mathbf{D}$. This situation $\operatorname{Re}(\mu) > 0$ for $\operatorname{Re}(\mu^s) < 0$, where μ matches μ^s, characterizes one or more unstable modes of $\mathbf{y}^s$ and hence a "spatial instability." This stability result may be different for each node 1. In case $\mu > 0$ the mode with number 1 may be activated.

The mode numbers 1 as well as the lumped parameter K take discrete values. We now extend them to the real numbers $\mathbb{R}$, and there use the notation $\tilde{1}$ and $\tilde{K}$. Then also the μ are parameterized continuously, $\tilde{\mu}_{\tilde{1}}(\mathbf{L}, \mathbf{D}, \Lambda)$. For given mode 1, the equations $\operatorname{Re}(\mu_1) = 0$, or $\operatorname{Re}(\tilde{\mu}_{\tilde{1}}) = 0$ define hypersurfaces $\mathcal{H}$ in the parameter space $\Pi := \{\mathbf{L}, \mathbf{D}, \Lambda\}$. Each mode 1 or $\tilde{1}$ has its own hypersurface, $\mathcal{H} = \mathcal{H}_{\tilde{1}}$ (compare Figure 3.19(a)). If we move in the parameter space Π along a curve $\mathcal{C}$, and cross $\mathcal{H}$, then the stability of the mode changes (Figure 3.19(b)). That is, a necessary condition for a bifurcation is passed. Accordingly, the hypersurfaces $\mathcal{H}$ are bifurcation surfaces. To each branch of the nonlinear problem there corresponds a mode of the linear eigenvalue problem. For $\tilde{\mu} \in \mathbb{R}$ the bifurcation is real, and is frequently called a *Turing instability* or *Turing bifurcation*. That is, Turing bifurcation describes the necessary criteria for local instability of the homogeneous state. Any perturbation of the homogeneous state may lead to amplification of the unstable modes (Figure 3.19(b)), they may grow and from a pattern. Ultimately, the nonlinear terms $\mathbf{f}$ decide what kind of pattern $\mathbf{y}$ emerges globally. —The above mechanism of pattern formation is widespread.

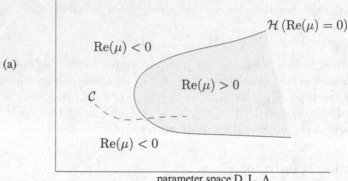

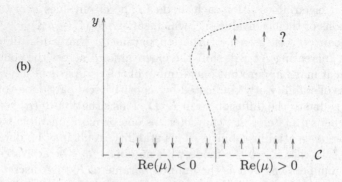

Fig. 3.19. Turing bifurcation, schematically

3.5.6 Special Case $n = 2$

For the special case $n = 2$ we use the notation

$$\mathbf{A} := \mathbf{f}_{\mathbf{y}}^{\mathrm{s}} = \begin{pmatrix} a_{11} & a_{12} \\ a_{21} & a_{22} \end{pmatrix}, \; \mathbf{D} = \begin{pmatrix} D_1 & 0 \\ 0 & D_2 \end{pmatrix}, \; \mathbf{B} := \mathbf{A} - K^2 \mathbf{D}.$$

Then, the vanishing determinant in equation (3.39) is simply the quadratic equation

$$\mu^2 - \mu \, \mathrm{trace}\, \mathbf{B} + \det \mathbf{B} = 0 \,,$$

with solution

$$\mu = \frac{1}{2} \left(\mathrm{trace}\, \mathbf{B} \pm \sqrt{(\mathrm{trace}\, \mathbf{B})^2 - 4 \det \mathbf{B}} \right). \tag{3.40}$$

Since K is part of $\mathbf{B}$, the eigenvalues $\mu = \mu_K$ depend on k_1, k_2, and thus on the mode number l_1, l_2. The equation (3.40) is the dispersion relation.

After these preparations we come back to the question, how the stability of $\mathbf{y}^{\mathrm{s}}$ may get lost. Note that in the absence of diffusion, $D_1 = D_2 = 0$, system

(3.34) reduces to the ODE system $\dot{\mathbf{y}} = \mathbf{f}(\mathbf{y}, \Lambda)$. Recall that we assume all eigenvalues of $\mathbf{A}$ have negative real parts. This criterion for temporal stability in the $n = 2$ setting is equivalent to the two requirements

$$\text{trace}\,\mathbf{A} < 0\,, \quad \det \mathbf{A} > 0\,.$$

Assuming that both inequalities hold means that any loss of stability of the no-pattern state $\mathbf{y}^s$ must be due to diffusion. The state $\mathbf{y}^s$ is unstable ("Turing unstable") if for some K

$$\text{Re}(\mu) > 0\,,$$

for μ from (3.40). The spatial instability can only happen when $\text{trace}\,\mathbf{B} > 0$ or $\det \mathbf{B} < 0$. Clearly, $\text{trace}\,\mathbf{B} > 0$ is not possible because $\text{trace}\,\mathbf{B} = \text{trace}\,\mathbf{A} - K^2(D_1 + D_2)$, $\text{trace}\,\mathbf{A} < 0$, and $D_1 + D_2 > 0$. Thus, spatial instability of $\mathbf{y}^s$ within the domain of temporal stability is equivalent to $\det \mathbf{B} < 0$, or to

$$s(K^2) := \det \mathbf{A} - K^2(D_1 a_{22} + D_2 a_{11}) + K^4 D_1 D_2 < 0\,. \tag{3.41}$$

Part of the further analysis is summarized in Exercise 3.9. If (3.41) holds for

$$K_{\min}^2 := \frac{1}{2}\left(\frac{a_{22}}{D_2} + \frac{a_{11}}{D_1}\right),$$

then there exists an *excitable band* $R_1 < K^2 < R_2$ such that for K^2 within this band a necessary criterion for spatial instability is satisfied. In particular those modes (l_1, l_2) may be activated with K close to $K_{\min}$. Then, for nearby parameters, one or more modes in

$$\sum_{l_1, l_2} \mathbf{a}_K \exp(\mu_K t) \cos\left(\frac{l_1 \pi}{L_1} x_1\right) \cos\left(\frac{l_2 \pi}{L_2} x_2\right) \tag{3.42}$$

have a positive μ, and may be activated by perturbations in $\mathbf{y}^s$. In case L_1, L_2 are prescribed fixed values we must assume that $s(K_{\min}^2) < 0$ is small enough such that at least one of the discrete values of K^2 is inside the excitable band $R_1 < K^2 < R_2$. Note that the exponential growth in (3.42) for some $\mu > 0$ has been derived by a linear analysis. Thus its validity is local—the exponential growth is restricted to a short-time period and to a small neighborhood of $\mathbf{y}^s$. The activated modes will eventually be dominated and bounded by the nonlinear terms of $\mathbf{f}$.

> In two (or three) spatial dimensions the minimum size that allows some pattern to develop depends on two (or three) directions. The aspect ratio L_1/L_2 of the spatial domain (rectangle Ω) determines the modes of which direction are activated first. On a narrow domain there may be a tendency that several modes in one direction are excited before the first mode in the other direction is activated. For example, for $L_1 \ll L_2$, several modes with $l_1 = 0$ and $l_2 = 1, 2, \ldots$ may be activated *before* the first pattern with $l_1 \geq 1$ is excited. Nice consequences and interpretations of the predominance of either stripes or spots on animal tails and other animal coats are found in [Ede88], [Mur89].

3.5.7 Numerical Computation

For arbitrary n the homogeneous states $\mathbf{y}^s$ are the constant solutions of the boundary-value problem

$$0 = \mathbf{D}\,\nabla^2 \mathbf{y} + \mathbf{f}(\mathbf{y}, \Lambda), \quad (\mathbf{n} \cdot \nabla)\mathbf{y} = \mathbf{0}.$$

The activated non-constant patterns bifurcate from $\mathbf{y}^s$. These bifurcations are of pitchfork type (generally without $\mathbf{Z}_2$-symmetry). The radii of the excitable band are the same for all dimensions of $\mathbf{x}$. Hence we can calculate the hypersurfaces $\mathcal{H}$ in the parameter space for the relatively simple ODE boundary-value problem with $x \in \mathbb{R}$

$$0 = \mathbf{D}\,\mathbf{y}'' + \mathbf{f}(\mathbf{y}, \Lambda), \quad \mathbf{y}'(0) = \mathbf{y}'(L) \doteq \mathbf{0}. \tag{3.43}$$

For related methods, see Chapter 6.

Example 3.7   Gierer-Meinhardt Reaction
A reaction of the Gierer-Meinhardt type is given by

$$\begin{aligned} u_t &= \nabla^2 u + \frac{u^2}{v} - bu \\ v_t &= D\nabla^2 v + u^2 - v \end{aligned} \tag{3.44}$$

(see [GiM72], [Mur89], [Ede88]). The variable u is an activator, and v is an inhibitor. Obviously, the nonzero homogeneous state is

$$u^s = \frac{1}{b}, \; v^s = \frac{1}{b^2}.$$

We study this reaction-diffusion system on an interval $0 \le x \le L$ (case $m = 1$), and choose $b = 0.5$ throughout, so $\mathbf{y}^s = (2, 4)$. Clearly, $D_1 = 1$, $D_2 = D$. The hypersurfaces $\mathcal{H}$ reduce to bifurcation curves in a (L, D)-plane. These curves have been calculated by solving the branching system that results from the boundary-value problem of the stationary situation $u_t = v_t = 0$. (The method will be explained in Section 6.2; compare also Example 6.1.) The result is shown in Figure 3.20. Each of the bifurcation curves corresponds to one mode number l. For small D, and for small values of L there will be no bifurcation with respect to D. The length L and the diffusion constant D must exceed some minimum value in order to let bifurcation/pattern happen. Specifically we show for $L = 10$ two branches of $u(x)$ solutions (Figure 3.21). The shapes of these profiles indicate the mode number l by the part of the (distorted) cos-profile shown in the figure. Computational results suggest that for large enough L and D, there is a branch for each l. The different patterns for $l = 1$ and $l = 2$ are illustrated in Figure 3.21. These patterns are still stationary, but no longer homogeneous in space. In Figure 3.21 the constant solution of the bifurcation off $u = 2$, and a turning point are visible.

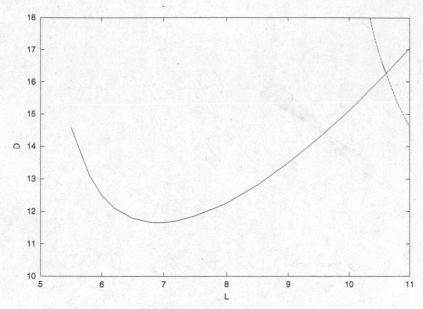

Fig. 3.20. Example 3.7, bifurcation curves $\text{Re}(\mu) = 0$ in the (L, D)-parameter plane. solid line: $l = 1$; dashed: $l = 2$. Above the curves, $\text{Re}(\mu) > 0$ holds, and the mode can be activated; below the curves the stationary state is stable.

Many systems in fluid dynamics, chemical reactions, biology, and other areas exhibit both spatial and temporal oscillations. Two well-known examples are the Brusselator [GlP71] (see Section 5.6.5) and the Belousov–Zhabotinskii reaction (often called "Oregonator") [FiN74], [GrHH78], [KoH73], [HaM75], [RiT82b]. For the Oregonator there is a pattern of spirals and rings that reflects spatial differences of concentrations; temporal oscillations are indicated by changes in color. The Kuramoto–Sivashinsky equation has been applied to model various problems, from combustion to fluid flow; references are in [KeNS90]. For a review on instabilities in chemical reactions, consult [Bor02].

Models in morphogenesis that exhibit regular patterns are discussed in [Kaz85]. For the Gray-Scott model of autocatalysis see [Pea93], [HaPT00]. Waves discussed in the context of pulsating fronts in combustion problems [MaS78] are related to Hopf bifurcation. Other bifurcation or wave phenomena occur in enzyme-catalyzed reactions [DoK86]. One means to analyze spatial patterns are cellular automata; see Section 1.4.3. For further hints and examples, refer to [GrHH78], [Hak77], [LeS85], [SwG81]. Obviously there is a relation between certain patterns and symmetry, see for instance [Hoy06] and the references therein. For an analysis of reaction-diffusion problems see [Fife79]; for numerical aspects consult [Mei00]. Beautiful illustrations of patterns on sea shells are shown in [Mei95].

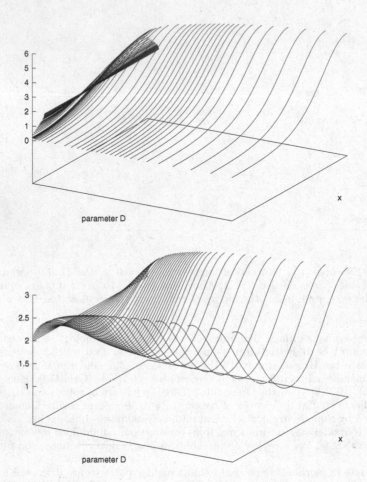

Fig. 3.21. Example 3.7, $m = 1$, $L = 10$, two branches of stationary solutions $u(x)$ for $0 \le x \le L$ and various values of the parameter D. upper figure: branch $(l = 1)$ emanating off $\mathbf{y}^s$ at $D = 15.1158$, for $11.5 \le D \le 73.6$; lower figure: branch $(l = 2)$ emanating at $D = 21.5415$, for $21.5 \le D \le 77.9$

3.6 Deterministic Risk

Traditionally, risk theory has been entirely based on probabilistic approaches. This works well for the essential tasks of insurance, which go along with large numbers of claims or premiums. But when a single risk is investigated, such as the operation of a chemical reactor or the behavior of an electric power generator, the laws of large numbers hardly help. The risky events we have in mind are related to loss of stability, to jumps in the state variables, or quite general to structural changes. Such phenomena are the causes of events that are often regarded as "failures." The underlying risk mechanism is bifurcation, and is basically deterministic. Hence, the bifurcation behavior is integral for a deterministic risk analysis.

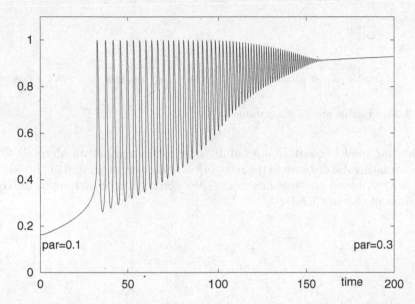

Fig. 3.22. Response of a system when a parameter is varied (Example 2.14 / 3.9)

When no model and no equation is known for a particular risk case, then the knowledge of bifurcation still helps in understanding failures in a qualitative way. Being aware, for example, of the phenomenon Hopf bifurcation, gives an immediate understanding of a scenario as shown in Figure 3.22, no matter whether this occurs in chemical engineering, or whether this is monitored in a biological system. But more insight is possible when a model is given, represented by a set of differential equations, for example. This is the theme of this book. (In some way, a system is not understood if no model is known.)

The fundamental role of bifurcation for the assessment of deterministic risk suggests to use the distance to the next bifurcation as a measure of risk. After having approximated this distance, it makes sense to define a *feasible range of parameters*, or a *risk area*. The bifurcation control or risk control then must take care that the parameters stay in their feasible ranges. This is illustrated in Figure 3.23 for the one-parameter case, and in Figure 3.24 for the two-parameter case. The λ^* indicate a guess on the closest bifurcation. Clearly, a prerequisite for the suggested risk analysis is the ability to find and approximate all relevant bifurcations. This assumption together with that

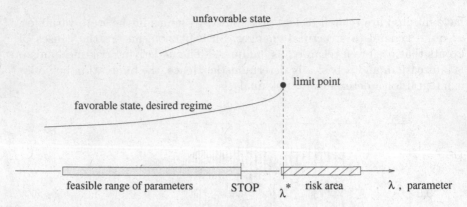

Fig. 3.23. Feasible areas for a parameter λ

of knowing model equations, are admittedly rather restrictive. Methods for approximating the distance to the next bifurcation were suggested in [Sey79c], and [Sey88], based on "test functions." We come back to such quantitative measures of risk in Chapter 5.

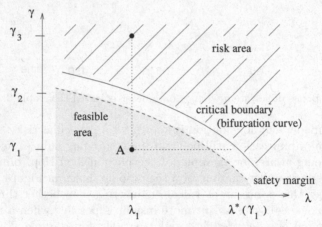

Fig. 3.24. Parameter chart of two parameters λ and γ. For γ_1 the critical threshold is at $\lambda^*(\gamma_1)$. For γ reaching γ_2 the safety margin shrinks to zero.

The bifurcations (that is, the risk cases) with respect to one parameter may depend on a second parameter. For two parameters, the loci of bifurcations form curves in the two-dimensional parameter chart (see Section 2.10). The current parameter combination or operation point marks a point in the parameter chart. If the two parameters are labelled λ and γ, the situation can be as in Figure 3.24. The risk investigator must estimate the distance between the current state of the real system (point $A = (\lambda_1, \gamma_1)$ in Figure 3.24) and the closest risk. To this end, the investigator calculates the bifurcation point, or the bifurcation curve. (Methods will be introduced in the following chapters.) Then it is easy to calculate the distance of the current parameter combination to the next bifurcation. Thereby one sees what other parameter needs to be changed when a critical parameter drifts towards a bifurcation/risk. A careful change of a second parameter may compensate for a dangerous drift of the first parameter. This allows to prevent risks. For instance, in the situation of Figure 3.24 an increase in λ must be compensated by decreasing the second parameter γ in order to keep clear of the safety margin along the bifurcation curve. Bifurcation curves or bifurcation surfaces separate feasible parameter areas from risk areas, up to a safety margin.

Example 3.8   Electric Power Generator

We shall illustrate methods, results, and phenomena by means of the model of Dobson and Chiang [DoC89], which we briefly list. The main variables are the reactive power demand $\lambda := Q_1$, the magnitude of the load voltage V, with phase angle δ, the generator voltage phase angle δ_m, and the rotor speed ω. The constants are taken from [WaAH92]. The model consists of the four ODEs

$$\dot{\delta}_m = \omega$$
$$M\dot{\omega} = -d_m\omega + P_m - E_mVY_m\sin(\delta_m - \delta)$$
$$K_{q\omega}\dot{\delta} = -K_{qv2}V^2 - K_{qv}V + Q(\delta_m, \delta, V) - Q_0 - Q_1$$
$$TK_{q\omega}K_{pv}\dot{V} = K_{p\omega}K_{qv2}V^2$$
$$+ (K_{p\omega}K_{qv} - K_{q\omega}K_{pv})V$$
$$+ K_{q\omega}(P(\delta_m, \delta, V) - P_0 - P_1)$$
$$+ K_{q\omega}(P(\delta_m, \delta, V) - P_0 - P_1)$$
$$- K_{p\omega}(Q(\delta_m, \delta, V) - Q_0 - Q_1)$$

with variables

$$P(\delta_m, \delta, V) = -E_0VY_0\sin(\delta) + E_mVY_m\sin(\delta_m - \delta)$$
$$Q(\delta_m, \delta, V) = E_0VY_0\cos(\delta) + E_mVY_m\cos(\delta_m - \delta) - (Y_0 + Y_m)V^2$$

and constants

$$M = 0.01464,\ Q_0 = 0.3,\ E_0 = 1.0,\ E_m = 1.05,\ Y_0 = 3.33,\ Y_m = 5.0,$$
$$K_{p\omega} = 0.4,\ K_{pv} = 0.3,\ K_{q\omega} = -0.03,\ K_{qv} = -2.8,\ K_{qv2} = 2.1,$$
$$T = 8.5,\ P_0 = 0.6,\ P_1 = 0.0,\ C = 12.0,\ P_m = 1.0,\ d_m = 0.05.$$

Figure 3.25 summarizes the basic solution behavior of this model. There is a turning point of the stationary states (critical parameter value $\lambda_0 = 2.61237$), and beyond that there are no such solutions. In this sense, approaching the turning point must be regarded as entering a risk situation. The lower half of the parabola-like curve consists of unstable states; they are not reached in a real experiment but leave a trace. But the situation of this model

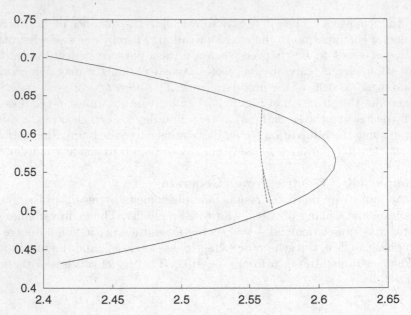

Fig. 3.25. Example 3.8, bifurcation diagram: load voltage V versus load λ; solid curve: stationary solutions, dashed curve: V_{min} of periodic solutions

is more subtle, which becomes clear when stability and periodic orbits are investigated. Figure 3.26 shows a detail of Figure 3.25; stable and unstable solutions are marked. The stability is lost in a Hopf point (hard loss of stability at $\lambda_0 = 2.55919$) before the turning point is reached. The emanating branch of periodic orbits undergoes a sequence of period doublings. One branch of the cascade of branches of periodic orbits is shown in Figure 3.26.

This example shows an interesting phenomenon: the periodic orbits are "cut" by the unstable stationary solutions of the lower branch in Figure 3.25, when they have grown enough such that they come into contact. Figure 3.27 shows phase diagrams of the metamorphosis of periodic orbits. The upper situation is in the bistable range not far from the Hopf point. The lower figure in Figure 3.27 shows an orbit shortly before it touches the lower stationary point (Su), where it is destroyed. In this sense, the cascade of branches of periodic orbits ends at the lower branch of stationary solutions. This emphasizes the fundamental role unstable states play in organizing dynamic behavior, and why they leave a trace in phase space. A phenomenon as reported here (illustrated by Figures 3.26 and 3.27) is sometimes called "blue-sky bifurcation," and is related to "crisis" [SoG92].

The risk background of Example 3.8 is illustrated by Figure 3.28. The load λ of the generator is simulated as slowly increasing through the time interval $0 \leq t < 70$, actually only to 68.67, the moment when the collapse sets in. Thereby the initial-value problem is integrated, and the result shows

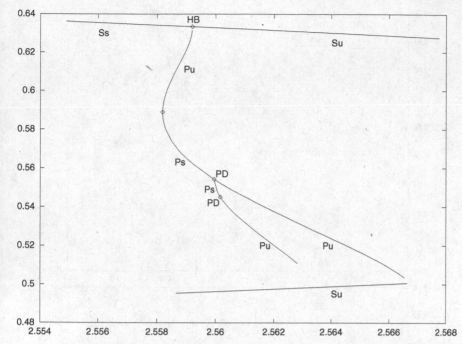

Fig. 3.26. Example 3.8, bifurcation diagram V versus λ, detail of Figure 3.25. notations: S: stationary, P: periodic, s: stable, u: unstable, HB: Hopf point, PD: period doubling

the collapsing of the generated voltage V. The same curve is shown twice, namely, as voltage $V(t)$ depending on time, and in the right-hand figure its projection to phase space.

Prototype Risk Index

The idea of a risk measure based on the distance d to the closest bifurcation suggests a *risk index* R of the type $R = \frac{1}{d}$. Figure 3.29 illustrates the basic idea in the situation of a single parameter λ. Let λ_1 denote the parameter value of the current operation point, and λ_0 the value of the closest bifurcation. Then $d = |\lambda_1 - \lambda_0|$, and a prototype of a risk index $R(\lambda_1)$ is $R(\lambda_1) := d^{-1}$. Then a large value of R indicates closeness of a risk. To be of practical relevance, the risk index must be properly scaled. Also uncertainties in the modeling (that is, in λ_0) and in the measurement of λ_1 must be considered. To this end, let $\Delta\lambda_1 > 0$ be the absolute error in λ_1, and $\Delta\lambda_0 > 0$ be the absolute error in λ_0 as compared to the (somewhat vague notion of the) bifurcation of the real-world problem. Typically, $\Delta\lambda_1$ can be neglected, $\Delta\lambda_1 = 0$. More involved is the role of $\Delta\lambda_0$, which is the result of several sources of error:

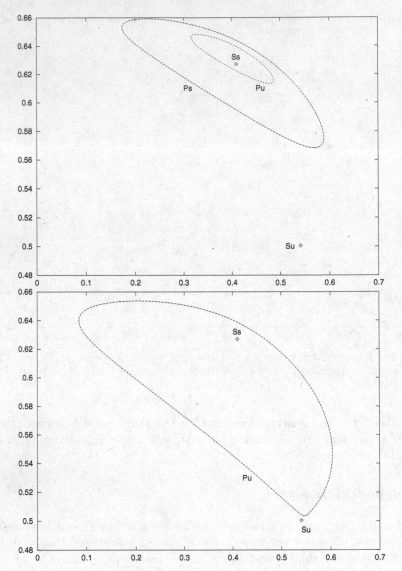

Fig. 3.27. Example 3.8, phase diagrams (projection); horizontal axis: δ_m, vertical axis: V; notations as in Figure 3.26; top: for $\lambda = 2.5588$, bottom: $\lambda = 2.5665$

(i)   inadequacy of the law represented by the model
(ii) errors in the coefficients of the model
(iii) computational error in approximating λ_0

Compared to the modeling errors due to (i) or (ii), the computational error (iii) usually is much smaller and mostly can be neglected. In many cases, it remains disclosed to what extent the model errors (i) and (ii) affect $\Delta\lambda_0$.

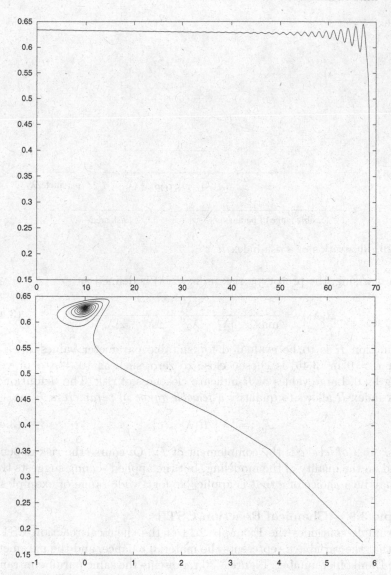

Fig. 3.28. Example 3.8, simulation of the voltage collapse: with the load parameter λ growing linearly from 2.5583 to 2.572. top: $V(t)$, bottom: phase diagram projection to (ω, V)-plane.

Then $\Delta\lambda_0$ can be set to an arbitrary symbolic number, such as 10^{-2}, or may be neglected again. Note that $\Delta\lambda_0$ compensates for all remaining sources of randomness in the formally deterministic model. Finally the data are scaled by λ_1 or λ_0. Since the scaling by the test point λ_1 has advantages in the two-dimensional situation, we scale by λ_1 for the sake of analogy.

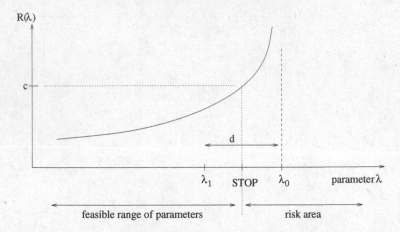

Fig. 3.29. Illustration of a risk index R

Summarizing, the prototype risk index $R(\lambda)$ is defined as

$$R(\lambda_1) := \frac{\lambda_1}{\max\{\epsilon, |\lambda_1 - \lambda_0| - \Delta\lambda_1 - \Delta\lambda_0\}} . \tag{3.45}$$

This function R is to be evaluated for suitable parameter values λ_1. The number $\epsilon > 0$ in (3.45) is chosen close to zero, such as 10^{-10}, to prevent dividing by 0. Large values of R indicate closeness of risk. The definition of the risk index R allows to quantify a *feasible range of parameters* $\mathcal{F}_c$ as

$$\mathcal{F}_c := \{\lambda \mid R(\lambda) < c\}. \tag{3.46}$$

The *risk area of level c* is the complement of $\mathcal{F}_c$. Of course the risk depends on c and on the quality of the modeling, but the applied scaling suggests that there may be a choice of c that is applicable for a wide range of examples.

Example 3.9   Chemical Reaction CSTR

This example continues the Example 2.14 of the chemical reaction CSTR. Recall that the variable u represents the material balance, and the parameter λ is the Damköhler number. Figure 3.30 represents the same parameter range as Figure 3.22, and shows a simulation of λ growing linearly from 0.1 to 0.3. Clearly, the risk index in the lower figure indicates the upcoming risks, which are Hopf bifurcations in this example. Approaching from $\lambda = 0.1$, or from $\lambda = 0.3$, and evaluating $R(\lambda)$ provides a measure of risk.

Early work on estimating the distance to the next bifurcation without having calculated the bifurcation point is [Sey79c], [Sey88]. Corresponding test functions will be discussed in Chapter 5. The role in a nonlinear sensitivity analysis was outlined in [Sey91]. An attempt to recommend a deterministic risk analysis to the (probabilistic) risk community was [Sey97]. There is ample literature on applications to voltage collapse of electric power systems, for example, [AbV84], [WaAH92], [WaG96], [Sou00], [MiCR00], [Sey01],

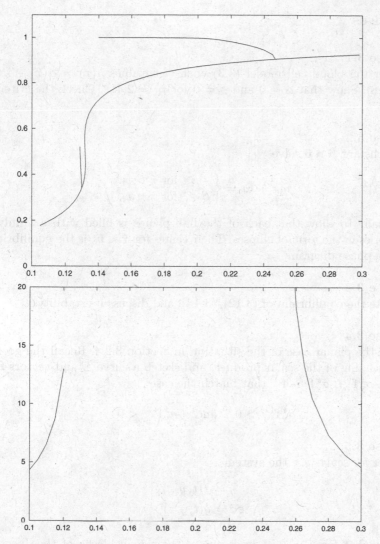

Fig. 3.30. Bifurcation diagram of the CSTR Example 3.9 and 2.14: top figure: stationary branch with two bifurcating branches of periodic solutions, connected by Hopf bifurcations. bottom figure: risk index $R(\lambda)$ of (3.45), with $\Delta\lambda_0 = \Delta\lambda_1 = 0$

[ChHW01], [HuCW02], [HuC03], [HuC04]. The aspect of calculating the distance to closest bifurcations is also discussed in [Dob93], [Dob03], [Sey04]. Applications in chemical engineering is found, for instance, in [FiMD91], [MaM05]. For stabilization of flight dynamics, see [AbL90].

Exercises

Exercise 3.1
Consider the blood cell model (3.3) with eigenvalues $\mu(\tau) = \alpha(\tau) + i\beta(\tau)$ from (3.8). Show that $\alpha = 0$ and $\beta \neq 0$ for $\tau_{\mathrm{H}} = 2.418$. Check the criterion $\frac{d\alpha}{d\tau} \neq 0$.

Exercise 3.2
For a constant $\beta > 0$ solve

$$m\ddot{x} + kx = \begin{cases} -\beta & \text{for } \dot{x} > v_b \\ \beta & \text{for } \dot{x} < v_b \end{cases}$$

analytically to show that each of the half planes is filled with a family of trajectories in the form of ellipses. Their center for $\dot{x} < v_b$ is the equilibrium. Sketch a phase diagram.

Exercise 3.3
Calculate the equilibrium of (3.12)/(3.14), and discuss its stability.

Exercise 3.4
Assume the planar case of the situation in Section 3.2.4. Recall the geometrical meaning of the scalar product, and sketch a curve Σ and vectors $\mathbf{f}^{(1)}$, $\mathbf{f}^{(2)}$, $\sigma_{\mathbf{y}}$, $\sigma_{\mathbf{y}}^{tr}\mathbf{f}^{(1)}$, $\sigma_{\mathbf{y}}^{tr}\mathbf{f}^{(2)}$, $\mathbf{f}^{\Sigma}$ that match the case

$$\sigma_{\mathbf{y}}^{tr}\mathbf{f}^{(1)} > 0 \quad \text{and} \quad \sigma_{\mathbf{y}}^{tr}\mathbf{f}^{(2)} < 0.$$

Exercise 3.5
Consider for scalar y, z the system

$$\dot{y} = f(t, y, z)$$
$$\varepsilon\dot{z} = g(t, y, z)$$

a) Show that for $\varepsilon \to 0$ at least one of the eigenvalues μ of the Jacobian matrix goes to infinity, $|\mu| \to \infty$.

b) For $\varepsilon = 0$, an implicit Euler step from t_0 to $t_0 + \Delta$ with step size Δ leads to the system of nonlinear equations

$$y - y_0 - \Delta \cdot f(t_0 + \Delta, y, z) = 0$$
$$g(t_0 + \Delta, y, z) = 0$$

to be solved for y, z. Investigate the Jacobian (iteration matrix of Newton's method) to show that $g_z \neq 0$ guarantees nonsingularity for small enough Δ.

Exercise 3.6

Analyze the model

$$\dot{u} = g(u) - v + I_{\mathrm{a}}$$
$$\dot{v} = bu - cv$$
$$g(u) := u(S - u)(u - 1)$$

a) Establish a sufficient criterion on S, b, c that guarantees a unique intersection point of the two null clines (compare Figure 3.17).

b) For $I_{\mathrm{a}} = 0$ sketch the flow field $(\dot{u}, \dot{v})$ in the (u, v)-plane for the case with three intersection points.

c) For $S = \frac{1}{4}$, $b = 1$, $c = 2$, $I_{\mathrm{a}} = \frac{145}{864}$ calculate the equilibrium $(u_{\mathrm{s}}, v_{\mathrm{s}})$ of the system.

d) For the parameter values of c) carry out numerical simulations.

Exercise 3.7

Given is the differential equation

$$\ddot{w} - w + \frac{3}{2}w^2 = 0$$

and a solution $w = 4e^t(1 + e^t)^{-2}$. Discuss stationary points and sketch solutions in a phase plane.

Exercise 3.8

Consider Fisher's model (Example 3.6) with $\zeta = 4$, $\lambda = 6$.

(a) Sketch a phase portrait and the profile $U(\xi)$.

(b) From the solution to the linearization at the saddle $(1, 0)$ derive an approximation $U_1(\xi) = 1 - e^{\delta\xi}$ that is close to $U(\xi)$ for large negative values of ξ.

(c) For the node $(0, 0)$ similarly calculate $U_0(\xi) = e^{\gamma\xi}$ as approximation to $U(\xi)$ for large values of ξ.

(d) Suggest boundary conditions $U(a)$ and $U(b)$ for the wave $U(\xi)$ in an interval $a < \xi < b$.

(e) In the (U, U')-plane, give a geometrical interpretation of (b), (c), and (d).

Exercise 3.9

Consider the linear stability analysis of the reaction-diffusion model (3.34) for $n = 2$.

(a) For

$$s(K^2) := \det \mathbf{A} - K^2(D_1 a_{22} + D_2 a_{11}) + K^4 D_1 D_2$$

show that spatial instability of $\mathbf{y}^{\mathrm{s}}$ is equivalent to $s < 0$.

(b) Prove that a necessary criterion for spatial instability is

$$D_1 a_{22} + D_2 a_{11} > 0.$$

4 Principles of Continuation

The diagrams in Chapter 2 are idealizations of the results one gets in practice. Numerical reality does not look quite as good as these diagrams. The smooth curves of the bifurcation diagrams are drawn from a skeleton of computed points. It requires—in addition to reliable computational methods—imagination and confidence to draw a good diagram from the relatively few points the computer gives us. How to select and construct the points is one concern, and the other is the correct interpretation, which may be summarized by drawing a smooth branch.

One can solve the equations for only a finite number of parameter values. Suppose that, because of limitations in time and money, we can find approximate solutions for only 90 parameter values $\lambda_1, \lambda_2, \ldots, \lambda_{90}$. We do not know exactly what solutions look like for any other parameter values than these 90. The situation is comparable to replacing a movie by a set of snapshots. Plotting the results means drawing a chain of dots (Figure 4.1b). Each dot represents one solution. Connecting the dots by a smooth curve gives a nice bifurcation diagram, but unfortunately no guarantee of correctness. One may be lucky, as in Figure 4.1a, when the interpolating curve represents the true parameter dependence. In such a situation the calculation of further solutions for intermediate parameter values will apparently not reveal more information. But in the gaps between consecutively calculated solutions there might be some interesting bifurcation.

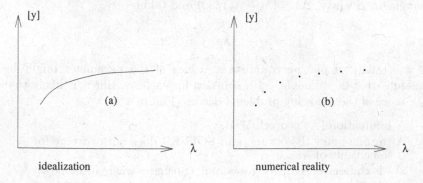

Fig. 4.1. Solutions in a bifurcation diagram

R. Seydel, *Practical Bifurcation and Stability Analysis*,
Interdisciplinary Applied Mathematics 5, DOI 10.1007/978-1-4419-1740-9_4,
© Springer Science+Business Media, LLC 2010

To focus on the construction of branches, we assume that any of the underlying equations can be solved by standard software of numerical analysis, provided a solution exists. Call this basic instrument "SOLVER." That is, SOLVER is a method at hand for solving either a system of algebraic equations, or an ODE boundary-value problem, or another relevant type of equation.

4.1 Principal Tasks

In a multiparameter problem we keep all parameters fixed except one (denoted, as usual, by λ). During the course of a parameter study, parameters may be switched and another parameter might become the bifurcation parameter. Suppose the parameter range of physical interest is

$$\lambda_a \leq \lambda \leq \lambda_b .$$

Let us calculate a first solution for $\lambda_1 = \lambda_a$. The calculation of this first solution is often the most difficult part. Initial guesses may be based on solutions to simplified versions of the underlying equations. Sometimes, choosing special values of the parameters (often $= 0$) leads to equations that are easily solved. In complicated cases, a hierarchy of simplifications must be gradually relaxed step by step, with the solution of each equation serving as an initial guess to the following more complicated equation. Solving a sequence of equations with diminishing degree of simplification until finally the "full" equation is solved is called *homotopy*. Homotopy will be explained in Section 4.3. For now we neglect possible trouble and assume that a solution to the full problem can be calculated for $\lambda_1 = \lambda_a$.

After having succeeded with λ_1, the next question is how to choose λ_2. Assume again that the available computer time allows us to compute no more than 90 solutions. We can try to get enough information by selecting an increment $\Delta\lambda$ (say, $\Delta\lambda = (\lambda_b - \lambda_a)/89$) and taking

$$\lambda_{j+1} = \lambda_j + \Delta\lambda, \quad j = 1, 2, \dots .$$

Such a strategy of picking equidistant values of the parameter might be successful—that is, we might get a solution for each λ_j. But it is likely that one or more of the following problems occurs (Figure 4.2):

★   A bifurcation (A) is overlooked.
★   A turning point (B) occurs, and SOLVER does not converge for some value of λ.
★ $\Delta\lambda$ is chosen too large or too small, causing a waste of money.

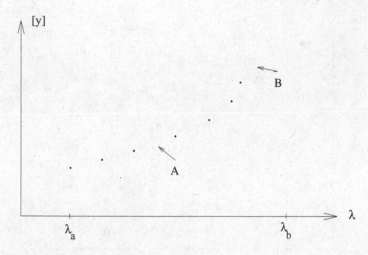

Fig. 4.2. See Section 4.1

Such situations are common. Figure 4.3 illustrates what kind of numerical methods are required. First we need a *continuation method*. A continuation method generates a chain of solutions. Continuation methods construct branches. Such methods are also called *branch tracing* or *path following*. In a good continuation procedure, reasonable candidates for the increment $\Delta\lambda$ are calculated adaptively. Using a continuation method, one can track a branch around a turning point. A device that indicates whether a bifurcation point is close or has been passed is needed. In many problems, stability must be checked. Finally, we need a method for switching branches. In Figure 4.3 assume the branch A/B/C has been calculated and switching to branch H/F/G is required. It suffices to calculate one solution on the "new" branch (Figure 4.3, F).

Then, by means of continuation methods, the entire branch (G/H) can be generated. The real situation underlying Figure 4.2 may look like that in Figure 4.3, for example. Detecting the bifurcation point (A) offers a straightforward way to reach λ_b (via H). A device passing the turning point (B) provides information about another part of the branch. Extending the path outside the "window" $\lambda_a \leq \lambda \leq \lambda_b$ may provide another way to reach λ_b (Figure 4.3 C, D, E). And following the new branch leftward from F for a sufficient distance may convince the investigator that this particular branch neither turns back (G) nor exhibits a bifurcation that leads back into the parameter range of interest.

Let us summarize the basic tools required for a parameter study. We need

(a) continuation methods with devices for detecting bifurcation and checking stability; and

(b) methods for switching from one branch to another with or without the option of calculating the bifurcation point itself.

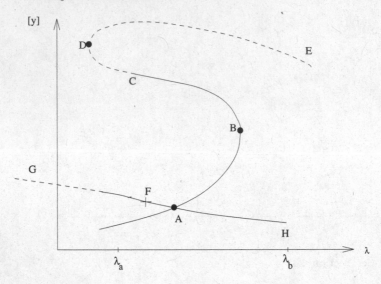

Fig. 4.3. See Section 4.1

The most time-consuming part of this is the use of the continuation method. Typically, 90 to 99% of the total effort consists in generating branches by the continuation method—that is, solving sequences of equations. In comparison, the handling of bifurcation points is less expensive.

We leave the bifurcation aspects to later chapters. This Chapter 4 introduces continuation principles. The system of nonlinear algebraic equations

$$0 = \mathbf{f}(\mathbf{y}, \lambda) \tag{4.1}$$

serves as a basis for the discussion. Here, $\mathbf{y}$ denotes an n-dimensional vector. No generality is lost by restricting attention to equation (4.1). Essentially, both ODEs and PDEs are approximated by such systems of nonlinear equations. Principles similar to those discussed for (4.1) can be applied directly to the differential equations.

> Papers introducing basic ideas of continuation were published in the 1960s by Haselgrove [Has61], Klopfenstein [Klo61], and Deist and Sefor [DeS67]. In the same period, continuation procedures were introduced in engineering and scientific applications. Today continuation is in wide use. In what follows we confine ourselves to "predictor–corrector" continuation methods, which have proved successful in applications. An introduction to simplicial continuation methods is beyond the scope of this book; for an exposition of this class of methods, and for further references see [AlG80], [AlG90].

4.2 Ingredients of Predictor–Corrector Methods

Equation (4.1) defines implicitly curves of solutions. As usual, we assume that solutions exist. In Section 2.5 we showed that a curve of solutions to equation (4.1) can be extended as long as the full-rank condition

$$\text{rank}(\mathbf{f_y}, \mathbf{f}_\lambda) = n \qquad (4.2)$$

is satisfied. If the function $\mathbf{f}$ is smooth, equation (4.2) guarantees a smooth branch of solutions. Under (4.2), pitchfork or transcritical bifurcations are ruled out; turning points can occur naturally.

Assume that at least one solution of equation (4.1) has been calculated. Let us denote this "first" solution on a branch by $(\mathbf{y}^1, \lambda_1)$. The continuation problem is to calculate the branch. Basically this amounts to calculating further solutions on the branch,

$$(\mathbf{y}^2, \lambda_2),\ (\mathbf{y}^3, \lambda_3), \dots,$$

until one reaches a target point, say at $\lambda = \lambda_b$. (The superscripts are not to be confused with exponents.)

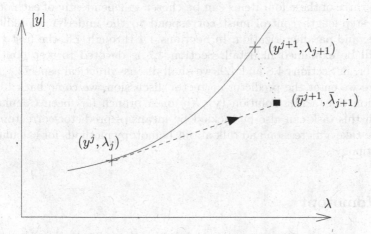

Fig. 4.4. Predictor–corrector principle

The jth continuation step starts from (an approximation of) a solution $(\mathbf{y}^j, \lambda_j)$ of equation (4.1) and attempts to calculate the solution $(\mathbf{y}^{j+1}, \lambda_{j+1})$ for the "next" λ, namely, λ_{j+1},

$$(\mathbf{y}^j, \lambda_j) \quad \longrightarrow \quad (\mathbf{y}^{j+1}, \lambda_{j+1}).$$

With *predictor–corrector* methods, the step $j \to j+1$ is split into two steps:

$$(\mathbf{y}^j, \lambda_j) \xrightarrow{\text{predictor}} (\bar{\mathbf{y}}^{j+1}, \bar{\lambda}_{j+1}) \xrightarrow{\text{corrector}} (\mathbf{y}^{j+1}, \lambda_{j+1})$$

(see Figure 4.4). In general, the predictor $(\bar{\mathbf{y}}, \bar{\lambda})$ is not a solution of equation (4.1). The predictor merely provides an initial guess for corrector iterations that home in on a solution of equation (4.1). In Figure 4.4 the path of the corrector iteration is indicated by the dotted line. The major portion of the work is either on the predictor step (resulting in an approximation $(\bar{\mathbf{y}}, \bar{\lambda})$ close to the branch) or on the corrector step (if a "cheaper" predictor has produced a guess far from the branch). The distance between two consecutively calculated solutions $(\mathbf{y}^j, \lambda_j)$ and $(\mathbf{y}^{j+1}, \lambda_{j+1})$ is called the *step length* or *step size*. Because this distance is not easy to measure—in particular when $\mathbf{y}^{j+1}$ is not yet known—we use these terms freely. For example, the length of the predictor step is called step length. In addition to equation (4.1), we need a relation that identifies the location of a solution on the branch. As we shall discuss in Section 4.5, this identification is related to the kind of *parameterization* strategy chosen to trace the branch.

Continuation methods differ, among other things, in the following:

(a) predictor,
(b) parameterization strategy,
(c) corrector, and
(d) step-length control.

The first three of these four items can be chosen independently of each other. But the step-length control must correspond to the underlying predictor, corrector, and parameterization. In Sections 4.4 through 4.6, the first three items will be explained in detail. Section 4.7 is devoted to step controls. Thereafter, in Sections 4.8 and 4.9, we shall discuss practical aspects.

Before we enter the predictor-corrector discussion, we come back to the assumption that a first solution $(\mathbf{y}^1, \lambda_1)$ on a branch has been calculated. Although this task can also be tackled by means of predictor-corrector methods, we take a digression and talk about homotopy methods for calculating *one* solution.

4.3　Homotopy

An important application of continuation is homotopy. The task of homotopy is to help calculating just one solution to a given equation. Because homotopy is not restricted to our context of bifurcation and stability, we denote the equation to be solved by

$$\mathbf{g}(\mathbf{y}) = \mathbf{0}\,.$$

For example, $\mathbf{g}(\mathbf{y}) := \mathbf{f}(\mathbf{y}, \lambda_1)$ for a prescribed value of λ_1 represents the problem of calculating the first solution on a branch. This application of homotopy is illustrated in Figure 4.5.

Assume that the equation $\mathbf{g}(\mathbf{y}) = \mathbf{0}$ is difficult to solve. This frequent difficulty is mostly due to a lack of "reasonable" initial guess for the solution. Recall that our equations are generally nonlinear and must be solved

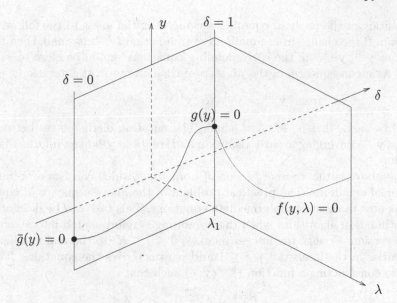

Fig. 4.5. Homotopy from $\delta = 0$ to $\delta = 1$ used as initial phase for finding a first solution on the branch

iteratively; see Section 1.5.1. That is, a sequence of approximations

$$\mathbf{y}^{(0)}, \mathbf{y}^{(1)}, \mathbf{y}^{(2)}, \ldots$$

is calculated such that the iterates $\mathbf{y}^{(\nu)}$ converge toward a solution. $\mathbf{y}^{(0)}$ is the initial guess. The iteration is terminated when for some M the iterate $\mathbf{y}^{(M)}$ satisfies a prescribed error tolerance. Typically, a termination criterion requires a small residual,

$$\|\mathbf{g}(\mathbf{y}^{(M)})\| < \epsilon,$$

for a given value of ϵ, and a small correction $\|\mathbf{y}^{(M)} - \mathbf{y}^{(M-1)}\|$.

Assume further that we also have an equation $\bar{\mathbf{g}}(\mathbf{y}) = \mathbf{0}$ that is easy to solve. The function $\bar{\mathbf{g}}$ is often obtained by simplifying $\mathbf{g}$. Homotopy constructs a chain of equations that are solved one at a time. The first equation is $\bar{\mathbf{g}}(\mathbf{y}) = \mathbf{0}$:

$$\mathbf{f}^{[0]}(\mathbf{y}) := \bar{\mathbf{g}}(\mathbf{y}) = \mathbf{0}$$

$$\mathbf{f}^{[1]}(\mathbf{y}) = \mathbf{0}$$

$$\vdots$$

$$\mathbf{f}^{[K-1]}(\mathbf{y}) = \mathbf{0}$$

$$\mathbf{f}^{[K]}(\mathbf{y}) := \mathbf{g}(\mathbf{y}) = \mathbf{0}.$$

The last equation is $\mathbf{g}(\mathbf{y}) = \mathbf{0}$ itself. This homotopy consists of $K+1$ equations to be solved in consecutive order. The idea of homotopy is to use in each step the solution of the previous equation serving as initial guess to the following equation. If the change from function $\mathbf{f}^{[j]}$ to function $\mathbf{f}^{[j+1]}$ is small, then the solutions $\mathbf{y}^{[j]}$, $\mathbf{y}^{[j+1]}$ of the corresponding equations should be close to each other. As a consequence, in the jth step of the homotopy, the iteration

$$\text{from } \mathbf{y}^{(0)} := \mathbf{y}^{[j]} \text{ to } \mathbf{y}^{(M_{j+1})} \approx \mathbf{y}^{[j+1]}$$

should be fast—that is, M_{j+1} is small. The notation distinguishes between iterates $\mathbf{y}^{(i)}$ converging toward a solution and solutions $\mathbf{y}^{[j]}$ of an intermediate equation.

The above is the *discrete* version of homotopy, which consists of a finite number of equations. The practical problem of the above sequence of equations is how to construct intermediate equations. This task can be delegated to continuation algorithms when the homotopy is reformulated into a *continuous* version. Change the integer index j, $0 \le j \le K$, to the real number δ that varies in the interval $0 \le \delta \le 1$ and parameterizes the equations. This leads to constructing a function $\mathbf{f}^{\mathrm{hom}}(\mathbf{y}, \delta)$ such that

$$\begin{aligned}
\mathbf{f}^{\mathrm{hom}}(\mathbf{y}, \delta) &= 0, \quad 0 \le \delta \le 1, \\
\mathbf{f}^{\mathrm{hom}}(\mathbf{y}, 0) &= \bar{\mathbf{g}}(\mathbf{y}), \quad \mathbf{f}^{\mathrm{hom}}(\mathbf{y}, 1) = \mathbf{g}(\mathbf{y}).
\end{aligned} \tag{4.3}$$

The continuous version (4.3) of homotopy parallels the discrete one in that the first equation to be solved (for $\delta = 0$) is the easy one, $\bar{\mathbf{g}}(\mathbf{y}) = \mathbf{0}$, and the last equation (for $\delta = 1$) is the difficult problem $\mathbf{g}(\mathbf{y}) = \mathbf{0}$. Between these two limiting equations there is a continuum of infinitely many equations. In applications of homotopy, only the solution to the final equation is of interest; the intermediate results (for $\delta < 1$) are discarded because they are not solutions to the original equation $\mathbf{g}(\mathbf{y}) = \mathbf{0}$. Identifying δ with λ, and $\mathbf{f}^{\mathrm{hom}}$ with $\mathbf{f}$, we see that the continuous version of homotopy, equation (4.3), is a special case of our problem to calculate branches. Hence, continuation algorithms can be applied to take care of constructing the chain of intermediate equations required to solve $\mathbf{g}(\mathbf{y}) = \mathbf{0}$. In this way the number K of equations is determined dynamically during the course of the homotopy.

It remains to discuss how to define the homotopy function $\mathbf{f}^{\mathrm{hom}}$. Examples of homotopy can be obtained by attaching a *factor* δ to those terms in the function $\mathbf{g}$ that are responsible for difficulties in calculating a solution. In this way, starting with $\delta = 0$, one first solves a simplified equation. A more systematic way of constructing a homotopy is to set

$$\mathbf{f}^{\mathrm{hom}}(\mathbf{y}, \delta) := \delta \mathbf{g}(\mathbf{y}) + (1 - \delta)\bar{\mathbf{g}}(\mathbf{y}). \tag{4.4}$$

The simple choice $\bar{\mathbf{g}}(\mathbf{y}) = \mathbf{y} - \mathbf{c}$, where $\mathbf{c}$ is any arbitrary vector, has been shown to work successfully [ChMY78]. Such a process of constructing a homotopy function $\mathbf{f}^{\mathrm{hom}}$ is called *embedding*; the equation $\mathbf{g}(\mathbf{y}) = \mathbf{0}$ is embedded into $\mathbf{f}^{\mathrm{hom}}$. For further references on homotopy, see [Kel78], [Wac78], [Wat79], [AlG90].

4.4   Predictors

We return to the basic framework of predictor-corrector methods, and begin with discussing predictors. Predictors can be divided into two classes:

(a) ODE methods, which are based on $\mathbf{f}(\mathbf{y}, \lambda)$ and its derivatives, and
(b) polynomial extrapolation, which uses only solutions $(\mathbf{y}, \lambda)$
   of equation (4.1).

4.4.1   ODE Methods; Tangent Predictor

Taking the differential of both sides of equation (4.1), one obtains

$$0 = \mathrm{d}\mathbf{f} = \mathbf{f_y}\,\mathrm{d}\mathbf{y} + \mathbf{f}_\lambda\,\mathrm{d}\lambda \tag{4.5}$$

and hence

$$\frac{\mathrm{d}\mathbf{y}}{\mathrm{d}\lambda} = -(\mathbf{f_y})^{-1}\mathbf{f}_\lambda\,.$$

Integrating this system starting from the initial values $(\mathbf{y}^1, \lambda_1)$, one obtains the branch on which $(\mathbf{y}^1, \lambda_1)$ lies. This procedure, proposed by Davidenko [Dav53], [Abb77], [Geo81a], fails at turning points because of the singularity of $\mathbf{f_y}$. In multiparameter problems this difficulty can be overcome by switching to an alternate parameter [AnM66].

Another way of overcoming the failure of the method at turning points is to change the parameter to the arclength s. Both $\mathbf{y}$ and λ are considered to be functions of the arclength parameter s: $\mathbf{y} = \mathbf{y}(s)$, $\lambda = \lambda(s)$. From equation (4.5) one obtains

$$0 = \mathbf{f_y}\frac{\mathrm{d}\mathbf{y}}{\mathrm{d}s} + \mathbf{f}_\lambda\frac{\mathrm{d}\lambda}{\mathrm{d}s}\,. \tag{4.6a}$$

The arclength s satisfies the relation

$$(\mathrm{d}y_1/\mathrm{d}s)^2 + \ldots + (\mathrm{d}y_n/\mathrm{d}s)^2 + (\mathrm{d}\lambda/\mathrm{d}s)^2 = 1\,. \tag{4.6b}$$

Equations (4.6a) and (4.6b) form an implicit system of $n + 1$ differential equations for the $n + 1$ unknowns

$$\mathrm{d}y_1/\mathrm{d}s, \ldots, \mathrm{d}y_n/\mathrm{d}s, \mathrm{d}\lambda/\mathrm{d}s\,.$$

A specific method for solving equation (4.6) will be described later.

An important ODE predictor is the tangent predictor. With the notation $z_i = i$th component of $\mathbf{z}$ and setting

$$z_i = \mathrm{d}y_i \ (1 \le i \le n), \ z_{n+1} = \mathrm{d}\lambda,$$

one derives formally from equation (4.5) the equation

$$(\mathbf{f_y}|\mathbf{f}_\lambda)\,\mathbf{z} = 0 \tag{4.7}$$

for a tangent $\mathbf{z}$. The $n \times (n+1)$ matrix $(\mathbf{f_y}|\mathbf{f}_\lambda)$ depends on $(\mathbf{y}, \lambda)$. The full-rank condition (4.2) guarantees a unique straight line tangent to the branch in $(\mathbf{y}, \lambda)$. The length and the orientation of a vector $\mathbf{z}$ on this tangent are still undetermined. A normalization of $\mathbf{z}$ must be imposed to give equation (4.7) a unique solution. One such normalizing equation is $\mathbf{c}^{tr}\mathbf{z} = 1$ for a suitable vector $\mathbf{c}$. Choose, for example, $\mathbf{c} := \mathbf{e}_k$,

$$\mathbf{e}_k^{tr}\mathbf{z} = z_k = 1\,,$$

where $\mathbf{e}_k$ is the $(n+1)$-dimensional unit vector with all elements equal to zero except the kth, which equals unity (recall that ^{tr} denotes transposure). Then the tangent $\mathbf{z}$ is the solution of the linear system

$$\left(\frac{\mathbf{f_y}|\mathbf{f}_\lambda}{\mathbf{e}_k^{tr}}\right)\mathbf{z} = \mathbf{e}_{n+1}\,. \tag{4.8}$$

Provided the full-rank condition $\mathrm{rank}(\mathbf{f_y}|\mathbf{f}_\lambda) = n$ holds along the whole branch, equation (4.8) has a unique solution at any point on the branch (k may have to be changed). In particular, the tangent can be calculated at turning points. The last row $\mathbf{e}_k^{tr}$ in the $(n+1)^2$ matrix of (4.8) can be replaced by any other vector $\mathbf{c}^{tr}$ such that the matrix is of rank $n+1$ (Exercises 4.1 and 4.2).

Several continuation methods—for example, those in [Kel77], [MeS78], [Rhe80], [RhB83]—are based on tangent predictors. In these methods the predictor point (initial approximation) for $(\mathbf{y}^{j+1}, \lambda_{j+1})$ is

$$(\bar{\mathbf{y}}^{j+1}, \bar{\lambda}_{j+1}) := (\mathbf{y}^j, \lambda_j) + \sigma_j\, \mathbf{z} \tag{4.9}$$

(see Figure 4.4). Here σ_j is an appropriate step length. This tangent predictor (4.9) can be considered a step of the Euler method for solving a differential equation that describes the branch. Hence, the discretization error of Euler's method [see equation (1.17)] applies here, and the tangent predictor (4.9) is of first order. That is, using the notation of the $(n+1)$-dimensional vector

$$\mathbf{Y} := \begin{pmatrix} \mathbf{y} \\ \lambda \end{pmatrix}, \tag{4.10}$$

the error of the tangent predictor $\bar{\mathbf{Y}}(\sigma)$ of (4.9) can be written as

$$\|\bar{\mathbf{Y}}(\sigma) - \mathbf{Y}(\sigma)\| = c\sigma^2 + O(|\sigma|^3)\,. \tag{4.11}$$

In equation (4.11), the solution $\mathbf{Y}(\sigma)$ of (4.1) corresponds to the predictor $\bar{\mathbf{Y}}(\sigma)$ in some unique way, as it will be discussed in the following sections.

We now return to differential equation (4.6), which implicitly defines the $n+1$ derivatives of $y_1, \ldots, y_n$, $y_{n+1} = \lambda$ with respect to arclength s. A solution of equation (4.6) can be obtained by the procedure proposed in [Kub76]. Note that the system equation (4.6a) is linear in its $n+1$ unknowns, while the arclength equation equation (4.6b) is nonlinear. Temporarily, we consider one of the components to be prescribed:

$$dy_m/ds \neq 0 \text{ for some fixed } m, \ 1 \leq m \leq n+1.$$

Then equation (4.6a) consists of n linear equations in n unknowns, which can be readily solved. The solutions of these equations depend linearly on dy_m/ds:

$$dy_i/ds = c_i \, dy_m/ds, \quad i = 1, 2, \ldots, m-1, m+1, \ldots, n+1.$$

[The coefficients c_i vary with $(\mathbf{y}, \lambda)$.] Substituting these expressions into equation (4.6b) yields dy/ds:

$$(dy_m/ds)^2 = (1 + c_1^2 + \ldots\ldots + c_{m-1}^2 + c_{m+1}^2 + \ldots + c_{n+1}^2)^{-1}.$$

This constitutes for each $(\mathbf{y}, \lambda)$ explicit expressions for the derivatives with respect to s.

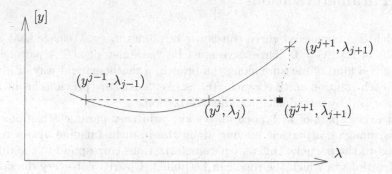

Fig. 4.6. Secant predictor

4.4.2 Polynomial Extrapolation; Secant Predictor

A polynomial in λ of degree ν that passes through the $\nu + 1$ points

$$(\mathbf{y}^j, \lambda_j), (\mathbf{y}^{j-1}, \lambda_{j-1}), \ldots, (\mathbf{y}^{j-\nu}, \lambda_{j-\nu})$$

can be constructed. One assumes that this polynomial provides an approximation to the branch in the region to be calculated next ($\lambda \approx \lambda_{j+1}$). The predictor consists in evaluating the polynomial at $\lambda = \lambda_{j+1}$.

The trivial predictor is the zeroth-order polynomial ($\nu = 0$), where the predictor point is the previous solution, $\bar{\mathbf{Y}}^{j+1} = \mathbf{Y}^j$, or

$$(\bar{\mathbf{y}}^{j+1}, \bar{\lambda}_{j+1}) := (\mathbf{y}^j, \lambda_j). \tag{4.12a}$$

The modified version

$$(\bar{\mathbf{y}}^{j+1}, \bar{\lambda}_{j+1}) := (\mathbf{y}^j, \lambda_{j+1}) \tag{4.12b}$$

is also used but is less favorable than (4.12a).

A polynomial of the first order ($\nu = 1$), the secant, is the alternative to the tangent in (4.9):

$$(\bar{\mathbf{y}}^{j+1}, \bar{\lambda}_{j+1}) := (\mathbf{y}^j, \lambda_j) + \sigma_j (\mathbf{y}^j - \mathbf{y}^{j-1}, \lambda_j - \lambda_{j-1}) \qquad (4.13)$$

(see Figure 4.6). Again, the quantity σ_j is an appropriate step length. Higher-order formulas ($\nu \geq 2$), some of which use values of previous tangents, are given in [Has61]. Frequently, however, lower-order predictors (tangent, secant) are in the long run less costly and therefore preferred. The discretization error of the secant predictor of (4.13) is of the same order as the tangent predictor, see equation (4.11).

4.5   Parameterizations

A branch is a connected curve consisting of points in $(\mathbf{y}, \lambda)$-space that are solutions of $\mathbf{f}(\mathbf{y}, \lambda) = \mathbf{0}$. This curve must be "parameterized." A parameterization is a kind of measure along the branch, a mathematical way of identifying each solution on the branch. Terms like "next" or "previous" solution are quantified by introducing a parameterization.

Not every (part of a) branch allows any arbitrary parameterization. To see this, imagine a particle moving along the branch. Imagine also a rope attached to the particle. Different parameterizations correspond to the different directions in which the rope can be pulled. Clearly, not every direction works. Pulling normal to the branch is the direction that does not work.

The most obvious parameter is the control variable λ. While this parameter has the advantage of having physical significance, it encounters difficulties at turning points, where the pulling direction is normal to the branch.

> But with a little extra effort a parameterization by λ can be sustained even at turning points. This can be accomplished by using the predictor $\bar{\mathbf{h}}$ of [Sey79a], [Sey81b] (cf. Definition 5.2 in Section 5.4.1). This predictor is perpendicular to the λ-axis and provides a way of jumping over the turning point and calculating a solution for the same value of λ on the opposite side of the branch (Figure 4.7). In the remainder of this section, parameterizations that do not need special provisions at turning points will be discussed.

4.5.1   Parameterization by Adding an Equation

Curves such as solution branches of equation (4.1) can be parameterized by *curve parameters*. A general curve parameter, call it γ, may have a geometrical rather than a physical meaning. A parameterization by γ means that solutions of $\mathbf{f}(\mathbf{y}, \lambda) = \mathbf{0}$ depend on γ:

$$\mathbf{y} = \mathbf{y}(\gamma), \ \lambda = \lambda(\gamma).$$

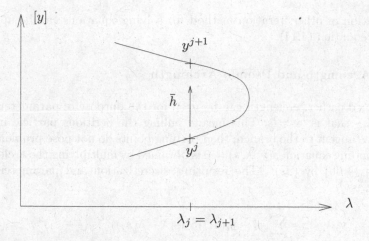

Fig. 4.7. Predictor passing a turning point

For a particular value of γ, the system $\mathbf{f}(\mathbf{y}, \lambda) = \mathbf{0}$ consists of n equations for the $n + 1$ unknowns $(\mathbf{y}, \lambda)$. If the parameterization is established by one additional scalar equation,

$$p(\mathbf{y}, \lambda, \gamma) = 0,$$

one can formulate an extended system

$$\mathbf{F}(\mathbf{Y}, \gamma) := \begin{pmatrix} \mathbf{f}(\mathbf{y}, \lambda) \\ p(\mathbf{y}, \lambda, \gamma) \end{pmatrix} = \mathbf{0}, \tag{4.14}$$

which consists of $n+1$ scalar equations for the $n+1$ unknowns $\mathbf{Y} = (\mathbf{y}, \lambda)$. The general setting of (4.14) includes all types of parameterization. For example, the option of taking λ as continuation parameter means choosing $\gamma = \lambda$ and

$$p(\mathbf{y}, \lambda) = \lambda - \lambda_{j+1}.$$

Other examples will follow soon. Solving equation (4.14) for specified values of γ yields information on the dependence $\mathbf{y}(\gamma), \lambda(\gamma)$. Attaching a normalizing or parameterizing equation is a standard device in numerical analysis; in the context of branch tracing, refer to [Riks72], [Kel77], [Sey79b], [Rhe80].

In equation (4.14) the parameterizing equation $p(\mathbf{y}, \lambda, \gamma) = 0$ is attached to the given system of equation (4.1). This allows one to apply a SOLVER such as a Newton method to equation (4.14); in this way the desired para-meterization is imposed on equation (4.1) automatically. Alternatively, one can apply the Newton method to equation (4.1) and impose an appropriate side condition on the iteration [Has61], [MeS78]. This latter approach will be described in Section 4.6. The formulation in (4.14) has the advantage of not being designed to be used with a special version of a Newton method. Any modification or other iteration method for solving equations can be applied to solve equation (4.14).

4.5.2   Arclength and Pseudo Arclength

As outlined earlier, arclength can be used for the purpose of parameterizing a branch—that is, $\gamma = s$. This means pulling the fictitious particle in the direction tangent to the branch; then turning points do not pose problems. A corresponding equation $p(\mathbf{y}, \lambda, s) = 0$ is obtained by multiplying the arclength equation (4.6b) by $(\mathrm{d}s)^2$. The resulting discretization and parameterizing equation is

$$0 = p_1(\mathbf{y}, \lambda, s) := \sum_{i=1}^{n}(y_i - y_i(s_j))^2 + (\lambda - \lambda(s_j))^2 - (s - s_j)^2. \qquad (4.15)$$

If $(\mathbf{y}^j, \lambda_j) = (\mathbf{y}(s_j), \lambda(s_j))$ is the solution previously calculated during continuation, equation (4.15) together with equation (4.1) fix the solution $(\mathbf{y}(s), \lambda(s))$ at discretized arclength distance $\Delta s = s - s_j$. Arclength parameterization has been used repeatedly; in particular, it was advocated by [Klo61], [Riks72], [Kub76], [Kel77], [DeK81], [ChK82]. In [Kel77] "pseudo arclength" was proposed—that is, for $0 < \zeta < 1$ either

$$0 = p_2(\mathbf{y}, \lambda, s) := \zeta \sum_{i=1}^{n}(y_i - y_i(s_j))^2 + (1 - \zeta)(\lambda - \lambda(s_j))^2 - (s - s_j)^2$$

or

$$0 = p_3(\mathbf{y}, \lambda, s) := \zeta \sum_{i=1}^{n}(y_i - y_i(s_j))\frac{\mathrm{d}y_i(s_j)}{\mathrm{d}s}$$
$$+ (1 - \zeta)(\lambda - \lambda(s_j))\frac{\mathrm{d}\lambda(s_j)}{\mathrm{d}s} - (s - s_j).$$

The tune factor ζ allows one to place different emphasis on $\mathbf{y}$ or λ. For $\zeta = 1$ this was suggested already by [Riks72]. The latter parameterization p_3 is motivated by the Taylor expansion,

$$\mathbf{y}(s) - \mathbf{y}(s_j) = (s - s_j)\frac{\mathrm{d}\mathbf{y}(s_j)}{\mathrm{d}s} + O(|s - s_j|^2).$$

The derivatives are evaluated at the previously calculated solution. The parameterization p_3 has the advantage of being linear in the increments $\Delta \mathbf{y}$, $\Delta \lambda$, and Δs.

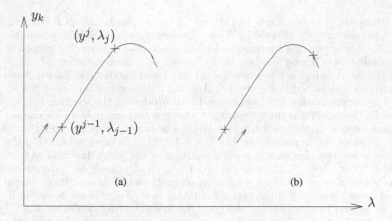

Fig. 4.8. Scenario where a failure is likely

4.5.3 Local Parameterization

Any of the components y_i $(i = 1, \ldots, n)$ can be admitted as a parameter, including $y_{n+1} = \lambda$. This leads to the parameterizing equation

$$p(\mathbf{y}, \eta) := y_k - \eta = 0 \,,$$

with an index k, $1 \leq k \leq n+1$, and a suitable value of η. The index k and the parameter $\eta = y_k$ are locally determined at each continuation step $(\mathbf{y}^j, \lambda_j)$ in order to keep the continuation flexible. For $k \leq n$, local parameterization means pulling the mentioned particle in a direction perpendicular to the λ-axis. Hence, an index k for local parameterization exists also at turning points. With $\mathbf{Y} = (\mathbf{y}, \lambda)$, equation (4.14) reduces to

$$\mathbf{F}(\mathbf{Y}, \eta; k) = \begin{pmatrix} \mathbf{f}(\mathbf{Y}) \\ y_k - \eta \end{pmatrix} = \mathbf{0} \,. \tag{4.16}$$

Such systems have been used in the context of branch tracing —for example, in [Abb78], [MeS78], [Sey79b], [Rhe80], [DeR81]. This kind of parameterization has been called *local parameterization* [RhB83]. At least two codes that implement a local parameterization strategy were developed [RhB83], [Sey83c], [Sey84].

It is easy to find a suitable index k and parameter value η. A continuation algorithm based on tangent predictors (4.8)/(4.9) determines k such that

$$|z_k| = \max\{|z_1|, \ldots, |z_n|, |z_{n+1}|\} \,. \tag{4.17}$$

This choice picks the component of the tangent $\mathbf{z}$ that is maximal. In the algorithm of [Sey84], which is based on the secant predictor (4.13), the index k is chosen such that $\Delta_r y_k$ is the maximum of all relative changes,

$$\Delta_r y_k = \max\{\Delta_r y_1, \ldots, \Delta_r y_{n+1}\} \,, \tag{4.18}$$

$$\text{with} \quad \Delta_r y_i := |y_i^j - y_i^{j-1}| / |y_i^j| \,.$$

The criteria of equations (4.17) and (4.18) establish indices k that can be expected to work effectively. These choices, however, do not guarantee that the corresponding continuation is especially fast; rather, the emphasis is on reliability. It may happen that a choice of k from equation (4.17) or (4.18) fails. A failure is likely when a turning point "arrives" suddenly. Such a scenario is depicted in Figure 4.8. Assume that a continuation step with local parameterization has been carried out, producing the solution $(\mathbf{y}^j, \lambda_j)$ close to the turning point. (Figure 4.8 shows a turning point with respect to y_k.) Assume further that the step control does not notice the change and still maintains y_k as parameter. In Figure 4.8, in both cases (a) and (b) the continuation fails, because it seeks a solution in the same direction characterized by $y_k^{j+1} > y_k^j$. In situation (a) the step size must be reduced too drastically in order to obtain a solution. A continuation procedure trying to avoid unnecessary cost interprets a reduction in step size below a certain limit as failure. In situation (b) of Figure 4.8, a successful calculation of $\mathbf{y}^{j+1}$ thereafter leads to a backward step of the continuation. As a result, the orientation of the branch tracing is reversed. In order to overcome a failure of a chosen k, it makes sense to have an alternative k_1 available. The choices (4.17)/(4.18) suggest choosing for k_1 the index that corresponds to the second largest value of the right-hand sides in (4.17)/(4.18). In case equation (4.16) with k from (4.17)/(4.18) does not lead to a solution, equation (4.16) is tried again with k_1. In this way, local parameterization is as reliable as any other method.

After an index k is fixed, a suitable parameter value η must be determined. A value of η depends on the index k, on the location of the current solution on the branch (i.e., on j), and on the desired step size σ,

$$\eta = \eta(k, j, \sigma).$$

A simple strategy relates the current step size $\eta - y_k^j$ to the previous step size $y_k^j - y_k^{j-1}$ by an adjustable factor ξ,

$$\eta = y_k^j + (y_k^j - y_k^{j-1})\xi. \tag{4.19}$$

An example for the adjustable factor ξ is given in Section 4.7.

The local parameterization concept can be generalized to a continuation that involves qualitative aspects. For example, a continuation by asymmetry is furnished by

$$p(\mathbf{y}, \lambda, \gamma) = y_k - y_m - \gamma \quad (\gamma \neq 0). \tag{4.20}$$

For this parameterization we refer to Section 5.6.6.

4.6 Correctors

In this section, we briefly discuss methods that are typically used to solve equation (4.1) or the extended systems (4.14) or (4.16). Because the latter systems are ready to be solved by the standard software SOLVER, we characterize this type of corrector only briefly. In the remainder, we shall concentrate

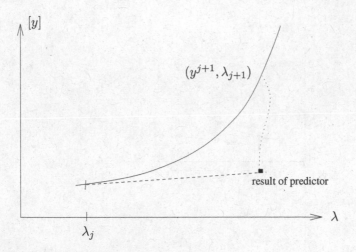

Fig. 4.9. Iteration of a corrector

on the previously mentioned alternative that establishes a parameterization by imposing a side condition on the iteration method.

First, we characterize correctors that result from solving equation (4.14); see Figure 4.9. The final limit $(\mathbf{y}^{j+1}, \lambda_{j+1})$ of the corrector iteration is determined by the parameterizing equation $p(\mathbf{y}, \lambda, \gamma) = 0$ in (4.14). The *path* of the iteration is not prescribed (dotted line in Figure 4.9). This path depends on the particular parameterization and on the kind of method implemented in SOLVER; the user does not care about the path. In principle, the same limit $(\mathbf{y}^{j+1}, \lambda_{j+1})$ can be produced by any parameterization if a proper step length is found. But in practice nothing is known about relations among y_i, λ, and arclength s. Consequently, different parameterizations usually lead to a different selection of solutions on the branch. Figure 4.10 illustrates correctors of the three parameterizations

   (a) by bifurcation parameter λ (step length $\Delta\lambda$),
   (b) local parameterization (step length Δy_k), and
   (c) arclength parameterization (step length Δs).

Imagine the dots in Figures 4.9 and 4.10 represent the corrector iteration. The upper and middle figures in Figure 4.10 reflect a choice of η in (4.16) (and hence of $\Delta\lambda$ or $\Delta\mathbf{y}$) that matches the predictor $\bar{\mathbf{y}}$. In other words, η and the step lengths of the chosen parameter are derived from the predictor step length σ. In this case the corrector iterations in Figures 4.10(a) and 4.10(b) can be expected to move more or less parallel to the axes. A corresponding identification of η with $\bar{y}_k$ is comfortable but not required. In general, relations between the step lengths σ and Δy_k are more involved.

Now we turn to the parameterization of the corrector iteration. To this end, consider one step of the Newton iteration, formulated for the $(n + 1)$-dimensional vector $\mathbf{Y} = (\mathbf{y}, \lambda)$,

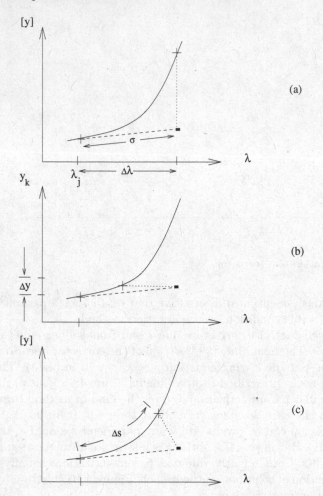

Fig. 4.10. Different parameterizations

$$\mathbf{f_y}\,\Delta\mathbf{y} + \mathbf{f_\lambda}\,\Delta\lambda = -\mathbf{f}(\mathbf{y}^{(\nu)},\lambda^{(\nu)})\,,$$
$$\mathbf{y}^{(\nu+1)} = \mathbf{y}^{(\nu)} + \Delta\mathbf{y}\,,\quad \lambda^{(\nu+1)} = \lambda^{(\nu)} + \Delta\lambda\,. \tag{4.21}$$

In equation (4.21) the first-order partial derivatives are evaluated at $(\mathbf{y}^{(\nu)}, \lambda^{(\nu)})$. The system of linear equations in (4.21) consists of n equations in $n+1$ unknowns $\Delta\mathbf{y}, \Delta\lambda$. The system can be completed by attaching one further equation that parameterizes the course of the corrector iteration. Following [Has61], [MeS78], one requests that the iteration be perpendicular to the tangent $\mathbf{z}$ from (4.8). Using the $(n+1)$ scalar product, this is written as

$$(\Delta\mathbf{y}^{tr}, \Delta\lambda)\,\mathbf{z} = 0\,. \tag{4.22}$$

Starting from the predictor,

$$(\mathbf{y}^{(0)}, \lambda^{(0)}) := (\bar{\mathbf{y}}^{j+1}, \bar{\lambda}_{j+1}),$$

equations (4.21) and (4.22) define a sequence of corrector iterations $\mathbf{Y}^{(\nu)}$

$$(\mathbf{y}^{(\nu)}, \lambda^{(\nu)}), \quad \nu = 1, 2, \ldots,$$

supposed to converge to a solution $(\mathbf{y}^{j+1}, \lambda_{j+1})$ of (4.1) (dots in Figure 4.4). This sequence eventually hits the branch in the solution that is the intersection of the curve defined by $\mathbf{f}(\mathbf{y}, \lambda) = \mathbf{0}$ and the hyperplane defined by equation (4.22). Hence, the parameterization of the branch is established by (4.22). In the parameterizing equation (4.22), the tangent vector $\mathbf{z}$ can be replaced by other vectors.

Now that the parameterization of (4.22) is characterized by a hyperplane in the $(n+1)$-dimensional $(\mathbf{y}, \lambda)$-space, we realize that the local parameterization concept can also be seen as defined by hyperplanes [see Figures 4.10(a) and 4.10(b)]. A straightforward analysis [Exercise 4.2(a)] reveals that conceptual differences among various parameterizations are not significant. In a practical implementation, however, differences become substantial. Note that equations (4.14) and (4.16) can be solved without requiring specific software. Available $\mathbf{LU}$-decompositions of $\mathbf{f_y}$ can also be used for the decomposition of $\mathbf{F_Y}$ (Exercise 4.3).

4.7 Step Controls

In simple problems the principles introduced above may work effectively without a step-length control. That is, constant step lengths are taken throughout, say with $\Delta s = 0.1$ in case of parameterization by arclength. If the step size is small enough, such a step strategy may be successful for a wide range of problems. But such results are often obtained in an inefficient manner, involving too many steps along "flat" branches. The step length should be adapted to the actual convergence behavior. Ultimately, it is the flexibility of a step control that decides whether a continuation algorithm works well. Because step controls must meet the features of the underlying combination predictor/parameterization/corrector, step-length algorithms are specifically designed.

Step controls can be based on estimates of the convergence quality of the corrector iteration. Corresponding measures of convergence are not easy to develop. [DeR81] showed that an extrapolation of convergence radii of previous continuation steps into the future is hardly possible. In the following, some general aspects are summarized.

As pointed out in [Sey77], [Sey84], step-length algorithms can be based on empirical arguments. Such a strategy is based on the observation that the total amount of work involved in a continuation depends on the average step length in a somewhat convex manner: The continuation is expensive both for very short steps (too many steps) and for very large steps (slow or no convergence of correctors). The costs of a continuation are moderate for a certain medium step length, which is related to an optimal number N_{opt} of iterations of a corrector. This number depends on the type of corrector and on the prescribed error tolerance ϵ. For example, with quasi-Newton correctors and $\epsilon = 10^{-4}$ the optimal number is about $N_{\text{opt}} \approx 6$. The aim is to adjust the step size so that each continuation step needs about N_{opt} corrector iterations. Let N_j denote the number of iterations needed to approximate the previous continuation step. Then a simple strategy is to reduce the step size in case $N_j > N_{\text{opt}}$ and to increase the step size in case $N_j < N_{\text{opt}}$. This can be done, for example, by updating the previous step size by multiplication with the factor

$$\xi := N_{\text{opt}}/N_j \, ;$$

compare equation (4.19). In case of local parameterization, this leads to the value

$$\eta := y_k^j + (y_k^j - y_k^{j-1})N_{\text{opt}}/N_j \tag{4.23}$$

to be prescribed in equation (4.16) for the current parameter y_k. This step control is easy to implement (Exercises 4.5 and 4.6). Such a strategy has become the core of several step-length algorithms.

In practice, the step-length correction factors ξ are bounded to a value $\bar{\xi}$ that is not too far away from 1. Choose, for instance,

$$\bar{\xi} := \begin{cases} 0.5 & \text{if } \xi < 0.5 \\ \xi & \text{if } 0.5 \le \xi \le 2 \\ 2 & \text{if } \xi > 2 \,. \end{cases} \tag{4.24}$$

In this way changes in step length are less significant, and thus the strategy becomes more careful. In addition to (4.23)/(4.24) further limitations of the step lengths are advisable in order to obtain enough information about the branch. A strategy that is easy to implement is to limit relative changes $\|\text{predictor}\|/\|\mathbf{Y}\|$ to a value not larger than, say, 0.1. In this way small details of branches most likely are resolved. The strategy (4.23)/(4.24) offers a robust way of step control that is based on few assumptions. Note that the factor ξ, or $\bar{\xi}$, is not a continuous function of the step length.

In [DeR81], [RhB83], the branch was locally replaced by a quadratic polynomial. This enables a derivation of convergence measures involving the curvature of the branch. Under more assumptions, continuous correction factors can be calculated that exploit the convergence of the corrector more closely. To this end, assume, for example,

(1) Newton's method is used,
(2) predictors are such that the Newton corrector converges quadratically, and
(3) convergence at two consecutive solutions is similar.

We give a flavor of related step-length procedures, in the lines of [Rhe80], [DeR81], [RhB83]. Denote the error of the νth Newton iteration by $e_\nu :=$ $\|\mathbf{Y}^{(\nu)} - \mathbf{Y}\|$. Then assumptions (1) and (2) lead to the asymptotic error model

$$\lim_{\nu \to \infty} \frac{e_{\nu+1}}{e_\nu^2} = \delta \qquad (4.25)$$

for some δ. Again the ultimate aim is to find a step length σ such that after N_{opt} corrector iterations a prescribed error tolerance ϵ for $e_{N_{\mathrm{opt}}}$ is reached. From equation (4.25) the asymptotic relation

$$\delta e_{\nu+1} = (\delta e_\nu)^2 \qquad (4.26)$$

is derived, which leads to

$$\delta \epsilon = \delta e_{N_{\mathrm{opt}}} = (\delta e_0)^{2^{N_{\mathrm{opt}}}} . \qquad (4.27)$$

By observing the iteration of Newton's method, a guess for δ can be obtained. With this δ, equation (4.26) serves as a convergence model of the corrector iteration. This allows one to calculate from equation (4.27) e_0, which serves as estimate of an optimal initial error e_{new}. Starting the corrector iteration from a predictor with this error e_{new} will require about N_{opt} iterations to reduce to ϵ. The error equation (4.11) of the predictor means to first order

$$e_{\mathrm{new}} = e_0 = c\sigma^2 .$$

The constant c depends on the solution, $c = c_j$. By assumption (3), $c_j = c_{j+1}$; hence,

$$c_{j+1} = \frac{e_{\mathrm{new}}}{\sigma_{\mathrm{new}}^2} = c_j = \frac{e_{\mathrm{last}}}{\sigma_{\mathrm{last}}^2}$$

is known from the previous step. This yields

$$\sigma_{\mathrm{new}} = \sigma_{\mathrm{last}} \sqrt{\frac{e_{\mathrm{new}}}{e_{\mathrm{last}}}} . \qquad (4.28)$$

The rough correction factor in (4.28) is subjected to safeguards such as provided by (4.24).

For references on continuation and on step control see [Kel77], [RhB83], [DeFK87], [ScC87], [Mac89], [RhRS90]. Aspects of large sparse systems are treated in [Chan84b], [HuiR90]. Calculating discrete solutions on the branch, as discussed above, is a secondary aim, derived from the primary task of approximating the branch. Concentrating on discrete solutions, as described in this chapter, has the tendency to neglect accuracy requirements for approximating the parts of the branch *between* the discrete solutions. Typically, accuracy control focuses on the discrete solutions. Step controls and continuation methods are possible that are designed for calculating branches [Neu93a]. Such methods calculate functions that approximate the branch within a prescribed error tolerance. For multiple-parameter continuation, see [AlS85], [Rhe88], [Hen02].

4.8   Practical Aspects

Readers who are more interested in phenomena and methods than in practical advice, are encouraged to finish reading Chapter 4 at this point.

So far we have discussed an idealized continuation carried out in an environment where enough computing time is available and no interference is expected. In many applications, however, a step control is affected by certain side conditions or restrictions. In our context of bifurcation and stability, the branch with all information is of interest, no detail should be missed. This situation affects a continuation strategy in a nontrivial way. Let us illustrate this with a typical example.

Assume that one is investigating a problem of periodic solutions and is concentrating on the stability. A main concern is to find a critical value of λ at which the stability is lost. One usually wants to study the situation close to a loss of stability (bifurcation) in more detail. This calls for substantially smaller step sizes. Therefore a step control needs to incorporate a "constraint."

The practical implications of this situation are significant. We may like to trace a branch close to the point of the loss of stability by prescribing a small step length in the physical parameter λ, not affected by any step control. The desire to choose equidistant values of λ temporarily renders superfluous any step-length algorithm and calls for a semiautomated and interactive way of tracing a branch.

Other examples of constraints of step-length controls are bounds for maximum step lengths, either measured in the predictor length, or in the difference of the parameter λ, or in the relative change of solutions. In many examples a window of reasonable values of λ or of $\mathbf{y}$ is known. Specifying such a range of admissible values allows the continuation procedure to halt when the window is left, thereby saving unnecessary expense of computing time.

Compromises between fully automating the continuation and taking into account the physical significance of λ are possible. For example, the variables y_i ($i = 1, \ldots, n$) can be treated differently from λ. This can be accomplished by making use of a weighted local parameterization [Sey84]. The choice of the index k [see (4.17) and (4.18)] is subjected to a suitable weighting factor strongly preferring λ. This strategy still enables passing turning points. Such a weighting, however, may slow down the turning, because it tends to stick too long to a parameterization by λ. Allowing more algorithmic complexity, step-length algorithms can be modified so that the values of λ chosen are as even as possible. Sophisticated continuation algorithms may suffer under profane restrictions.

A common situation in running continuations is a halt followed by a restart. The solution calculated last [say, $(\mathbf{y}^j, \lambda_j)$] must be kept available for a subsequent restart of the continuation. In case the continuation is based on a tangent predictor, equations (4.8) and (4.9), a guess $(\bar{\mathbf{y}}^{j+1}, \bar{\lambda}_{j+1})$ for the

next solution is readily obtained, provided the tangent z and information on
the last predictor length σ have been saved. For a reasonable restart, some
information on the history of the completed part of the continuation must
be available. Apart from (y^j, λ_j), the data of σ_j, the index k in the case of
local parameterization and the last actual step length must be saved. Then
the new continuation can benefit from what was calculated before. Whether
the tangent z should be saved or recalculated upon restart will depend on
the amount of available storage.

We discuss the matter of storage for the secant predictor equation (4.13),
which is based on two previous solutions. Sufficient data for generating a
new predictor are available when two previous solutions are stored, but this
strategy may encounter difficulties. Typically, a parameter study is split into
sequences of various continuation runs. Storing all terminating solutions can
easily produce a flood of data. Storing two solutions of each termination
leads to double the amount of data. One alternative is to restart a continuation with the trivial predictor equation (4.12a). This choice usually requires
a starting step length that is drastically smaller than the terminating step
size that may have taken advantage of secant predictors. It is worth starting
as follows: In a preiteration, a zeroth step is calculated with a small step
length of, say, $\Delta\lambda = \lambda\,10^{-3}$. The resulting auxiliary solution only serves to
generate a secant, now enabling us to proceed with the terminating step
length of the previous part of a continuation run.

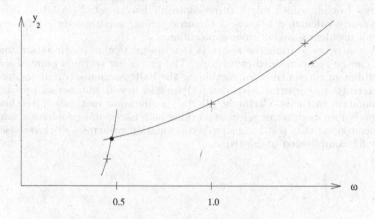

Fig. 4.11. Branch jumping for the Duffing equation (2.9)

The reader might have the impression that a nontrivial predictor (such as
a tangent or a secant) always leads to better results than just starting from
the previous solution [trivial predictor, equation (4.12)]. In fact, nontrivial
predictors generally result in faster convergence of the subsequent corrector.
But this convergence is not always toward the desired solution. As an example, we reconsider the Duffing equation (2.9). A continuation along the upper
branch with decreasing values of the exciting frequency ω produced results
as depicted in Figure 4.11. In this example, a secant predictor provided a
guess close to the emanating branch. As a result, the corrector converged
to this branch. This example shows that a predictor such as equation (4.9)

or (4.13) may lead to an undesired branch switching. This *branch jumping* did not arise with a weighting of the bifurcation parameter λ [Sey84]. More examples of branch jumping will be given in Section 4.9.

The extrapolation provided by tangent and secant predictors speeds up the continuation because it makes larger steps possible. But larger steps reduce the sensitivity. In bifurcation problems it is not uncommon to have two different continuation runs tracing the same part of the branch reveal a "different" bifurcation behavior. For example, in Figure 4.15 below, a run with extrapolation based on a secant or tangent predictor obtained the indicated results, whereas the trivial predictor equation (4.12) probably would have traced the left branch with the turning point. In branching problems no predictor is generally "better" than others. Instead, in many models it might be recommended that continuations be run with another predictor strategy. We conclude that a continuation algorithm should offer the option of switching between extrapolation yes or no —that is, between an optimistic view and a more pessimistic view.

The continuation approach does not guarantee that all possible solutions can be located in a specific example. In particular, chances to detect isolated branches are small. For some problems it might be worthwhile to solve the governing equations for a random choice of parameters and to check whether an obtained solution is part of the already constructed branching diagram. In case a "new" solution is found, the branch-tracing algorithm will follow the new branch, which might turn out to be disconnected (isolated) from the already calculated branches. To some extent, a parameter study of a difficult problem is a venturous exploration.

A consequence from the above is that no particular continuation method can be recommended exclusively. The preceding sections pointed out that different alternatives for combining the basic elements (predictor, parameterization, corrector, step control) lead to a vast number of possible continuation methods. Certainly, all the continuation methods have their merits. For an engineer or scientist working on a particular problem, it may be encouraging that relatively simple continuation principles work satisfactorily for complicated problems.

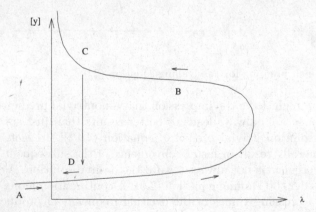

Fig. 4.12. Branch jumping I

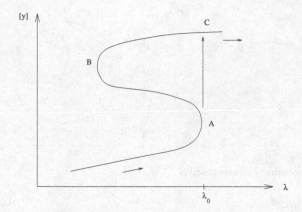

Fig. 4.13. Branch jumping II

4.9 What Else Can Happen

In the preceding section, we briefly discussed some common situations. The interpretation was straightforward. Many other phenomena are difficult to interpret. In this section we sketch some examples that encourage the reader to double-check results carefully. In particular we return to the frequent problem of undesired jumping between branches. Such branch jumping is easily overlooked.

Figure 4.12 depicts a branch whose upper part grows to infinity for λ approaching zero. Suppose a continuation starts at point A with tracing of the branch to the right (see the arrows). After rounding the turning point, the upper part B can be traced using decreasing values of λ. At C the continuation procedure should stop, because "infinite" values of **y** are not likely to make sense. Instead of stopping there, a jump to the lower part of the branch (D) may take place. This jump might be interpreted as indicating a closed path and is in a sense a "failure." A similar situation is shown in Figure 4.13. Tracing a hysteresis for increasing values of λ, one can see results that behave like a physical experiment. The continuation may not detect the turning point (at A) quickly enough and may try to calculate a solution for $\lambda > \lambda_0$. In this situation, SOLVER may fail to converge, or the iteration may jump to the other part of the branch (C). In the second case, the hysteresis phenomenon is often overlooked, especially if it is small.

In Figures 4.12 and 4.13 the question "Is the continuation procedure giving us the same branch or a different branch?" arises. This common question is illustrated in Figure 4.14. Suppose we are tracking a closed branch (left of Figure 4.14) in a counterclockwise direction starting from A (symbol +). At B we change the symbol to "o." Arriving again at parameter values in a neighborhood of A, we ask whether we are generating the same branch or a different branch (right side of Figure 4.14). In practice it can be difficult to answer this question.

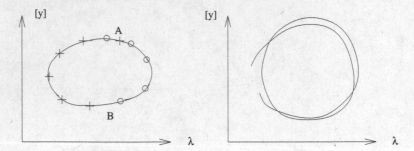

Fig. 4.14. Is the same branch traced?

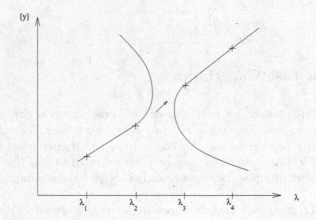

Fig. 4.15. Branch jumping III

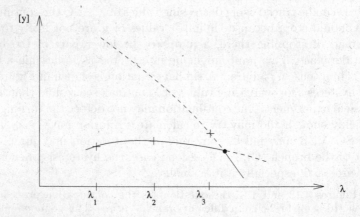

Fig. 4.16. Branch jumping IV

Let us give more examples of the same dilemma. In Figure 4.15 a situation where the continuation procedure jumps across a gap is depicted; the "+" symbols represent consecutively calculated solutions. Another example is depicted in Figure 4.16. This figure shows a transcritical bifurcation with an exchange of stability. Three consecutive solutions obtained during the continuation are marked. The instability of the solution at λ_3 may cause the algorithm to search for a bifurcation in the wrong interval $\lambda_2 \le \lambda \le \lambda_3$.

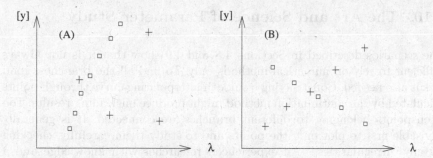

Fig. 4.17. Do the branches intersect?

Another question that is sometimes not easy to answer is whether two branches intersect (in a bifurcation point). Figure 4.17 shows two such situations. The solutions of one branch are indicated by crosses, the solutions of the other branch by squares. In such situations, clarity is achieved by investigating the branches in more detail. If there are bifurcation points on either branch with the same value of $(\mathbf{y}_0, \lambda_0)$, we can be sure that both branches intersect. An alternative is to interpolate the values of neighboring solutions on each branch and compare the results. Close to a bifurcation, the solution values of both branches must match. Situation (A) in Figure 4.17 may be a pitchfork bifurcation, and (B) might be no bifurcation at all.

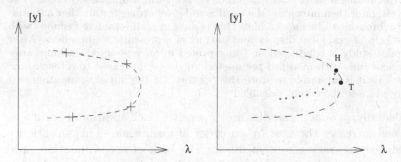

Fig. 4.18. Large steps (left) have skipped a detail (right)

Finally, we give an example of how bifurcation can be overlooked (Figure 4.18). In a first run with coarse step sizes, a turning point is passed. All the calculated stationary solutions ($+$ in the left figure) are unstable. The same branch is then recalculated with smaller step sizes. During this second run, a short branch of stable equilibria bounded by a Hopf bifurcation (H) and the turning point (T) is discovered. There is an emanating branch of stable periodic solutions. The results suddenly look completely different. This example shows how results are affected by step sizes in continuation methods. Unfortunately, one rarely knows beforehand how small a "good" step size must be.

This section has discussed diagrams without reference to specific examples. It should be stressed that the difficulties adduced are common.

4.10   The Art and Science of Parameter Study

The scenarios described in Sections 4.8 and 4.9 show that it is not always sufficient to rely on numerical methods only. To find all the branches, good luck is also needed. Constructing a smooth interpolating curve through points calculated by the continuation method might produce misleading results. Too often points belonging to different branches are connected. It is generally advisable first to plot only the points and to study them carefully, checking for any "irregularities." An experienced researcher with knowledge and a good imagination may find an interpretation that is different from that of a machine. Although it is possible to construct a "black box" that takes care of almost everything, its use may have some drawbacks. First, a fully automated parameter study carries no proof of correct interpretation. Second, it is not easy to write algorithms that decide, for instance, whether "it is the same branch" or not. A machine can easily waste time in closed-loop branches or regions that are of no interest. To make a decision, a fully automated black box would require accuracy greater than that needed when the human mind is doing the interpreting.

> It may be wise to confine automatic procedures to limited tasks such as to calculating a short continuation or carrying out a branch switching. The investigator then interprets the results and gives orders for another automatic subroutine to be carried out. This is done in an interactive fashion, with the help of graphical display and output of representative numbers. After a pilot study is conducted and interpreted in an interactive fashion, subsequent runs with modified parameters might be delegated to a black box. For this it is advisable to store the "history" of the initial semiautomated procedure to the extent possible.

Generally, parameter studies are expensive. Each additional parameter to be varied increases the cost by an order of magnitude. This situation calls for reduced accuracy requirements for the bulk of a continuation process. Intermittently, results should be confirmed by recalculating selected solutions with higher accuracy. In this way the overall cost of branch tracing can be reduced significantly.

The reader may now have the impression that a parameter study is an art, one that requires some creativity and imagination. This is partly justified, but the methods described in this chapter and in the following chapters show that a parameter study is also a science.

Exercises

Exercise 4.1
Use notation of (4.10) and assume full rank, $\text{rank}(\mathbf{f_Y}) = n$.

(a) Show that a tangent $\mathbf{z}$ to the branch is uniquely determined by

$$\mathbf{f_Y z} = 0, \quad \mathbf{z}^{tr}\mathbf{z} = 1, \quad \det\begin{pmatrix} \mathbf{f_Y} \\ \mathbf{z}^{tr} \end{pmatrix} > 0.$$

(b) The tangent vector $\mathbf{z}$ is a function of $\mathbf{f_Y}$. To show this, consider

$$G(\mathbf{f_Y}) := \begin{pmatrix} \mathbf{f_Y z} \\ \frac{1}{2}\mathbf{z}^{tr}\mathbf{z} - \frac{1}{2} \end{pmatrix},$$

and apply the implicit function theorem.

Exercise 4.2
The $(n+1)^2$ Jacobian $\mathbf{J}$ associated with the Newton method that corresponds to equation (4.14) or equations (4.21) and (4.22) is composed of matrix $\mathbf{A} := \mathbf{f_Y}$, vector $\mathbf{b} := \mathbf{f_\lambda}$, a vector $\mathbf{c}$, and a scalar ζ,

$$\mathbf{J} = \begin{pmatrix} \mathbf{A} & \mathbf{b} \\ \mathbf{c}^{tr} & \zeta \end{pmatrix}.$$

Note also that the matrix in equation (4.8) is of that "bordered" form.

(a) What are the vectors $\mathbf{c}$ and the scalars ζ in the cases of local parameterization (4.16), pseudo arclength parameterization p_3, and the parameterization (4.22)?

(b) Suppose $\mathbf{A}$ is nonsingular. Show that $\mathbf{J}$ is nonsingular if $\zeta \neq \mathbf{c}^{tr}\mathbf{A}^{-1}\mathbf{b}$.

(c) Suppose $\mathbf{A}$ has a one-dimensional null space, $\mathbf{b} \notin \text{range}\mathbf{A}$, $\mathbf{c} \notin \text{range}(\mathbf{A}^{tr})$. Show that $\mathbf{J}$ is nonsingular.

Exercise 4.3
Consider the linear system of equations

$$\mathbf{J}\begin{pmatrix} \mathbf{z} \\ \xi \end{pmatrix} = \begin{pmatrix} \mathbf{r} \\ \gamma \end{pmatrix}$$

with the matrix $\mathbf{J}$ from Exercise 4.2 involving $\mathbf{c}$, $\mathbf{b}$, ζ, and some given right-hand side $(\mathbf{r}, \gamma)^{tr}$. Assume $\mathbf{A}$ is decomposed into $\mathbf{A} = \mathbf{LU}$.

(a) Show that a decomposition of $\mathbf{J}$ into

$$\mathbf{J} = \begin{pmatrix} \mathbf{L} & \mathbf{0} \\ \mathbf{d}^{tr} & 1 \end{pmatrix}\begin{pmatrix} \mathbf{U} & \mathbf{a} \\ \mathbf{0}^{tr} & \delta \end{pmatrix}$$

is obtained by

$$\mathbf{U}^{tr}\mathbf{d} = \mathbf{c}, \quad \mathbf{La} = \mathbf{b}, \quad \delta = \zeta - \mathbf{d}^{tr}\mathbf{a}.$$

(b) Derive an algorithm to calculate the solution $(\mathbf{z}, \xi)$ of the system of linear equations.

Exercise 4.4
Programming a continuation algorithm based on the secant predictor equation (4.13) is easy when three arrays are used. Design a continuation algorithm requiring only two arrays.

Exercise 4.5
In designing a continuation algorithm, one does not allow the step length to become arbitrarily large. Hence, the absolute value of a maximal step size should be prescribed as input data. The following questions may stimulate one to think about a practical realization:

(a) Does a limitation in the step length affect the choice of k in the local parameterization criteria equation (4.17) or (4.18)?
(b) Design a step-length control that relates the bound on the step length to the λ component only.

Exercise 4.6
Design a continuation algorithm implementing elements of your choice. The following should be implemented: a predictor, a parameterization strategy, a corrector ("CALL SOLVER" suffices), and a step-length control. [Hint: A simple but successful algorithm is obtained by combining equations (4.13), (4.16), and (4.18) for k and k_1, and (4.23).]

5 Calculation of the Branching Behavior of Nonlinear Equations

Equipped now with some knowledge about continuation, we assume that we are able to trace branches. We take for granted that the entire branch can be traced, provided one solution on that branch can be found. In this chapter we address problems of locating bifurcation points and switching branches. Essential ideas and methods needed for a practical bifurcation and stability analysis are presented.

The basic principles discussed here apply to both ODE boundary-value problems and systems of algebraic equations. Although some of the ideas were first introduced for the more complicated case of boundary-value problems, we begin with the simpler situation of systems of algebraic equations

$$\mathbf{f}(\mathbf{y}, \lambda) = \mathbf{0} \,. \tag{5.1}$$

First we concentrate on varying one bifurcation parameter (λ) only. The results can be used in analyzing multiparameter models. In Section 5.9, models with two bifurcation parameters are treated. Methods for boundary-value problems are deferred to Chapter 6.

5.1 Calculating Stability

In this section we interpret equation (5.1) as an equation that defines the equilibria of an ODE system $\dot{\mathbf{y}} = \mathbf{f}(\mathbf{y}, \lambda)$. Consequently, gain and loss of stability are crucial aspects of bifurcation. The principles of local stability were introduced in Chapter 1. Recall that the n^2 Jacobian matrix $\mathbf{J} = \mathbf{f_y}$ is evaluated at the solution $(\mathbf{y}, \lambda)$, the stability of which is to be investigated. The n eigenvalues

$$\mu_k(\lambda) = \alpha_k(\lambda) + i\beta_k(\lambda) \quad (k = 1, \ldots, n) \tag{5.2}$$

of the Jacobian vary with λ. Linearized or local stability is ensured, provided all real parts α_k are negative. The basic principles are straightforward. But carrying them out by calculating the Jacobian $\mathbf{J}$ and the eigenvalues μ is not as simple.

R. Seydel, *Practical Bifurcation and Stability Analysis*,
Interdisciplinary Applied Mathematics 5, DOI 10.1007/978-1-4419-1740-9_5,
© Springer Science+Business Media, LLC 2010

5.1.1 Accuracy Problems

Let us discuss first how the eigenvalues are affected by the accuracy in evaluating the Jacobian $\mathbf{J}$. This problem splits into two subproblems. The first is
how the first-order partial derivatives are evaluated; the second is *where* they
are evaluated—that is, for which argument(s) $(\mathbf{y}, \lambda)$. The question of how to
evaluate $\mathbf{J}$ was discussed in Section 1.5.1. Recall that $\mathbf{J}$ can be approximated
by numerical differentiation, updated by rank-one approximations, or evalua-
ted analytically. The latter produces the most accurate results but is usually
limited to simple examples. Often the solution algorithm ("SOLVER") for
equation (5.1) is based on a quasi-Newton method. Because an approximate
Jacobian is calculated on each step of a quasi-Newton method, we have an
approximation $\bar{\mathbf{J}}$ available on the last step. The quality of this Jacobian is
strongly affected by the history of the iteration. The initial guess and the
rank-one update strategy both have significant influence on $\bar{\mathbf{J}}$. For instance,
inaccuracies in $\bar{\mathbf{J}}$ are caused by the following factors:

⋆ SOLVER needs only one iteration. $\bar{\mathbf{J}}$ is then evaluated at the initial
  guess, which may not be close to a solution of $\mathbf{f}(\mathbf{y}, \lambda) = \mathbf{0}$.
⋆ The continuation step length is "large." Rank-one updates to $\bar{\mathbf{J}}$ converge more slowly to the true Jacobian than the iterates $(\mathbf{y}^{(\nu)}, \lambda^{(\nu)})$
  converge to the solution $(\mathbf{y}, \lambda)$. In general, the quality of $\bar{\mathbf{J}}$ deteriorates
  with increasing step length.
⋆ Predictor equation (4.12b) is used instead of (4.12a). Hence, the initial approximation for the Newton method "leaves" the branch. The
  resulting Jacobians $\bar{\mathbf{J}}$ are frequently worse than when equation (4.12a)
  is used.

The question is how useful eigenvalues of a poor approximation $\bar{\mathbf{J}}$ of $\mathbf{J} = \mathbf{f_y}$
can be. Experience shows that the eigenvalues based on $\bar{\mathbf{J}}$ are often good
enough. Close to critical situations such as loss of stability, however, the eigenvalues and thus $\bar{\mathbf{J}}$ must be calculated more accurately. Here the real part
of a critical eigenvalue is close to zero, and the relative error may be unaccep-
table even though the absolute error might be tolerable. Occasionally, smaller
continuation steps improve the accuracy of $\bar{\mathbf{J}}$. Higher demands for accuracy
call for an additional approximation of the Jacobian by numerical differentiation at the final Newton iterate. The advantage in efficiency gained by
approximating $\mathbf{J}$ by rank-one updates can be lost when accurate eigenvalues
are required.

5.1.2 Applying the QR Method

Software mentioned in Section 1.5.3 and Appendix A.9 is available for calculating the eigenvalues. The widely used QR method calculating all eigenvalues
of $\bar{\mathbf{J}}$ is relatively expensive. In fact, the cost of carrying out a Newton step
and of evaluating the eigenvalues by the QR method are of the same order

of magnitude $O(n^3)$. Hence, for large systems (n large) as it typically arises when PDEs are discretized into large ODE systems, the use of the QR method is prohibitive. For such problems only a few selected eigenvalues are calculated; this will be described at the end of this section. One alternative that often provides information about stability is simulation of the transient behavior by numerically integrating the ODE system associated with equation (5.1). This simulation was described in Section 1.3 (Algorithm 1.8). The disadvantage is a lack of knowledge how to choose z and t_f. The advantage is a more global character of the results. For the special case of symmetric Jacobians an inexpensive method for approximating the eigenvalue with real part closest to zero has been proposed in [Sim75]. In case the QR method is applied, costs can be saved when stability is not checked for each solution but rather for selected solutions only.

TABLE 5.1.   Table of eigenvalues.

Parameter	μ_1	μ_2	μ_3	μ_4
λ_1	0.82	$-5.1+0.4$i	$-5.1-0.4$i	-17.3
λ_2	0.02	$-5.9+0.43$i	$-5.9-0.43$i	-17.1
λ_3	-0.51	$-6.5+0.45$i	$-6.5-0.45$i	-16.8
$\vdots$		$\vdots$		

We now show how eigenvalues monitored during a continuation procedure can be used to draw practical conclusions about bifurcation behavior. Consider Table 5.1 for $n = 4$ equations, in which the four eigenvalues μ_1, μ_2, μ_3, and μ_4 are listed for consecutive values of the branching parameter λ. For clarity, the eigenvalues in the table are ordered in the same way for all λ, although this situation cannot be expected from the QR method in "raw" output data. The QR method frequently changes the positions of the eigenvalues, causing difficulties in tracing "the same" eigenvalue. As the table shows, one eigenvalue changes the sign of its real part for a value of λ between λ_2 and λ_3 (first column). The question is whether the crossing of the imaginary axis is real or complex. That is, the investigator must decide whether a stationary bifurcation or a Hopf bifurcation has been passed. Although the three calculated values of μ_1 are all real, one might expect a tiny subinterval between λ_2 and λ_3 with complex values of μ_1. This situation is difficult to survey when there are many eigenvalues (n large). In case of uncertainty, it is advisable to retrace the branch in the critical subinterval, using a small step length. The specific situation illustrated by the table above is easy to resolve. In order to become complex, the eigenvalue μ_1 needs a real-valued partner with which to coalesce. The numbers taken by μ_2 and μ_3 obviously rule out μ_2 and μ_3 as candidates; they are partners of each other. The remaining candidate μ_4 does not show any desire to join with μ_1. Thus, the example clearly indicates that a real crossing by μ_1 of the imaginary axis is as certain as numerical results can be.

5.1.3 Alternatives

As mentioned above, the QR method is too costly for large n. In most applications of a stability analysis, the QR method delivers more than is needed. All n eigenvalues are calculated, and the accuracy is high. In a context where the only concern is local stability, we mainly need to know the *sign of the eigenvalue with maximum real part*. In most cases this information can be easily obtained by the inverse power method, see equations (1.23) and (1.24). As a typical situation, assume that during the course of a continuation the eigenvalue μ with maximum real part together with its eigenvector $\mathbf{z}$ have been calculated for each j and $(\mathbf{y}_j, \lambda_j)$. For the path (μ, λ) of the eigenvalue in the three-dimensional $(\mathrm{Re}(\mu), \mathrm{Im}(\mu), \lambda)$-space, the predictor-corrector principles discussed in Chapter 4 apply. Hence, a secant predictor can be used to deliver an approximation $\bar{\mu}$. To this end replace in equation (4.13) $\mathbf{y}$ by μ. As an initial guess $\mathbf{z}^{(0)}$ for the inverse iteration of equation (1.23), use the eigenvector of the last solution, which corresponds to the trivial predictor equation (4.12a). Note that the matrix $\mathbf{A} - \bar{\mu}\mathbf{I}$ in equation (1.23) is complex for complex $\bar{\mu}$. Hence, a complex factorization is required. After a few iterations, limits $\mathbf{z}$ and $\tilde{\mu}_1$ are obtained, and equation (1.24) gives the required μ.

The above strategy is not complete; several complications arise. First, an initialization is needed that calculates μ for the first solution. Second, a test is required to make sure that μ is the eigenvalue with maximum real part. The eigenvalue that is traced by the predictor-corrector principle can lose its dominance. As mentioned in Section 1.5.3, there are iterative methods that approximate the eigenvalue *largest in magnitude*, which is a different dominance than that of the dominant real part required for our stability. By means of the *generalized Cayley transform*

$$\mathbf{C}(\mathbf{A}) := (\mathbf{A} - a_1\mathbf{I})^{-1}(\mathbf{A} - a_2\mathbf{I}) \tag{5.3}$$

for real a_1 and a_2 $(a_1 > a_2,\ a_1$ not eigenvalue), the eigenvalues μ of $\mathbf{A}$ are mapped to the eigenvalues θ of $\mathbf{C}$ dominant in magnitude (Exercise 5.1). Applying Arnoldi iteration, one calculates the θ_1 dominant in magnitude and then the eigenvalue μ_1 dominant in real part via

$$\mu = \frac{a_1\theta - a_2}{\theta - 1};$$

see [GaMS91], [Neu93b]. In this way the dominant eigenvalue μ of the Jacobian of the first solution is calculated. The second difficulty, a change in dominance, is checked by a *verification step* [Neu93b], which only needs to be performed if $\mathrm{Re}(\mu) < 0$. Assume that μ is a candidate for the dominant eigenvalue calculated by a secant predictor and an inverse iteration. Then the specific Cayley transformation

$$\mathbf{C} := \mathbf{A}^{-1}(\mathbf{A} - 2\operatorname{Re}(\mu)\mathbf{I}) \tag{5.4}$$

maps any eigenvalue with larger real part than $\operatorname{Re}(\mu)$ outside the unit circle of the complex plane. A few Arnoldi iterations show where the convergence drifts, answering whether μ is dominant or not. Using such methods, [Neu93b] has shown that the costs of calculating the stability is less than the costs of calculating the solutions for all n.

5.2 Bifurcation Test Functions

During branch tracing a bifurcation must not be overlooked. Suppose that a chain of solutions

$$(\mathbf{y}^1, \lambda_1),\ (\mathbf{y}^2, \lambda_2),\ldots$$

to $\mathbf{f}(\mathbf{y}, \lambda) = \mathbf{0}$ has been approximated and that a bifurcation point is straddled by $(\mathbf{y}^{j-1}, \lambda_{j-1})$ and $(\mathbf{y}^j, \lambda_j)$. How can we recognize from the data from these solutions whether a bifurcation $(\mathbf{y}_0, \lambda_0)$ is close? The required information is provided by a *test function* $\tau(\mathbf{y}, \lambda)$, which is evaluated during the branch tracing. A bifurcation is indicated by a zero of τ—that is, a bifurcation test function (or branching test function) satisfies the property

$$\tau(\mathbf{y}_0, \lambda_0) = 0 \tag{5.5}$$

and is continuous in a sufficiently large interval that includes λ_0. A change of sign of τ at $(\mathbf{y}_0, \lambda_0)$ will be convenient. The application of τ is illustrated in Figure 5.1. The upper part of the figure shows several solutions calculated on one branch. During these computations, the values of a bifurcation test function are monitored (lower part of figure). Because it is unlikely to hit the zero (or the bifurcation, respectively) exactly, it makes sense to check for a change of sign of τ. A bifurcation is most likely passed if the criterion

$$\tau(\mathbf{y}^{j-1}, \lambda_{j-1})\,\tau(\mathbf{y}^j, \lambda_j) < 0 \tag{5.6}$$

is satisfied. Figure 5.1 depicts an ideal situation with clearly isolated zeros of τ; step length $|\lambda_{j-1} - \lambda_j|$ and accuracy of τ together ensure a smooth interpolating curve. It is difficult to verify a bifurcation via criterion (5.6) when step lengths are too large or the computed τ exhibits artificial oscillations that are the result of inaccurate evaluation.

Some examples of test functions are immediately obvious, others are defined in an artificial way. We delay a discussion of the latter until later. A natural choice of τ is the maximum of all real parts of the eigenvalues $\alpha_k + \mathrm{i}\beta_k$ of the Jacobian $\mathbf{J}$,

$$\tau := \max\{\alpha_1, \ldots, \alpha_n\}. \tag{5.7}$$

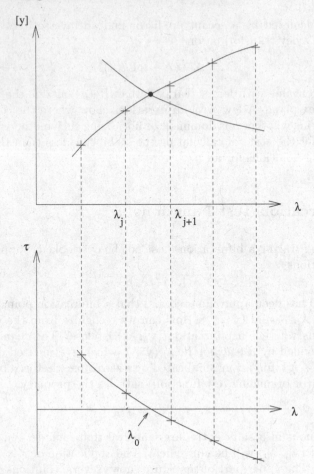

Fig. 5.1. Values of a test function τ straddling a zero

This choice has the advantage of being physically meaningful because $\tau < 0$ guarantees local stability. τ is continuous, provided $\mathbf{f}(\mathbf{y}, \lambda)$ is continuously differentiable. However, τ from equation (5.7) does not need to be smooth.

Example 5.1   Electric Power Generator
For an example of the schematic diagram of Figure 5.1, we return to Example 3.8. In Figure 5.2 the correspondence between branch and test function $\tau(\lambda)$ is shown for the model of an electric power system. Clearly, the test function indicates the location of the turning point.   □

The test function equation (5.7) does not signal turning points in which two unstable half-branches coalesce. The same holds for bifurcations that connect unstable branches only. At such bifurcation points, all real parts are negative, and hence τ from equation (5.7) is nonzero. This minor problem

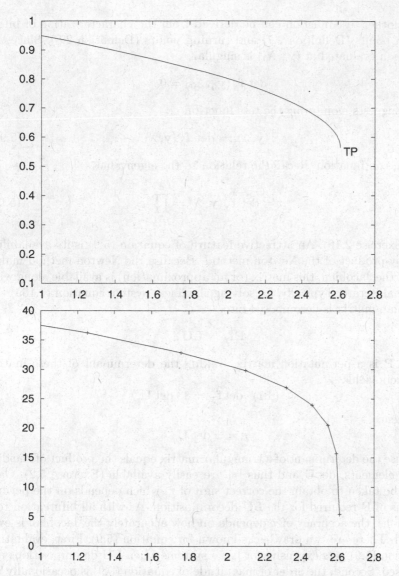

Fig. 5.2. Example 5.1 / 3.8; top: branch with turning point TP (corresponds to Figure 3.25), load voltage V versus load λ; bottom: test function $\tau(\lambda)$ for TP

can be overcome by modifying equation (5.7) to

$$\tau = \alpha_k \quad \text{with } |\alpha_k| = \min\{|\alpha_1|, \dots |\alpha_n|\}; \tag{5.8}$$

this was used in [Sim75]. A strong advantage of equation (5.7), compared with the test function to be discussed next, is that it also indicates Hopf bifurcations. Problems concerning accuracy and cost in evaluating this τ were investigated in the preceding section.

Another test function can be derived from the characterization of bifurcation points (Definition 2.7) and turning points (Definition 2.8). Since the Jacobian evaluated at $(\mathbf{y}_0, \lambda_0)$ is singular,

$$\det \mathbf{f_y}(\mathbf{y}_0, \lambda_0) = 0\,.$$

This suggests monitoring the test function

$$\tau(\mathbf{y}, \lambda) := \det \mathbf{f_y}(\mathbf{y}, \lambda) \qquad\qquad (5.9a)$$

during continuation. Recall the relation to the eigenvalues

$$\tau = \det \mathbf{f_y}(\mathbf{y}, \lambda) = \prod_{j=1}^{n} \mu_j \qquad\qquad (5.9b)$$

(see Exercise 2.10). An attractive feature of equation (5.9) is its availability as a by-product of the Newton method. Because the Newton method makes use of the Jacobian, this matrix (or an approximation) is available along with the final iterate. Typically, for solving the linear system equation (1.15a) the Jacobian matrix is decomposed into

$$\mathbf{Pf_y} = \mathbf{LU}\,.$$

Here, $\mathbf{P}$ is a permutation matrix. Taking the determinant of these matrix products yields

$$(\pm 1)\cdot \det \mathbf{f_y} = 1 \cdot \det \mathbf{U}$$

and hence

$$\tau = \pm \det \mathbf{U}\,.$$

Because the determinant of a triangular matrix equals the product of the diagonal elements, $\det \mathbf{U}$ and thus $|\tau|$ are easily available (Exercise 5.2). Care must be taken to obtain the correct sign of τ, which depends on the permutations of $\mathbf{P}$ required for the $\mathbf{LU}$ decomposition. As with all bifurcation test functions, the accuracy of τ depends on how accurately the Jacobian is evaluated. There are two drawbacks known for equation (5.9). First, evaluating equation (5.9) gets expensive for large systems when $\mathbf{LU}$ decompositions are not used. Second, the order of magnitude of equation (5.9) is occasionally inconvenient. That is, a loss of significance due to scaling problems may occur. In fact, rescaling all components f_i by multiplying with a factor ζ leads to a factor ζ^n in the determinant. Hence, it can be difficult to decide whether a particular value of this τ is "large" or "small." Alternative test functions, which apparently do not suffer from these disadvantages, were introduced in [Abb78] and [Sey79a]; the latter will be discussed in Section 5.4.3. A test function specifically for Hopf bifurcation points is

$$\tau_{\text{Hopf}} := \prod_{1 \le k < j \le n} (\mu_j + \mu_k)\,.$$

Other test functions are based on Hurwitz determinants [Khi90], [Yu05], or on bialternate products [GuMS97], [Kuz98].

5.3 Indirect Methods for Calculating Bifurcation Points

In this and the following section, we report on general methods for calculating bifurcation points, deferring special methods for special purposes to Section 5.7. For reasons that will become apparent later, we distinguish between indirect and direct methods.

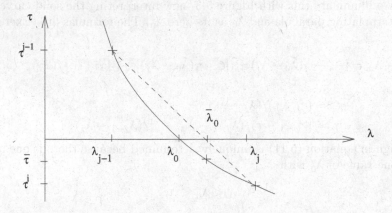

Fig. 5.3. Approximating a zero of τ

5.3.1 Interpolation with Test Functions

Standard methods for calculating bifurcation points benefit from the data obtained during the continuation procedure, completed by values of a bifurcation test function. With τ_j denoting the value of a test function evaluated at $(\mathbf{y}^j, \lambda_j)$, the data

$$(\mathbf{y}^j, \lambda_j, \tau_j), \qquad j = 1, 2, \ldots$$

are available. Assume that a bifurcation has been located in the subinterval $\lambda_{j-1} < \lambda < \lambda_j$ by noticing a change of the sign of τ [see criterion (5.6)]. It is a straightforward calculation to interpolate the data in order to obtain an approximation of the zero of τ (see Figure 5.3). To this end, assume that τ has exactly one zero λ_0 in the subinterval under consideration. Then an approximation $\bar{\lambda}_0$ is obtained by calculating the zero of the interpolating straight line (dashed in Figure 5.3). The formula is

$$\bar{\lambda}_0 = \lambda_{j-1} + (\lambda_j - \lambda_{j-1}) \tau_{j-1}/(\tau_{j-1} - \tau_j). \qquad (5.10)$$

In many applications this approximation is close enough to λ_0, because the continuation step length $|\lambda_j - \lambda_{j-1}|$ is typically small when bifurcations are investigated. The formula of equation (5.10) is well defined since we are based on $\tau_{j-1}\,\tau_j < 0$.

In some cases it is necessary to improve the approximation or get information on the error $|\bar\lambda_0 - \lambda_0|$. In order to achieve this objective, the data of at least one additional solution are required. Three solutions with three values of τ enable the approximation of τ by an interpolating parabola. In order to save work, one may take the data obtained at λ_{j-2}. The results will become better when an additional solution for a value of λ close to the expected bifurcation is approximated. The natural candidate is $\lambda = \bar\lambda_0$. Solving $\mathbf{f}(\mathbf{y}, \lambda) = \mathbf{0}$ for this parameter value provides a third value $\bar\tau$ of the test function. We illuminate this with Figure 5.3, now interpreting the solid curve as the interpolating parabola and λ_0 as its zero λ_0^*. The formulas are (Exercise 5.3)

$$\lambda_0^* - \bar\lambda_0 = \{-\zeta - \gamma(\bar\lambda_0 - \lambda_j) \pm ([\zeta + \gamma(\bar\lambda_0 - \lambda_j)]^2 - 4\gamma\bar\tau)^{1/2}\}/(2\gamma) \quad (5.11)$$

with

$$\zeta := (\tau_j - \bar\tau)/(\lambda_j - \bar\lambda_0)\,,$$
$$\gamma := [(\tau_{j-1} - \bar\tau)/(\lambda_{j-1} - \bar\lambda_0) - \zeta]/(\lambda_{j-1} - \lambda_j)\,.$$

The sign in equation (5.11) is uniquely determined because there is one and only one value of λ_0^* with

$$\lambda_{j-1} < \lambda_0^* < \lambda_j\,.$$

The relation equation (5.11) is used in two ways. First, this formula provides a corrector to be added to λ_0 in order to obtain a better approximation. Second, assuming that the second approximation λ_0^* is by far better than the first approximation $\bar\lambda_0$,

$$|\lambda_0^* - \lambda_0| \ll |\bar\lambda_0 - \lambda_0|\,,$$

equation (5.11) offers an estimate of the error of the first approximation $\bar\lambda_0$. To this end, substitute λ_0 for λ_0^*. In this way, no error estimate for $\lambda_0 - \lambda_0^*$ is obtained. But finding a small value of $|\lambda_0^* - \bar\lambda_0|$ by (5.11) suggests that $|\lambda_0 - \lambda_0^*|$ is much smaller. On the other hand, a large value of (5.11) ("large" compared with the interval length $|\lambda_j - \lambda_{j-1}|$) serves as a hint that the accuracy is not sufficient. In such a case the continuation steps have usually been too large. The procedure can be repeated, then resembling the Muller method for calculating zeros [StB80]. For the topic of *root finding* of a scalar function $\tau(\lambda)$ consult [PrFTV89] or [KaMN89]. Convergence problems due to an almost singular Jacobian may occur when $\mathbf{f}(\mathbf{y}, \lambda) = \mathbf{0}$ is solved for a λ too close to λ_0.

After an approximation $\bar\lambda_0$ (or λ_0^*, respectively) has been obtained, the next step is to approximate the corresponding solution $\mathbf{y}$ of $\mathbf{f}(\mathbf{y}, \lambda) = \mathbf{0}$. This can be done most economically by applying linear interpolation also to

$\mathbf{y}$, rather than by solving the equation $\mathbf{f}(\mathbf{y}, \lambda) = \mathbf{0}$. The approximation is accomplished by

$$\bar{\mathbf{y}}_0 = \mathbf{y}^{j-1} + (\mathbf{y}^j - \mathbf{y}^{j-1}) \tau_{j-1} / (\tau_{j-1} - \tau_j). \tag{5.12}$$

The approximation $(\bar{\mathbf{y}}_0, \bar{\lambda}_0)$ obtained in this way is accurate enough in most practical applications. The cost of the interpolation formulas is negligible, compared with the cost of solving $\mathbf{f}(\mathbf{y}, \lambda) = \mathbf{0}$. Note that equations (5.10) and (5.12) are also valid for extrapolation: a guess on the next bifurcation is obtained when equation (5.10) is evaluated even before τ changes sign. This use of equation (5.10) is illustrated in Figure 5.4. The information that a λ_0 is close can be used in advance for deciding whether the continuation step length is to be reduced. In general, the accuracy of the guess λ_0 improves as the bifurcation is approached. There are, however, exceptions to this rule: note that the criterion $\tau_{j-1}\tau_j < 0$ is not met when equation (5.10) is used for extrapolation. Hence, the guess λ_0 may suffer severely from inaccurate τ values. In particular, an evaluation of equation (5.10) does not make sense when $\tau_{j-1} \approx \tau_j$.

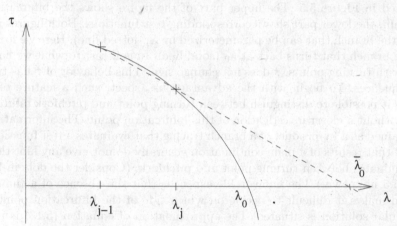

Fig. 5.4. Extrapolation on a zero of τ

5.3.2 Approximation of Turning Points

So far, we have obtained approximations of a bifurcation point on a part of a branch that can be parameterized by λ. In a similar fashion, approximations of turning points can be based on τ. Before going into detail, we stress that there are more economical methods that do not require test functions. We postpone this for a moment and briefly outline the application of test functions. Because at a turning point a bifurcation test function reflects the behavior of the branch, the curve (τ, λ) also turns back in λ_0. This means that τ crosses the λ-axis at λ_0 perpendicularly, and justifies modeling the shape of τ locally by the parabola

$$\lambda - \lambda_0 = \xi \tau^2 .$$

This ansatz function allows computing an approximation $\bar{\lambda}_0$ based on only two solutions $(\mathbf{y}^{j-1}, \lambda_{j-1})$, $(\mathbf{y}^j, \lambda_j)$ with associated values of τ_{j-1} and τ_j. As is easily seen (Exercise 5.7), the relation

$$\bar{\lambda}_0 = (\lambda_j \tau_{j-1}^2 - \lambda_{j-1} \tau_j^2)/(\tau_{j-1}^2 - \tau_j^2) \tag{5.13}$$

establishes an approximation $\bar{\lambda}_0$ of the value λ_0 of a turning point.

> Note that an evaluation of equation (5.13) should be avoided if $|\tau_{j-1}| \approx |\tau_j|$ holds. For such values of τ_{j-1} and τ_j, equation (5.13) is not well defined; $\bar{\lambda}_0$ then reacts sensitively to inherent errors in τ. If $|\tau_{j-1}|$ and $|\tau_j|$ are close, one set of data (say, $\mathbf{y}^j, \lambda_j, \tau_j$) may be replaced by corresponding values of another solution, aiming at $|\tau_j| < \frac{1}{2}|\tau_{j-1}|$. Based on the approximation of equation (5.13), approximate values for $\bar{\mathbf{y}}_0$ can be obtained (Exercise 5.7).

Applying the above approximation formula is easy, provided the bifurcation manifests itself by a change of sign of the test function. But there are bifurcations that do not do us this favor. Consider a pitchfork bifurcation as depicted in Figure 5.5. The upper part of the figure shows the bifurcation diagram, the lower part shows corresponding test functions. Both figures include the branch that can be parameterized by λ (dotted line). Here we focus on the branch that turns back at λ_0 (solid line). In contrast to what we have seen with turning points, τ does not change sign. This behavior of τ has two consequences. To begin with the advantageous aspect, such a feature of τ makes it possible to distinguish between turning point and pitchfork bifurcation without a closer investigation of this bifurcation point. The information is obtained as a by-product of a branch tracing that evaluates a test function. Recall that results of a plain continuation generally do not give any hint that discriminates between turning point and pitchfork. (Consider the dots in Figure 5.5 unknown.) The unfavorable aspect is that the absence of a change of sign makes it difficult to decide on which side of the bifurcation point a particular solution is situated. The approximation of equation (5.13) is not valid here. One must instead prefer equation (5.10).

Now, we address the alternative, where turning points are approximated most efficiently without evaluating a particular test function. Instead, one can exploit the relation $d\lambda/d\mathbf{y} = 0$ that characterizes a turning point as a local maximum or minimum in λ. Transferring this feature to an interpolating inverse polynomial leads to simple formulas based on $(\mathbf{y}, \lambda)$ values only (see Exercise 5.9).

5.3.3 Inverse Interpolation

The approximation formulas introduced so far are convenient in that they involve only two or three solutions $(\mathbf{y}^j, \lambda_j)$ together with τ_j. We now turn to approximating bifurcation points when more solutions are involved. One

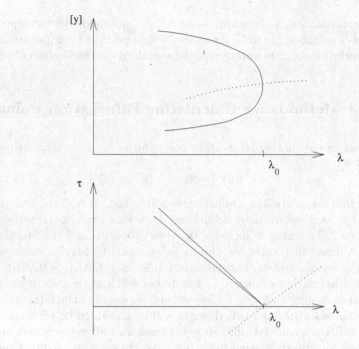

Fig. 5.5. Test function at a pitchfork, schematically

either is content with the solutions available from continuation, or one calculates additional solutions close to the bifurcation point. A powerful tool that
can be applied is *inverse interpolation.* Assume we have four to six solutions
to $\mathbf{f}(\mathbf{y}, \lambda) = \mathbf{0}$ available with corresponding values of τ in a range where τ
changes sign and where its derivative with respect to λ does not vanish. For
good results, choose the solutions in such a way that about half have positive
values of τ and the remaining have negative values of τ. Then consider the
interpolating polynomial $\lambda = p(\tau)$ with

$$\lambda_j = p(\tau_j)$$

for the chosen data λ_j, τ_j. Evaluating this polynomial for $\tau = 0$ gives an
approximation for λ_0 that is generally extremely accurate, provided the values
of τ_j are accurate enough and the step lengths are not too large. The practical
evaluation of $p(0)$ is done by applying Neville's method. (This recursion is
described in any textbook of numerical analysis.) It is worthwhile to append
a corresponding algorithm to any continuation method that provides values
of a test function (see Exercise 5.8). The simple interpolations of equations
(5.10) and (5.11) can be considered special cases of inverse interpolation.

It has become customary to call methods that approximate λ_0 by interpolation "indirect methods." Indirect methods are based mostly on data

generated by continuation. Before addressing "direct" methods, let us emphasize that indirect methods are to be recommended in practical computations.
With small overhead, they are appropriate for standard demands of accuracy.

5.4 Direct Methods for Calculating Bifurcation Points

A *direct method* for calculating bifurcation points formulates a suitable equation

$$\mathbf{F}(\mathbf{Y}) = \mathbf{0} \qquad (5.14)$$

in such a way that its solution is a bifurcation point. To this end, the original
system $\mathbf{f}(\mathbf{y}, \lambda) = \mathbf{0}$ is enlarged by additional equations that characterize a
bifurcation point. The vector $\mathbf{Y}$ includes $(\mathbf{y}, \lambda)$ as subvector and the function
$\mathbf{F}$ includes $\mathbf{f}$. A direct method solves the single equation (5.14), for example,
by just calling standard software of numerical analysis ("CALL SOLVER").
This might raise hopes that eventually a researcher will first be able to calculate all bifurcation points and then decide whether to connect the bifurcation
points with branches. Although this thought will be shown to be an illusion,
it illustrates well the principal difference between an indirect method and
a direct method for approximating branch points. By concept, a direct method is an iterative process because it exploits the iterative approach of the
numerical method in SOLVER.

By the definition of a test function $\tau(\mathbf{y}, \lambda)$, the enlarged system of equations

$$\mathbf{F}(\mathbf{Y}) := \mathbf{F}(\mathbf{y}, \lambda) = \begin{pmatrix} \mathbf{f}(\mathbf{y}, \lambda) \\ \tau(\mathbf{y}, \lambda) \end{pmatrix} = \mathbf{0} \qquad (5.15)$$

is suitable for calculating bifurcation points. Approaches based on equation
(5.15) were proposed by [Riks72], [Kub75], [Abb78], [KuM79], [PöS82]. As an
example, we apply the determinant equation (5.9). This leads to the extended
system

$$\begin{pmatrix} \mathbf{f}(\mathbf{y}, \lambda) \\ \det \mathbf{f}_{\mathbf{y}}(\mathbf{y}, \lambda) \end{pmatrix} = \mathbf{0}. \qquad (5.16)$$

Both turning points and stationary bifurcation points are among the solutions. Solving equation (5.16) by standard Newton method is restricted to
small n because the determinant is required analytically, which is difficult to
obtain.

> Efficient methods based on equations (5.15) and (5.16) have been pro-
> posed by Abbott [Abb77], [Abb78]. His approach is based on the method of
> Brown and Brent [Bre73], which turns out to be suitable for equations [see
> equation (5.15)] with the feature that $\mathbf{f}_{\mathbf{y}}$ is available analytically whereas
> a derivative of τ is not available. In [Abb78], three choices of the scalar
> equation $\tau(\mathbf{y}, \lambda) = 0$ are presented; one of them is equation (5.16).

5.4.1 The Branching System

Another prototype of direct methods was introduced by [Sey77]. We describe
the basic method in some detail for two reasons. First, a specific aspect of the
method will give rise to a class of branch-switching methods and to new test
functions. Second, this method can be modified in various ways to adapt its
performance to special applications. This method has served as the basis for
many direct methods that have been used to calculate bifurcations. Special
versions for special purposes will be discussed in Section 5.7; the present
section is confined to the original version of [Sey77], [Sey79c].

The linearization of $\mathbf{f}(\mathbf{y}, \lambda) = \mathbf{0}$ about $\mathbf{y}$ is the system of n equations

$$\mathbf{f_y}(\mathbf{y}, \lambda)\,\mathbf{h} = \mathbf{0}. \tag{5.17}$$

The vector $\mathbf{h}$ consists of n components. We know (see Section 2.5) that the
Jacobian matrix $\mathbf{J}(\mathbf{y}, \lambda) = \mathbf{f_y}(\mathbf{y}, \lambda)$ is singular if evaluated at a bifurcation
point $(\mathbf{y_0}, \lambda_0)$. An equivalent criterion for $\mathbf{f_y}$ being singular is that the linearization equation (5.17) has a nontrivial solution $\mathbf{h_0} \neq \mathbf{0}$. The feature "$\mathbf{h}$
nontrivial" can be written as an equation $\zeta(\mathbf{h}) = \xi$, where ζ is a functional
that assigns a nonzero value ξ to $\mathbf{h}$. The argument $(\mathbf{y}, \lambda)$ in equation (5.17)
must solve $\mathbf{f}(\mathbf{y}, \lambda) = 0$, which can be attached to equation (5.17). It has been
shown that the three characteristic features $\mathbf{f} = \mathbf{0}$, $\mathbf{f_y}\mathbf{h} = \mathbf{0}$, $\zeta(\mathbf{h}) = \xi$ can be
coupled to one large system with as many equations as unknowns:

$$\begin{pmatrix} \mathbf{f}(\mathbf{y}, \lambda) \\ \mathbf{f_y}(\mathbf{y}, \lambda)\,\mathbf{h} \\ \zeta(\mathbf{h}) - \xi \end{pmatrix} = \mathbf{0}. \tag{5.18}$$

For each $(\mathbf{y_0}, \lambda_0)$ that makes $\mathbf{f_y}$ singular there is an $\mathbf{h_0}$ such that the combined
vector $(\mathbf{y_0}, \lambda_0, \mathbf{h_0})$ solves the extended system of equation (5.18). Hence, the
relevant bifurcation points are among the solutions of equation (5.18). In
fact, a large class of bifurcation points can be calculated by making use of
equation (5.18). The functional $\zeta(\mathbf{h})$ is simply chosen to be a scalar product
$\zeta(\mathbf{h}) = \mathbf{r}^{tr}\mathbf{h}$ with a suitably chosen vector $\mathbf{r}$. One possible choice is $\mathbf{r} = \mathbf{h}$,
which amounts to $\|\mathbf{h}\| = 1$. This functional was used in the early days of
extended systems [Sey77]. The practical choice of $\mathbf{r}$ is a unit vector, $\mathbf{r} = \mathbf{e}_k$.
That is, $\zeta(\mathbf{h}) = \xi$ requires the kth component of $\mathbf{h}$ to take a certain prescribed
value ξ,

$$h_k = \xi. \tag{5.19}$$

Because equation (5.17) can be multiplied by any nonzero number without
changing the quality of the linearization, we are free to impose any nonzero
value of ξ. For the sake of a reasonable scaling of variables, we choose $\xi = 1$.
To transform the extended system of equation (5.18) into the form of equation
(5.14), define the vector $\mathbf{Y}$ of $2n + 1$ components by

$$\mathbf{Y} := \begin{pmatrix} \mathbf{y} \\ \lambda \\ \mathbf{h} \end{pmatrix}. \tag{5.20}$$

The system $\mathbf{F}$ of $2n + 1$ scalar functions and equations is given by

$$\mathbf{F}(\mathbf{Y}) := \begin{pmatrix} \mathbf{f}(\mathbf{y}, \lambda) \\ \mathbf{f_y}(\mathbf{y}, \lambda)\,\mathbf{h} \\ h_k - 1 \end{pmatrix} = \mathbf{0}. \tag{5.21}$$

Referring to its general applicability, Seydel's system (5.21) has also been called a *branching system.*

Solving equation (5.21) by such standard methods as Newton methods requires a reasonable initial guess $(\bar{\mathbf{y}}, \bar{\lambda}, \bar{\mathbf{h}})$ to start the iterative process. That is, the iteration that generates a sequence of vectors $\mathbf{Y}^{(1)}, \mathbf{Y}^{(2)}, \ldots$ must be equipped with an initial approximation

$$\mathbf{Y}^{(0)} = \bar{\mathbf{Y}} = \begin{pmatrix} \bar{\mathbf{y}} \\ \bar{\lambda} \\ \bar{\mathbf{h}} \end{pmatrix}$$

that is close enough to the expected solution $\mathbf{Y}_0$ to $\mathbf{F}(\mathbf{Y}) = \mathbf{0}$. Without a close initial approximation, success of the Newton method is doubtful. Hence a direct method based on equation (5.21) requires a precursive phase that generates $\bar{\mathbf{Y}}$. The easy part is to find $(\bar{\mathbf{y}}, \bar{\lambda})$, provided one has located the bifurcation point. One possibility is to take the current solution of $\mathbf{f}(\mathbf{y}, \lambda) = \mathbf{0}$ at which a change of sign for a bifurcation test function was detected. This choice of $(\bar{\mathbf{y}}, \bar{\lambda})$ can be replaced by results of an approximation $(\bar{\mathbf{y}}_0, \bar{\lambda}_0)$ that is obtainable by the interpolation outlined in the preceding subsection. The problem remains to calculate a vector $\bar{\mathbf{h}}$, depending on $(\bar{\mathbf{y}}, \bar{\lambda})$, such that $\bar{\mathbf{h}} \approx \mathbf{h}_0$.

This problem is solved in [Sey77] based on the following principle: The linearized system

$$\mathbf{f_y}(\bar{\mathbf{y}}, \bar{\lambda})\,\mathbf{h} = \mathbf{0} \tag{5.22}$$

has no nontrivial solution, because the Jacobian $\mathbf{J}(\bar{\mathbf{y}}, \bar{\lambda}) = \mathbf{f_y}(\bar{\mathbf{y}}, \bar{\lambda})$ evaluated at $(\bar{\mathbf{y}}, \bar{\lambda}) \neq (\mathbf{y}_0, \lambda_0)$ is nonsingular. Removing one of the n scalar equations of equation (5.22) produces a solvable system. It is convenient to replace the removed equation by the normalizing equation $h_k = 1$, which enforces $\mathbf{h} \neq \mathbf{0}$. The process can be described formally as follows: Let us denote by l the index of the removed equation and define the n^2 matrix

$$\bar{\mathbf{J}}_{lk} := \mathbf{J}_{lk}(\bar{\mathbf{y}}, \bar{\lambda}) := (\mathbf{I} - \mathbf{e}_l\mathbf{e}_l^{tr})\,\mathbf{J}(\bar{\mathbf{y}}, \bar{\lambda}) + \mathbf{e}_l\mathbf{e}_k^{tr}. \tag{5.23}$$

As above, the n-vector $\mathbf{e}_i$ denotes the ith unit vector. The matrix $\bar{\mathbf{J}}_{lk}$ essentially consists of $\mathbf{J}(\bar{\mathbf{y}}, \bar{\lambda})$, except that the lth row is replaced by $\mathbf{e}_k^{tr}$. With these notations, the method of producing a solvable linear system leads to

$$\bar{\mathbf{J}}_{lk}\mathbf{h} = \mathbf{e}_l \,. \tag{5.24}$$

Definition 5.2 $\bar{\mathbf{h}}$ *is the solution of equation (5.24),* $\bar{\mathbf{J}}_{lk}\bar{\mathbf{h}} = \mathbf{e}_l,$ $\bar{\mathbf{J}}_{lk}$ *from equation (5.23).*

Modifications are possible, working with the matrix $\mathbf{f_y}(\bar{\mathbf{y}}, \bar{\lambda})$ instead of $\mathbf{J}_{lk}$ (Exercise 5.12). The vector $\bar{\mathbf{h}}$ varies with $(\bar{\mathbf{y}}, \bar{\lambda})$ and changes with the choice of the indices l and k. Remarks on the choices of l and k will follow later. An approximation to $\mathbf{h}_0$ can also be calculated by inverse iteration, observing that $\mathbf{h}_0$ is the eigenvector of $\mathbf{f_y}(\mathbf{y}_0, \lambda_0)$ for the eigenvalue 0.

Let us make sure that the $\bar{\mathbf{h}}$ defined as the solution of equation (5.24) meets the requirement of being close to $\mathbf{h}_0$. The vector $\bar{\mathbf{h}}$ qualifies as approximation if

$$\operatorname{rank} \mathbf{J}(\mathbf{y}_0, \lambda_0) = n - 1 \,,$$

which is satisfied by turning points and (simple) bifurcation points. In this case of rank deficiency one, indices l and k can be found such that $\mathbf{J}_{lk}(\mathbf{y}_0, \lambda_0)$ is nonsingular, see Section 5.5. The following continuity property has been established (for a proof, see [Sey79c]):

Theorem 5.3 *Assume* $(\mathbf{y}_0, \lambda_0, \mathbf{h}_0)$ *solves equation (5.18) or (5.21),* $\mathbf{f}$ *is two times continuously differentiable, and* $\mathbf{J}_{lk}(\mathbf{y}_0, \lambda_0)$ *is nonsingular. Then, in the limit,* $\bar{\mathbf{h}}$ *from Definition 5.2 approaches* $\mathbf{h}_0$ *if* $(\bar{\mathbf{y}}, \bar{\lambda})$ *approaches* $(\mathbf{y}_0, \lambda_0)$.

The direct method can be summarized as follows:

Algorithm 5.4   Method for Calculating Bifurcation Points. *Let a bifurcation point be located between the continuation steps* $(\mathbf{y}^{j-1}, \lambda_{j-1})$ *and* $(\mathbf{y}^j, \lambda_j)$ *with values* τ_{j-1} *and* τ_j *of the bifurcation test function. Suppose* τ *is continuous in an interval that includes* λ_0, λ_{j-1}, λ_j. *Evaluate equations (5.10) and (5.12) and choose* $(\bar{\mathbf{y}}, \bar{\lambda}) = (\bar{\mathbf{y}}_0, \bar{\lambda}_0)$ *as an initial approximation. Calculate* $\bar{\mathbf{h}}$, *for instance, by establishing matrix* $\mathbf{J}_{lk}$ *from equation (5.23) and by solving equation (5.24). Call SOLVER to solve equation (5.21), starting from* $(\bar{\mathbf{y}}, \bar{\lambda}, \bar{\mathbf{h}})$.

The main disadvantage of this direct method is that the number $2n+1$ of scalar equations to be solved in equation (5.21) is essentially doubled. Because solvers work with matrices (as Jacobians) of this size, the amount of storage required to solve the branching system with standard software is about four times as high as in performing an indirect method. This restricts the use of Algorithm 5.4 to small or moderate values of n. Another typical disadvantage of direct methods is that for the setup of $\mathbf{F}$ analytical expressions for the first-order derivatives in $\mathbf{f_y}$ are calculated and implemented in an appropriate procedure. The disadvantage of the high storage requirements can occasionally be removed by exploiting the special block structure that underlies equation (5.21) (discussed in Section 5.7.1). In many cases the matrix $\mathbf{f_y}$ can be set up by means of symbolic software. One must be aware that in solving equation (5.21) by Newton-type methods, second-order derivatives of

f are involved in the Jacobian of **F**. Although this adds to the complexity, the convenience of solving equation (5.21) is hardly affected because the numerical differentiation device of a Newton method takes care of the derivatives. In spite of storage problems with high dimensions n, the method of Algorithm 5.4 has also been applied to PDE examples [Sey79c], [RoPHR84]. Calculating bifurcations by solving equation (5.21) by calling SOLVER requires little overhead.

Convergence depends on the type of bifurcation. For Hopf bifurcation, equation (5.21) has to be modified; see Section 5.7.4. Let us assume that equation (5.21) is solved by Newton-like methods. Because Algorithm 5.4 takes care of a reasonable initial guess, global convergence can be expected to be fast. That is, it is likely that $\bar{\mathbf{Y}}$ is already in the domain of attraction. Concerning the local convergence, one must distinguish between the turning point and other bifurcation points. At a turning point the Jacobian of **F** is nonsingular (Exercise 5.22), which guarantees fast local convergence. For a bifurcation point as it is defined in Definition 2.7 the Jacobian of **F** is singular, because its upper part consists of $(\mathbf{f_y}|\mathbf{f}_\lambda)$ with $\mathbf{f}_\lambda \in \text{range}(\mathbf{f_y})$. Hence, one may expect locally slow (linear) convergence.

Surprisingly, convergence has shown to be good enough for many practical applications. That is, for average demands of accuracy (four to five significant digits), Newton-like methods show a similar convergence behavior for the nonsingular case of a turning point and the singular case of a bifurcation point. In particular, the convergence in the λ-component is satisfactory. Extensive experiments on a computer with about fourteen decimal digits have shown that the iteration switches from fast to slow convergence after an average of eight digits accuracy is obtained (in λ). Summarizing, we may say that the basic Algorithm 5.4 shows fast convergence to turning points for all accuracies and to bifurcation points for modest accuracies.

Further remarks on the singularity: Convergence behavior of Newton or Newton-like methods at solutions with singular Jacobians has been investigated repeatedly, see [Red78], [DeKK83], [KeS83], [Gri85] and the references therein. Although in general the convergence to such singular solutions must be expected to be slow, there are certain exceptions to that "rule." Werner [Wer84] found an explanation for the surprisingly good convergence behavior of Newton-like methods that iterate on the branching system (5.21) in order to approximate a bifurcation point: if one reduces the domain of **y** and **h** to certain subdomains (symmetric and antisymmetric vectors, to be introduced in Section 5.6.4), then the Jacobian of **F** is guaranteed to be nonsingular. Apparently the method for constructing an initial approximation (Algorithm 5.4) establishes an iteration that sticks closely enough to these subdomains. This may explain why the convergence behavior of the general direct method in Algorithm 5.4 has been satisfactory. For critical demands on accuracy see the methods in Subsection 5.7.2.

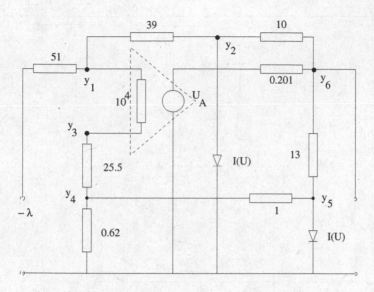

Fig. 5.6. Trigger circuit of Section 5.4.2; the boxes represent resistances

5.4.2 An Electrical Circuit

In [PöS82] an electrical circuit is presented (Figure 5.6). Several resistors are connected with an operational amplifier in which a voltage U_A is induced. The amplifier is modeled by

$$U_A(U) = 7.65 \ \text{arctan} \ (1962U),$$

two diodes are modeled by

$$I(U) = 5.6 \cdot 10^{-8}(e^{25U} - 1);$$

resistances can be found in Figure 5.6. The voltages at the nodes are denoted by y_i, $i = 1, \ldots, 6$. The input voltage is λ, the output voltage is y_6. The currents satisfy the first law of Kirchhoff. Hence, at the six nodes the following six equations hold:

$$
\begin{aligned}
f_1 &= (y_1 - y_3)/10000 + (y_1 - y_2)/39 + (y_1 + \lambda)/51 = 0, \\
f_2 &= (y_2 - y_6)/10 + I(y_2) + (y_2 - y_1)/39 = 0, \\
f_3 &= (y_3 - y_1)/10000 + (y_3 - y_4)/25.5 = 0, \\
f_4 &= (y_4 - y_3)/25.5 + y_4/0.62 + y_4 - y_5 = 0, \\
f_5 &= (y_5 - y_6)/13 + y_5 - y_4 + I(y_5) = 0, \\
f_6 &= (y_6 - y_2)/10 - [U_A(y_3 - y_1) - y_6]/0.201 + (y_6 - y_5)/13 = 0.
\end{aligned}
\tag{5.25}
$$

Upon varying the input voltage λ, one observes jumps of the output voltage. As expected, the jumps can be explained by hysteresis (compare Figure 5.7).

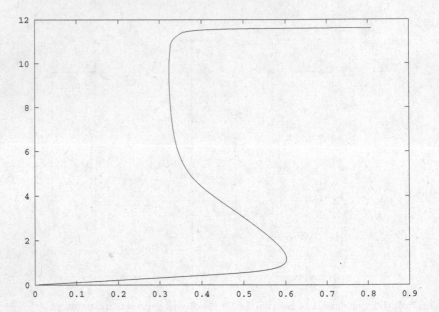

Fig. 5.7. Trigger equation (5.25), output voltage y_6 versus input voltage λ

A level of high voltage ($y_6 > 11$) and a low-voltage level are connected by a branch of unstable solutions. In short, the circuit exhibits *trigger* behavior.

We demonstrate the branching system with this example. To arrange for **F**, the first-order partial derivatives of $\mathbf{f}(\mathbf{y}, \lambda)$ are required. Because most expressions are linear, the calculation of the Jacobian is easy. The only nonlinear terms arise in f_2, f_5, and f_6 . The nonzero entries c_{ij} of the Jacobian are

$$c_{11} = 10^{-4} + \tfrac{1}{39} + \tfrac{1}{51} \,,$$
$$c_{12} = c_{21} = -\tfrac{1}{39} \,,$$
$$c_{13} = c_{31} = -10^{-4} \,,$$
$$c_{22} = \tfrac{1}{10} + \tfrac{1}{39} + 1.4 \cdot 10^{-6} \exp(25\, y_2) \,,$$
$$c_{26} = c_{62} = -\tfrac{1}{10} \,,$$
$$c_{33} = 10^{-4} + \tfrac{1}{25.5} \,,$$
$$c_{34} = c_{43} = -\tfrac{1}{25.5} \,,$$
$$c_{44} = 1 + \tfrac{1}{25.5} + \tfrac{1}{0.62} \,,$$
$$c_{45} = c_{54} = -1 \,,$$
$$c_{55} = 1 + \tfrac{1}{13} + 1.4 \cdot 10^{-6} \exp(25\, y_5) \,,$$
$$c_{56} = c_{65} = -\tfrac{1}{13} \,,$$
$$c_{61} = -c_{63} = 15009.3/[1 + 3849444(y_3 - y_1)^2]/0.201 \,,$$
$$c_{66} = \tfrac{1}{10} + \tfrac{1}{0.201} + \tfrac{1}{13} \,.$$

TABLE 5.2.   Two turning points.

y_1	0.049367	0.23578
y_2	0.54736	0.66297
y_3	0.049447	0.237598
y_4	0.049447	0.237602
y_5	0.12920	0.62083
y_6	1.16602	9.6051
λ	0.60185	0.32288

With the notations

$$y_7 = \lambda, \; y_8 = h_1, \ldots, \; y_{13} = h_6,$$

the six equations of the linearized problem $\mathbf{f_y}(\mathbf{y}, \lambda)\mathbf{h} = \mathbf{0}$ can be written as

$$\sum_{j=1}^{6} c_{ij} h_j = 0 \qquad (i = 1, \ldots, 6).$$

It is possible but tedious to write the equations

$$f_7 = 0, \ldots, f_{12} = 0$$

down explicitly. We omit that, emphasizing that this kind of formula is conveniently organized by carrying out multiplications matrix∗vector (Exercise 5.13). In order to apply a direct method, the Jacobian is implemented in a separate procedure (with the name COEFF, say, and output c_{ij}). As is typical for a wide range of examples, the Jacobian is *sparse*—that is, mostly zero. In the above example of a specific electric circuit, one evaluation of COEFF takes even fewer arithmetic operations than evaluating the function $\mathbf{f}$ of equation (5.25) itself, provided the constants are implemented as decimals. But notice how much effort is needed before COEFF is constituted. Here again symbolic formula-manipulating programs offer support.

The equations of the linearization are

$$f_{6+i} = \sum_{j=1}^{6} c_{ij} y_{6+1+j} = 0, \quad i = 1, \ldots, 6.$$

The normalizing equation $h_k = 1$ with $k = 1$ is

$$f_{13} = y_8 - 1 = 0.$$

This completes the branching system. This system of thirteen equations requires a good initial approximation, as proposed earlier in this section. We omit further details and merely present the two obtained solutions to the branching system in Table 5.2. The $\mathbf{h}$-components $(y_8, \ldots, y_{13})$ are not listed.

The results are rounded to five digits behind the decimal point. (The value of y_6 at the turning point is sensitive to the number of digits in the computation.) There is no value of the input voltage λ such that a value of the output voltage y_6 exists in the range

$$1.1660 < y_6 < 9.6051 .$$

The critical input voltages λ_0 where the bistability manifests itself by jumps are

$$0.60185 \text{ and } 0.32288$$

(compare Figure 5.7).

5.4.3 A Family of Test Functions

We go back to the description of the direct method in Algorithm 5.4 and in particular review the calculation of the initial approximation $\bar{\mathbf{h}}$. Remember that the lth equation of the linearized system

$$\mathbf{f_y}(\bar{\mathbf{y}}, \bar{\lambda}) \, \mathbf{h} = \mathbf{0}$$

has been removed in order to produce a solvable system. After having calculated $\bar{\mathbf{h}}$ by solving equation (5.24), one can calculate the value of the removed component:

$$\mathbf{e}_l^{tr} \mathbf{f_y}(\bar{\mathbf{y}}, \bar{\lambda}) \, \bar{\mathbf{h}} .$$

This number measures the distance to a bifurcation point $(\mathbf{y}_0, \lambda_0)$ because it is zero for $(\bar{\mathbf{y}}, \bar{\lambda}) = (\mathbf{y}_0, \lambda_0)$ and nonzero for $(\bar{\mathbf{y}}, \bar{\lambda}) \neq (\mathbf{y}_0, \lambda_0)$. Thus, the variable

$$\tau_{lk}(\mathbf{y}, \lambda) := \mathbf{e}_l^{tr} \mathbf{f_y}(\mathbf{y}, \lambda) \, \mathbf{h} \qquad (5.26)$$

with $\mathbf{h}$ from

$$[(\mathbf{I} - \mathbf{e}_l \mathbf{e}_l^{tr}) \, \mathbf{f_y}(\mathbf{y}, \lambda) + \mathbf{e}_l \mathbf{e}_k^{tr}] \, \mathbf{h} = \mathbf{e}_l$$

is a bifurcation test function. The matrix in brackets is $\mathbf{J}_{lk}$, which was shown to be easily established. Hence, formally, the test function (5.26) can be written

$$\tau_{lk} = \mathbf{e}_l^{tr} \, \mathbf{J} \, \mathbf{J}_{lk}^{-1} \, \mathbf{e}_l .$$

The test functions τ_{lk} can be evaluated following the above definition by carrying out the two steps:

(a) Solve the linear system $\mathbf{J}_{lk} \mathbf{h} = \mathbf{e}_l$, and
(b) evaluate the scalar product of $\mathbf{h}$ with the lth row of the Jacobian $\mathbf{J}$.

Alternatively, τ and $\bar{\mathbf{h}}$ can be calculated using the original Jacobian $\mathbf{f_y}$ [Sey91]. It easy to show that the above τ is theoretically equivalent to the τ calculated via the vector $\mathbf{v}$ defined by

$$\mathbf{f_y}(\bar{\mathbf{y}}, \bar{\lambda})\, \mathbf{v} = \mathbf{e}_l .\tag{5.27}$$

Then

$$\tau = \frac{1}{v_k}\tag{5.28}$$

holds. Using this version of τ one can benefit from the **LU**-decomposition available from the calculation of $(\mathbf{y}, \lambda)$ by Newton's method.

Because $\mathbf{h}$ depends on the choice of the indices l and k, equation (5.26) defines quite a number of test functions τ_{lk}. The requirement that $\mathbf{J}_{lk}(\mathbf{y}_0, \lambda_0)$ must be nonsingular hardly restricts the choice of l and k. One may think of picking l and k in such a way that the nonsingularity of $\mathbf{J}_{lk}(\mathbf{y}_0, \lambda_0)$ becomes most pronounced. A related criterion (proposed in [Käs85]) is based on the Householder decomposition of this matrix. Such a strategy faces the inherent difficulty that $\mathbf{J}_{lk}$ is evaluated at $(\bar{\mathbf{y}}, \bar{\lambda}) \neq (\mathbf{y}_0, \lambda_0)$. A nonsingularity of $\mathbf{J}_{lk}(\bar{\mathbf{y}}, \bar{\lambda})$ does not ensure the nonsingularity of $\mathbf{J}_{lk}(\mathbf{y}_0, \lambda_0)$. Fortunately, in most practical applications an arbitrarily chosen pair of indices works well, say, with $l = k = 1$. In examples with very large values of n with sparse or banded Jacobians, l and k should be chosen that $\mathbf{J}_{lk}$ retains its structure. Then the special iterative solvers that solve the linear system of a Newton step can also be applied for calculating $\mathbf{h}$. Another alternative chooses l and k such that the number of arithmetic operations are minimal [Sey91]; see also Exercise 5.12c. The relation between the indices l, k and the nonsingularity of $\mathbf{J}_{lk}$ will be analyzed in section 5.5.3. The test function τ of this section was used in equation (5.15) to calculate "generalized" turning points by a direct method [GrR84a]. For test functions making use of symmetry, see [Wer93].

As pointed out by [Sey91], the test function monitors how the component f_l varies when $\mathbf{y}$ is varied in direction $\mathbf{h}$,

$$\tau_{lk} = \lim_{\varepsilon \to 0} \frac{f_l(\mathbf{y} + \varepsilon \mathbf{h}, \lambda)}{\varepsilon}\tag{5.29}$$

This relation shows that if order ε perturbations in the solution are made, the function is perturbed by an amount $\tau\varepsilon$. Hence, this test function provides a specific nonlinear sensitivity analysis of $\mathbf{f}(\mathbf{y}, \lambda) = \mathbf{0}$.

The test function of this section has shown great potential in detecting critical buses of electrical power systems. The test function serves as performance index that indicates a possible collapse of power. For references consult [WaG96], [Sou00], [HuCW02], [HuC04].

We conclude the introduction of the test function τ_{lk} by giving two practical hints:

(a) $\mathbf{J}_{lk}(\mathbf{y}, \lambda)$ can be singular for some $(\bar{\mathbf{y}}, \bar{\lambda})$, although $\mathbf{J}_{lk}(\mathbf{y}_0, \lambda_0)$ is nonsingular. This singularity of $\mathbf{J}_{lk}$ causes a pole of the test function τ_{lk}. For sufficiently small continuation step size, a pole can be discovered by incorporating a third value of τ in criterion (5.6): Require

$$|\tau(\mathbf{y}^{j-2}, \lambda_{j-2})| > |\tau(\mathbf{y}^{j-1}\lambda_{j-1})|$$

to hold in addition to the change of sign.
(b) Formulas establishing a calculation of the derivative $d\tau/d\lambda$ of τ were derived. These formulas are costly because they require numerical approximation of the second-order derivatives of $\mathbf{f}(\mathbf{y}, \lambda)$ with respect to $\mathbf{y}$ [Sey79c]. A method of how to evaluate the gradient of τ was proposed in [GrR84a]. The test functions were further developed in [Gov00].

Example 5.5

We reconsider the example of the electrical circuit in Figure 5.6, with voltages satisfying the system of equations (5.25). Upon tracing the lower branch (Figure 5.7, $y_6 < 1.16$), test functions τ_{11} and τ_{42} are calculated. The choice of these indices was random. The discrete values of the test functions are connected by smooth interpolating curves. The result is shown in Figure 5.8. Both test functions signal the presence of the turning point. The figure also illustrates that the characteristic behavior of various test functions may differ even though the roots are identical.   □

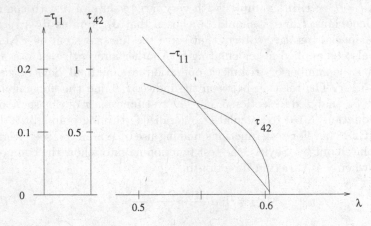

Fig. 5.8. Two test functions of Example 5.5.

5.4.4 Direct Versus Indirect Methods

We are going to discuss differences between direct and indirect methods. The following example will serve as an illustration.

Example 5.6   Temperature Distribution in a Reacting Material
In [BaW78] the partial differential equation

$$\frac{\partial\Theta}{\partial t} = \nabla^2\Theta + \lambda\exp\frac{\Theta}{1 + \epsilon\Theta} \tag{5.30}$$

is investigated. The variable

$$\Theta = (T - T_{\mathrm{a}})E/(RT_{\mathrm{a}}^2)$$

is a normalized temperature, with

T: actual temperature,
T_a: surface temperature,
R: gas constant,
E: Arrhenius activation energy,
$\epsilon = RT_a/E$,
λ: nondimensional lumped parameter, and
$\nabla^2\Theta$: Laplacian (with respect to dimensionless space variable).

Equation (5.30) models the temperature within a material with exothermic reaction. We confine ourselves to the steady-state situation in a unit square $0 \le x_1 \le 1, 0 \le x_2 \le 1$,

$$-\nabla^2\Theta = \lambda \exp \frac{\Theta}{1 + \epsilon\Theta} \quad \text{inside the square,}$$

$$\Theta = 0 \qquad \text{on the boundary.}$$

(5.31)

Employing the standard five-point star for discretizing the Laplacian on a uniform square mesh with N^2 interior grid points

$$\left(\frac{i}{N+1}, \frac{j}{N+1}\right), \quad i,j = 1,2\ldots,N$$

yields a system $\mathbf{f}(\mathbf{y}, \lambda) = \mathbf{0}$ of N^2 equations (Exercise 5.14).

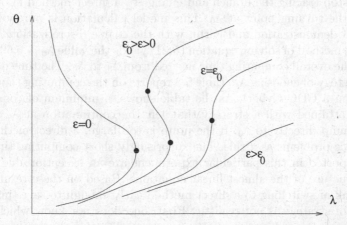

Fig. 5.9. Branching structure of Example 5.6 depending on ϵ

Calculating solutions for various values of λ and ϵ produces bifurcation diagrams as depicted in Figure 5.9. For $\epsilon = 0$ there is one turning point only; for $\epsilon > 0$ the branch turns back in a second turning point. Eventually, if ϵ is increased to ϵ_0, the turning points collapse in a hysteresis point. The critical value λ_0 of the lower-temperature turning point has a physical meaning: If λ is increased past λ_0, a jump to higher temperature occurs and the material

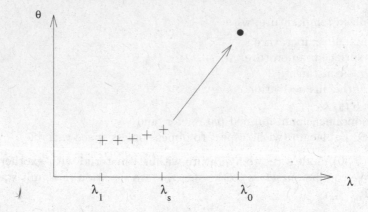

Fig. 5.10. Experiment: Jump to a turning point

ignites. For $\epsilon > \epsilon_0$ there is no such jump. The value of ϵ_0 depends on the geometry [BaW78], [BaW81]. □

This example may serve to illustrate characteristic differences between direct and indirect methods. We confine ourselves to the case $\epsilon = 0$, which has been treated repeatedly in the literature (for example, [Mey75], [Sim75], [Abb78], [MoS80]). For the numerical experiment to be described next, see Figure 5.10. We run a continuation starting with the (small) value of $\lambda = \lambda_1$. At λ_s we stop tracing the branch and change to a direct method in order to calculate the turning point at λ_0. This model calculation is purely for the purpose of demonstration and is run with the coarse discretization $N = 3$. The direct method of solving equation (5.21) yields the value $\lambda_0 = 6.6905$. We measure the overall computing time needed from λ_1 to λ_0. The time depends on the strategy of choosing λ_s. Table 5.3 reports on the computing times (run long ago on a CDC Cyber). As the table shows, a minimum of computing time was attained with a strategy that left the continuation at $\lambda_s \approx 4$ in order to jump directly to λ_0. If the jump is too large, a direct method has convergence problems as a rule. The surprisingly short computing times for $\lambda_s \approx 1$ reported in this particular experiment are an exception due to the simple structure of the almost linear equations. Based on these results, one might think of switching to a direct method early enough to save time. But this kind of strategy is not realistic. First, one does not know which value of λ_s is "fast." Second, nothing is known about the solution structure in the gap between λ_s and λ_0. Usually this lack of information cannot be tolerated. Hence, a direct method should not replace part of the continuation. Instead, a direct method is applied *after* at least rudimentary results have been obtained by an indirect method.

Another aspect that distinguishes between indirect and direct methods is the effort required to obtain bifurcation points in high accuracy. If the iteration of a direct method converges, then any additional iteration step im-

TABLE 5.3.   Computing times of the experiment in Figure 5.10.

λ_s	6.67	6.6	6.2	5.2	4.05	3.07	2.07	1.07	0.27
Seconds	0.94	0.8	0.67	0.65	0.59	0.65	0.73	0.39	0.32

proves the accuracy significantly. In contrast, indirect methods on the average have more difficulty in achieving highly accurate approximations of bifurcation points. But this does not mean that the accuracy is required. Table 5.4 gives the values of λ_0 for four values of N (results partly from [Abb78]). For instance, the effort of applying a direct method in case $N = 7$ is worthless because the discretization error is large. In fact, any continuation offers an approximation to λ_0 with an accuracy smaller than this discretization error. The mere checking of a graphical output of a corresponding bifurcation diagram produces enough accuracy. Even for $N = 15$ the discretization error is still larger than the accuracy that can be obtained by an indirect method as an easy by-product of a continuation (see Exercises 5.8 to 5.10, for instance). It is futile in this particular example to apply a direct method such as that of Algorithm 5.4.

TABLE 5.4.   Example 5.6, $\epsilon = 0$.

N	λ_0
3	6.6905 ...
7	6.7833 ...
15	6.8080865 ...
23	6.80811698 ...

Some tentative guidelines on the use of direct or indirect methods appear to be appropriate. The present section introduces direct methods that are basic for many different kind of bifurcation points. Because an introduction into variants and further developments is deferred to Section 5.7, we restrict our remarks to features inherent in all direct methods.

A judgment on what kind of method may be preferred is strongly affected by the demands for accuracy. In case one wants to calculate bifurcation points with high accuracy, it may be worthwhile to implement analytical expressions for the Jacobian in order to apply a direct method. A direct method does not replace part of a continuation, but it typically improves the guess produced during branch tracing by an indirect method. The direct method need not be applied when the medium accuracy achieved by the indirect method is sufficient.

There are two classes of problems for which it does not seem to make sense to request small error tolerances for $|\lambda_0 - \bar{\lambda}_0|$. First, equations resulting from discretization are subjected to discretization errors that typically exceed the error $|\lambda_0 - \bar{\lambda}_0|$ of a readily available indirect method. The second class of problems not requiring accurate approximations $\bar{\lambda}_0$ is the calculation of bifurcation points that are unstable to perturbations. In this context we mentioned simple bifurcation points and hysteresis points; their calculation in generic one-parameter problems can be regarded as an ill-posed problem. Hence, for the approximation of bifurcation points during branch tracing and of bifurcation points in equations resulting from discretization, indirect methods are sufficient. For all cases, indirect methods are well qualified and are to be recommended.

So far we have made a sharp distinction between indirect and direct methods. Essentially our "direct" methods have met two criteria:

(a) The iteration is not confined to walk along the branch.
(b) Equations in standard form have been formulated and the solution is delegated to standard software.

This understanding is by no means obvious. Relaxing criterion (b) qualifies a wide range of methods as direct methods or semidirect methods. In a way, then, even indirect methods might be formulated as "direct" methods, because (a) does not forbid walking along the branch. We therefore take the position that a direct method should meet both criteria (a), (b). A free interpretation of what "delegate to standard software" means is a compromise between a meaningless and a too-restrictive definition of direct methods.

5.5 Analyzing the Singular Matrix

Many of the characteristic features of a bifurcation can be realized by closer analyzing derivatives of $\mathbf{f}$, evaluated at the bifurcation point $(\mathbf{y}_0, \lambda_0)$. Readers not interested in this mathematical background may like to skip this section. Assume again $\mathbf{f} \in \mathcal{C}^2$. In what follows, the derivatives evaluated at the bifurcation will be denoted with the superscript ^{0}, as in

$$\mathbf{f}_{\mathbf{y}}^0 := \mathbf{f}_{\mathbf{y}}(\mathbf{y}_0, \lambda_0), \quad \mathbf{f}_{\lambda}^0 := \mathbf{f}_{\lambda}(\mathbf{y}_0, \lambda_0).$$

5.5.1 Simple Rank Deficiency

We assume a simple rank deficiency of $\mathbf{f}_{\mathbf{y}}^0$ —that is,

$$\mathrm{null}(\mathbf{f}_{\mathbf{y}}^0) = \mathrm{span}(\mathbf{h}), \quad \mathrm{null}(\mathbf{f}_{\mathbf{y}}^{0^{tr}}) = \mathrm{span}(\mathbf{g}). \tag{5.32}$$

For the zero eigenvalue, $\mathbf{h}$ is the right eigenvector, and $\mathbf{g}$ is the left eigenvector:

$$\mathbf{f}_{\mathbf{y}}^0 \mathbf{h} = \mathbf{0}, \quad \text{and} \quad \mathbf{g}^{tr} \mathbf{f}_{\mathbf{y}}^0 = \mathbf{0}^{tr}.$$

Equation (5.32) guarantees $\mathbf{g}^{tr}\mathbf{h} \neq 0$, and without loss of generality we assume $\mathbf{g}^{tr}\mathbf{h} = 1$. As a consequence of

$$\text{range}\,(\mathbf{A}) = (\text{null}\,(\mathbf{A}^{tr}))^{\perp}$$

(see Appendix A.2), the relation

$$\text{range}\,(\mathbf{f}_{\mathbf{y}}^0) = \{\mathbf{y} \in \mathbb{R}^n \mid \mathbf{y}^{tr}\mathbf{g} = 0\}$$

holds. The necessary criterion for a simple stationary bifurcation

$$\mathbf{f}_{\lambda}^0 \in \text{range}\,\mathbf{f}_{\mathbf{y}}^0$$

can be written as

$$\mathbf{g}^{tr}\mathbf{f}_{\lambda}^0 = 0\,.$$

This implies that the inhomogeneous system of linear equations

$$\mathbf{f}_{\mathbf{y}}^0\mathbf{v} + \mathbf{f}_{\lambda}^0 = \mathbf{0}$$

has a solution $\tilde{\mathbf{v}}$, which is not unique since all $\mathbf{v}(\alpha) := \alpha\mathbf{h} + \tilde{\mathbf{v}}$ solve the equation. Among these $\mathbf{v}(\alpha)$ there is exactly one vector orthogonal to $\mathbf{g}$ (show $\alpha = -\mathbf{g}^{tr}\tilde{\mathbf{v}}/\mathbf{g}^{tr}\mathbf{h}$). Consequently, a unique vector $\mathbf{v}$ exists solving

$$\mathbf{f}_{\mathbf{y}}^0\mathbf{v} + \mathbf{f}_{\lambda}^0 = \mathbf{0} \quad \text{with} \quad \mathbf{g}^{tr}\mathbf{v} = 0\,. \tag{5.33}$$

Note that $\mathbf{h}$ is part of the solution of the branching system equation (5.21), $\mathbf{h} = \mathbf{h}_0$.

Let $(\mathbf{y}(s), \lambda(s))$ be a smooth branch passing through the bifurcation point, with $(\mathbf{y}(s_0), \lambda(s_0)) = (\mathbf{y}_0, \lambda_0)$. The parameter s stands for arclength or any other local parameter (see Section 4.5). In the bifurcation point the tangent defined by

$$\left(\frac{d\mathbf{y}}{ds}, \frac{d\lambda}{ds}\right) \tag{5.34}$$

satisfies the equation

$$\mathbf{f}_{\mathbf{y}}^0\frac{d\mathbf{y}}{ds} + \mathbf{f}_{\lambda}^0\frac{d\lambda}{ds} = \mathbf{0}\,. \tag{5.35}$$

Equations (5.33) and (5.35) and the property $\mathbf{f}_{\mathbf{y}}^0\mathbf{h} = \mathbf{0}$ imply

$$\frac{d\mathbf{y}}{ds} = \gamma_0\mathbf{v} + \gamma_1\mathbf{h}\,, \tag{5.36}$$

a linear combination of $\mathbf{v}$ and $\mathbf{h}$. The two constants γ_0, γ_1 satisfy

$$\gamma_0 = \frac{d\lambda}{ds}, \quad \gamma_1 = \mathbf{g}^{tr}\frac{d\mathbf{y}}{ds} \tag{5.37}$$

(Exercise 5.16).

5.5.2 Second-Order Derivatives

In order to derive equations for γ_0 and γ_1, we differentiate the identity (4.6a)

$$0 = \mathbf{f_y}(\mathbf{y}(s), \lambda(s)) \frac{d\mathbf{y}(s)}{ds} + \mathbf{f_\lambda}(\mathbf{y}(s), \lambda(s)) \frac{d\lambda(s)}{ds}$$

with respect to s (Exercise 5.17). Evaluating the resulting expressions at the bifurcation point and indicating derivatives with respect to s by primes,

$$\mathbf{y}' = \frac{d\mathbf{y}}{ds}, \quad \lambda' = \frac{d\lambda}{ds},$$

one has

$$\mathbf{f_y^0}\mathbf{y}'' + \mathbf{f_\lambda^0}\lambda'' = -[\mathbf{f_{yy}^0}\mathbf{y}'\mathbf{y}' + 2\mathbf{f_{y\lambda}^0}\mathbf{y}'\lambda' + \mathbf{f_{\lambda\lambda}^0}\lambda'\lambda']. \tag{5.38}$$

The first term on the left-hand side is clearly in the range of $\mathbf{f_y^0}$. For the second term, this holds for the case of a bifurcation with $\mathbf{f_\lambda^0} \in \text{range}\,(\mathbf{f_y^0})$, or $\mathbf{g}^{tr}\mathbf{f_\lambda^0} = 0$. Then the right-hand side must be in this range too,

$$\mathbf{g}^{tr}\,[\mathbf{f_{yy}^0}\mathbf{y}'\mathbf{y}' + 2\mathbf{f_{y\lambda}^0}\mathbf{y}'\lambda' + \mathbf{f_{\lambda\lambda}^0}\lambda'\lambda'] = 0.$$

Inserting equations (5.36/37) and using the notations

$$\begin{aligned}
a &:= \mathbf{g}^{tr}\mathbf{f_{yy}^0}\,\mathbf{h}\,\mathbf{h}, \\
b &:= \mathbf{g}^{tr}(\mathbf{f_{yy}^0}\mathbf{v} + \mathbf{f_{y\lambda}^0})\,\mathbf{h}, \\
c &:= \mathbf{g}^{tr}(\mathbf{f_{yy}^0}\mathbf{v}\mathbf{v} + 2\mathbf{f_{y\lambda}^0}\mathbf{v} + \mathbf{f_{\lambda\lambda}^0})
\end{aligned} \tag{5.39}$$

yields the quadratic equation

$$a\gamma_1^2 + 2b\gamma_1\gamma_0 + c\gamma_0^2 = 0 \tag{5.40}$$

for the two unknowns γ_0, γ_1. For a positive discriminant

$$b^2 - ac > 0, \tag{5.41}$$

this has two distinct pairs of solutions (γ_0, γ_1), which via (5.37)

$$\begin{aligned}
\lambda' &= \gamma_0 \\
\mathbf{y}' &= \gamma_0\mathbf{v} + \gamma_1\mathbf{h}
\end{aligned} \tag{5.42}$$

yields two different tangents to branches passing $(\mathbf{y}_0, \lambda_0)$. This is the scenario of exactly two different branches intersecting in the bifurcation point transversally. Accordingly, the criterion (5.41) replaces the fourth criterion in Definition 2.7. The condition (5.41) serves as criterion for a simple bifurcation point; see, for instance [Kel77], [Moo80], [BrRR81], [WeS84]. The results of the derivation can be summarized as follows:

Theorem 5.7 *Assume* $\mathbf{f} \in C^2$, $\mathbf{f}(\mathbf{y}_0, \lambda_0) = \mathbf{0}$, null $(\mathbf{f_y^0}) = \mathbf{h}$, range $(\mathbf{f_y^0}) =$ null $(\mathbf{g})$, $\mathbf{g}^{tr}\mathbf{h} = 1$, $\mathbf{f_\lambda^0} \in$ range $(\mathbf{f_y^0})$, $\mathbf{v}$ *is the solution of (5.33), and* $b^2 - ac > 0$. *Then exactly two branches intersect in* $(\mathbf{y}_0, \lambda_0)$ *transversally.*

Now consider the case of one of the two branches being vertical in the sense

$$\begin{pmatrix} \mathbf{y}'(s_0) \\ \lambda'(s_0) \end{pmatrix} = \begin{pmatrix} \mathbf{h} \\ 0 \end{pmatrix},$$

which amounts to the case of a pitchfork bifurcation. So $\gamma_0 = 0$, and (5.40) reduces to $a\gamma_1^2 = 0$. This shows that $a = 0$ is a criterion for pitchfork bifurcation.

For a turning point, on the other hand, we start again from (5.38). Now $\mathbf{f_\lambda^0} \notin$ range $(\mathbf{f_y^0})$, and $\mathbf{g}^{tr}\mathbf{f_\lambda^0} \neq 0$. From (5.35) $\lambda' = \frac{d\lambda}{ds} = 0$ (in σ_0 , clear from the geometry). Hence

$$\lambda'' = -\frac{\mathbf{g}^{tr}\mathbf{f_{yy}^0}\mathbf{y}'^2}{\mathbf{g}^{tr}\mathbf{f_\lambda^0}}.$$

Since by (5.35) $\mathbf{f_y^0}\mathbf{y}' = \mathbf{0}$, the vector $\mathbf{y}'$ is a scalar multiple of $\mathbf{h}$, $\mathbf{y}' = \beta\mathbf{h}$ for a nonzero β, and

$$\lambda'' = -\beta^2 \frac{\mathbf{g}^{tr}\mathbf{f_{yy}^0}\mathbf{hh}}{\mathbf{g}^{tr}\mathbf{f_\lambda^0}.} \tag{5.43}$$

This implies

$$a \neq 0 \iff \lambda''(\sigma_0) \neq 0,$$

and we realize that $a \neq 0$ is equivalent for a turning point not being degenerated into a hysteresis point. The fourth condition of Definition 2.8 of a turning point, can be replaced by the condition $a \neq 0$.

We conclude the exploration of the second-order derivatives by a note on its computation. This can be based on

$$\mathbf{f_y}(\mathbf{y} + \varepsilon\mathbf{u}, \lambda) = \mathbf{f_y}(\mathbf{y}, \lambda) + \varepsilon\mathbf{f_{yy}}(\mathbf{y}, \lambda)\mathbf{u} + t.h.o.$$

Multiplication with $\mathbf{v}$ leads to

$$\mathbf{f_{yy}}(\mathbf{y}, \lambda)\mathbf{uv} = \lim_{\varepsilon \to 0} \frac{1}{\varepsilon}(\mathbf{f_y}(\mathbf{y} + \varepsilon\mathbf{u}, \lambda)\mathbf{v} - \mathbf{f_y}(\mathbf{y}, \lambda)\mathbf{v})$$

For a choice of ε, the coefficients a, b, c in (5.39) can be calculated in this way, provided $\mathbf{f_y}$ is implemented.

5.5.3 How to Get it Nonsingular

In Sections 5.4.1 and 5.4.3 we have introduced the branching system and test functions using indices k and l such that certain matrices $\mathbf{J}_{lk}$ are nonsingular. Now we have the means to discuss this more closely. In what follows, for consistency and ease of notation, we denote the singular matrix by $\mathbf{J}$, which represents the square Jacobian $\mathbf{f}_{\mathbf{y}}^0$. Assume again a simple rank deficiency as in (5.32), with two nonzero vectors $\mathbf{h}$ and $\mathbf{g}$ such that $\mathbf{J}\mathbf{h} = \mathbf{0}$ and $\mathbf{g}^{tr}\mathbf{J} = \mathbf{0}$. A bordering matrix as in Excrcise 4.2 is

$$\mathbf{B} := \begin{pmatrix} \mathbf{J} & \mathbf{b} \\ \mathbf{c}^{tr} & \zeta \end{pmatrix} \text{ with } \mathbf{b}, \mathbf{c} \in \mathbb{R}^n, \zeta \in \mathbb{R}.$$

Theorem 5.8 *For a matrix $\mathbf{J}$ with* rank $(\mathbf{J}) = n-1$ *the following statements are equivalent:*
  (1) $\mathbf{B}$ *is nonsingular*
  (2) $\mathbf{b} \notin$ range $(\mathbf{J})$ *and* $\mathbf{c} \notin$ range $(\mathbf{J}^{tr})$
  (3) $\mathbf{g}^{tr}\mathbf{b} \neq 0$ *and* $\mathbf{c}^{tr}\mathbf{h} \neq 0$
  (4) $(\mathbf{I} - \mathbf{b}\mathbf{b}^{tr})\mathbf{J} + \mathbf{b}\mathbf{c}^{tr}$ *has full rank n*
  (for the latter equivalence assume in addition $\mathbf{b}^{tr}\mathbf{b} = 1$.)

Proof: see Appendix A8.

Application
Definition 5.2 defines $\mathbf{h}$ [the embedded version $\bar{\mathbf{h}} = \mathbf{h}(\mathbf{y}, \lambda)$] via

$$\mathbf{J}_{lk}\mathbf{h} = \left((\mathbf{I} - \mathbf{e}_l\mathbf{e}_l^{tr})\mathbf{f}_{\mathbf{y}}(\mathbf{y}, \lambda) + \mathbf{e}_l\mathbf{e}_k^{tr}\right)\mathbf{h} = \mathbf{e}_l.$$

That is, in Theorem 5.8 the vector $\mathbf{b}$ corresponds to $\mathbf{e}_l$, and $\mathbf{c}$ to $\mathbf{e}_k$, and the criterion transforms to

$$\mathbf{g}^{tr}\mathbf{b} \neq 0 \Leftrightarrow g_l \neq 0; \quad \mathbf{c}^{tr}\mathbf{h} \neq 0 \Leftrightarrow h_k \neq 0.$$

So we have:

Corollary 5.9 *Assume* rank $(\mathbf{f}_{\mathbf{y}}(\mathbf{y}_0, \lambda_0)) = n - 1$.
The matrix $\mathbf{J}_{lk}$ is nonsingular in $(\mathbf{y}_0, \lambda_0)$ if and only if $g_l \neq 0$, $h_k \neq 0$.

  How likely is the criterion $\mathbf{g}^{tr}\mathbf{b} \neq 0$, $\mathbf{c}^{tr}\mathbf{h} \neq 0$ (or $g_l \neq 0$, $h_k \neq 0$) fulfilled? Both range$(\mathbf{J})$ and range$(\bar{\mathbf{J}})$ are $(n-1)$-dimensional hyperspaces. So, for arbitrary chosen $\mathbf{b}$, $\mathbf{c}$, a violation of the criterion is highly unprobable. This backs the claim of Section 5.4 that the choice of indices l and k is not problematic. A slightly more involved alternative is to choose $\mathbf{b} = \mathbf{g}$ and $\mathbf{c} = \mathbf{h}$.

5.6   Branch Switching

In this section we consider methods for calculating (at least) one solution on the emanating branch. We assume the following typical situation: some solutions $\mathbf{y}(\lambda)$ to $\mathbf{f}(\mathbf{y}, \lambda) = \mathbf{0}$ have been calculated—call two of them $\mathbf{y}(\lambda_1)$ and $\mathbf{y}(\lambda_2)$. Further, suppose a bifurcation has been roughly located by an indirect method that travels along the branch. It makes sense to assume that at least one solution (say, $\mathbf{y}(\lambda_1)$) is situated somewhat close to the bifurcation $(\mathbf{y}_0, \lambda_0)$. We denote a solution on the emanating branch by $\mathbf{z}$— that is, for some λ the equation $\mathbf{f}(\mathbf{z}, \lambda) = \mathbf{0}$ is satisfied (see Figure 5.11). As was mentioned earlier, branch switching means to calculate one "emanating solution" $(\mathbf{z}, \lambda)$. This "first solution" on the emanating branch then serves as the starting point for a subsequent tracing of the entire branch. As indicated in Figure 5.11, we assume that the calculated branch can be parameterized by λ. The other situation in which the calculated solutions are points on a branch that cannot be parameterized by λ will be discussed at the end of this section.

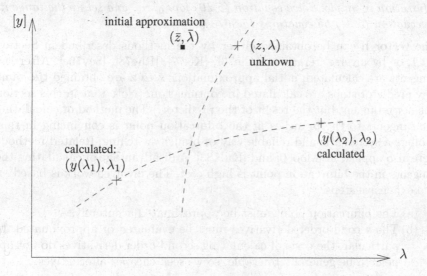

Fig. 5.11. Scenario of branch switching

Methods for switching branches consist of two phases: First, an initial guess $(\bar{\mathbf{z}}, \bar{\lambda})$ is constructed, and then an iteration is established that should converge to the new branch. That is, branch switching can be seen as a matter of establishing suitable predictors and correctors.

5.6.1 Constructing a Predictor via the Tangent

A first idea is to calculate the tangent to the emanating branch in the bi-furcation point. This requires a rather accurate calculation of the bifurcation point $(\mathbf{y}_0, \lambda_0)$ and of the derivatives of $\mathbf{f}(\mathbf{y}, \lambda)$ in the bifurcation point. The tangents are given by (5.42). The procedure of calculating tangents is part of Section 5.5.2, consult also [Kel77].

One of the two tangents corresponds to the calculated branch and need not be recalculated by this analysis. The existence of this real tangent implies a nonnegative value $b^2 - ac$ in equation (5.41). For pitchfork bifurcations, the analysis of this Subsection is not needed because approximations to both tangents are known, see Section 5.5.2. We summarize the method to calculate tangents in the bifurcation point in Algorithm 5.10.

Algorithm 5.10   Calculation of Tangents in $(\mathbf{y}_0, \lambda_0)$. *Calculate the bifurcation point $(\mathbf{y}_0, \lambda_0)$ including the vector $\mathbf{h} = \mathbf{h}_0$. Calculate $\mathbf{g}$ and $\mathbf{v}$ from (5.33), the second-order derivatives of $\mathbf{f}$ in $(\mathbf{y}_0, \lambda_0)$, and evaluate a, b, c from equation (5.39). Check $\mathbf{g}^{tr}\mathbf{f}_\lambda^0 = 0$ and equation (5.41) for whether the bifurcation is simple. Solve equation (5.40) for γ_0, γ_1, and set up the tangents in equation (5.34) by equations (5.36/37).*

The vector $\mathbf{h}$ can be obtained either by the methods described in Section 5.4.1, or by inverse iteration [Ruhe73], [Kel77], [Rhe78], [Sey79c]. After the tangents are calculated, initial approximations $\bar{\mathbf{z}}$ to $\mathbf{z}$ are obtained the same way predictor steps are calculated in continuation. (Note that in this section $\mathbf{z}$ is not a tangent but the result of the predictor.) The method of calculating the tangents to the branches in the bifurcation point is convincing in that it offers a systematic and reliable way of branch switching. Related methods were also applied in [Moo80] and [KuK83]. A disadvantage of calculating the tangents in the bifurcation point is high cost. The involved work is based on two expensive steps:

(a) The bifurcation point must be approximated accurately.
(b) The second-order derivatives must be evaluated or approximated. In particular, the costs of calculating second-order derivatives do not appear to be generally tolerable, so we seek cheaper alternatives.

5.6.2 Predictors Based on Interpolation

Two branches intersecting transversely in a bifurcation point $(\mathbf{y}_0, \lambda_0)$ define a plane that is spanned by the two tangents to the branches in $(\mathbf{y}_0, \lambda_0)$. It would be nice to have a vector $\mathbf{d}$ in that plane, perpendicular to the λ-axis. If the curvature of the branches is small enough, this vector $\mathbf{d}$ can be regarded as an approximation to the difference between the branches, provided one knows an appropriate length of $\mathbf{d}$. In other words, define

$$\mathbf{d}(\bar{\lambda}) = \mathbf{z}(\bar{\lambda}) - \mathbf{y}(\bar{\lambda})$$

as this difference, where $\mathbf{f}(\mathbf{y}, \bar{\lambda}) = \mathbf{f}(\mathbf{z}, \bar{\lambda}) = \mathbf{0}$. Here, for ease of demonstration, both branches are taken as parameterized by λ. For $\bar{\lambda}$ close to λ_0, the vector $\mathbf{d}$ is almost parallel to the plane that is spanned by the tangents in $(\mathbf{y}_0, \lambda_0)$. If one knew this vector $\mathbf{d}$, one could easily switch directly from one branch to the other without taking the detour of calculating $(\mathbf{y}_0, \lambda_0)$. Such an approach, if possible, would be less affected by unfoldings of the bifurcation due to perturbations. The assumption that the distance $\mathbf{d} = \mathbf{z} - \mathbf{y}$ is obtainable without calculating the bifurcation point is not realistic, but a major part of the program can be realized effectively. Related ideas have been introduced in [Sey81b], [Sey83a], [Sey83b].

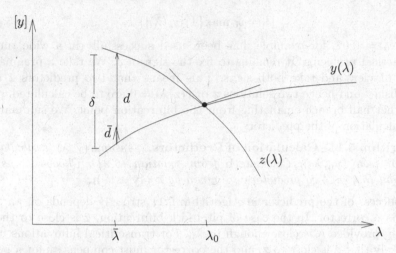

Fig. 5.12. Distance d between two branches

The problem of finding an approximation to $\mathbf{d}$ can be split into two subproblems (Figure 5.12):

(a) Find a direction $\bar{\mathbf{d}}$ pointing to the other branch, where $\bar{\mathbf{d}}$ is normalized by $\bar{d}_k = 1$ (say).
(b) Find the distance δ between the branches, measured in the kth component,

$$\delta = z_k - y_k \, .$$

Such variables $\bar{\mathbf{d}}$ and δ establish a predictor by

$$\bar{\mathbf{z}} := \mathbf{y} + \delta \bar{\mathbf{d}} \, . \tag{5.44}$$

Notice that $\delta, \bar{\mathbf{d}}, \mathbf{y}$, and $\bar{\mathbf{z}}$ depend on λ.

In equation (5.24) we have introduced a variable $\bar{\mathbf{h}}$ as a solution of a system of linear equations (Definition 5.2). This vector $\bar{\mathbf{h}}$ serves as direction $\bar{\mathbf{d}}$. We know from Section 5.4.1 that $\bar{\mathbf{h}} \approx \mathbf{h}_0$ for $\bar{\lambda} \approx \lambda_0$. The vector $\mathbf{h}_0$ lies in the plane that is spanned by the two tangents in the bifurcation point, and

$\bar{\mathbf{h}}_0$ is perpendicular to the λ-axis. This holds for the majority of bifurcation problems. Hence, we can consider the easily available $\bar{\mathbf{h}}$ as an approximation to subproblem (a).

Finding the distance δ that fits the current value of the parameter $\lambda = \bar{\lambda}$ (Figure 5.12) appears to be impossible, at least for acceptable cost (no second-order derivatives, no accurate calculation of the bifurcation point). This difficulty can be shifted to the corrector, where it is much easier to cope with. This postponing and mitigation of difficulties can be accomplished by exchanging the role of δ and λ. Instead of thinking about $\delta = \delta(\lambda)$, we prescribe δ and let the corrector take care of how to calculate $\lambda = \lambda(\delta)$. The value

$$|\delta| = \delta_0 \max\{1, |y_k(\bar{\lambda})|\} \qquad (5.45)$$

with $\delta_0 = 0.02$, for example, has been used successfully in a wide range of practical problems. It remains to fix the sign of δ. We take a pragmatic point of view and take both signs. This means that two predictors $\bar{\mathbf{z}}$ are established and hence two solutions $\mathbf{z}$ of $\mathbf{f}(\mathbf{z}, \lambda) = 0$ are to be calculated, one on either half-branch emanating from the bifurcation point. We summarize the calculation of the predictor:

Algorithm 5.11 Calculation of Predictors. *Assume* $(\bar{\mathbf{y}}, \bar{\lambda})$ *is calculated not far from* $(\mathbf{y}_0, \lambda_0)$. *Calculate* $\bar{\mathbf{h}}$ *from equation (5.24). Choose* $|\delta|$ *as in equation (5.45). Two predictors are given by* $\bar{\mathbf{z}} = \bar{\mathbf{y}} \pm |\delta|\bar{\mathbf{h}}$.

The success of the predictor in Algorithm 5.11 strongly depends on an appropriate corrector. In the case of pitchfork bifurcation, $\bar{\mathbf{z}}$ is close to the $\mathbf{z}$ branch, provided $\bar{\lambda}$ is close enough to λ_0. For transcritical bifurcations it is not likely that $\bar{\mathbf{z}}$ is close to $\mathbf{z}$, and the corrector must compensate for a poor predictor. This situation calls for some care in designing a corrector (see next subsection).

Calculation of the predictor (Algorithm 5.11) is based on results obtained at one solution $(\bar{\mathbf{y}}, \bar{\lambda})$ only. In practice, at least two solutions $(\mathbf{y}^1, \lambda_1)$ and $(\mathbf{y}^2, \lambda_2)$ are available close to the bifurcation. Interpolating the data of two such solutions enables a calculation of better approximations to $\mathbf{h}_0$. The formulas are elementary; most of them were used in the context of indirect methods (Section 5.3). The interpolation requires the values of a bifurcation test function τ. Linear interpolation yields the formulas

$$\begin{aligned} \xi &:= \tau(\mathbf{y}^1, \lambda_1)/[\tau(\mathbf{y}^1, \lambda_1) - \tau(\mathbf{y}^2, \lambda_2)], \\ \bar{\lambda}_0 &:= \lambda_1 + \xi(\lambda_2 - \lambda_1), \\ \bar{\mathbf{y}}_0 &:= \mathbf{y}^1 + \xi(\mathbf{y}^2 - \mathbf{y}^1). \end{aligned} \qquad (5.46a)$$

Let $\mathbf{h}^1$ and $\mathbf{h}^2$ be the vectors defined by equation (5.24), that is,

$$[(\mathbf{I} - \mathbf{e}_l\mathbf{e}_l^{tr})\,\mathbf{f_y}(\mathbf{y}^i, \lambda_i) + \mathbf{e}_l\mathbf{e}_k^{tr}]\,\mathbf{h}^i = \mathbf{e}_l \quad \text{for } i = 1, 2 \qquad (5.46b)$$

$$\bar{\mathbf{h}}_0 := \mathbf{h}^1 + \xi(\mathbf{h}^2 - \mathbf{h}^1)$$
$$\bar{\mathbf{z}} := \bar{\mathbf{y}}_0 \pm |\delta|\bar{\mathbf{h}}_0 \qquad (5.46c)$$

with $|\delta|$ from equation (5.45). These formulas establish an approximation $(\bar{\mathbf{y}}, \bar{\lambda}_0)$ of the bifurcation $(\mathbf{y}_0, \lambda_0)$ and two predictors $\bar{\mathbf{z}}$ to emanating solutions. Equations (5.46) do not require that $(\mathbf{y}_0, \lambda_0)$ be straddled by $(\mathbf{y}^1, \lambda_1)$, $(\mathbf{y}^2, \lambda_2)$. The results of (5.46) are better approximations than those in Algorithm 5.11 provided the bifurcation test function τ is smooth enough in the interval that includes $\lambda_0, \lambda_1, \lambda_2$. Hence, the continuation must provide $(\mathbf{y}^1, \lambda_1), (\mathbf{y}^2, \lambda_2)$ close enough to $(\mathbf{y}_0, \lambda_0)$ such that linear interpolation makes sense. The indices l and k are the same indices used in the preceding section.

Let us stress the importance of the vectors $\mathbf{h}$ defined in equations (5.24) or (5.46), introduced in [Sey77],

$$\mathbf{h} = \mathbf{h}(\mathbf{y}, \lambda, l, k).$$

They provide a readily available basis for both a family of test functions and a guess at emanating solutions. The latter does not require calculation of the bifurcation point itself.

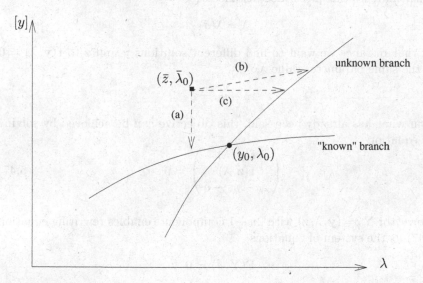

Fig. 5.13. See Section 5.6.3.

5.6.3 Correctors with Selective Properties

We now turn to the problem of how to proceed to a solution to $\mathbf{f}(\mathbf{y}, \lambda) = \mathbf{0}$, starting from $(\bar{\mathbf{z}}, \bar{\lambda}_0)$ [Sey81b], [Sey83a], [Sey83b]. An iteration toward the known branch (Figure 5.13a) must be avoided. Situation (a) frequently happens when $\bar{\lambda} = \lambda_0$ is kept fixed and Newton iterations on $\mathbf{f}(\mathbf{y}, \lambda) = \mathbf{0}$ are carried out. A convergence back to the known branch is considered a failure. In the case of pitchfork bifurcation, the probability that no emanating solution exists for $\lambda = \bar{\lambda}_0$ is 50 percent. Hence, the seemingly obvious option of parameterizing the emanating branch by λ is ruled out. We require parameterizations or equations that have a greater affinity for the emanating branch than for the known branch. Such *selective properties* of a corrector are badly needed because the predictor $(\bar{z}, \bar{\lambda}_0)$ might be a poor initial approximation only. In this subsection, we establish a corrector that proceeds parallel to the known branch [(b) in Figure 5.13], and a corrector that iterates parallel to the λ-axis [(c) in Figure 5.13]. Here the term "parallel" is motivating rather than strict. In Section 5.6.6 we discuss correctors that exploit qualitative differences between the branches.

The initial approximation $\bar{z}$ is obtained by assuming a certain distance δ between the branches—that is, for example, that for the kth component of the initial guess the relation

$$\bar{z}_k - \bar{y}_k = \delta$$

holds. Proceeding parallel to the known $\mathbf{y}$ branch means finding the value of λ that matches this prescribed distance δ,

$$\lambda = \lambda(\delta).$$

For that distance we want to find different solutions $\mathbf{y}$ and $\mathbf{z}$ to $\mathbf{f}(\mathbf{y}, \lambda) = \mathbf{0}$ for the same parameter value λ,

$$z_k - y_k = \delta.$$

From what has already been said, this objective can be achieved by solving the equation

$$\begin{pmatrix} \mathbf{f}(\mathbf{y}, \lambda) \\ \mathbf{f}(\mathbf{z}, \lambda) \\ z_k - y_k - \delta \end{pmatrix} = \mathbf{0}. \tag{5.47}$$

The vector $\mathbf{Y} := (\mathbf{y}, \lambda, \mathbf{z})$ with $2n + 1$ components enables rewriting equation (5.47) as the system of equations

$$\mathbf{F}(\mathbf{Y}, \delta) = \mathbf{0}$$

in standard form, which can be solved by standard software (CALL SOLVER). For $\delta \neq 0$, each solution $\mathbf{Y}$ includes a solution on the emanating branch. The initial approximation required for solving equation (5.47) is

$$\mathbf{Y}^{(0)} := \begin{pmatrix} \bar{\mathbf{y}} \\ \lambda \\ \bar{\mathbf{z}} \end{pmatrix}. \tag{5.48}$$

Solving equation (5.47) is the *method of parallel computation*. Carrying it out with standard software [i.e, without exploiting the structure of (5.47)] must be considered expensive because the number of equations is essentially doubled. This leads to storage requirements that can not be tolerated for larger values of n. In addition, computing time increases with the number of equations. But the method of parallel computation is attractive when an analytic expression for y_k is known and inserted. Then the method includes only $n + 1$ scalar equations.

As an alternative, we ask for the value of λ such that a solution $\mathbf{y}$ of $\mathbf{f}(\mathbf{y}, \lambda) = \mathbf{0}$ has the same value as $\bar{\mathbf{z}}$ in the kth component. This can be seen as proceeding parallel to the λ-axis when in Figure 5.13 the ordinate chosen is $[\mathbf{y}] = y_k$. Such a $\mathbf{y}$ is calculated by solving

$$\begin{pmatrix} \mathbf{f}(\mathbf{y}, \lambda) \\ y_k - \bar{z}_k \end{pmatrix} = \mathbf{0}, \quad \text{initial guess } \begin{pmatrix} \mathbf{y} \\ \lambda \end{pmatrix} := \begin{pmatrix} \bar{\mathbf{z}} \\ \lambda_0 \end{pmatrix}. \qquad (5.49)$$

Because $\bar{\mathbf{z}}$ depends on δ, solutions of equation (5.49) can be seen as being parameterized by δ. Equation (5.49) is a special version of equation (4.16), which establishes local parameterization in the context of continuation. Again the equation is of standard form and amenable to SOLVER. Solving equation (5.49) is efficient because only $n + 1$ variables are involved. But there is no guarantee that a solution of equation (5.49) lies on the emanating branch. Rather, there may be a solution on the known branch with the same value of $\bar{z}_k$. Hence, the selective properties of equation (5.49) are not fully satisfactory; much depends on the quality of the predictor. But because equation (5.49) is more selective than the ordinary parameterization by λ, and because the predictor from equation (5.46) is close to an emanating solution at least for the pitchfork case, equation (5.49) is recommended. In a large number of examples tested so far, the predictor/corrector combination equations (5.46) and (5.49), has been successful in switching branches. A branch-switching approach with better selective properties will be described next.

5.6.4 Symmetry Breaking

Functions $\mathbf{f}(\mathbf{y}, \lambda)$ often have specific properties. For example, $\mathbf{f}$ may be odd in $\mathbf{y}$,

$$\mathbf{f}(-\mathbf{y}, \lambda) = -\mathbf{f}(\mathbf{y}, \lambda). \qquad (5.50)$$

Such a relation may hold for some components only, compare Exercise 2.19. The relation of equation (5.50) is a special case of the equivariance condition equation (2.30),

$$\mathbf{f}(\mathbf{S}\mathbf{y}, \lambda) = \mathbf{S}\mathbf{f}(\mathbf{y}, \lambda). \qquad (5.51)$$

Specifically, if the matrix $\mathbf{S}$ obeys $\mathbf{S} \neq \mathbf{I}$, $\mathbf{S}^2 = \mathbf{I}$, then equation (5.51) descri- bes $\mathbf{Z}_2$ symmetry. A common type of transformation that satisfies equation

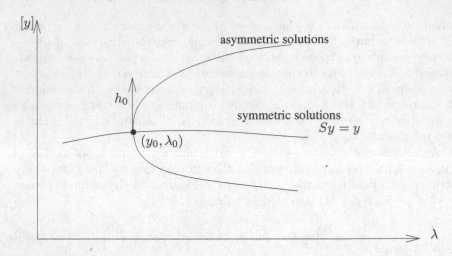

Fig. 5.14. Symmetry-breaking pitchfork

(5.51) is a permutation; we shall handle a corresponding example in the following subsection.

To be more specific, assume that $(\mathbf{y}_0, \lambda_0)$ is a simple bifurcation point, where $\mathbf{y}_0$ is symmetric and $\mathbf{h}_0$ is *antisymmetric*,

$$\mathbf{Sy}_0 = \mathbf{y}_0, \quad \mathbf{Sh}_0 = -\mathbf{h}_0. \tag{5.52}$$

Then the bifurcation point is of the pitchfork type and the emanating branch is asymmetric [BrRR81], [WeS84]. The tangent to the emanating branch in $(\mathbf{y}_0, \lambda_0)$ is given by $(\mathbf{h}_0, 0)$. This explains why the the predictor, Algorithm 5.11 and equation (5.46), provides a close approximation to an emanating solution in the pitchfork situation. The symmetry breaking is summarized in Figure 5.14. As we shall see, the frequent case of symmetry breaking has consequences both for branch switching and for the calculation of bifurcation points.

5.6.5 Coupled Cell Reaction

In [GlP71] the trimolecular reaction

$$A \rightleftharpoons X$$

$$2X + Y \rightleftharpoons 3X$$

$$B + X \rightleftharpoons Y + D$$

$$X \rightleftharpoons E$$

was introduced. The concentrations of the chemicals A, B, D, E are assumed to remain at a constant level. Taking all direct kinetic constants equal to one, and neglecting the reverse processes, the reaction is described by the reaction-diffusion problem

$$\frac{\partial X}{\partial t} = A + X^2Y - BX - X + D_1\frac{\partial^2 X}{\partial x^2},$$
$$\frac{\partial Y}{\partial t} = BX - X^2Y + D_2\frac{\partial^2 Y}{\partial x^2}. \tag{5.53}$$

The variable x measures the length of the reactor. In what follows, we shall concentrate on the kinetic equations—that is, we neglect the diffusion terms $(D_1 = D_2 = 0)$; for the impact of diffusion in this equation, see Section 6.1. Although equation (5.53) does not seem to have much chemical relevance [Sch85], this model has become famous under the name "Brusselator," after the city in which the model was created. With various choices of boundary conditions, coefficients, and extensions, the Brusselator has been investigated often. A slightly more realistic modification of equation (5.53) allows diffusion of the initial components A and B [JaHR83]. For any nonlinear effect one may be interested in, there appears to be an appropriate variant of the Brusselator.

Here we consider the steady-state situation ($\dot{X} = \dot{Y} = 0$) in three connected reaction cells with coefficients $A = 2$, $B = 6$ chosen in [KuK83]. The branching parameter λ is taken as a coupling coefficient. The corresponding coupled cell reaction is described by six equations, $f_i(y_1, \ldots, y_6, \lambda) = 0$, $i = 1, \ldots, 6$,

$$2 - 7y_1 + y_1^2y_2 + \lambda(y_3 - y_1) = 0,$$
$$6y_1 - y_1^2y_2 + 10\lambda(y_4 - y_2) = 0,$$
$$2 - 7y_3 + y_3^2y_4 + \lambda(y_1 + y_5 - 2y_3) = 0,$$
$$6y_3 - y_3^2y_4 + 10\lambda(y_2 + y_6 - 2y_4) = 0, \tag{5.54}$$
$$2 - 7y_5 + y_5^2y_6 + \lambda(y_3 - y_5) = 0,$$
$$6y_5 - y_5^2y_6 + 10\lambda(y_4 - y_6) = 0.$$

Here, y_1 and y_2 stand for X and Y in the first cell, and so on.

An inspection of equation (5.54) reveals that an exchange

$$y_1 \leftrightarrow y_5, \quad y_2 \leftrightarrow y_6$$

yields the same equations. After a permutation of the order of equations, we arrive back at equation (5.54). This process of exchanging is described by the transformation

$$\mathbf{Sy} := \begin{pmatrix} 0 & 0 & 0 & 0 & 1 & 0 \\ 0 & 0 & 0 & 0 & 0 & 1 \\ 0 & 0 & 1 & 0 & 0 & 0 \\ 0 & 0 & 0 & 1 & 0 & 0 \\ 1 & 0 & 0 & 0 & 0 & 0 \\ 0 & 1 & 0 & 0 & 0 & 0 \end{pmatrix} \begin{pmatrix} y_1 \\ y_2 \\ y_3 \\ y_4 \\ y_5 \\ y_6 \end{pmatrix}. \tag{5.55}$$

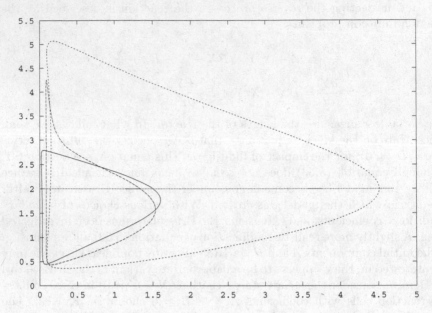

Fig. 5.15. Equation (5.54), y_1 versus λ

We realize that the transformation (5.55) and $\mathbf{f}(\mathbf{y}, \lambda)$ from equation (5.54) satisfy the equivariance condition equation (5.51), and $\mathbf{S}^2 = \mathbf{I}$ holds. That is, equation (5.54) is $\mathbf{Z}_2$ symmetric (Exercise 5.18). Hence, in this example, we expect pitchfork bifurcations with symmetry breaking.

As inspection of equations (5.53) and (5.54) shows, there is a stationary solution for all λ,

$$X = A, \quad Y = \frac{B}{A}.$$

In equation (5.54) this solution is given by

$$\mathbf{y} = (2, 3, 2, 3, 2, 3)^{tr}. \tag{5.56}$$

We carry out a numerical analysis of equation (5.54) for the parameter range $0 < \lambda < 5$. (Results were calculated using BIFPACK, see Appendix A.9.) The resulting bifurcation diagram is depicted in Figure 5.15, and a detail for small values of λ is shown in Figure 5.16. As these figures show, there are a great number of turning points, pitchfork bifurcations, and transcritical bifurcations. The bifurcation points are listed in Table 5.5 (rounded to six digits). These points were calculated by Algorithm 5.4. In addition, Hopf bifurcations exist.

Figure 5.17 condenses the bifurcation behavior of Figures 5.15 and 5.16 in a qualitative way. The nontrivial branches are connected with the trivial branch (5.56) via bifurcation points 2, 7, 13, 16. Because only two are pitchfork bifurcations, the way symmetries are broken or preserved is especially

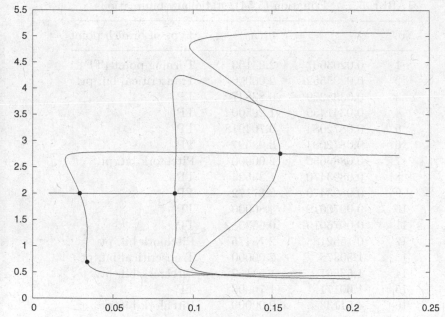

Fig. 5.16. Equation (5.54), y_1 versus λ, detail of Figure 5.15.

interesting. The trivial branch reflects a "higher" regularity than the symmetry condition with (5.55) because

$$y_1 = y_3 = y_5, \quad y_2 = y_4 = y_6 \tag{5.57}$$

holds. In Figure 5.17, symmetric solutions [with respect to equation (5.54)] are indicated by solid lines, and branches of asymmetric solutions are dashed, or dotted. The symmetry defined by the matrix **S** of (5.55) is preserved at the transcritical bifurcation points 2 and 13, but at these points the regularity responsible for equation (5.57) is broken. Symmetry breaking of the $\mathbf{Z}_2$ symmetry takes place at the pitchfork bifurcations 7, 12, 14, 16. The difference between the transcritical bifurcations and the pitchfork bifurcations is reflected in the $\mathbf{h}_0$ vectors: At the pitchfork points, these vectors are antisymmetric,

$$h_1 = -h_5, \quad h_2 = -h_6,$$

while at the transcritical bifurcation points 2 and 13, the $\mathbf{h}_0$ vectors remain symmetric,

$$h_1 = h_5, \quad h_2 = h_6.$$

This different bifurcation behavior can be discovered during the tracing of the symmetric branches by inspecting the approximations $\bar{\mathbf{h}}_0$.

TABLE 5.5.   Equation (5.54), stationary bifurcations.

No.	λ_0	$y_1(\lambda_0)$	Type of branch point
1	0.0203067	2.64153	Turning point (TP)
2	0.0295552	2.00000	Transcritical bif. pt.
3	0.0335279	0.838479	TP
4	0.0341459	1.265004	TP
5	0.0872084	0.764918	TP
6	0.0872084	3.98597	TP
7	0.0886656	2.00000	Pitchfork bif. pt.
8	0.0893470	1.35524	TP
9	0.0893470	2.87172	TP
10	0.0976019	4.80000	TP
11	0.0976019	0.653982	TP
12	0.155218	2.75475	Pitchfork bif. pt.
13	1.50378	2.00000	Transcritical bif. pt.
14	1.53561	1.57162	Pitchfork bif. pt.
15	1.60171	1.76993	TP
16	4.51133	2.00000	Pitchfork bif. pt.

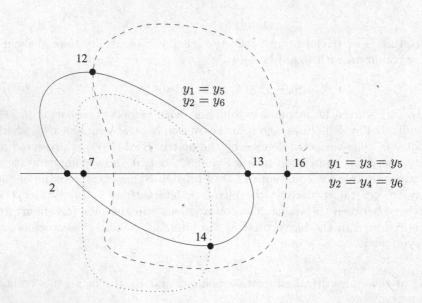

Fig. 5.17. Equation (5.54), schematic bifurcation; for numbers of bifurcations see Table 5.5.

5.6.6 Parameterization by Irregularity

The common scenario of symmetry breaking can be exploited for numerical purposes [Sey83b]. Instead of trying to find numerically a symmetry relation $\mathbf{S}$ with $\mathbf{f}(\mathbf{S}\mathbf{y}, \lambda) = \mathbf{S}\mathbf{f}(\mathbf{y}, \lambda)$, the numerical approach described next concentrates on readily available *symptoms* of a symmetry condition. That is, the components of $\mathbf{y}$ are investigated with regard to regularities. For example, a related algorithm easily detects whether two components of a current solution $\mathbf{y}$ are identical,

$$y_i = y_m.$$

In case two such indices i and m are found, the next step is to check the vector $\bar{\mathbf{h}}_0$ to determine whether its corresponding components are distinct. In this way a symmetry breaking can be detected.

The details are as follows: After having located a symmetry breaking (or *regularity breaking*) with the two indices i and m, the equation

$$\begin{pmatrix} \mathbf{f}(\mathbf{y}, \lambda) \\ y_i - y_m - \gamma \end{pmatrix} = \mathbf{0} \tag{5.58a}$$

with

$$\gamma := \delta(\bar{h}_{0i} - \bar{h}_{0m}) \tag{5.58b}$$

is solved, starting from the initial approximation

$$\mathbf{Y}^{(0)} := \begin{pmatrix} \bar{\mathbf{z}} \\ \bar{\lambda}_0 \end{pmatrix}. \tag{5.58c}$$

By the construction in equation (5.46), this initial approximation obeys the attached equation $y_i - y_m - \gamma = 0$. Equation (5.58a) forms a system $\mathbf{F}(\mathbf{Y}, \gamma) = \mathbf{0}$ or $\mathbf{F}(\mathbf{Y}, \delta) = \mathbf{0}$ consisting of $n + 1$ components. The parameter γ measures the asymmetry. Equation (5.58a) enables parameterizing the emanating branch by asymmetry; for $\gamma = 0$, one obtains a symmetric solution.

Comparing the corrector established by equations (5.58a–c) to the corrector equation (5.49) shows that both are equally efficient. The additional checking for asymmetry required for equation (5.58) can be automated at negligible cost. Concerning selectivity, equation (5.58) is preferable by far because it excludes solutions that are regular in a sense such as $y_i = y_m$. The branch-switching approach, equations (5.46) and (5.58), is reliable, fast, and easy to implement. The process of checking solutions for a symmetry of the type $y_i = y_m$ can be extended to other kinds of regularities (such as $y_i = 0$ for some i). Equation (5.58) is then be adapted accordingly. The method advocated here can be regarded as *parameterization by irregularity*.

5.6.7 Other Methods

Apart from the methods based on tangents (Section 5.6.1) and the methods based on interpolation combined with selective correctors (Sections 5.6.2, 5.6.3, and 5.6.6), other methods have been proposed—for example, the methods of [Kel77], [Lan77], [Sch77], [Rhe78], [Chan84a], [Hui91]. Other approaches were based on simplicial principles (see [PeP79], [AlG80] and the references therein). Instead of describing these approaches, we turn our attention to another technique that can be practical.

The idea is to perturb the system $\mathbf{f}(\mathbf{y}, \lambda)$ in a way that destroys the underlying regularity that is intrinsic to the current bifurcation. In literature this approach is referred to as *unfolding*. The problem is to find a parameter value γ_0 in $\mathbf{f}$ that is critical for the underlying regularity. This value is then varied to $\gamma \neq \gamma_0$ so that the regularity is destroyed and the bifurcation disappears. The parameter may occur naturally in $\mathbf{f}$, or it can be added artificially. Frequently, just adding γ to the first component (say) helps,

$$f_1(y, \lambda) + \gamma \, ;$$

in this situation $\gamma_0 = 0$. By choosing either $\gamma > \gamma_0$ or $\gamma < \gamma_0$, branches are connected in such a way that branch switching is carried out by standard branch tracing (cf. Figure 2.47). Branch switching by unfolding is not always easy to apply. Not knowing in advance where to impose the perturbation and what values of γ to choose makes it difficult.

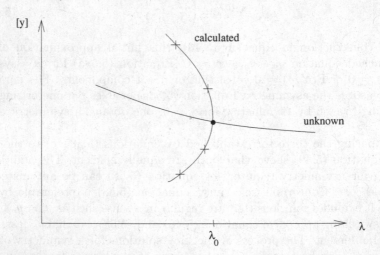

Fig. 5.18. Here, branch switching is easy

So far, we have discussed methods for branch switching under the assumption that some solutions were calculated on a branch that can be parameterized by λ. Now we comment briefly on the other situation, in which branch switching starts from a branch that cannot be parameterized by λ close to

a bifurcation. We visualize this scenario as a pitchfork bifurcation and assume that solutions on the "vertical" branch are available (Figure 5.18). For
switching branches in this situation no particular method is required. One
starts from an approximation $(\bar{\mathbf{y}}_0, \bar{\lambda}_0)$ to the bifurcation point, which can be
obtained, for instance, by an indirect method. Then solve $\mathbf{f}(\mathbf{y}, \lambda) = \mathbf{0}$ for

$$\lambda = \bar{\lambda}_0 \pm |\delta| \tag{5.59}$$

for a small value of $|\delta|$ chosen as in equation (5.45), say. Because on one
side of λ_0 locally only the unknown branch exists, this approach is selective
for one sign in equation (5.59). In general, the appropriate sign becomes
apparent when the shape of the "vertical" branch is inspected. For example,
in a situation like that depicted in Figure 5.18, the positive sign must be
chosen in equation (5.59). This branch switching can be incorporated into
equation (5.49), choosing $k = n + 1$.

5.7 Methods for Calculating Specific Bifurcation Points

In Sections 5.3 and 5.4, we discussed indirect and direct methods for calculating bifurcation points of a general type. To simplify, one could say that any
of these methods can calculate without change both bifurcation and turning
points. Several of these methods are recommended for their favorable relation
overhead (required software) versus applicability.

Allowing specific methods for specific purposes, methods that are more
efficient can be constructed. Many such methods have been proposed, most
of them direct methods. Several of the approaches are modifications or extensions of the methods described earlier. In particular, the branching system
equations (5.18) or (5.21) has been a starting point for further developments.

Systems of equations that are solved to obtain specific bifurcation points
have been called *determining equations*. This reflects interest in finding different equations for different types of bifurcation points. Such a desire was
indicated in [GrR83], which announced a system for Hopf points that does
not admit turning points as solutions. From such a remark and from the
expression "determining equation," one may envision being able to have as
many different systems as types of bifurcation points, each equation being
selective in that other bifurcation points are not admitted as solutions. This
is fascinating from a theoretical point of view, but it does not appear to
be practical, because the overhead is significant, and the current type of bifurcation point is usually *not known in advance*. In practice there is some
uncertainty in applying a specific apparatus. It is desirable that a method
perform reasonably well in a variety of phenomena.

On the other hand, improvements in the efficiency of a specialized method can be so significant that it is worthwhile to discuss related ideas. This is the object of this section. For additional literature, consult for example, [ClST00].

5.7.1 A Special Implementation for the Branching System

Moore and Spence [MoS80] proposed a significant storage reduction that takes into account the structure of the branching system equation (5.21),

$$0 = \begin{pmatrix} \mathbf{f}(\mathbf{y}, \lambda) \\ \mathbf{f_y}(\mathbf{y}, \lambda)\mathbf{h} \\ h_k - 1 \end{pmatrix} = \mathbf{F} \begin{pmatrix} \mathbf{y} \\ \lambda \\ \mathbf{h} \end{pmatrix} = \mathbf{F}(\mathbf{Y}).$$

The reduction is based on the assumption that the classical Newton method is used for solving $\mathbf{F}(\mathbf{Y}) = \mathbf{0}$. We report on this implementation in detail because similar approaches can also be used in other instances.

Each iteration step of the Newton method must solve a linear system of the form

$$\begin{pmatrix} \mathbf{f_y} & \mathbf{f_\lambda} & \mathbf{0} \\ \mathbf{f_{yy}h} & \mathbf{f_{y\lambda}h} & \mathbf{f_y} \\ \mathbf{0}^{tr} & 0 & \mathbf{e}_k^{tr} \end{pmatrix} \begin{pmatrix} \Delta\mathbf{y} \\ \Delta\lambda \\ \Delta\mathbf{h} \end{pmatrix} = -\mathbf{F} \begin{pmatrix} \mathbf{y} \\ \lambda \\ \mathbf{h} \end{pmatrix}. \tag{5.60}$$

Here $\Delta\mathbf{y}$, $\Delta\lambda$, $\Delta\mathbf{h}$ are the current unknown correction vectors (the symbol Δ is no factor). Without loss of generality, the index k is chosen as $k = 1$. The matrix in block structure is the Jacobian of $\mathbf{F}$; the arguments $(\mathbf{y}, \lambda)$ are suppressed. With the nomenclature

$$\mathbf{J} := \mathbf{f_y}(\mathbf{y}, \lambda), \quad \mathbf{B} := \mathbf{f_{yy}}(\mathbf{y}, \lambda)\mathbf{h}, \quad \mathbf{d} := \mathbf{f_{y\lambda}}(\mathbf{y}, \lambda)\mathbf{h},$$

equation (5.60) can be written as two subsystems:

$$\mathbf{J}\Delta\mathbf{y} + \mathbf{f_\lambda}\Delta\lambda = -\mathbf{f} \tag{5.61a}$$

and

$$\mathbf{J}\Delta\mathbf{h} + \mathbf{d}\Delta\lambda + \mathbf{B}\Delta\mathbf{y} = -\mathbf{Jh}. \tag{5.61b}$$

The first component of $\Delta\mathbf{h}$ is known, $\Delta h_1 = 0$. Following [MoS80], we first solve subsystem equation (5.61a) and then use the results in solving the second system equation (5.61b). Because equation (5.61a) has one unknown more than scalar equations, one of the $n + 1$ unknowns

$$\Delta\lambda, \Delta y_1, \ldots, \Delta y_n$$

is temporarily regarded as being known. Choosing Δy_1 as such a parameter takes advantage of $h_1 = 1$ and $\Delta h_1 = 0$, and allows us to replace the first column of $\mathbf{J}$ in the left-hand sides of both subsystems by a suitable vector. This can be seen as follows: Rewriting equation (5.61) yields

$$\mathbf{J}[\Delta\mathbf{y} - \Delta y_1\mathbf{h}] + \mathbf{f_\lambda}\Delta\lambda = -\mathbf{f} - \Delta y_1\mathbf{Jh} \tag{5.62a}$$

and

$$\mathbf{J}\Delta\mathbf{h} + \mathbf{f}_\lambda\Delta\lambda = -\mathbf{Jh} - \mathbf{d}\Delta\lambda - \mathbf{B}\Delta\mathbf{y} + \mathbf{f}_\lambda\Delta\lambda. \qquad (5.62b)$$

Equation (5.62a) is obtained from equation (5.61a) by subtracting $\Delta y_1 \mathbf{Jh}$. This produces a zero in the first component of the bracketed vector. Now in both the equations (5.62a) and (5.62b), $\Delta\lambda$ can take the position of the first component of the unknown vectors on the left side—that is, the vector $\mathbf{f}_\lambda$ replaces the first column of $\mathbf{J}$. Let us introduce the unknown vectors $\Delta\mathbf{a}$, $\Delta\mathbf{b}$ by

$$\Delta\mathbf{a} := \begin{pmatrix} \Delta\lambda \\ \Delta y_2 - \Delta y_1 h_2 \\ \vdots \\ \Delta y_n - \Delta y_1 h_n \end{pmatrix}, \quad \Delta\mathbf{b} := \begin{pmatrix} \Delta\lambda \\ \Delta h_2 \\ \vdots \\ \Delta h_n \end{pmatrix} \qquad (5.63)$$

and the square matrix $\mathbf{A}$ by

$$\mathbf{A} := (\,\mathbf{f}_\lambda|(\mathbf{J})\,) ,$$

where the notation $(\mathbf{J})$ indicates the remaining part of $\mathbf{J}$. With this notation, the left-hand sides of equations (5.62a) and (5.62b) can be written as $\mathbf{A}\Delta\mathbf{a}$ and $\mathbf{A}\Delta\mathbf{b}$ with the same matrix $\mathbf{A}$.

The solution of equation (5.62) is obtained by splitting equation (5.62a) into two systems of equations for two unknown vectors $\mathbf{c}^{(1)}$, $\mathbf{c}^{(2)}$,

$$\mathbf{A}\mathbf{c}^{(1)} = -\mathbf{f},$$
$$\mathbf{A}\mathbf{c}^{(2)} = -\mathbf{Jh}.$$

The solution $\Delta\mathbf{a}$ is given by linear combination,

$$\Delta\mathbf{a} = \mathbf{c}^{(1)} + \Delta y_1 \mathbf{c}^{(2)}.$$

This constitutes $\Delta\lambda, \Delta y_2, \ldots, \Delta y_n$ in dependence of Δy_1

$$\Delta\lambda = c_1^{(1)} + \Delta y_1 c_1^{(2)} \qquad (5.64)$$

and

$$\Delta y_i = c_i^{(1)} + \Delta y_1 (c_i^{(2)} + h_i) \quad \text{for } i = 2, \ldots, n.$$

These expressions are substituted into equation (5.62b), leading to the equation

$$\mathbf{A}\Delta\mathbf{b} = \mathbf{r}^{(1)} + \Delta y_1\, \mathbf{r}^{(2)}$$

with two vectors $\mathbf{r}^{(1)}$ and $\mathbf{r}^{(2)}$. We take $\Delta\mathbf{b}$ in the form

$$\Delta\mathbf{b} = \mathbf{c}^{(3)} + \Delta y_1\, \mathbf{c}^{(4)}$$

and solve two further systems of linear equations (Exercise 5.20), ending with the relation

$$\Delta\lambda = c_1^{(3)} + \Delta y_1\, c_1^{(4)} \qquad (5.65)$$

together with relations for Δh_i ($i = 2, ..., n$). The two equations (5.64) and
(5.65) give the final values of $\Delta\lambda$, Δy_1, which establishes Δy_i, Δh_i for $i > 1$.
As proved in [MoS80], the final system of two equations for $\Delta\lambda$, Δy_1 is nonsingular at turning points. Apart from calculating second-order derivatives, only
one **LU** decomposition per step is required, because the four linear systems
involve the same matrix **A**.

To summarize, this special implementation of the Newton method requires
less storage than solving the branching system equation (5.21) by standard
software. The method is designed for turning points. No quasi-Newton method is possible in this particular implementation. The size of n should be
such that **LU** decompositions are convenient.

> In the light of the limited value of direct methods in case the equations
> result from discretizations (see Section 5.4.4), the latter requirement does
> not impose a further restriction. Note the dichotomy that the savings of this
> implementation are most pronounced for large n resulting from discretizations, the case in which a direct method is most questionable. In [MoS80],
> 10 or 11 digits are calculated, although because of inherent discretization
> errors only four digits are accurate.

5.7.2 Regular Systems for Simple Bifurcation Points

As seen from the matrix structure in (5.60), and as mentioned in Section
5.4.1, the Jacobian of the branching system equation (5.21) is singular at a
simple bifurcation point. Following a suggestion of Moore [Moo80], [GrR84a],
the singularity can be removed by adding a suitable vector to **f**. To this end
define

$$\tilde{\mathbf{f}}(\mathbf{y}, \lambda, \tilde{\lambda}) := \mathbf{f}(\mathbf{y}, \lambda) + \tilde{\lambda}\mathbf{r} \qquad (5.66)$$

for a parameter $\tilde{\lambda}$ and a vector **r** that satisfies

$$\mathbf{r} \notin \text{ range } (\mathbf{f}_\mathbf{y}(\mathbf{y}_0, \lambda_0) | \mathbf{f}_\lambda(\mathbf{y}_0, \lambda_0)) .$$

One considers $\tilde{\lambda}$ a new bifurcation parameter and sets $\tilde{\mathbf{y}} := (\mathbf{y}, \lambda)$. For the
modified function $\tilde{\mathbf{f}}$

$$\partial\tilde{\mathbf{f}}/\partial\tilde{\lambda} = \mathbf{r} \notin \text{ range } (\partial\tilde{\mathbf{f}}/\partial\tilde{\mathbf{y}})$$

holds. Hence, the bifurcation point of $\mathbf{f}(\mathbf{y}, \lambda) = \mathbf{0}$ has become a turning point
of $\tilde{\mathbf{f}}(\tilde{\mathbf{y}}, \tilde{\lambda}) = \mathbf{0}$. Because this modified equation has a rectangular Jacobian
with n rows and $n+1$ columns, its singularity is characterized by the existence
of a *left* null vector **g**,

$$\mathbf{0} = \mathbf{g}^{tr}(\partial\tilde{\mathbf{f}}/\partial\tilde{\mathbf{y}}) ,$$

rather than by the right null vector **h**, as in the branching system equation
(5.21). This is equivalent to the two equations

$$\mathbf{g}^{tr}\mathbf{f}_\mathbf{y} = \mathbf{0}^{tr}, \quad \mathbf{g}^{tr}\mathbf{f}_\lambda = 0 .$$

Collecting these equations together with $\tilde{\mathbf{f}}(\tilde{\mathbf{y}}, \tilde{\lambda}) = \mathbf{0}$ and a scaling condition $\mathbf{g}^{tr}\mathbf{r} = 1$ leads to the system

$$\begin{pmatrix} \mathbf{f}(\mathbf{y}, \lambda) + \tilde{\lambda}\mathbf{r} \\ (\mathbf{f}_{\mathbf{y}}(\mathbf{y}, \lambda))^{tr}\mathbf{g} \\ \mathbf{g}^{tr}\mathbf{f}_{\lambda}(\mathbf{y}, \lambda) \\ \mathbf{g}^{tr}\mathbf{r} - 1 \end{pmatrix} = \mathbf{0} \qquad (5.67a)$$

for the $2n + 2$ unknowns $\mathbf{y}, \lambda, \tilde{\lambda}$, and $\mathbf{g}$. The ambiguity of the choice of $\mathbf{r}$ can be removed by identifying $\mathbf{r}$ with $\mathbf{g}$:

$$\mathbf{r} = \mathbf{g}. \qquad (5.67b)$$

This choice qualifies, since $\mathbf{g} \notin \text{range } (\mathbf{f}_{\mathbf{y}}^{0}|\mathbf{f}_{\lambda}^{0})$ holds as required. (Assume the contrary, which immediately leads to a contraction.) Note that for $\mathbf{r} = \mathbf{e}_k$ the equation (5.67a) is equivalent to the branching system of equation (5.21) for $\tilde{\mathbf{f}}(\tilde{\mathbf{y}}, \tilde{\lambda})$. Solutions of equation (5.67) are called "imperfect bifurcations" [Moo80]. The Jacobian of equation (5.67) is regular in a (simple) bifurcation point. Solving equation (5.67) yields for the artificial perturbation parameter $\tilde{\lambda}$ the value $\tilde{\lambda} = 0$. Then obviously $\mathbf{f}(\mathbf{y}, \lambda) = \mathbf{0}$, $\mathbf{f}_{\mathbf{y}}$ is singular, and $\mathbf{f}_{\lambda} \in \text{range } \mathbf{f}_{\mathbf{y}}$.

The above approach ignores possible symmetry. Regularity or symmetry in bifurcation can be exploited numerically. Following a proposal of Abbott [Abb77], [Abb78], regularity can be achieved by replacing suitable components of $\mathbf{f}$ by symmetry relations. Werner and Spence in [WeS84] treat the case of a $\mathbf{Z}_2$ symmetry-breaking bifurcation, which amounts to a pitchfork bifurcation. The approach starts from our branching system $\mathbf{F}(\mathbf{Y}) = \mathbf{0}$ in equation (5.21). $\mathbf{Z}_2$ symmetry is assumed; compare Section 2.11. Defining the subdomains of $\mathbb{R}^n$ of symmetric or antisymmetric vectors by

$$\mathbb{R}_s^n := \{\mathbf{y} \mid \mathbf{S}\mathbf{y} = \mathbf{y}\},$$
$$\mathbb{R}_a^n := \{\mathbf{h} \mid \mathbf{S}\mathbf{h} = -\mathbf{h}\},$$

it can be shown that the Jacobian $\mathbf{F}_{\mathbf{Y}}$ is nonsingular if restricted to symmetric $\mathbf{y} \in \mathbb{R}_s^n$, and to antisymmetric $\mathbf{h} \in \mathbb{R}_a^n$. This restriction allows one to set up an implementation of the branching system that requires only $n + 1$ scalar equations [WeS84].

5.7.3 Methods for Turning Points

The literature on the calculation of turning points is rich; for early references and first comparisons, see [MeR82]. The methods of using the branching-system (5.21) and that of Abbott have already been mentioned. Pönisch and Schwetlick proposed two algorithms [PöS81], [PöS82]. An approach of Rheinboldt is described in [Rhe82]. Griewank and Reddien [GrR84a] describe a direct method for the turning point case, which is based on our test function

τ from equation (5.26) in Section 5.4.3. With this τ equation (5.15) is solved; a method to obtain the derivatives of τ is given [GrR84a]. As [Käs85] shows, differences in numerical performance of various direct methods do not appear to be significant.

5.7.4 Methods for Hopf Bifurcation Points

Hopf points are characterized by a simple pair of purely imaginary eigenvalues $\pm i\beta$ of the Jacobian $\mathbf{f_y}(\mathbf{y_0}, \lambda_0)$ (see Section 2.6). Hence, the equation

$$\mathbf{f_y}(\mathbf{y_0}, \lambda_0)\mathbf{w} = i\beta\mathbf{w} \qquad (5.68)$$

holds for a nonzero complex vector $\mathbf{w} = \mathbf{h} + i\mathbf{g}$. This complex system is equivalent to two real systems for the vectors of the real part $\mathbf{h}$ and the imaginary part $\mathbf{g}$,

$$\mathbf{f_y}\mathbf{h} = -\beta\mathbf{g}, \quad \mathbf{f_y}\mathbf{g} = \beta\mathbf{h}. \qquad (5.69)$$

In order to normalize $\mathbf{w}$, impose as in (5.21)

$$w_k = 1,$$

which is equivalent to two real normalizing conditions. Collecting these equations yields the determining system

$$\begin{pmatrix} \mathbf{f}(\mathbf{y}, \lambda) \\ \mathbf{f_y}(\mathbf{y}, \lambda)\mathbf{h} + \beta\mathbf{g} \\ \mathbf{f_y}(\mathbf{y}, \lambda)\mathbf{g} - \beta\mathbf{h} \\ h_k - 1 \\ g_k \end{pmatrix} = \mathbf{0}, \qquad (5.70)$$

with an index k, $1 \le k \le n$. The system of equations (5.70) can be solved to calculate a Hopf bifurcation point [Jep81], [HoK84a]. In case of stationary bifurcation, $\beta = 0$, and the linear subsystem for $\mathbf{g}$ is redundant. This shows that equation (5.70) is the complex formulation of the branching system (5.21). Solving equation (5.70) by calling standard software is advisable for small or moderate values of n only, since equation (5.70) consists of $(3n + 2)$ scalar equations. An efficient implementation of equation (5.70) is proposed in [GrR83]. A variant of (5.70) works with the squared matrix $(\mathbf{f_y}(\mathbf{y}, \lambda))^2$ and requires fewer equations (see [HoK84a], [Roo85], [Wer96], and Exercise 5.21).

These methods for Hopf points are direct methods. The branching system in the corresponding ODE form can also be applied (solving (7.23), see Section 7.6.2). The above direct methods have analogs in the ODE/PDE situation (see Section 6.7). Branch-switching from stationary branch to periodic branch is relegated to Section 7.6.3 because we shall require elements of boundary-value problems that will be introduced in the following chapter. A sophisticated algorithm has been developed by Hassard [Has80] (see also [HaKW81]). Other algorithms are based on bialternate products, see [Ful68], [GuMS97]. We emphasize the effective indirect methods described earlier (Section 5.3) that are readily applied to Hopf bifurcations.

5.7.5 ' Further Bibliographical Remarks

Today, numerical bifurcation has reached a level of sophistication. Initially, the numerical treatment of bifurcations was dominated by implementations of analytic methods. That is, computers were used as machines that amplified the analytical potential of researchers. Frequently, the applicability was restricted by the assumption that a branch $y(\lambda)$ be known. Since the validity of analytical methods is often only local, such "numerical" approaches were not convincing. Beginning in the late 1970s, numerical methods were suggested that broke with traditional approaches in that they were based on ideas not used previously for analytical treatment. A prototypical example for the new approaches is the branching system (5.18/5.21), pioneered in [Sey77, Sey79]. This approach makes use of the ability to solve complex nonlinear systems of equations iteratively, which can not be done with analytical methods. Following these lines, also the computation of emanating solutions can dispense with traditional analytical methods of calculating emanating solutions only locally.

Many papers have been devoted to the finite-dimensional situation of Eq.(5.1) which requires "only" linear algebra. An influencial paper was [Kel77]. The methods of [Sey77, Sey79a] including the basic branching system were first formulated for the more difficult infinite-dimensional case of boundary-value problems, and subsequently simplified and specialized to the finite-dimensional case [Sey79c]. The idea of regularization due to Moore [Moo80] (see (5.66)), and the attempt to set up minimal extended systems have further inspired the field. Many of the related papers have been quoted in this Chapter. The notion of bifurcation test functions goes back to [Sey77, Sey79].

Today, for most bifurcation points a defining system is available. As mentioned earlier, the Jacobian $\mathbf{F_Y}$ of the basic branching system equation (5.21) is singular if evaluated at bifurcation points. This feature holds a fortiori for singularities that are even less generic. For many types of singularities of $\mathbf{f_y}$ regularizations of the branching system have been suggested. Typically, the effort increases with the deficiency of the singularity; extended systems become larger and more complicated. We restrict ourselves to an extended system suitable for the calculation of hysteresis points. But since for a reasonable formulation of the problem two parameters are required, we postpone the system to Section 5.9.

We conclude this section with some hints on further methods. There are approaches that do not work with any derivatives [Geo81b], [Kea83]. The repeatedly occurring extended systems can be solved by specific block-elimination methods [Kel83], [Chan85]. Deflation techniques as suggested in [BrG71] can be helpful for branch-switching purposes [Abb77]. Relations between different approaches for calculating turning points were investigated in [Schw84]. In [Beyn84], [Men84], [Beyn85] the calculation of multiple bifurcation points is treated. Related methods in a general setting were proposed in [AlB88], [Yam83]; these papers include multiparameter problems. Other references in numerical bifurcation are [Pön85], [MiR90], [RoDS90], [JeSC91], [SeSKT91], [Kuz98], [Gov00].

5.8   Concluding Remarks on One-Parameter Problems

The reader may now have the (correct) impression that there are many methods for a numerical bifurcation analysis. Instead of attempting to report on all variants and details, we have restricted ourselves to basic principles. Nevertheless, the preceding sections may have been confusing for those who are just looking for help with practical stability and bifurcation analysis. Hence, we offer some preliminary advice.

For the location of turning points, no particular method will be needed. A simple device like that proposed by Exercises 5.8 through 5.10 can be readily implemented in any continuation algorithm. In this way a turning point is approximated with reasonable accuracy. The situation is not complicated, because it is difficult to miss a turning point during branch tracing.

A location of bifurcation points is much more complex. Test functions have to be involved, and they are expensive when based on eigenvalues. In the context of branch tracing, one must be aware that the approximation of stationary bifurcation points in high accuracy is an ill-posed problem, because the bifurcation is destroyed by a perturbation of the underlying regularity. Therefore, indirect methods as described in Section 5.3 are appropriate. Occasions in which a specific direct method is preferable seem to be rare. For the application of some package consult Appendix A.9. As far as branch switching is concerned, the methods of Sections 5.6.2, 5.6.3, and 5.6.6 are easy to implement. The question of what method is "best" can hardly be answered, but such questions are not too important because many aspects of a bifurcation treatment just do not occur in a two-parameter setting.

5.9   Two-Parameter Problems

So far we have studied one-parameter families of equations $\mathbf{f}(\mathbf{y}, \lambda) = \mathbf{0}$. Here no method for calculating hysteresis points, isola centers, or multiple bifurcation points is required (see Section 2.10). This situation changes when a two-parameter family of equations is studied,

$$\mathbf{f}(\mathbf{y}, \lambda, \gamma) = \mathbf{0},$$

where organizing centers are most important. For example, consider the hysteresis center $(\lambda_0^{HY}, \gamma_0^{HY})$, in which two fold lines coalesce. Figure 5.19 shows the part of a (λ, γ)-parameter chart that contains two critical boundaries, (1) and (2), of parameter combinations for which turning points exist. These bifurcation curves characterize a hysteresis phenomenon [see Figures 2.45 and 2.46 and the discussion of equation (2.28)]. Figure 5.19 illustrates three principles for calculating an organizing center such as a hysteresis point.

(i) Calculate the center *point* $(\lambda_0^{HY}, \gamma_0^{HY})$ directly.

(ii) Calculate the bifurcation *curves* (1) and (2) and trace them until they coalesce.

(iii) Calculate the solution *manifold*, for instance, as follows: Choose (λ_1, γ_1) and start a standard continuation varying one parameter (say, λ) and keeping the other parameter (γ) fixed in a systematic fashion, as illustrated by Figure 5.19b. The critical boundaries (1) and (2) can be generated by interpolation.

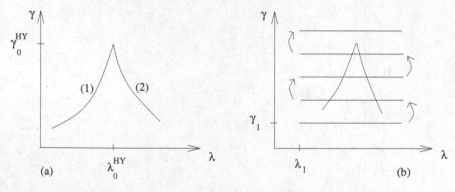

Fig. 5.19. Searching for an organizing center

Before mentioning specific methods, let us discuss differences between the above three approaches. Approach (i) is fully direct and produces only the one particular solution of the organizing center, here for parameter value $(\lambda_0^{HY}, \gamma_0^{HY})$. On its own, this approach does not provide sufficient information. Approach (ii) may be called semidirect because each turning point on the bifurcation curves (1) or (2) can be calculated by a direct method. The results of approach (ii) are highly informative because it produces an approximation to the hysteresis point and shows where the range of hysteresis extends. Finally, approach (iii) is fully indirect and produces a great amount of information on the solution structure. With (iii), an approximation of the hysteresis point is obtained as a by-product, because the main objective is to approximate the manifold (surface) defined by $\mathbf{f}(\mathbf{y}, \lambda, \gamma) = \mathbf{0}$. In case turning points are calculated by indirect methods, the illustration of a class-ii approach resembles Figure 5.20. Both (ii) and (iii) can apply one-parameter strategies. The amount of work in applying (ii) is proportional to the number of curves of critical boundaries. A fully direct method (class i) cannot

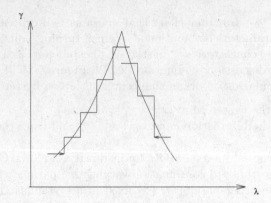

Fig. 5.20. Fully indirect tracing of bifurcation curves

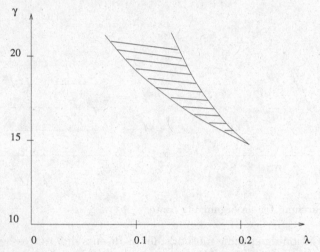

Fig. 5.21. Example 6.3. Equation (6.16), parameter chart with two turning-point curves, $\beta = 0.4$

be seen as an alternative to a class-ii or class-iii approach because the latter are required anyway in order to calculate global information. This situation resembles the discussion "direct versus indirect methods" in Section 5.4.4.

Next we briefly mention specific methods falling into the above three classes. We begin with class (ii). Methods for tracing critical boundaries are given by suitable direct methods, as they are described in the previous sections. We illustrate this by means of Seydel's method [Sey79a], which was described in Section 5.4.1. The extended branching system equation (5.21) depends on the second parameter,

$$\mathbf{F}(\mathbf{Y}, \gamma) = \begin{pmatrix} \mathbf{f}(\mathbf{y}, \lambda, \gamma) \\ \mathbf{f_y}(\mathbf{y}, \lambda, \gamma)\mathbf{h} \\ h_k - 1 \end{pmatrix} = \mathbf{0}. \tag{5.71}$$

A standard continuation with respect to γ traces critical boundaries. Since the specific branch of bifurcations defined by equation (5.71) may itself encounter a turning point with respect to γ, the continuation methods of Chapter 4 are applied—that is, an additional parameterizing equation is added. We see that for tracing of bifurcation curves no extra numerical method is needed if the branching system is used.

As an example for results produced by this approach we consider a specific catalytic reaction with two parameters λ and γ. The equations will be given in equation (6.15) in Section 6.2; here we focus on the two-parameter aspect of solutions. Figure 5.21 shows two critical boundaries that were produced by this bifurcation curve tracing due to [Sey79a].

Other methods for calculating turning points can be applied similarly, including indirect methods (Figure 5.20). Further methods of class (ii) are proposed in [Rhe82], [SpW82], [SpJ84]. In general, such methods are not restricted to the handling of hysteresis points but also apply to the calculation of bifurcations and isola centers. This situation emphasizes the importance of being able to calculate turning points. The calculation of bifurcation points (gap centers), isola centers, and hysteresis centers can be reduced to repeated calculations of turning points. Analogously as equation (5.71) traces turning points, a curve of Hopf points is calculated by using equation (5.70) with a second parameter γ included.

Direct procedures of class (i) for calculating hysteresis and Hopf points have been proposed by Roose in [RoP85], [KüST87]. To give an example, the extended system for hysteresis points is

$$\begin{pmatrix} \mathbf{f}(\mathbf{y}, \lambda, \gamma) \\ \mathbf{f_y}(\mathbf{y}, \lambda, \gamma)\,\mathbf{h} \\ h_k - 1 \\ \mathbf{g}^{tr}\mathbf{f_y}(\mathbf{y}, \delta, \gamma) \\ g_l - 1 \\ \mathbf{g}^{tr}\mathbf{f_{yy}}(\mathbf{y}, \lambda, \gamma)\,\mathbf{h}\,\mathbf{h} \end{pmatrix} = \mathbf{0}. \tag{5.72}$$

The last component is the coefficient a in equation (5.39), $\mathbf{g}$ is left eigenvector; for details see [RoP85]. Equation (5.72) illustrates that the branching-system equation (5.21) is an integral part of related defining systems. Equations defining organizing centers have been discussed in [Beyn84], [Kun88], [DeRR90], [GrR90], [Kun91], [Kuz98], [Gov00]. For the two-parameter setting, a defining system such as equation (5.72) is of class (i). Note that for a problem involving three or more variable parameters this equation becomes a class-ii method.

Finally, we come to approach (iii), which approximates a part of the manifold defined by $\mathbf{f}(\mathbf{y}, \lambda, \gamma) = \mathbf{0}$. Considering a scalar measure $[\mathbf{y}]$ of solutions

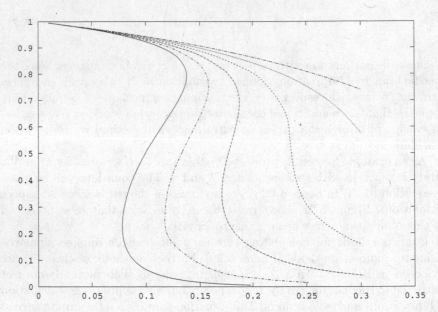

Fig. 5.22. Equation (6.16), bifurcation diagram $y_1(0)$ versus λ for $\beta = 0.4$, and $\gamma = 20$ (left profile), $\gamma = 18, 14, 12$, and $\gamma = 10$ (right profile)

to $\mathbf{f}(\mathbf{y}, \lambda, \gamma) = \mathbf{0}$, the results form a surface in the three-dimensional $([\mathbf{y}], \lambda, \gamma)$ space. Figure 5.19b suggests how to apply a standard one-parameter continuation. In this way, cuts through the surface are obtained. Accuracy can be improved by varying the spacing. Figure 5.22 illustrates results as they are obtained by approximating the surface in the way indicated by Figure 5.19b. Figure 5.22 again refers to the mentioned catalytic reaction and thus matches Figure 5.21. In the approach indicated by Figure 5.19b or 5.22, it is helpful to take advantage of calculated results on the previous parallel line in the parameter chart. Alternatives to the simple parallel slicing construct a triangulation on the surface. Related efficient approaches have been developed by Allgower and Schmidt [AlS85] and Rheinboldt [Rhe88], see also [ChCJ05]. Triangulation for the computation of bifurcation surfaces is investigated in [StGSF08].

Exercises

Exercise 5.1

(a) Show that the eigenvalues θ_k of $\mathbf{C}$ and μ_k of $\mathbf{A}$ $(k = 1,\ldots,n)$ of the Cayley transformation equation (5.3) satisfy

$$\mu = \frac{a_1\theta - a_2}{\theta - 1}, \qquad \theta = \frac{\mu - a_2}{\mu - a_1}.$$

(b) Show $|\theta_{1,2}| > 1$, $|\theta_k| \le 1$ for $k \ge 3$ where $\mu_k = \alpha_k + i\beta_k$, $\alpha_j \ge \alpha_k$ for $j < k$, $\beta_1 \ne 0$, $\alpha_1 > \alpha_3$, and a_1, a_2 are chosen such that $\alpha_3 \le \frac{1}{2}(a_1 + a_2) < \alpha_1$.
(c) What is the application of the Cayley transformation for $a_1 = 0$, $a_2 = 2\mathrm{Re}(\mu)$?

Exercise 5.2
Establish a way to determine the sign of $\det \mathbf{P}$, where $\mathbf{P}$ is a permutation matrix.

Exercise 5.3
Verify equation (5.11) using the ansatz

$$p(\lambda) = \bar{\tau} + \zeta(\lambda - \bar{\lambda}_0) + \gamma(\lambda - \bar{\lambda}_0)(\lambda - \lambda_j)$$

for the interpolating polynomial.

Exercise 5.4
Suppose criterion (5.6) holds and the subinterval $\lambda_{j-1} < \lambda < \lambda_j$ has exactly one zero of a continuous $\tau(\lambda)$. Show that there is exactly one of the two values of λ_0^* [furnished by equation (5.11)] in this subinterval.

Exercise 5.5
Equation (5.11) may suffer from *cancellation* (loss of significant digits due to subtraction of nearly equal numbers). Establish a formula that does not suffer from cancellation.

Exercise 5.6
Design a flowchart of an algorithm that exploits data $(\mathbf{y}^j, \lambda_j, \tau_j)$ provided by continuation. The objective is to locate bifurcations and calculate the approximation λ_0. You are encouraged to program your algorithm on a computer and extend it by making use of equation (5.11).

Exercise 5.7
Approximation of turning points $(\mathbf{y}_0, \lambda_0)$ by means of test functions. Verify equation (5.13). Rewrite equation (5.13) in the way equation (5.10) is written. Establish an approximation to $\mathbf{y}_0$ based on the assumption that the components of $\mathbf{y}$ locally behave like parabolas.

Exercise 5.8
(Project, see Section 5.3.3.) Construct an algorithm that approximates λ_0 by inverse interpolation. Design the algorithm so that it can be easily attached to any continuation procedure that also provides values of a bifurcation test function.

Exercise 5.9
(Project.) As explained in Section 5.3.2, turning points can be approximated most effectively without making use of any test function. This approach requires three or more solutions $(\mathbf{y}^j, \lambda_j)$, $j = 1, 2, \ldots$, calculated in the neighborhood of a current turning point $(\mathbf{y}_0, \lambda_0)$. For convenience we restrict ourselves to three solutions, interpreting y and y^j as any component (the first, say) of these vectors.

Exercise 5.10
(Project.) Design an algorithm that interrupts the continuation in order to approximate a turning point more accurately. To this end, use the formulas of Exercise 5.9 and calculate an additional solution using equation (4.16) with a value of η obtained from $\bar{\mathbf{y}}_0$. After a solution is calculated, a better approximation to the turning point can be obtained in one of two ways. For one, apply the formula in Exercise 5.9 based on three solutions. Replace the solution that is furthest from the current guess by the new approximation. As an alternative, use a polynomial of the next higher degree. After the approximation is obtained, the process can be repeated in an iterative way. One or two iterations of this simple scheme are usually sufficient to produce an accurate approximation to a turning point.

Exercise 5.11
(An academic example, following [CrR71].) Consider the equations

$$f_1(y_1, y_2, \lambda) = y_1 + \lambda y_1(y_1^2 + y_2^2 - 1) = 0,$$
$$f_2(y_1, y_2, \lambda) = 10 y_2 - \lambda y_2(1 + 2y_1^2 + y_2^2) = 0.$$

(a) Establish the branching system for this example.
(b) For the branch $y_1 = ((\lambda - 1)/\lambda)^{1/2}$, $y_2 = 0$ calculate the matrices $\mathbf{J}_{11}$ and $\mathbf{J}_{22}$.
(c) Show that τ_{11} and τ_{22} signal distinct branch points.

Exercise 5.12
Consider the system

$$\begin{pmatrix} \mathbf{f}(\mathbf{y}, \lambda) \\ (\mathbf{I} - \mathbf{e}_l \mathbf{e}_l^{tr})\mathbf{f}_\mathbf{y}(\mathbf{y}, \lambda)\mathbf{h} \\ h_k - 1 \\ p(\mathbf{y}, \lambda, \mathbf{h}) \end{pmatrix} = \mathbf{0}.$$

Solutions are parameterized by the scalar equation $p(\mathbf{y}, \lambda, \mathbf{h}) = 0$ in one of two ways:

(1) $p_1(\mathbf{y}, \lambda, \mathbf{h}) = \lambda - \eta$,

(2) $p_2(\mathbf{y}, \lambda, \mathbf{h}) = \mathbf{e}_l^{tr} \mathbf{f_y}(\mathbf{y}, \lambda)\mathbf{h} - \eta$.

The solutions depend on η, $\mathbf{y}(\eta)$, $\lambda(\eta)$, and $\mathbf{h}(\eta)$. The parameterization p_2 establishes an *embedding* of the branching system [Sey77]. This idea gives rise to modifications of the calculation of an initial approximation.

(a) For which choices of parameterization does one obtain the initial approximation $\bar{\mathbf{h}}$ from equation (5.24) and the branching system equation (5.21)?

(b) How can this be used to construct a homotopy (continuation) for solving the branching system?

(c) Assume that the inverse of the Jacobian exists. Show that $\bar{\mathbf{h}}$ from equation (5.24) can be defined as a solution of the inhomogeneous equation

$$\mathbf{f_y}(\bar{\mathbf{y}}, \bar{\lambda})\mathbf{h} = \mathbf{z}, \quad \mathbf{z} = \mathbf{e}_l (\mathbf{e}_k^{tr} \mathbf{f_y}^{-1} \mathbf{e}_l)^{-1}.$$

(d) Is it possible to dispense with indices l and k?

(e) Assume that an **LU** decomposition of the Jacobian is calculated. Show that for $l = n$ the calculation of $\mathbf{h}$ requires only one backsubstitution. What result does one obtain for $l = k = n$?

Exercise 5.13

(a) Design an algorithm that constitutes the function $\mathbf{F}$ of equation (5.21). Input: n, $\mathbf{Y}$; assume that procedures FCN [calculates $\mathbf{f}(\mathbf{y}, \lambda)$] and COEFF [calculates $\mathbf{f_y}(\mathbf{y}, \lambda)$] are available.

(b) Modify the algorithm so that the known component h_k is saved and not recalculated as part of the branching system.

Exercise 5.14

Derive the N^2 equations that are the discrete analog of the Dirichlet problem (5.31). Denote the approximation to

$$\Theta\left(\frac{i}{N+1}, \frac{j}{N+1}\right)$$

by u_{ij}, and approximate the associated Laplacian by

$$(N+1)^2(-4u_{ij} + u_{i-1,j} + u_{i+1,j} + u_{i,j-1} + u_{i,j+1})$$

(why?). Take $n = N^2$ and

$$y = (u_{11}, \dots, u_{1N}, u_{21}, \dots, u_{NN}),$$

and constitute the n equations of $\mathbf{f}(\mathbf{y}, \lambda) = \mathbf{0}$.

Exercise 5.15

Suppose that $(\mathbf{y}_0, \lambda_0)$ is a bifurcation point with equation (5.32). Show that there is a unique vector $\mathbf{v}$ such that equation (5.33) holds.

Exercise 5.16
Show that equation (5.37) holds.

Exercise 5.17
Derive the quadratic in equation (5.40) in detail.

Exercise 5.18
Consider another symmetry of the Brusselator equations (5.54): Take the three cells in a triangular configuration, each cell is coupled to the two others. To this end, add in the coupling terms of the first and second equation $y_5 - y_1$ and $y_6 - y_2$, respectively. Similarly, add in the coupling terms of the fifth and the sixth equations $y_1 - y_5$ and $y_2 - y_6$. Realize that the coupling terms, which adapt for diffusion among the three cells, are now identical. Symmetries of these modified equations are described by 3×3 block matrices with 2×2 identity matrices or 2×2 zero blocks. Check that the matrix in equation (5.55) is of this type. Realize that the symmetry group consists of all permutations of these 3×3 block matrices.

Exercise 5.19
Assume that solutions $\mathbf{y}$ to $\mathbf{f}(\mathbf{y}, \lambda) = \mathbf{0}$ have been approximated with an accuracy such that the absolute error in each component of $\mathbf{y}$ is bounded by ϵ_1 ($\epsilon_1 = 10^{-4}$, for instance). Assume further that the error in $\bar{\mathbf{h}}_0$ [from equation (5.46)] is bounded by ϵ_2 (e.g., $\epsilon_2 = 50\epsilon_1$). Design an algorithm that checks $\mathbf{y}$ and $\bar{\mathbf{h}}_0$ for symmetry/regularity breaking. Based on the result of this investigation, an appropriate corrector equation must be chosen. To this end, unify equations (5.49) and (5.59a) by introducing the general $(n+1)$st component

$$y_i - \zeta y_m - \gamma = 0.$$

Fix $i, m, \zeta,$ and γ, depending on the outcome of the regularity check, in such a way that equation (5.49) is carried out in case no regularity breaking is detected.

Exercise 5.20
Carry out in detail the special implementation of the Newton method as indicated by equations (5.60) through (5.65). In particular, calculate the vectors $\mathbf{r}^{(1)}$ and $\mathbf{r}^{(2)}$ and give a solution $\Delta\lambda$, Δy_1 to equations (5.64) and (5.65).

Exercise 5.21
Find another determining equation for Hopf points. (Hint: Combine the two systems of equation (5.69) in a suitable way.)

Exercise 5.22

Show that for a turning point $(\mathbf{y}_0, \lambda_0)$ a solution $(\mathbf{x}_1, \mu, \mathbf{x}_2)$ of

$$\mathbf{f}_\mathbf{y}^0 \mathbf{x}_1 + \mathbf{f}_\lambda^0 \mu = 0$$

$$\mathbf{f}_{\mathbf{yy}}^0 \mathbf{h}_0 \mathbf{x}_1 + \mathbf{f}_{\mathbf{y}\lambda}^0 \mathbf{h}_0 \mu + \mathbf{f}_\mathbf{y}^0 \mathbf{x}_2 = 0$$

$$\mathbf{e}_k^t \mathbf{x}_2 = 0$$

must be the trivial one ($\mathbf{x}_1 = \mathbf{x}_2 = \mathbf{0}$, $\mu = 0$).

Hint: Use from Section 5.5 $\mathbf{g}^{tr} \mathbf{f}_\mathbf{y}^0 = \mathbf{0}$ and $\mathbf{g}^{tr} \mathbf{f}_\lambda^0 \neq 0$. Start with the first equation, argue $\mathbf{x}_1 = \beta \, \mathbf{h}_0$, and substitute into the second equation.

6 Calculating Branching Behavior of Boundary-Value Problems

The main topic of this chapter is the calculation of branching behavior of the one-parameter family of two-point boundary-value problems

$$\mathbf{y}' = \mathbf{f}(t, \mathbf{y}, \lambda), \quad \mathbf{r}(\mathbf{y}(a), \mathbf{y}(b)) = \mathbf{0} . \tag{6.1}$$

As usual, the variable $\mathbf{y}(t)$ consists of n scalar functions $y_1(t), \ldots, y_n(t)$. The right-hand side $\mathbf{f}(t, \mathbf{y}, \lambda)$ is a vector function; the boundary conditions [the second equation (6.1)] consist of n scalar equations,

$$r_1(y_1(a), \ldots y_n(a), y_1(b), \ldots, y_n(b)) = 0$$
$$\vdots \qquad\qquad \vdots$$
$$r_n(y_1(a), \ldots, y_n(a), y_1(b), \ldots, y_n(b)) = 0$$

The independent variable t $(a \leq t \leq b)$ need not be time; accordingly, the derivative with respect to t is denoted by a prime $'$ rather than a dot: $\mathbf{y}' = \mathrm{d}\mathbf{y}/\mathrm{d}t$. The bifurcation parameter λ can occur in the boundary conditions:

$$\mathbf{r}(\mathbf{y}(a), \mathbf{y}(b), \lambda) = \mathbf{0} .$$

However, because the methods discussed in this chapter are not affected by the dependence of $\mathbf{r}$ on λ, the notation $\mathbf{r}(\mathbf{y}(a), \mathbf{y}(b))$ of equation (6.1) will be retained.

The methods for boundary-value problems resemble methods explained in the previous chapter. Following [Sey77], our approach stays in the infinite-dimensional space of ODEs and lets standard software take care of the transition to the finite-dimensional world of numerical approximation. Most of the methods discussed in this chapter are not dependent on any particular solution procedure or software. As an alternative to the methods set out in this chapter, the methods of the preceding chapter can be applied, provided equation (6.1) is discretized first. As an example of such a discretization, see the PDE model Example 5.6 (in Section 5.4.4), which was transformed into a system of algebraic equations.

In addition to ODE boundary-value problems, we shall study certain classes of PDEs, mainly to investigate stability. Solutions of equation (6.1) can be stationary states of certain classes of PDEs, the long-term temporal behavior of which is of interest.

R. Seydel, *Practical Bifurcation and Stability Analysis*,
Interdisciplinary Applied Mathematics 5, DOI 10.1007/978-1-4419-1740-9_6,
© Springer Science+Business Media, LLC 2010

6.1 Enlarged Boundary-Value Problems

Specific tasks (such as continuation or branch switching) can be reformulated into boundary-value problems of the general type

$$\mathbf{Y}' = \mathbf{F}(t, \mathbf{Y}), \quad \mathbf{R}(\mathbf{Y}(a), \mathbf{Y}(b)) = \mathbf{0}. \tag{6.2}$$

Typically, equation (6.2) includes equation (6.1) as a subsystem. Because solvers for two-point boundary-value problems of type (6.2) are well established, we consider that the specific task is solved whenever we succeed in

(1) formulating an equation of type (6.2) that is equivalent to the specific task, and

(2) constructing a reasonable initial guess to the solution.

In this section we use two examples to explain elementary ways of constructing a boundary-value problem of the general type (6.2).

The most widely applicable enlarged boundary-value problem is

$$\begin{pmatrix} \mathbf{y} \\ \lambda \end{pmatrix}' = \begin{pmatrix} \mathbf{f}(t, \mathbf{y}, \lambda) \\ 0 \end{pmatrix}, \quad \begin{pmatrix} \mathbf{r}(\mathbf{y}(a), \mathbf{y}(b)) \\ y_k(a) - \eta \end{pmatrix} = \mathbf{0}. \tag{6.3}$$

The system (6.3) of dimension $n+1$ is of type (6.2), with vector function $\mathbf{Y} = (\mathbf{y}, \lambda)$. Equation (6.3) is applied in the same way the related equation (4.16) is applied for continuation of systems of equations. In particular, equation (6.3) is appropriate for passing turning points. Prescribing a suitable index k and a value of η enables the calculation of λ as a dependent variable. The "trivial" differential equation $\lambda' = 0$ characterizes λ as constant. Solving equation (6.3) automatically yields a value of λ that matches the underlying k and η. Concerning the choice of k and η, the strategy outlined in Section 4.5.3 is applied. If one of the n boundary conditions (r_j, say) is of the form of an initial condition for y_ν

$$r_j = y_\nu(a) - \text{ constant } = 0,$$

the index k must be different from ν in order to avoid a contradiction. A strategy such as equation (4.18), evaluated for $\mathbf{y}(a)$, prevents k from conflicting with an initial condition. Choosing $k = n + 1$ in equation (6.3), the parameter λ becomes the control parameter.

Example 6.1   Brusselator with Diffusion
We reconsider the Brusselator model [see equation (5.53) in Section 5.6.5] now without neglecting the diffusion. Concentrating on the steady-state situation $\dot{X} = \dot{Y} = 0$ yields two differential equations of the second order, namely,

$$0 = A + X^2 Y - BX - X + D_1 \frac{\partial^2 X}{\partial x^2},$$
$$0 = BX - X^2 Y + D_2 \frac{\partial^2 Y}{\partial x^2}. \tag{6.4}$$

These ordinary differential equations describe the spatial dependence of the two chemicals X and Y along a reactor with length $L, 0 \le x \le L$. We impose fixed boundary conditions,

$$X = A \qquad \text{for } x = 0, \ x = L,$$
$$Y = B/A \qquad \text{for } x = 0, \ x = L.$$

Scaling the independent variable according to $t := x/L$ and writing $\lambda := L^2$ lead to the system of four ODEs of the first order

$$
\begin{aligned}
y_1' &= y_2, \\
y_2' &= -\lambda[A + y_1^2 y_3 - (B+1)y_1]/D_1, \\
y_3' &= y_4, \\
y_4' &= -\lambda[By_1 - y_1^2 y_3]/D_2,
\end{aligned}
\tag{6.5a}
$$

with boundary conditions

$$y_1(0) = y_1(1) = A, \quad y_3(0) = y_3(1) = B/A \tag{6.5b}$$

(Exercise 6.1). We adopt the constants

$$D_1 = 0.0016, \qquad D_2 = 0.008, \qquad A = 2, \qquad B = 4.6$$

from [KuRM78]. As equation (6.5a) shows, the independent variable t in equation (6.1) need not occur explicitly in the right-hand side; the right-hand side of equation (6.5a) is of the form $\mathbf{f}(\mathbf{y}, \lambda)$.

Because the boundary conditions in equation (6.5b) include two initial conditions (imposed on y_1 and y_3), the index k in equation (6.3) can take only the values $k = 2$ or $k = 4$. In order to extend equation (6.5) to the system of equation (6.3), define $y_5 = \lambda$ and attach the differential equation and boundary condition

$$y_5' = 0, \quad y_k(0) = \eta. \tag{6.5c}$$

Writing the boundary conditions in the form used in equations (6.1), (6.2), or (6.3) yields

$$
\begin{aligned}
y_1(0) - A &= 0, \\
y_1(1) - A &= 0, \\
y_3(0) - B/A &= 0, \\
y_3(1) - B/A &= 0, \\
y_k(0) - \eta &= 0.
\end{aligned}
\tag{6.5d}
$$

This concludes the formal preparations that transform the original problem, equation (6.4), into the standard forms of equation (6.1) or (6.2).

After having implemented the right-hand side of equation (6.5a) together with $y_5' = 0$ in a routine for $\mathbf{F}$, and equation (6.5d) in a routine for $\mathbf{R}$,

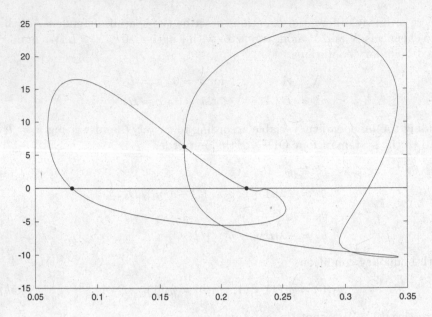

Fig. 6.1. Example 6.1, bifurcation diagram with two nontrivial branches; $y_2(0)$ versus L. Three bifurcation points are marked. (There are more branches than illustrated here.)

the next step is to call SOLVER in order to calculate solutions and trace branches. The specific Brusselator model equation (6.5) has a great number of solutions [KuRM78]; some are shown in the bifurcation diagram of Figure 6.1. One nontrivial bifurcation point is found in Figure 6.1 for $L = 0.1698$, $y_2(0) = 6.275$. The closed branches have been calculated by using the $k - \eta$ strategy in equations (4.18), (4.19), or (4.23).   □

The above Brusselator has served as a first example illustrating the transformation from a particular model into the two-point boundary-value problem, equation (6.1) or (6.3), in standard form. The second example represents a different class of solutions—namely, the time-dependent periodic solutions.

Example 6.2   Forced Duffing Equation
We reconsider the Duffing equation (2.9) introduced in Section 2.3,

$$\ddot{u} + 0.04\dot{u} - 0.2u + 8u^3/15 = 0.4\cos\omega t\,.$$

The harmonic forcing term on the right-hand side provokes a response of the system (beam, electric current). We confine ourselves to *harmonic oscillations* $u(t)$, which have the same period T as the excitation,

$$T = \frac{2\pi}{\omega}\,. \tag{6.6}$$

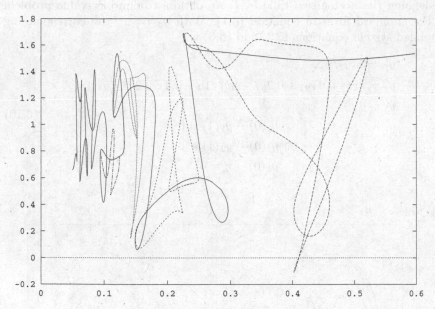

Fig. 6.2. Example 6.2, bifurcation diagram, branches of harmonic solutions, $u(0)$ versus ω

Hence, harmonic oscillations obey the boundary conditions

$$u(0) = u(T), \quad \dot{u}(0) = \dot{u}(T).$$

(For subharmonic oscillations, see Section 7.7.)

Because the period T varies with ω, the integration interval $0 \le t \le T$ must be adapted whenever ω is changed. This is inconvenient when nodes of a numerical solution procedure must also be adapted, so we transform the integration interval to unit length, thereby shifting the dependence on ω to the right-hand side of the differential equation. The normalized time $\tilde{t}$ satisfies

$$T\tilde{t} = t, \quad 0 \le \tilde{t} \le 1. \tag{6.7}$$

The transformation

$$y_1(\tilde{t}) := u(t), \quad y_2(\tilde{t}) := \dot{u}(t) \tag{6.8}$$

leads to the first-order system

$$\begin{aligned} y_1' &= Ty_2, \\ y_2' &= T(-0.04y_2 + 0.2y_1 - 8y_1^3/15 + 0.4\cos 2\pi\tilde{t}) \end{aligned} \tag{6.9}$$

with boundary conditions

$$\begin{aligned} y_1(0) - y_1(1) &= 0, \\ y_2(0) - y_2(1) &= 0. \end{aligned}$$

Redefining the normalized time by t, we obtain a boundary-value problem of the standard form in equation (6.1). With $y_3 := \omega$, the corresponding extended system equations (6.2) and (6.3) is

$$y_1' = 2\pi y_2/y_3 \,,$$
$$y_2' = 2\pi(-0.04y_2 + 0.2y_1 - 8y_1^3/15 + 0.4\cos 2\pi t)/y_3 \,,$$
$$y_3' = 0 \,,$$

$$\begin{aligned} y_1(0) - y_1(1) &= 0 \,, \\ y_2(0) - y_2(1) &= 0 \,, \\ y_k(0) - \eta &= 0 \,. \end{aligned}$$

(6.10)

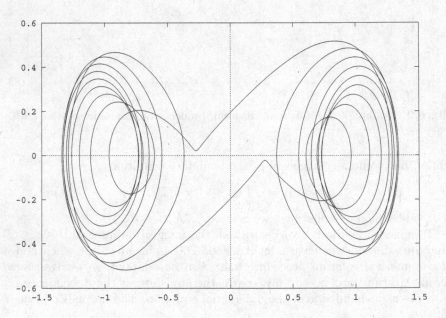

Fig. 6.3. Example 6.2, $(u, \dot{u})$-plane for $\omega = 0.053$, symmetric oscillation

Harmonic solutions can be calculated by solving equation (6.10). With appropriate choices for k and η, the bifurcation diagram in Figure 2.7 results. The branching behavior is rich for small values of ω. As Figure 6.2 shows, there are many branches, turning points, and bifurcation points for values of the parameter $\omega < 0.5$. The solid curve in this branching diagram represents oscillations with phase diagrams being symmetric with respect to the origin (Figure 6.3). For decreasing ω, each loop of this "main" branch attaches a further wiggle to the oscillation. The wiggles indicate that the small-amplitude oscillation around any of the stable equilibria (Section 2.3) takes much time to collect enough energy before a transition to the other attracting basin is possible; compare the phase diagram in Figure 6.3 and

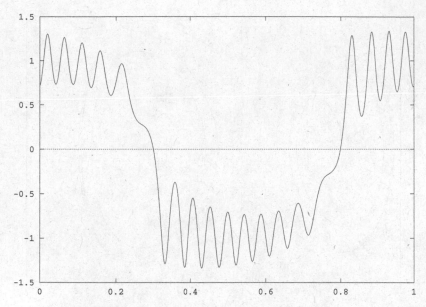

Fig. 6.4. Example 6.2, $u(t)$ for $\omega = 0.053$, t is a normalized time, $t = 1$ corresponds to $T = 2\pi/\omega = 118.55$

the time dependence in Figure 6.4 ($\omega = 0.053$). At first glance, the many wiggles might give the impression that such solutions are not harmonic; but they do have period T.

Each of the closed branches in Figure 6.2 (dashed and dotted curves) is attached to the main branch via two pitchfork bifurcations. Accordingly, the phase diagrams of these "secondary" branches are asymmetric with respect to the origin; Figure 6.5 shows one such phase plot. The closed branch drawn in a dashed line (Figure 6.2) appears to have sharp corners, and one might expect difficulties in tracing this particular branch. The plotted behavior refers only to the dependence $y_1(0)$ versus λ; the corresponding graph of $y_2(0)$ behaves more smoothly (see Figure 6.6). Accordingly, a local parameterization does not encounter difficulties in finding an appropriate index k. The closed branch depicted in Figure 6.6 illustrates how difficult it is to decide when the tracing of the branch is completed. As outlined in Section 4.9, it is not easy to implement a criterion that recognizes safely that a part of a branch is already known. Note that all of the Figures 6.2 through 6.6 depict harmonic solutions. □

For periodic oscillations it is interesting to know the amplitude. The amplitude can often be measured sufficiently accurately from related plots (as in Figures 6.3, 6.4, and 6.5). If more accuracy is required, one must resort to numerical approximations. The amplitude of an oscillation can be calculated with high accuracy as part of the solution procedure [BeS80]. As an alternative, the amplitude can be approximated after the solution is cal-

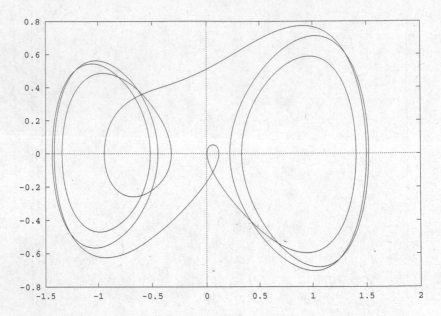

Fig. 6.5. Example 6.2, $(u, \dot{u})$-plane for $\omega = 0.12861$, asymmetric oscillation

culated. To this end, a spline that approximates the current solution can be constructed. Evaluating the zeros of the derivative of the spline yields the values of the amplitude (Exercise 6.3). A table of 20 bifurcation points of the above Duffing equation with accurate values of the amplitude can be found in [BeS80]. The above Duffing equation also exhibits *subharmonic* oscillations; then the period of the system is an integral multiple of the period of the external frequency. The period-adding behavior has been investigated in [EnL91]. Note that periodic oscillations can also be calculated by methods other than applying shooting. For example, Fourier expansion methods have shown good results. For an application to bifurcation, see [Del92].

6.2 Calculation of Bifurcation Points

The calculation of bifurcation points in boundary-value problems follows along the lines outlined in the previous chapter. We establish a general direct method and deduce a bifurcation test function on the ODE level.

The linearization of equation (6.1) with respect to $\mathbf{y}$ is the boundary-value problem

$$\begin{aligned}
\mathbf{h}' &= \mathbf{f_y}(t, \mathbf{y}, \lambda)\mathbf{h}, \\
\mathbf{A}\mathbf{h}(a) &+ \mathbf{B}\mathbf{h}(b) = \mathbf{0}.
\end{aligned} \tag{6.11}$$

The vector-valued function $\mathbf{h}(t)$ corresponds to the $\mathbf{h}$ of Chapter 5. $\mathbf{A}$ and $\mathbf{B}$ are the n^2 matrices of the linearization of the boundary conditions,

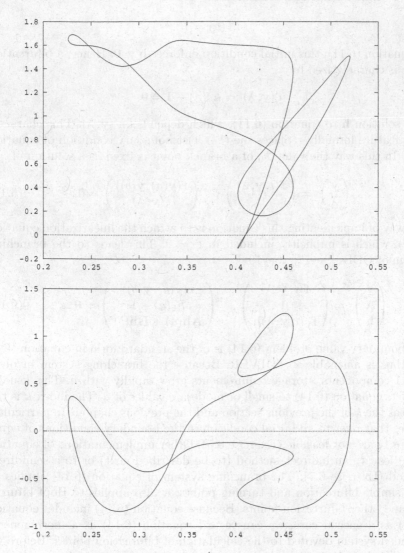

Fig. 6.6. Example 6.2: Closed branch u versus ω (top), $\dot{u}$ versus ω (bottom)

$$\mathbf{A} := \frac{\partial \mathbf{r}(\mathbf{y}(a), \mathbf{y}(b))}{\partial \mathbf{y}(a)}, \quad \mathbf{B} := \frac{\partial \mathbf{r}(\mathbf{y}(a), \mathbf{y}(b))}{\partial \mathbf{y}(b)}. \tag{6.12}$$

Note that $\mathbf{A}$ and $\mathbf{B}$ in general vary with $\mathbf{y}$. But, since $\mathbf{A}$ and $\mathbf{B}$ are frequently constant matrices, we suppress the dependence on $\mathbf{y}$ for the sake of convenience. Evaluated at a bifurcation point $(\mathbf{y}_0, \lambda_0)$, the linearized problem in equation (6.11) has a nontrivial solution $\mathbf{h} \neq \mathbf{0}$. For $(\mathbf{y}, \lambda) \neq (\mathbf{y}_0, \lambda_0)$, the only solution to equation (6.11) is $\mathbf{h} = \mathbf{0}$. Along the lines set out in the previous chapter, we impose the equation

$$h_k(a) = 1$$

on equation (6.11); this initial condition enforces $\mathbf{h} \neq \mathbf{0}$. Hence, a bifurcation point is characterized by

$$\zeta(\mathbf{y}, \lambda) := h_k(a) - 1 = 0$$

for a solution $\mathbf{h}$ to equation (6.11), which depends on $(\mathbf{y}, \lambda)$. This characteristic feature formally replaces the $(n+1)$st boundary condition of equation (6.3). In this way the value λ_0 of a branch point is fixed as a solution of

$$\begin{pmatrix} \mathbf{y} \\ \lambda \end{pmatrix}' = \begin{pmatrix} \mathbf{f}(t, \mathbf{y}, \lambda) \\ 0 \end{pmatrix}, \quad \begin{pmatrix} \mathbf{r}(\mathbf{y}(a), \mathbf{y}(b)) \\ \zeta(\mathbf{y}, \lambda) \end{pmatrix} = \mathbf{0}. \qquad (6.13)$$

One way of implementing this equation is to attach the linearization equation (6.11), which is implicitly included in $\zeta = 0$. This leads to the branching system of ODEs [Sey77], [Sey79a],

$$\begin{pmatrix} \mathbf{y} \\ \lambda \\ \mathbf{h} \end{pmatrix}' = \begin{pmatrix} \mathbf{f}(t, \mathbf{y}, \lambda) \\ 0 \\ \mathbf{f_y}(t, \mathbf{y}, \lambda)\mathbf{h} \end{pmatrix}, \quad \begin{pmatrix} \mathbf{r}(\mathbf{y}(a), \mathbf{y}(b)) \\ h_k(a) - 1 \\ \mathbf{A}\mathbf{h}(a) + \mathbf{B}\mathbf{h}(b) \end{pmatrix} = \mathbf{0}. \qquad (6.14)$$

The boundary-value problem (6.14) is of the standard form in equation (6.2) and thus is amenable to SOLVER. Because the branching system involves $2n+1$ components, storage requirements grow rapidly with n. This usually restricts equation (6.14) to small or moderate values of n. The index k is the same as the k of the previous section and the previous chapter. In particular, indices that refer to an initial condition in the boundary conditions of equation (6.1) are not feasible (Exercise 6.5). Other implementations of equation (6.13) lead to an indirect method (to be described next) or to a semidirect method (Exercise 6.7). The branching system of equation (6.14) applies to both simple bifurcation and turning points; it also applies to Hopf bifurcation and other bifurcation points. Because equation (6.14) includes equation (5.21) as a special case, we can regard equation (6.14) as a most general extended system devoted to the calculation of bifurcation points. Before we discuss such topics as the calculation of an initial guess, we study an example.

Example 6.3   Catalytic Reaction
Heat and mass transfer within a porous catalyst particle can be described by the boundary-value problem

$$\frac{d^2 y}{dx^2} + \frac{a}{x}\frac{dy}{dx} = \vartheta^2 y^m \exp\left[\frac{\gamma\beta(1-y)}{1+\beta(1-y)}\right]. \qquad (6.15)$$

The meaning of the variables (see [HlMK68]) is

y: dimensionless concentration;
x: dimensionless coordinate;

a: coefficient defining shape of particle ($a = 0$: flat plate; $a = 1$: cylindrical; $a = 2$: spherical);

ϑ: Thiele modulus;

γ: dimensionless energy of activation; and

β: dimensionless parameter of heat evolution.

The boundary conditions are

$$\frac{dy(0)}{dx} = 0, \quad y(1) = 1.$$

We choose the simple case of a flat particle ($a = 0$) with first-order reaction ($m = 1$) and take as a bifurcation parameter the modified Thiele modulus $\lambda := \vartheta^2$.

The resulting first-order system is

$$\begin{aligned} y_1' &= y_2, \\ y_2' &= \lambda y_1 e(y_1), \\ y_1(1) &= 1, \quad y_2(0) = 0. \end{aligned} \tag{6.16}$$

Here $e(y)$ serves as an abbreviation of the exponential term,

$$e(y_1) = \exp\left[\gamma\beta(1 - y_1)/(1 + \beta(1 - y_1))\right].$$

The Jacobian matrix $\mathbf{f_y}$ associated with equation (6.16) is

$$\mathbf{f_y}(t, \mathbf{y}, \lambda) = \begin{pmatrix} 0 & 1 \\ \lambda\left[e(y_1) + \frac{de}{dy_1}y_1\right] & 0 \end{pmatrix}.$$

We write $y_3 = \lambda$, $y_4 = h_1$, $y_5 = h_2$ and obtain for the linearized system

$$h_1' = y_4' = h_2 = y_5,$$

$$h_2' = y_5' = y_3 y_4\left[e(y_1) + \frac{de}{dy_1}y_1\right].$$

Because the boundary conditions are linear (Exercise 6.6), the boundary conditions of the linearization can be written down immediately as

$$h_1(1) = 0, \quad h_2(0) = 0.$$

Collecting all parts, one obtains the branching system

$$\begin{aligned} y_1' &= y_2, \\ y_2' &= y_1 y_3 e(y_1), \\ y_3' &= 0, \\ y_4' &= y_5, \\ y_5' &= y_3 y_4 e(y_1)[1 - y_1\gamma\beta(1 + \beta(1 - y_1))^{-2}], \\ y_1(1) &= 1, \quad y_2(0) = 0, \\ y_4(1) &= 0, \quad y_5(0) = 0, \quad y_4(0) = 1 \ \ (k = 1). \end{aligned} \tag{6.17}$$

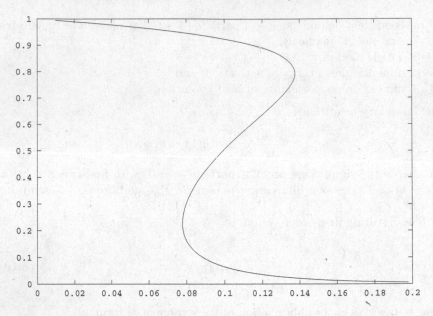

Fig. 6.7. Example 6.3, $y_1(0)$ versus λ for $\gamma = 20$, $\beta = 0.4$

For $\gamma = 20$, $\beta = 0.4$, and positive values of λ, there is a branch that exhibits hysteresis behavior; see the bifurcation diagram in Figure 6.7. During the course of the continuation, the two turning points have been calculated by solving the branching system (6.17). The resulting values are

$$\lambda_0 = 0.07793\,, \quad y_0(0) = 0.2273\,,$$
$$\lambda_0 = 0.13756\,, \quad y_0(0) = 0.7928\,.$$

For values of λ between these two critical values, there are three different concentrations (see Figure 6.8). Only the upper and lower profiles are stable.
□

Section 5.9 showed how the hysteresis behavior varies with γ (see Figures 5.21 and 5.22). The critical boundaries of Figure 5.21 are obtained by continuation of the branching system of equation (6.17) with respect to γ or by indirect methods (see Section 5.9). For completeness we remark that for $\lambda < 0$ another branch can be found. These solutions are not physical because concentrations become negative or larger than one. From a mathematical point of view, however, the new branch is interesting because it branches off to infinity for $\lambda \to 0$.

Next we discuss how to calculate initial approximations, and how to define bifurcation test functions. Consider the situation that $(\bar{\mathbf{y}}, \bar{\lambda})$ is a solution to equation (6.1), calculated not far from a bifurcation point $(\mathbf{y}_0, \lambda_0)$. The linearized boundary-value problem equation (6.11) has only the trivial solution if evaluated about $(\bar{\mathbf{y}}, \bar{\lambda}) \neq (\mathbf{y}_0, \lambda_0)$. As in the methods described in Section

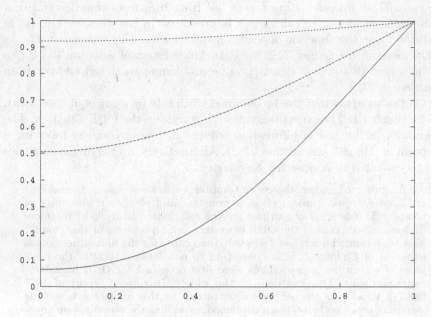

Fig. 6.8. Example 6.3, $y_1(x)$ for $\lambda = 0.1$, three different solutions

5.4.1, the lth boundary condition (for an index l, $1 \leq l \leq n$) is removed to obtain a solvable system,

$$\mathbf{h}' = \mathbf{f_y}(t, \bar{\mathbf{y}}, \bar{\lambda})\mathbf{h} ,$$
$$(\mathbf{I} - \mathbf{e}_l\mathbf{e}_l^{tr})(\mathbf{A}\mathbf{h}(a) + \mathbf{B}\mathbf{h}(b)) = \mathbf{0} , \quad h_k(a) = 1 . \tag{6.18}$$

We denote the solution to equation (6.18) by $\bar{\mathbf{h}}$. This vector function serves as the concluding part of the initial approximation $(\bar{\mathbf{y}}, \bar{\lambda}, \bar{\mathbf{h}})$ of the solution $(\mathbf{y}_0, \lambda_0, \mathbf{h}_0)$ to the branching system of equation (6.14). One way of calculating $\bar{\mathbf{h}}$ is to solve the boundary-value problem

$$\begin{pmatrix} \mathbf{y} \\ \lambda \\ \mathbf{h} \end{pmatrix}' = \begin{pmatrix} \mathbf{f}(t, \mathbf{y}, \lambda) \\ 0 \\ \mathbf{f_y}(t, \mathbf{y}, \lambda)\mathbf{h} \end{pmatrix} , \quad \begin{pmatrix} \mathbf{y}(a) - \bar{\mathbf{y}}(a) \\ \lambda - \bar{\lambda} \\ (\mathbf{I} - \mathbf{e}_l\mathbf{e}^{tr})(\mathbf{A}\mathbf{h}(a) + \mathbf{B}\mathbf{h}(b)) \\ h_k(a) - 1 \end{pmatrix} = \mathbf{0} , \quad (6.19)$$

which differs from equation (6.14) only in the boundary conditions. Solving equation (6.19) is not problematic because it is a linear problem; the nonlinear part is not effective because its solution $\bar{\mathbf{y}}$ is already calculated. In the following section, we see how $\bar{\mathbf{h}}$ can be approximated more efficiently.

By construction, the value of the removed boundary condition is a bifurcation test function τ,

$$\tau := \mathbf{e}_l^{tr}(\mathbf{A}\mathbf{h}(a) + \mathbf{B}\mathbf{h}(b)) . \tag{6.20}$$

One possibility for calculating τ is to use $\mathbf{h}(a), \mathbf{h}(b)$ from equation (6.19); a more economical way to calculate τ is presented in the next section. As in Chapter 5, the test function depends on the choice of the indices l and k. As in the method in Exercise 5.12, the branching system of equation (6.14) can be embedded in a set of boundary-value problems parameterized by τ from equation (6.20).

On the accuracy that can be obtained efficiently by using equation (6.14), see Section 5.4.1. The same observations are valid in the ODE situation. The singularity in the case of bifurcation points can be overcome by following a proposal in [Moo80] (see Section 5.7.2). Alternatively, underlying symmetries can be exploited to remove the singularity.

> As indicated earlier, the entire Chapter 6 can be seen as a special case of Chapter 6; just consider all y_i as constant and identify $\mathbf{r}$ with the $\mathbf{f}$ of Chapter 5. Numerical algorithms for the bifurcation analysis of the more difficult infinite case of the ODE boundary-value problems of this chapter have been treated much less frequently than those for the finite-dimensional situation of Chapter 5. It is interesting to note that several of the basic ideas of algorithms and methods were first proposed for ODE boundary-value problems. This was done in the widely distributed report of 1977 [Sey77], which was the author's dissertation. In this work, the branching system of equation (6.14) was introduced, as well as the test function equation (6.20), the vector $\bar{\mathbf{h}}$, and the step control equation (4.23). In the same year (1977) the influential paper of Keller [Kel77] was published, and the dissertation of Abbott [Abb77] was completed. These two papers considered the finite-dimensional case. Several years of generalizations, specializations, and applications followed, and many further methods have been proposed. Today, the numerical treatment of bifurcations, and continuation methods can be considered as being on a sophisticated level.

6.3   Stepping Down for an Implementation

Although from a theoretical point of view $\mathbf{h}$ and τ can be satisfactorily defined and calculated via ODE boundary-value problems, this approach is not practical. Notice that equation (6.19) represents a boundary-value problem of the double size. In order to construct an efficient implementation, we temporarily step down from the ODE level to a finite-dimensional approximation. The vector function $\mathbf{h}(t)$ can be approximated for discrete values of t_j, which will suffice for both branch switching and calculating τ.

Information required for evaluating $\mathbf{h}$ and τ is hidden in the data calculated during the approximation of a solution $\mathbf{y}$ to equation (6.1) or (6.3). It is possible to reveal this information, thereby benefiting from data that are available without extra cost. Different methods incorporated in SOLVER produce data in different ways. Accordingly, the analysis of revealing $\mathbf{h}$ and τ depends on the type of method applied to solve equation (6.1) or (6.3). To discuss an example, let us assume that SOLVER is based on multiple shooting. A corresponding analysis has been carried out in [Sey79a]; the results are as follows.

Assume that $\mathbf{y}$ was calculated by multiple shooting. In a shooting implementation in condensed form, an n^2 iteration matrix, which is the core of the internal Newton iteration, is established [StB80]. This iteration matrix is

$$\mathbf{E} := \mathbf{A} + \mathbf{B}\mathbf{G}_{m-1} \cdot \ldots \cdot \mathbf{G}_1 \,. \tag{6.21}$$

The matrices $\mathbf{A}$ and $\mathbf{B}$ are those of the linearized boundary conditions equation (6.12). The matrices $\mathbf{G}_j$ are defined as

$$\mathbf{G}_j = \mathbf{G}(t_{j+1}) \,, \tag{6.22}$$

where $\mathbf{G}(t)$ solves the matrix initial-value problem

$$\mathbf{G}' = \mathbf{f}_{\mathbf{y}}(t, \mathbf{y}, \lambda)\mathbf{G} \,, \quad \mathbf{G}(t_j) = \mathbf{I} \,.$$

Here $a = t_1 < t_2 < \ldots < t_m = b$ are the nodes of multiple shooting ($m = 2$ for simple shooting). When SOLVER terminates, the internal iteration matrix $\mathbf{E}$ and the matrices $\mathbf{G}_j$ can be made available without extra cost. As in the method in Section 5.4.1, we replace the lth row of $\mathbf{E}$ by the kth unit vector to obtain the matrix $\mathbf{E}_{lk}$. This can be formally written as

$$\mathbf{E}_{lk} := (\mathbf{I} - \mathbf{e}_l\mathbf{e}_l^{tr})\mathbf{E} + \mathbf{e}_l\mathbf{e}_k^{tr} \,; \tag{6.23}$$

compare equation (5.23). The matrix $\mathbf{E}_{lk}$ depends on the solution $(\mathbf{y}, \lambda)$ or $(\bar{\mathbf{y}}, \bar{\lambda})$, at which $\mathbf{E}$ was evaluated. If the indices l and k are chosen in a way such that $\mathbf{E}_{lk}$ is nonsingular (not problematic), then equation (6.18) has a unique solution $\mathbf{h}(t)$. The initial vector $\mathbf{h}(a)$ is the solution of the linear equation

$$\mathbf{E}_{lk}\,\mathbf{h}(a) = \mathbf{e}_l \,. \tag{6.24}$$

Test functions $\tau = \tau_{lk}$ are given by the scalar product of $\mathbf{h}(a)$ with the lth row of $\mathbf{E}$,

$$\tau_{lk} = \mathbf{e}_l^{tr}\,\mathbf{E}\,\mathbf{h}(a) \,. \tag{6.25}$$

Because SOLVER outputs $\mathbf{y}$ at the discrete shooting nodes, $\mathbf{y}(t_j)$, it is sufficient to calculate $\mathbf{h}$ at the same points t_j. This is furnished by

$$\mathbf{h}(t_j) = \mathbf{G}_{j-1}\mathbf{h}(t_{j-1}) \quad \text{for } j = 2, \ldots, m \,. \tag{6.26}$$

Summarizing, the bifurcation test function τ can be calculated by solving one linear system, equation (6.24), and evaluating one scalar product, equation (6.25). This amount of work compares favorably to solving boundary-value problem equation (6.19). Because the matrices $\mathbf{A}$, $\mathbf{B}$, and $\mathbf{G}_j$ are approximated by numerical differentiation, the iteration matrix $\mathbf{E}$ is subjected to discretization errors. This affects the accuracy of $\mathbf{h}$ and τ. The remarks in Section 5.1 are valid here too—that is, the accuracy of τ depends on the strategy of rank-one approximations in SOLVER and on the history of previous continuation steps. Further simplifications are possible that exploit decompositions of $\mathbf{E}$.

6.4 Branch Switching and Symmetry

We end the excursion to a finite-dimensional approximation and come back
to the ODE level. Concerning branch switching, approaches similar to those
described earlier are valid. We confine ourselves to a brief list of suitable
boundary-value problems; for motivations and illustrating figures, see Section
5.6.

Predictors $\bar{\mathbf{z}}$ can be based on interpolation,

$$\bar{\mathbf{z}}(t) := \bar{\mathbf{y}}_0(t) \pm |\delta| \bar{\mathbf{h}}_0(t) \,. \tag{6.27}$$

Here $\bar{\mathbf{y}}_0(t)$ and $\bar{\mathbf{h}}_0(t)$ are linear interpolations based on the values of $\mathbf{y}$ and
$\mathbf{h}$ obtained at two solutions not far from a bifurcation point. Equation (5.46)
holds true for boundary-value problems; in equation (5.46b) the Jacobian $\mathbf{f}_{\mathbf{y}}$
is replaced by the matrix $\mathbf{E}$. The distance δ is given by equation (5.45); in
this formula, y_k is evaluated at $t = a$. If $\mathbf{h}$ is approximated as described in
the previous section, the predictor $\bar{\mathbf{z}}$ is given only at the discrete nodes t_j.
This is no disadvantage because SOLVER does not require more.

The boundary-value problem that establishes a parallel computation of
two distinct solutions is written in standard form, equation (6.2), with $\mathbf{Y} =
(\mathbf{y}, \lambda, \mathbf{z})$,

$$\mathbf{Y}' = \mathbf{F}(t, \mathbf{Y}) = \begin{pmatrix} \mathbf{f}(t, \mathbf{y}, \lambda) \\ 0 \\ \mathbf{f}(t, \mathbf{z}, \lambda) \end{pmatrix} ,$$

$$\mathbf{R}(\mathbf{Y}(a), \mathbf{Y}(b), \delta) = \begin{pmatrix} \mathbf{r}(\mathbf{y}(a), \mathbf{y}(b)) \\ z_k(a) - y_k(a) - \delta \\ \mathbf{r}(\mathbf{z}(a), \mathbf{z}(b)) \end{pmatrix} = \mathbf{0} \,. \tag{6.28}$$

The initial approximation is $(\bar{\mathbf{y}}_0, \bar{\lambda}_0, \bar{\mathbf{z}})$. We shall prefer equations of the simple dimension $n+1$ to the dimension $2n+1$ of equation (6.28). As mentioned
in Chapter 5, the method of parallel computation is most attractive when an
analytic expression for $\mathbf{y}$ is known and inserted. Another possibility to obtain
a "small" system is given by equation (6.3), initial approximation and η are

$$(\mathbf{y}, \lambda) = (\bar{\mathbf{z}}, \bar{\lambda}_0), \quad \eta = \bar{z}_k(a) \,. \tag{6.29}$$

Better selective properties are obtained by exploiting symmetry breaking.
To classify and investigate symmetries in the solution, consider a *reflection*
$\tilde{y}_i(t)$ of $y_i(t)$. For example,

$$\tilde{y}_i(t) := y_i(a + b - t)$$

defines the reflection with respect to the line $t = \frac{1}{2}(a + b)$. We define the
component y_i to be symmetric if

$$\tilde{y}_i(t) = y_i(t)$$

holds for all t in the interval $a \leq t \leq b$. The type of symmetry depends on the reflection and can be different in the components of the vector $\mathbf{y}$. We shall say that a boundary-value problem *supports* symmetry if it is solved by $\tilde{\mathbf{y}}$ whenever it is solved by $\mathbf{y}$.

In particular, two kinds of symmetries occur frequently in bifurcation problems: even functions (6.30a) and odd functions (6.30b):

$$y_i(t) = \tilde{y}_i(t) \qquad \text{for } \tilde{y}_i(t) := y_i(a + b - t), \tag{6.30a}$$

$$y_i(t) = \tilde{y}_i(t) \qquad \text{for } \tilde{y}_i(t) := -y_i(a + b - t). \tag{6.30b}$$

In equation (6.30b) the reflection is with respect to the point $(y, t) = (0, \frac{1}{2}(a + b))$. Note that a component y_i can be odd while another component of the same solution is even. Two other types of symmetry are characteristic for bifurcations of periodic solutions with period $b - a$:

$$y_i(t) = \tilde{y}_i(t) \qquad \text{for } \tilde{y}_i(t) := -y_i\left(t + \tfrac{b-a}{2}\right), \tag{6.30c}$$

$$y_i(t) = \tilde{y}_i(t) \qquad \text{for } \tilde{y}_i(t) := y_i\left(t + \tfrac{b-a}{2}\right). \tag{6.30d}$$

A periodic solution of type (6.30d) actually has a period $\frac{1}{2}(b - a)$.

For a practical evaluation of symmetry, we first check a current solution to see whether its components are symmetric. Because it is not practical to check the above criteria for all t in the interval $a \leq t \leq b$, we confine ourselves to $t = a$. This establishes necessary criteria for the onset of symmetries. For a small error tolerance ϵ_1 (not smaller than the error tolerance of SOLVER) we check the criteria

$$|y_i(b) - y_i(a)| < \epsilon_1, \tag{6.31a}$$

$$|y_i(b) + y_i(a)| < \epsilon_1, \tag{6.31b}$$

$$|y_i(a) + y_i\left(\tfrac{a+b}{2}\right)| < \epsilon_1, \tag{6.31c}$$

$$|y_i(a) - y_i\left(\tfrac{a+b}{2}\right)| < \epsilon_1. \tag{6.31d}$$

If for one index i (say k) one of the inequalities holds, we assume a symmetry of the corresponding type in (6.30). The next question is whether this symmetry is broken in the component z_k of the emanating solution. Because the emanating solution $\mathbf{z}$ is yet unknown, we resort to the predictor $\bar{\mathbf{z}}$ from equation (6.27). For a symmetric $\bar{\mathbf{z}}$, the following value of γ is zero:

$$
\begin{aligned}
\gamma &:= \bar{z}_k(b) - \bar{z}_k(a) && \text{in case (6.31a)},\\
\gamma &:= \bar{z}_k(b) + \bar{z}_k(a) && \text{in case (6.31b)},\\
\gamma &:= \bar{z}_k(a) + \bar{z}_k\left(\tfrac{a+b}{2}\right) && \text{in case (6.31c)},\\
\gamma &:= \bar{z}_k(a) - \bar{z}_k\left(\tfrac{a+b}{2}\right) && \text{in case (6.31d)}.
\end{aligned}
\tag{6.32}
$$

Hence, asymmetry of $\bar{\mathbf{z}}$ manifests itself by

$$|\gamma| > \epsilon_2; \tag{6.33}$$

here we choose an error tolerance ϵ_2 slightly larger than ϵ_1. Numerical experience indicates that the test criteria (6.31) through (6.33) reliably detect a symmetry breaking. This success is to be expected, because $\bar{z}$ is close to an emanating solution in case of pitchfork bifurcations—that is, in particular the $\mathbf{Z}_2$ symmetry breaking can be detected in advance during branch tracing. This requires monitoring both τ and $\mathbf{h}$. The above test for symmetry breaking is easily automated. Note that the test is based on solutions instead of on equations.

Assume now that for $i - k$ one of the criteria in equation (6.31) holds and that equation (6.33) is satisfied. For each kind of symmetry breaking there is a corresponding boundary-value problem with an appropriate selective boundary condition:

$$\begin{pmatrix} \mathbf{y} \\ \lambda \end{pmatrix}' = \begin{pmatrix} \mathbf{f}(t,\mathbf{y},\lambda) \\ 0 \end{pmatrix}, \quad \begin{pmatrix} \mathbf{r}(\mathbf{y}(a),\mathbf{y}(b)) \\ r_{n+1} \end{pmatrix} = \mathbf{0} \qquad (6.34)$$

with $\gamma \neq 0$ from equation (6.32) and the boundary condition

$$\begin{aligned} r_{n+1} &:= y_k(b) - y_k(a) - \gamma & \text{in case (6.31a)}, \\ r_{n+1} &:= y_k(b) + y_k(a) - \gamma & \text{in case (6.31b)}, \\ r_{n+1} &:= y_k(a) + y_k\left(\tfrac{a+b}{2}\right) - \gamma & \text{in case (6.31c)}, \\ r_{n+1} &:= y_k(a) - y_k\left(\tfrac{a+b}{2}\right) - \gamma & \text{in case (6.31d)}. \end{aligned}$$

The initial approximation is $(\bar{z}, \bar{\lambda}_0)$. In cases (6.31c) and (6.31d) the boundary-value problem is a three-point boundary-value problem; a transformation to a two-point boundary-value problem is possible [Sey83a]. Case (6.30d) refers to period doubling; this can be seen by setting $T = \frac{1}{2}(b-a)$. The corresponding bifurcation means there is a transition from period T to period $2T$. This kind of bifurcation will be discussed in Chapter 7.

The above numerical test for symmetry breaking dispenses with an analytical investigation of equation (6.1), which can be cumbersome (Exercise 6.10). There is no guarantee that the criteria in equation (6.31) detect all the relevant regularities. For example, equation (6.31b) fails if equation (6.30b) is generalized to

$$y_i(a + b - t) + \vartheta = \vartheta - y_i(t)$$

for a nonzero value of ϑ. Fortunately, if such a y_i is solution of a second-order differential equation, the derivative or antiderivative is even and equation (6.31a) works.

Example 6.4   Superconductivity Within a Slab

In [Odeh67] a model of a superconducting slab in a parallel magnetic field is discussed. The Ginzburg–Landau equations are equivalent to the boundary-value problem of two second-order ODEs

$$\begin{aligned} \Theta'' &= \Theta(\Theta^2 - 1 + \lambda\Psi^2)\kappa^2, \\ \Psi'' &= \Theta^2\Psi \end{aligned} \qquad (6.35)$$

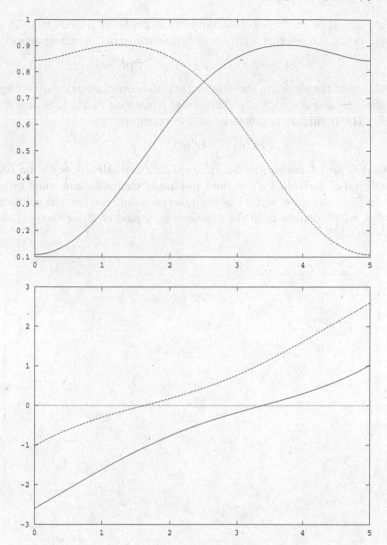

Fig. 6.9. Example 6.4; antisymmetric solutions for $\lambda = 0.6647$. $\Theta(t)$ and $\tilde{\Theta}(t)$ (top), $\Psi(t)$ and $\tilde{\Psi}(t)$ (bottom)

for $0 \leq t \leq d$,

$$\Theta'(0) = \Theta'(d) = 0, \quad \Psi'(0) = \Psi'(d) = 1.$$

The meaning of the variables is

    Θ: an order parameter that characterizes different superconducting states;

    Ψ: potential of the magnetic field;

    λ: square of external field;

    d: thickness of the slab ($d = 5$); and

    κ: Ginzburg–Landau parameter ($\kappa = 1$).

Investigation of the above boundary-value problem reveals that it supports symmetry. To see this, define the "antisymmetric" counterparts

$$\tilde{\Theta}(t) := \Theta(d-t), \quad \tilde{\Psi}(t) := -\Psi(d-t)$$

and realize that the signs are the same after differentiating twice with respect to t. Hence, $\tilde{\Theta}$ and $\tilde{\Psi}$ satisfy the differential equations (6.35) if Θ and Ψ do. Analyzing the boundary conditions, as, for example,

$$\tilde{\Theta}'(0) = -\Theta'(d) = 0\,,$$

shows that $\tilde{\Theta}$ and $\tilde{\Psi}$ satisfy the boundary conditions. Hence, equation (6.35) supports even Θ and odd Ψ. This does *not* imply that solutions must be even or odd. In fact, equation (6.35) has asymmetric solutions. See the solution in Figure 6.9, which depicts both the solutions Θ, Ψ and $\tilde{\Theta}, \tilde{\Psi}$ for the parameter value $\lambda = 0.6647$.

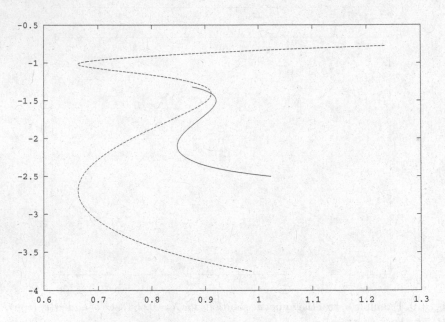

Fig. 6.10. Example 6.4, bifurcation diagram, $y_3(0)$ versus λ

In order to calculate solutions, we rewrite the boundary-value problem as a first-order system. Denoting $y_1 = \Theta$, $y_2 = \Theta'$, $y_3 = \Psi$, and $y_4 = \Psi'$ yields

$$
\begin{aligned}
y_1' &= y_2\,,\\
y_2' &= y_1(y_1^2 - 1 + \lambda y_3^2)\,,\\
y_3' &= y_4\,,\\
y_4' &= y_1^2 y_3\,,\\
y_2(0) &= y_2(5) = 0\,, \quad y_4(0) = y_4(5) = 1\,.
\end{aligned}
\tag{6.36}
$$

Solutions are summarized in the bifurcation diagram in Figure 6.10. A branch of asymmetric solutions branches off a branch of symmetric solutions at a pitchfork bifurcation point. The data of the pitchfork point and four turning points (TP) are listed in Table 6.1. The values in that table have been calculated by solving the branching system. The solution profiles of the bifurcation point are shown in Figure 6.11; notice the symmetries. This bifurcation point is straddled between the two turning points with asymmetric solutions, which are depicted in Figure 6.9. Compare Figures 6.9 and 6.11 to see how this straddling becomes evident in the solution profiles.   □

TABLE 6.1.   Example 6.4, bifurcations.

λ_0	$\Theta_0(0)$	$\Psi_0(0)$	Type
0.66294	0.08996	−2.687	TP, asymmetric sol.
0.66294	0.8461	−1.016	TP, asymmetric sol.
0.91196	0.5853	−1.407	Pitchfork bif. point
0.92070	0.5214	−1.496	TP, symmetric sol.
0.84769	0.1923	−2.095	TP, symmetric sol.

The boundary-value problem equation (6.34) is designed to calculate asymmetric solutions. For the sake of completeness, we mention that boundary-value problems can be reformulated in a way that admits only symmetric solutions. We illustrate this possibility by means of the above example. Symmetric solutions to equation (6.35) satisfy

$$\Theta'\left(\tfrac{d}{2}\right) = 0, \quad \Psi\left(\tfrac{d}{2}\right) = 0$$

(see Figure 6.11). This allows one to reduce the integration interval from $0 \le t \le d$ to

$$0 \le t \le \frac{d}{2}, \quad \text{or, generally,} \quad a \le t \le \frac{a+b}{2}.$$

Replacing the previous boundary conditions at $t = d$ by the midpoint conditions leads to the new boundary conditions

$$\Theta'(0) = 0, \quad \Psi'(0) = 1, \quad \Theta'\left(\tfrac{d}{2}\right) = 0, \quad \Psi\left(\tfrac{d}{2}\right) = 0.$$

Solving a corresponding boundary-value problem yields the symmetric solutions supported by this example. Note that for this modified boundary-value problem the symmetry-breaking bifurcation *does not exist*.

Let us discuss the matter in more general terms. The "space" of symmetric functions is a (small) subspace of the space of all functions. The smaller space reflects certain assumptions that are imposed on the underlying model (here it is the assumption of symmetry). Relaxing a restricting assumption

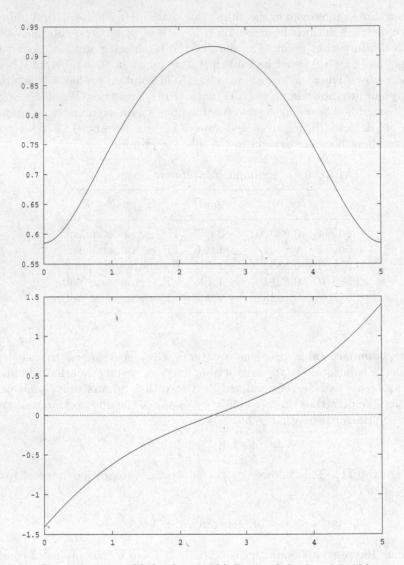

Fig. 6.11. Example 6.4, $\Theta(t)$ (top) and $\Psi(t)$ (bottom) for $\lambda = 0.91196$

means enlarging the space of admissible functions. As a result, the bifurcation behavior may become richer. Another example of this phenomenon will be presented in Example 7.6 in Section 7.4.1.

6.5   Trivial Bifurcation

Before modern computers and powerful numerical methods became generally available, a substantial part of the bifurcation research concentrated on problems where one *basic solution* $\mathbf{y}^B$ is known analytically for all λ. This drastic restriction is theoretically equivalent to the assumption

$$\mathbf{f}(t, \mathbf{0}, \lambda) = \mathbf{0}, \quad \mathbf{r}(\mathbf{0}, \mathbf{0}) = \mathbf{0}. \tag{6.37}$$

This can be seen by applying the transformation

$$\mathbf{y}^* = \mathbf{y} - \mathbf{y}^B$$

to the original problem (Exercise 6.10). We refer to bifurcations from the trivial solution $\mathbf{y} = \mathbf{0}$ as *trivial bifurcations* or *primary bifurcations*.

In principle, all the methods described so far apply to trivial bifurcation; some of the relevant methods simplify significantly. The branching system equation (6.14) reduces to

$$\begin{pmatrix} \mathbf{h} \\ \lambda \end{pmatrix}' = \begin{pmatrix} \mathbf{f}_\mathbf{y}(t, \mathbf{0}, \lambda)\mathbf{h} \\ 0 \end{pmatrix}, \quad \begin{pmatrix} \mathbf{A}\mathbf{h}(a) + \mathbf{B}\mathbf{h}(b) \\ h_k(a) - 1 \end{pmatrix} = \mathbf{0}. \tag{6.38}$$

In contrast to equation (6.14), this boundary-value problem of the simple dimension must be considered an efficient tool for calculating bifurcation points, at least as far as storage requirements are concerned. The Jacobian $\mathbf{f}_\mathbf{y}$, however, is still required analytically. Indirect methods simplify drastically in the case of trivial bifurcations [Sey81b]; they are a recommended alternative to solving equation (6.38). Note that the techniques reported here and in Section 6.2 are readily applied to linear *eigenvalue problems*—namely, to the calculation of eigenvalues and eigenvectors.

The assumption in equation (6.37) often enables a simple analytical evaluation of bifurcation points. Although we generally put emphasis on numerical methods, let us take a break and carry out such a hand calculation.

Example 6.5
Consider the simple second-order boundary-value problem

$$-u'' + u^3 = \lambda u, \quad u(0) = u(\pi) = 0. \tag{6.39}$$

Clearly, equation (6.37) is satisfied. For the left-hand operator $F(u) := -u'' + u^3$ the linearization is obtained from

$$F(u + v) - F(u) = -v'' + 3u^2v + v^2(3u + v).$$

The linear part is $-v'' + 3u^2v$. Hence, the linearization around $u = 0$ is the eigenvalue problem

$$-v'' = \lambda v, \quad v(0) = v(\pi) = 0.$$

Because both $\cos(\sqrt{\lambda}t)$ and $\sin(\sqrt{\lambda}t)$ solve the linearized differential equation, and the general solution of the equation must be of the form

$$v(t) = c_1 \cos \sqrt{\lambda}t + c_2 \sin \sqrt{\lambda}t.$$

The constants and eigenvalues are determined by the boundary conditions. We obtain $c_1 = 0$ and $\sin(\sqrt{\lambda}t) = 0$, which fixes the eigenvalues

$$\lambda = j^2, \quad j = 1, 2, 3, \ldots.$$

In this example, all the eigenvalues are bifurcation points. Simple eigenvalues of the linearization are always bifurcation points of the nonlinear problem. We remark in passing that the emanating solutions of equation (6.39) can be calculated analytically in terms of elliptic functions [Sey75].   □

We close this section by investigating a more demanding example:

Example 6.6   Buckling of a Rod
Consider a rod of length L that is subject to end loading (load $P > 0$). The shape of the rod is described by the tangent angle $u(x)$ as a function of the material points x on the rod in its originally straight position [Olm77]. The variation of the angle with x is zero at the ends of the rod,

$$u'(0) = u'(L) = 0. \tag{6.40a}$$

A deformation of the rod depends on the pressure P, on the stiffness K of the rod material, and on compressibility effects. Neglecting the latter effects, the tangent angle satisfies the differential equation

$$u'' + \frac{P}{K} \sin(u) = 0. \tag{6.40b}$$

This is the famous Euler buckling problem [Eul52]; a finite-element analog is discussed in Exercise 2.14. Euler showed that the undeformed state $u = 0$ is the only solution for P/K less than the first eigenvalue of the linearization,

$$\frac{P}{K} < \left(\frac{\pi}{L}\right)^2.$$

A bent state is stable for values of P/K exceeding this critical value. Solutions in closed form can be found in [Sta71].

A rod with nonlinear compressibility properties is discussed in [Olm77]; the differential equation is

$$u'' + \frac{P}{K} \exp[C(P \cos u(x)) - C(0)] \sin u = 0 \tag{6.41}$$

with $C(\zeta)$ chosen to be

$$C(\zeta) = \frac{1}{5} \tanh[5(3 - 5\zeta + 2\zeta^2)].$$

In order to make it fit into the analytical framework of [Olm77], this problem was simplified by fixing a unit load $P = 1$. This means varying the stiffness K or the parameter

$$\lambda := \frac{\exp(-C(0))}{K} = \frac{0.81873\ldots}{K}$$

in the boundary-value problem

$$u'' + \lambda \exp[C(\cos(u))]\sin(u) = 0, \quad u'(0) = u'(L) = 0. \tag{6.42}$$

We follow this simplification in order to obtain comparable results; numerical methods can handle the original problem, equation (6.41), taking $\lambda = P$ as the parameter.

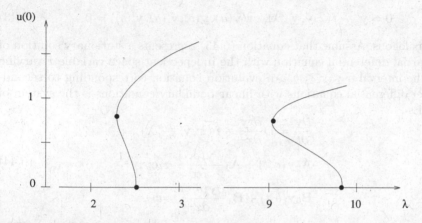

Fig. 6.12. Equation (6.42), bifurcation diagram, $u(0)$ versus λ

The boundary-value problem equation (6.42) has been solved for $L = 2$. Figure 6.12 shows a bifurcation diagram. The eigenvalues of the linearization (Exercise 6.15) can be calculated numerically by solving equation (6.38). Alternatively, approximations $\bar{\lambda}_0, \bar{h}_0$ are obtained effectively by the method of Section 6.3, which simplifies significantly in case of trivial bifurcation. Starting from these approximations, the emanating branch is traced by means of the boundary-value problem (6.3). This reveals that the nontrivial branches locally branch off to the left, but globally the branches "soon" turn back to the right. The related turning points are easily obtained as by-products of the continuation; the methods indicated in Exercises 5.7 through 5.9 apply without change to boundary-value problems. The turning point with the smallest value of λ is

$$\lambda_0 = 2.242, \quad u(0) = 0.7293.$$

□

6.6   Testing Stability

This section looks at the stability of solutions to boundary-value problems. Recall the situation encountered with nonlinear equations

$$0 = f(y, \lambda).$$

If this equation represents equilibria of a system of ODEs, the dynamic behavior is governed by the *evolution equation*

$$\dot{y} = \frac{dy}{dt} = f(y, \lambda).$$

The situation of second-order ODE boundary-value problems

$$0 = y'' - f(x, y, y', \lambda), \quad r(y(a), y(b), y'(a), y'(b)) = 0 \qquad (6.43)$$

is analogous. Assume that equation (6.43) represents a stationary solution of a partial differential equation with the independent space variable x varying in the interval $a \le x \le b$. An evolution equation corresponding to second-order differential equations with linear boundary conditions is the system of PDEs

$$\frac{\partial y}{\partial t} = D \frac{\partial^2 y}{\partial x^2} + f\left(x, y, \frac{\partial y}{\partial x}, \lambda\right),$$

$$A_0 y(a, t) + A_1 \frac{\partial y(a, t)}{\partial x} = c_a, \qquad (6.44)$$

$$B_0 y(b, t) + B_1 \frac{\partial y(b, t)}{\partial x} = c_b.$$

Prominent examples of this class are reaction-diffusion systems, see Section 3.5.3. Solutions to equation (6.44) depend on both time t and space variable x, $y(x, t)$ for $a \le x \le b$, $t \ge 0$. The matrices D, A_0, A_1, B_0, and B_1 are of size n^2, and c_a and c_b are n-vectors, fixing linear boundary conditions for $x = a$ and $x = b$. Often, the matrices are diagonal. With

$$y' = \frac{\partial y}{\partial x}, \quad y'' = \frac{\partial^2 y}{\partial x^2},$$

we write the ODE problem that represents the steady states of equation (6.44) in a form similar to equation (6.43),

$$0 = D y'' + f(x, y, y', \lambda),$$
$$A_0 y(a) + A_1 y'(a) = c_a, \qquad (6.45)$$
$$B_0 y(b) + B_1 y'(b) = c_b.$$

The Brusselator model is an example for the pair of equations (6.44) and (6.45). The full PDE problem of the Brusselator is listed in equation (5.53); the steady-state problem is in equation (6.4).

6.6.1 Elementary Approaches

It makes sense to investigate the stability of a solution $\mathbf{y}_s(x)$ of the steady-state problem equation (6.45) that has the evolution equation (6.44) as background. Clearly, $\mathbf{y}(x,t) = \mathbf{y}_s(x)$ is a solution to the full system in equation (6.44) for all t. The question is whether nearby solutions approach the steady state or leave its neighborhood. Testing the stability of a steady state is cumbersome [HeP81], [PeOH81], [JeR82], [RoH83], [KuH84a], [Sey85a]. In Section 3.5.4 we have discussed a linear stability analysis for homogeneous solutions of reaction-diffusion problems. In what follows we describe some numerical strategies.

One method for testing a steady state for stability simulates the dynamic behavior. To this end, $\mathbf{y}_s(x)$ is taken as an initial profile for $t = 0$. Then equation (6.44) is solved numerically as an initial-value problem starting from

$$\mathbf{y}(x,0) = \mathbf{y}_s(x)$$

or from an artificially perturbed $\tilde{\mathbf{y}}_s$. Initial deviations eventually die out or grow. Stability is indicated when $\mathbf{y}(x,t)$, for increasing t, converges to $\mathbf{y}_s(x)$ or remains close. On the other hand, instability is signaled when $\mathbf{y}(x,t)$ takes values that are significantly distinct from the initial profile. The difficulty with simulation lies in the uncertainty about how to perturb $\tilde{\mathbf{y}}_s$ and which value of t_f to carry out the integration.

A second method is based on a semidiscretization in the space variable —namely, the *method of lines*. This can also be used for a simulation. But now our aim is to set up a framework for an eigenvalue analysis. For example, perform a simple discretization using the standard difference quotients

$$\begin{aligned}
\frac{\partial^2 \mathbf{y}(x)}{\partial x^2} &\approx \frac{\mathbf{y}(x+\Delta) - 2\mathbf{y}(x) + \mathbf{y}(x-\Delta)}{\Delta^2}, \\
\frac{\partial \mathbf{y}(x)}{\partial x} &\approx \frac{\mathbf{y}(x+\Delta) - \mathbf{y}(x-\Delta)}{2\Delta}
\end{aligned} \tag{6.46}$$

for a small increment Δ. This is easily organized by imposing an equidistant grid on the interval $a \le x \le b$, using N interior grid points,

$$\Delta := \frac{b-a}{N+1}, \quad x_j := a + j\Delta \quad \text{for } j = 0, 1, \dots, n+1. \tag{6.47}$$

The components $y_i(x,t)$ of $\mathbf{y}(x,t)$ are approximated along the discrete "lines" $x = x_j$. This gives rise to nN scalar functions

$$y_i(x_j, t),$$

which depend on time only. Introducing an nN-vector $\mathbf{Y}$ by

$$Y_\nu(t) := y_i(x_j, t) \text{ with } \nu = (j-1)n + i, \quad i = 1, \dots, n, \ j = 1, \dots, N \tag{6.48}$$

and inserting the difference quotients (6.46) into equation (6.44) yields an ordinary differential equation $\dot{\mathbf{Y}} = \mathbf{F}(\mathbf{Y}, \lambda)$. The generation of $\mathbf{F}(\mathbf{Y}, \lambda)$ is best done by computer; some care in implementing the boundary conditions is required (Exercise 6.17). After the ODE system in standard form is constituted, eigenvalue-based methods for checking the stability apply [Neu93a].

This second approach may be attractive because it is more deterministic than simulation is, but there are many difficulties to overcome. An equidistant grid with a constant spacing Δ is not adequate for many applications. Steep gradients in the $\mathbf{y}_s(x)$ profile call for a nonequidistant grid. Also, to keep the dimension small, difference quotients of an order higher than the second-order discretizations equation (6.46) are preferred. Such improvements lead to highly complex algorithms. In any case, the dimension nN will be large, and the evaluation of the eigenvalues of the Jacobian of $\mathbf{F}$ is mostly too expensive. For alternatives, see Section 5.1. One must frequently resort to a small value of N, hoping that the stability of the stationary PDE solution is reflected by a coarse ODE approximation.

Summarizing, we point out that it requires much effort to test the stability of a solution that was the straightforward result of some ODE solver. Consequently, during continuation stability is tested only occasionally, in order to save on cost.

6.6.2 Inertial Manifolds

The above approach has set up a system of ODEs to approximate the PDE solution. The better the approximation, the more trustworthy a prediction of stability or bifurcation will be. The *method of inertial manifolds* has been applied successfully for a bifurcation and stability analysis [Foi88], [Tém90], [KeJT91], [JoT92], [MaJ92]. This class of methods handles, for example, evolution equations of the type

$$u_t + Au + F(u) = 0 \,. \tag{6.49}$$

Here A is a linear differentiation operator such as $Au = -\gamma u_{xx}$ and $F(u)$ stands for other terms. An example of this type of equation is the reaction-diffusion problem

$$u_t - \gamma u_{xx} + u^3 - u = 0 \tag{6.50}$$

with boundary conditions $u(0, t) = u(\pi, t) = 0$. The linear operator Au acts on an appropriate function space that has a complete orthonormal basis consisting of eigenfunctions $w_1, w_2, w_3, \ldots$ of A, observing the given boundary conditions. Example 6.5 has revealed for the operator $Au = -\gamma u_{xx}$ of equation (6.50) the eigenfunctions $w_k(x) = \sin kx$.

An approximation to u is sought in the form of the truncated series

$$p(x, t) := \sum_{j=1}^{l} a_j(t) w_j(x) \,. \tag{6.51}$$

Let us denote by P_l the projection onto $\mathrm{span}\{w_1, ..., w_l\}$, $p = P_l u$, which implies $q := u - p = (\mathrm{id} - P_l)u$. This allows one to split equation (6.49) into two equations by multiplying with P_l and the complementary $(\mathrm{id} - P_l)$,

$$
\begin{aligned}
p_t + Ap + P_l F(p + q) &= 0\,, \\
q_t + Aq + (\mathrm{id} - P_l)F(p + q) &= 0\,.
\end{aligned}
\tag{6.52}
$$

The first part in equation (6.52) is an equation for the l lower modes, and the second equation takes care of the infinite number of higher modes. Traditionally, Galerkin-type methods discard the higher modes completely and solve the finite-dimensional part

$$
p_t + Ap + P_l F(p) = 0\,;
$$

no correction term is applied in the argument of F. Frequently, this truncation is too severe to accurately predict the long term behavior of $u(t)$. Inertial manifold methods solve

$$
p_t + Ap + P_l F(p + \psi_m(p)) = 0\,,
\tag{6.53}
$$

constructing an *approximate inertial manifold* $\bar{q} = \psi_m(p)$ such that $\bar{q}$ captures the essence of the truncated dynamics. The function $\bar{q} = \psi_m(p)$ serves as a correction term in the argument of F, $u \approx p + \bar{q}$. Constructing $\psi_m(p)$ can be accomplished by taking $m - l$ further eigenfunctions $w_{l+1}, ..., w_m$ into consideration, for $m > l$. This defines a projection Q onto $\mathrm{span}\{w_{l+1}, ..., w_m\}$, $\bar{q} = Qu$. For convenience, we rewrite $\bar{q}$ as q. The truncated dynamics $q = \psi_m(p)$ is then defined by

$$
q_t + Aq + QF(p + q) = 0\,.
\tag{6.54}
$$

A crude approximation for q will suffice. Take, for example, an implicit Euler integration step, with

$$
q_t \approx \frac{q - q_0}{\Delta t}\,,
$$

and start from the trivial approximation $q_0 = 0$, which corresponds to the traditional Galerkin approach. We obtain an implicit equation for q

$$
q + \Delta t\, Aq + \Delta t\, QF(p + q) = 0\,,
$$

which can be written

$$
q = -(I + \Delta t\, A)^{-1} \Delta t\, QF(p + q)\,.
$$

This equation is regarded a fixed point equation; the fixed point iteration converges for suitably chosen Δt. Carrying out only one iteration, starting again from $q = 0$, yields as approximate inertial manifold

$$
q = \psi_m(p) := -\Delta t\,(1 + \Delta t\, A)^{-1} QF(p)\,.
\tag{6.55}
$$

This establishes the higher modes q ($\bar{q}$) as functions of the lower modes p.

For $m \to \infty$, $\psi_m(p)$ approaches the *inertial manifold* $\psi(p)$, and Q approaches $\mathrm{id} - P_l$. Hence, equation (6.54) approximates the second equation in equation (6.52), and the function $q = \psi_m(p)$ in equation (6.53) can be seen as a correction term that improves the accuracy of this equation. Typically, equation (6.53) represents a system of l ODEs for the functions $a_j(t)$. For instance, for $l = 2$, $m = 4$ the above approach means to express the higher modes a_3, a_4 by the lower ones, $a_3 = a_3(a_1, a_2)$, $a_4 = a_4(a_1, a_2)$. The reader is encouraged to try this on equation (6.50). The inertial manifold methods, when applicable, have shown remarkably accurate results using a smaller l than required for traditional Galerkin-type methods.

6.7 Hopf Bifurcation in PDEs

The stability of solutions to ODE boundary-value problems can be lost in similar ways, as is common with solutions to equations $\mathbf{f}(\mathbf{y}, \lambda) = \mathbf{0}$. The way a critical eigenvalue crosses the imaginary axis is crucial [JoS72], [CrR77].

Consider the PDE in equation (6.44) and a solution $\mathbf{y}_s(x)$ to the steady-state equations (6.45). Let us denote the difference between $\mathbf{y}_s(x)$ and a solution $\mathbf{y}(x, t)$ to equation (6.44) by $\tilde{\mathbf{d}}$,

$$\mathbf{y}(x, t) = \mathbf{y}_s(x) + \tilde{\mathbf{d}}(x, t).$$

(For a constant $\mathbf{y}_s$ see Section 3.5.4.) Substituting $\mathbf{y}$ into the PDE in equation (6.44) leads to

$$\frac{\partial \tilde{\mathbf{d}}}{\partial t} = \mathbf{D}\mathbf{y}_s'' + \mathbf{D}\tilde{\mathbf{d}}'' + \mathbf{f}(x, \mathbf{y}_s + \tilde{\mathbf{d}}, \mathbf{y}_s' + \tilde{\mathbf{d}}', \lambda).$$

Linearizing $\mathbf{f}$ about $\mathbf{y}_s, \mathbf{y}_s'$ yields the linearized PDE

$$\frac{\partial \mathbf{d}}{\partial t} = \mathbf{D}\mathbf{d}'' + \mathbf{f}_\mathbf{y}(x, \mathbf{y}_s, \mathbf{y}_s', \lambda)\mathbf{d} + \mathbf{f}_{\mathbf{y}'}(x, \mathbf{y}_s, \mathbf{y}_s', \lambda)\mathbf{d}',$$

$$\mathbf{A}_0\mathbf{d}(a, t) + \mathbf{A}_1\mathbf{d}'(a, t) = \mathbf{0}, \qquad (6.56)$$

$$\mathbf{B}_0\mathbf{d}(b, t) + \mathbf{B}_1\mathbf{d}'(b, t) = \mathbf{0}.$$

For $\mathbf{y} \approx \mathbf{y}_s$, the linear PDE in equation (6.56) reflects the dynamics of the solutions to equation (6.44) that are close to $\mathbf{y}_s(x)$. This analysis parallels the ODE stability analysis outlined in Section 1.2 and in Section 5.7.4. Compare also the analysis of pattern formation in Section 3.5.4. The ansatz

$$\mathbf{d}(x, t) = e^{\mu t}\mathbf{w}(x),$$

substituted into equation (6.56), yields the ODE eigenvalue problem

$$\mu\mathbf{w} = \mathbf{D}\mathbf{w}'' + \mathbf{f_y}(x, \mathbf{y}_s, \mathbf{y}_s', \lambda)\mathbf{w} + \mathbf{f}_{\mathbf{y}'}(x, \mathbf{y}_s, \mathbf{y}_s', \lambda)\mathbf{w}',$$

$$\mathbf{A}_0\mathbf{w}(a) + \mathbf{A}_1\mathbf{w}'(a) = \mathbf{0}, \tag{6.57}$$

$$\mathbf{B}_0\mathbf{w}(b) + \mathbf{B}_1\mathbf{w}'(b) = \mathbf{0}.$$

For complex μ and $\mathbf{w}$, this boundary-value problem is equivalent to two boundary-value problems for real variables. The two real systems are obtained by substituting

$$\mu = \alpha + \mathrm{i}\beta, \quad \mathbf{w}(x) = \mathbf{h}(x) + \mathrm{i}\mathbf{g}(x)$$

into equation (6.57). After collecting real and imaginary parts we have

$$\mathbf{D}\mathbf{h}'' + \mathbf{f_y}(x, \mathbf{y}_s, \mathbf{y}_s', \lambda)\mathbf{h} + \mathbf{f}_{\mathbf{y}'}(x, \mathbf{y}_s, \mathbf{y}_s', \lambda)\mathbf{h}' = \alpha\mathbf{h} - \beta\mathbf{g},$$

$$\mathbf{D}\mathbf{g}'' + \mathbf{f_y}(x, \mathbf{y}_s, \mathbf{y}_s', \lambda)\mathbf{g} + \mathbf{f}_{\mathbf{y}'}(x, \mathbf{y}_s, \mathbf{y}_s', \lambda)\mathbf{g}' = \beta\mathbf{h} + \alpha\mathbf{g},$$

$$\begin{aligned}
\mathbf{A}_0\mathbf{h}(a) + \mathbf{A}_1\mathbf{h}'(a) &= \mathbf{0}, \\
\mathbf{B}_0\mathbf{h}(b) + \mathbf{B}_1\mathbf{h}'(b) &= \mathbf{0}, \\
\mathbf{A}_0\mathbf{g}(a) + \mathbf{A}_1\mathbf{g}'(a) &= \mathbf{0}, \\
\mathbf{B}_0\mathbf{g}(b) + \mathbf{B}_1\mathbf{g}'(b) &= \mathbf{0}.
\end{aligned} \tag{6.58}$$

This coupled system consists of $2n$ second-order ODEs with boundary conditions. The difficulty with the ODE eigenvalue problem in equation (6.58) is that it has an infinite number of eigenvalues. Hence, it can hardly be calculated whether, for instance, all eigenvalues have negative real parts. Bifurcation points are characterized in the usual way: The α and β depend on λ, and for one of the α we have $\alpha(\lambda_0) = 0$ at the parameter value λ_0 of a bifurcation point. In the real case, $\beta(\lambda_0) = 0$, we encounter a turning point or a stationary bifurcation point. In the imaginary case, $\beta(\lambda_0) \neq 0$, we have a Hopf bifurcation.

This situation enables us to set up direct methods for calculating bifurcation points. As in the branching system in equation (6.14), we keep the solution $\mathbf{y}_s(x)$ available by attaching equation (6.45) to equation (6.58). The α terms in equation (6.58) can be dropped because $\alpha(\lambda_0) = 0$. For the stationary bifurcation, the β terms are also dropped. In the latter case, the two systems in equation (6.58) are identical, and one can be removed (the $\mathbf{g}$ system, say). Imposing a normalizing equation such as $h_k(a) = 1$, and attaching the trivial differential equation $\lambda' = 0$, reproduces the branching system in equation (6.14), here written for second-order ODEs. This shows that the bifurcation points calculated by equation (6.14) are candidates for gain or loss of stability. We conclude that it is sufficient to test stability only once on either side of a bifurcation point.

The direct method for calculating Hopf points is now easily set up. Because $\beta(\lambda_0) \neq 0$ is an unknown constant, we set up the trivial differential equation $\beta' = 0$. Imposing a normalizing condition on $\mathbf{g}(a)$ and collecting all equations, one has

$$\mathbf{D}\mathbf{y}'' + \mathbf{f}(x, \mathbf{y}, \mathbf{y}', \lambda) = \mathbf{0},$$

$$\mathbf{D}\mathbf{h}'' + \mathbf{f_y}(x, \mathbf{y}, \mathbf{y}', \lambda)\mathbf{h} + \mathbf{f_{y'}}(x, \mathbf{y}, \mathbf{y}', \lambda)\mathbf{h}' = -\beta\mathbf{g},$$
$$\mathbf{D}\mathbf{g}'' + \mathbf{f_y}(x, \mathbf{y}, \mathbf{y}', \lambda)\mathbf{g} + \mathbf{f_{y'}}(x, \mathbf{y}, \mathbf{y}', \lambda)\mathbf{g}' = \beta\mathbf{h},$$
$$\lambda' = 0, \quad \beta' = 0,$$
$$\mathbf{A}_0\mathbf{y}(a) + \mathbf{A}_1\mathbf{y}'(a) = \mathbf{c}_a, \quad \mathbf{B}_0\mathbf{y}(b) + \mathbf{B}_1\mathbf{y}'(b) = \mathbf{c}_b, \qquad (6.59)$$
$$\mathbf{A}_0\mathbf{h}(a) + \mathbf{A}_1\mathbf{h}'(a) = \mathbf{0}, \quad \mathbf{B}_0\mathbf{h}(b) + \mathbf{B}_1\mathbf{h}'(b) = \mathbf{0},$$
$$\mathbf{A}_0\mathbf{g}(a) + \mathbf{A}_1\mathbf{g}'(a) = \mathbf{0}, \quad \mathbf{B}_0\mathbf{g}(b) + \mathbf{B}_1\mathbf{g}'(b) = \mathbf{0},$$
$$h_k(u) = 1, \quad g_k(a) = 1.$$

In order to use standard software for ODEs, each of the second-order systems is transformed into a system of $2n$ first-order ODEs. Hence, altogether, equation (6.59) represents a first-order boundary-value problem consisting of $6n + 2$ differential equations. Numerical results with $n = 2$ are reported in [KuH84a]. Severe problems in finding reasonable initial approximations were indicated.

The above approach can be generalized to PDEs with more than one space variable. For ease of notation, we suppress boundary conditions and write down the equations for the PDE

$$\frac{\partial \mathbf{y}}{\partial t} = \nabla^2 \mathbf{y} + \mathbf{f}(\mathbf{x}, \mathbf{y}, \lambda).$$

Here the independent space variable $\mathbf{x}$ is two-dimensional or three-dimensional. For Hopf bifurcation, the system that corresponds to equation (6.59) is based on

$$\nabla^2 \mathbf{y} + \mathbf{f}(\mathbf{x}, \mathbf{y}, \lambda) = \mathbf{0},$$
$$\nabla^2 \mathbf{h} + \mathbf{f_y}(\mathbf{x}, \mathbf{y}, \lambda)\mathbf{h} = -\beta\mathbf{g}, \qquad (6.60)$$
$$\nabla^2 \mathbf{g} + \mathbf{f_y}(\mathbf{x}, \mathbf{y}, \lambda)\mathbf{g} = \beta\mathbf{h}.$$

The system reduces in the case of stationary bifurcation ($\beta = 0$). A general reference on numerical bifurcation in reaction-diffusion equations is [Mei00].

Example 6.7   Nonstationary Heat and Mass Transfer
A nonstationary heat and mass transfer inside a porous catalyst particle can be described by the two parabolic PDEs [KuH84a], [RoH83]:

$$\mathrm{Lw}\frac{\partial c}{\partial t} = \frac{\partial^2 c}{\partial x^2} - \frac{\lambda c}{\gamma\beta}\exp\left(\frac{\Theta}{1 + \Theta/\gamma}\right),$$
$$\frac{\partial \Theta}{\partial t} = \frac{\partial^2 \Theta}{\partial x^2} + \lambda c \cdot \exp\left(\frac{\Theta}{1 + \Theta/\gamma}\right) \qquad (6.61a)$$

with boundary conditions

$$\frac{\partial c(0, t)}{\partial x} = \frac{\partial \Theta(0, t)}{\partial x} = 0,$$
$$c(1, t) = 1, \quad \Theta(1, t) = 0. \qquad (6.61b)$$

The meaning of the variables is

$c(x, t)$: concentration;
$\Theta(x, t)$: temperature;
λ: Frank–Kamenetskii parameter;
γ: activation energy $(= 20)$;
β: heat evolution parameter $(= 0.2)$; and
Lw: Lewis number $(= 10)$.

The steady state satisfies the boundary-value problem

$$c'' = \frac{\lambda c}{\gamma \beta} \exp\left(\frac{\Theta}{1 + \Theta/\gamma}\right),$$

$$\Theta'' = -\lambda c \cdot \exp\left(\frac{\Theta}{1 + \Theta/\gamma}\right),$$

$$c'(0) = 0, \quad c(1) = 1, \quad \Theta'(0) = 0, \quad \Theta(1) = 0.$$

After evaluating the Jacobian $\mathbf{f_y}$, the remaining parts of equation (6.59) can be easily written down. Denoting the exponential terms

$$E_0 := \exp\left(\frac{\Theta}{1 + \Theta/\gamma}\right),$$

$$E_1 := \frac{c}{(1 + \Theta/\gamma)^2} E_0,$$

one obtains

$$h_1'' - \frac{\lambda}{\gamma \beta} E_0 h_1 - \frac{\lambda}{\gamma \beta} E_1 h_2 = -\beta \cdot \text{Lw} \cdot g_1,$$

$$h_2'' + \lambda E_0 h_1 + \lambda E_1 h_2 = -\beta g_2,$$

$$g_1'' - \frac{\lambda}{\gamma \beta} E_0 g_1 - \frac{\lambda}{\gamma \beta} E_1 g_2 = \beta h_1,$$

$$g_2'' + \lambda E_0 g_1 + \lambda E_1 g_2 = \beta h_2,$$

$$h_1'(0) = h_1(1) = h_2'(0) = h_2(1) = 0,$$

$$g_1'(0) = g_1(1) = g_2'(0) = g_2(1) = 0.$$

The coupling with the steady state is hidden in the exponential terms. Writing down the 14 first-order differential equations explicitly is beyond the scope of this book. Such systems are seldom written down explicitly because they can be conveniently set up in a computer working with n^2-matrices and subvectors. In order to give an impression of how the system equation (6.59) can be organized, we write down the components of the resulting extended vector $\mathbf{Y}$:

$$Y_1 = c, \quad Y_2 = c', \quad Y_3 = \Theta, \quad Y_4 = \Theta',$$

$$Y_5 = h_1, \quad Y_6 = h_1', \quad Y_7 = h_2, \quad Y_8 = h_2',$$

$$Y_9 = g_1, \quad Y_{10} = g_1', \quad Y_{11} = g_2, \quad Y_{12} = g_2',$$

$$Y_{13} = \lambda, \quad Y_{14} = \beta.$$

$\square$

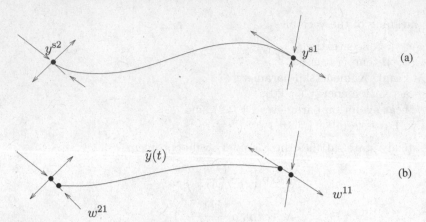

Fig. 6.13. Heteroclinic orbit and its approximation

6.8 Computation of Heteroclinic and Homoclinic Orbits

The computation of heteroclinic or homoclinic orbits leads to boundary-value problems of a specific kind. The derivation of such boundary-value problems is the topic of this section. For an illustration of a homoclinic orbit see Figure 2.29, and Figure 6.13a depicts a heteroclinic orbit. We assume an autonomous system of ODEs

$$\dot{\mathbf{y}} = \mathbf{f}(\mathbf{y})$$

and postpone possible parameters until later. To explain basic ideas, we start with a simple planar scenario, where a heteroclinic orbit of the differential equation connects two saddles $\mathbf{y}^{s1}, \mathbf{y}^{s2}$. Figure 6.13a shows a related phase portrait, including the lines of the eigenspaces.

It takes an infinite time for a particle to follow the ideal heteroclinic trajectory from one stationary solution to the other, $-\infty < t < \infty$. In practice, a trajectory can only be approximated for a finite time interval of length T. Since we have assumed an autonomous ODE, the interval of the truncated problem can be taken as $0 \leq t \leq T$. The aim is to calculate an orbit $\bar{\mathbf{y}}(t)$ that approximates the heteroclinic orbit in that $\mathbf{y}(0)$ is close to $\mathbf{y}^{s1}$, and $\mathbf{y}(T)$ is close to $\mathbf{y}^{s2}$, for a large prescribed value of T. Since the heteroclinic orbit enters or leaves the stationary solutions tangentially to the eigenspaces, it is reasonable to try to find an orbit $\bar{\mathbf{y}}(t)$ that starts and ends on the relevant eigenspace close to the saddles (Figure 6.13b). This amounts to formulate boundary conditions

$$\bar{\mathbf{y}}(0) = \mathbf{y}^{s1} + \varepsilon_1 \mathbf{w}^{11} \,,$$
$$\bar{\mathbf{y}}(T) = \mathbf{y}^{s2} + \varepsilon_2 \mathbf{w}^{21} \,.$$

(6.62)

In equation (6.62), $\mathbf{w}^{11}$ is the vector tangent to the unstable manifold of $\mathbf{y}^{s1}$, $\mathbf{w}^{21}$ defines the stable manifold of $\mathbf{y}^{s2}$, and ε_1 and ε_2 measure the distances from the start and end points of $\bar{\mathbf{y}}$ to the saddles. The trajectory $\bar{\mathbf{y}}$ can be seen as being parameterized by $\varepsilon_1, \varepsilon_2$, which is a parameterization by "irregularity" or "discrepancy" following the lines of Section 5.6.6 and Section 6.4. We expect ε_1 and ε_2 to be close to 0 for large T.

In order that a solution of the differential equation with boundary conditions equation (6.62) exists, the problem must have enough *free parameters* to be flexible. The number n_p of necessary free parameters depends on n and the type of the stationary solutions $\mathbf{y}^{s1}, \mathbf{y}^{s2}$. We arrange the free parameters in a parameter vector $\Lambda = (\lambda_1, ..., \lambda_{n_p})$ and shall see what n_p might be. That is, we study a differential equation of the form $\mathbf{y}' = \mathbf{f}(\mathbf{y}, \Lambda)$. Transforming the integration interval to $0 \leq t \leq 1$ leads to

$$\mathbf{y}' = T\mathbf{f}(\mathbf{y}, \Lambda). \tag{6.63}$$

For this differential equation we need boundary conditions for $\mathbf{y}(0)$ and $\mathbf{y}(1)$ in the spirit of equation (6.62). For general stationary solutions, the boundary conditions are more involved, taking into account higher-dimensional stable manifolds and unstable manifolds. We describe this more general setting next. For ease of notation, we drop the overbar on the approximation $\bar{\mathbf{y}}$.

Assume that the Jacobian $\mathbf{f}_{\mathbf{y}}(\mathbf{y}^{s1}, \Lambda)$ has n_1 real positive eigenvalues μ_{1k} with n_1 eigenvectors $\mathbf{w}^{1k}$ ($k = 1, ..., n_1$). For an easy setting suppose (as in [DoKK91]) that the remaining $n - n_1$ eigenvalues have negative real parts. Then the unstable eigenspace belonging to $\mathbf{y}^{s1}$ consists of all linear combinations

$$\mathbf{y}^{s1} + \sum_{k=1}^{n_1} c_{1k}\mathbf{w}^{1k},$$

for coefficients $c_{11}, ..., c_{1n_1}$. Demanding the start point $\mathbf{y}(0)$ to be on this eigenspace gives the boundary condition for $\mathbf{y}(0)$ that replaces the planar saddle situation in equation (6.62). For this saddle, $n_1 = 1$, and c_{11} is included in ε_1. This outlines the general idea.

For the other stationary solution $\mathbf{y}^{s2}$ we assume similar conditions to describe the stable manifold: Let n_2 be the number of real negative eigenvalues μ_{2k} and eigenvectors $\mathbf{w}^{2k}$ of the matrix $\mathbf{f}_{\mathbf{y}}(\mathbf{y}^{s2}, \Lambda)$, $k = 1, ..., n_2$. The other $n - n_2$ eigenvalues are assumed to be "unstable," having positive real parts. We collect the defining equations:

$$\mathbf{f}(\mathbf{y}^{s1}, \Lambda) = \mathbf{0}, \tag{6.64a}$$

$$\mathbf{f}(\mathbf{y}^{s2}, \Lambda) = \mathbf{0}, \tag{6.64b}$$

$$\mathbf{f}_{\mathbf{y}}(\mathbf{y}^{s1}, \Lambda)\mathbf{w}^{1k} = \mu_{1k}\mathbf{w}^{1k}, \quad k = 1, ..., n_1, \tag{6.64c}$$

$$\mathbf{f}_{\mathbf{y}}(\mathbf{y}^{s2}, \Lambda)\mathbf{w}^{2k} = \mu_{2k}\mathbf{w}^{2k}, \quad k = 1, ..., n_2, \tag{6.64d}$$

$$\|\mathbf{w}^{1k}\| = 1, \quad k = 1, ..., n_1, \tag{6.64e}$$

$$\|\mathbf{w}^{2k}\| = 1, \quad k = 1, ..., n_2. \tag{6.64f}$$

These equations define the unstable manifold of $\mathbf{y}^{s1}$ and the stable manifold of $\mathbf{y}^{s2}$. The equations (6.64e/f) normalize the eigenvectors to unit length, using the Euclidian norm; other normalizing conditions may also be used. The following equations formulate the boundary conditions, which specify $\mathbf{y}(0)$ to lie on the unstable manifold of $\mathbf{y}^{s1}$, and $\mathbf{y}(1)$ to arrive on the stable manifold of $\mathbf{y}^{s2}$:

$$\mathbf{y}(0) = \mathbf{y}^{s1} + \varepsilon_1 \sum_{k=1}^{n_1} c_{1k} \mathbf{w}^{1k}, \tag{6.65a}$$

$$\mathbf{y}(1) = \mathbf{y}^{s2} + \varepsilon_2 \sum_{k=1}^{n_2} c_{2k} \mathbf{w}^{2k}. \tag{6.65b}$$

To give ε_1 and ε_2 a unique meaning, the coefficients c_{1k} and c_{2k} must be normalized,

$$\sum_{k=1}^{n_1} c_{1k}^2 = 1, \tag{6.65c}$$

$$\sum_{k=1}^{n_2} c_{2k}^2 = 1. \tag{6.65d}$$

Finally, a phase condition is required, because the differential equation (6.63) is autonomous. This topic will be explained in Section 7.1. As an example of a phase condition one may choose

$$f_1(\mathbf{y}(0), \Lambda) = 0. \tag{6.66}$$

The equations (6.64), (6.65), and (6.66) consist of

$$n + n + nn_1 + nn_2 + n_1 + n_2 + n + n + 1 + 1 + 1$$

scalar equations. By substituting $\mathbf{y}^{s1}$ and $\mathbf{y}^{s2}$ from equation (6.65a/b) into equation (6.64) $2n$ of the equations can be replaced. Additional n of the equations are needed as boundary conditions for the differential equation for $\mathbf{y}(t)$ in equation (6.63). The remaining number of equations is

$$n + 3 + (n + 1)(n_1 + n_2). \tag{6.67}$$

The unknown variables are the reals $\Lambda_1, ..., \Lambda_{n_p}$, $c_{11}, ..., c_{1n_1}$, $c_{21}, ..., c_{2n_2}$, ε_1, ε_2, $\mu_{11}, ..., \mu_{1n_1}$, $\mu_{21}, ..., \mu_{2n_2}$, the vectors $\mathbf{w}^{1k}$ ($k = 1, ..., n_1$) and $\mathbf{w}^{2k}$ ($k = 1, ..., n_2$). Altogether we have

$$2 + n_p + (n + 2)(n_1 + n_2) \tag{6.68}$$

scalar variables. The counting of the constraints in equation (6.67) and of the variables in equation (6.68) does not include $\mathbf{y}$, for which we have the differential equation with boundary conditions. The numbers in equations (6.67) and (6.68) must match, which leads to the relation

$$1 + n = n_1 + n_2 + n_p. \qquad (6.69)$$

This equation fixes the number n_p of free parameters needed to define an orbit as illustrated in Figure 6.13b. In order to have a *branch* of such orbits, another free parameter is needed. This introduces our λ, and the differential equation is then

$$\mathbf{y}' = T\mathbf{f}(\mathbf{y}, \Lambda, \lambda), \qquad (6.70)$$

where Λ consists of $n + 1 - n_1 - n_2$ free parameters.

The above involved set of equations is solved numerically. If analytic information on stationary solutions and eigenvectors is available, the number of equations in equation (6.67) can be reduced. It is straightforward to reduce the above method such that homoclinic orbits are approximated (Exercise 6.18). For references on this and related methods on global bifurcation, see for example [Beyn90a], [Beyn90b], [Kuz90], [DoKK91], [FrD91], [ChK94], [Moo95], [BaLS96], [Kuz98], [OlCK03]. Test functions and other techniques described in this book are extended and applied frequently.

Example 6.8   Fisher's Model

This equation was discussed in Section 3.5.1; see Example 3.6. The ODE system ($n = 2$) is

$$y_1' = y_2,$$
$$y_2' = -\lambda y_2 - \zeta y_1 (1 - y_1).$$

The heteroclinic orbit starts from the saddle $(y_1, y_2) = (1, 0)$; we have $n_1 = 1$. The other stationary solution $(0, 0)$ is for $\lambda > 2\sqrt{\zeta}$ a stable node, hence $n_2 = 2$. The count of free parameters n_p of equation (6.69) is $n_p = 0$. No free parameter is required for a heteroclinic orbit, $\zeta > 0$ and $\lambda > 2\sqrt{\zeta}$ can be fixed constants. A branch of heteroclinic orbit is obtained when one parameter, say λ, is varied. □

Exercises

Exercise 6.1

(a) Derive equations (6.5a) and (6.5b) from equation (6.4).
(b) What are the boundary conditions that replace those in equation (6.5b) when *zero flux* is considered,

$$\frac{\partial X}{\partial x} = \frac{\partial Y}{\partial x} = 0 \quad \text{for } x = 0, x = L?$$

(c) Write down explicitly the functions f_i and r_i $(i = 1, \ldots, 4)$ from equation (6.1) that define equations (6.5a) and (6.5b).

(d) Calculate all first-order partial derivatives of f_i and r_i with respect to y_j $(i, j = 1, \ldots, 4)$ and arrange them in matrices. Notice that the derivatives of $\mathbf{r}$ include $2n^2$ entries.

(e) Which constant solution satisfies equations (6.5a) and (6.5b)? Where is this solution found in Figure 6.1?

Exercise 6.2
Derive in detail boundary-value problem equation (6.10). Start from the original Duffing equation and use equations (6.7) and (6.8).

Exercise 6.3
(Project.) Assume that a periodic oscillation has been calculated with a vector of initial values $\mathbf{y}(0)$ and period T. The right-hand side of the differential equation is $\mathbf{f}$. Suppose further that N values $\mathbf{y}(t_j)$ at equidistant nodes t_j are required (for plotting purposes, say), together with an approximation of the amplitude. Design and implement an algorithm that calculates the required approximations. To this end, modify an integration routine with step control (as, for instance, an RKF method; see Appendix A.6) as follows: Each function evaluation of $\mathbf{f}$ is to be used to calculate locally cubic splines [StB80], [DaB08]. Establish formulas that calculate the extrema of the splines. Use the splines also for calculating the desired approximations of $\mathbf{y}(t_j)$; in this way the step control need not be manipulated.

Exercise 6.4
Establish the linearized boundary-value problems in equation (6.11) for the Brusselator in equation (6.5) and the Duffing equation in equation (6.9).

Exercise 6.5
Show that a boundary condition $r_j = y_\nu(a) -$ constant $= 0$ implies $h_\nu(a) = 0$.

Exercise 6.6
Boundary conditions are frequently linear, $\mathbf{r} = \tilde{\mathbf{A}}\mathbf{y}(a) + \tilde{\mathbf{B}}\mathbf{y}(b) - \mathbf{c}$, with an n-vector $\mathbf{c}$ and matrices $\tilde{\mathbf{A}}$ and $\tilde{\mathbf{B}}$.

(a) What are the boundary conditions of the linearization?

(b) What are $\tilde{\mathbf{A}}$, $\tilde{\mathbf{B}}$, and $\mathbf{c}$ in the case of periodic boundary conditions?

Exercise 6.7
(Laborious Project.) Construct a direct method for calculating bifurcation points as follows: Implement a special shooting code that solves equation (6.13), exploiting equation (6.24) and $\mathbf{E} \cdot \mathbf{h}(a) = \mathbf{0}$ internally.

Exercise 6.8
For each of the cases (6.30a) through (6.30d), sketch an example.

Exercise 6.9
Assume that $u(t)$ and $v(t)$ are two scalar variables defined for $-1 \le t \le 1$. $u(t)$ is supposed to be odd, and $v(t)$ is even. What can be said about $u', v', u'', v'', u^2, uv$?

Exercise 6.10
Show that the Brusselator model equation (6.4) supports even solutions. (Hint: Substitute $X^* = X - A$, $Y^* = Y - B/A$, and investigate the resulting system in X^*, Y^*.)

Exercise 6.11
Show that the Duffing equation (6.9) supports a symmetry of the type in equation (6.30c).

Exercise 6.12
Consider a periodic solution of a second-order oscillator [such as the Duffing equation in equation (6.9)] and its graphical representation in a phase plane. What can be said about the graphs of $u(t)$ and $\tilde{u}(t)$ when there is a symmetry of the type of equation (6.30c)?

Exercise 6.13
Find the trivial solution(s) of the superconductivity example in equation (6.35).

Exercise 6.14
For the boundary-value problem

$$-u'' + u^3 = \lambda u, \quad u(0) = u(\pi), \quad u'(0) = u'(\pi)$$

calculate the eigenvalues of the linearization about $u = 0$.

Exercise 6.15
Consider the buckling rod problem in equations (6.40a) and (6.40b). Calculate the eigenvalues of the linearization about $u = 0$. Show that the linearization of equation (6.42) has the same eigenvalues.

Exercise 6.16
Consider stationary solutions ($u_t = 0$) of the Kuramoto–Sivashinsky equation (3.30). Show that bifurcations from the trivial solution $u \equiv 0$ can only take place for $\gamma = 4j^2$, $j = 1, 2, 3, \ldots$.

Exercise 6.17
Assume that the scalar dependent variable $u(x, t)$ satisfies the boundary conditions of equation (6.44) (consider A_0, A_1, B_0, B_1 to be scalars). Suppose further that the standard difference quotients in equation (6.46) with equidistant spacing are applied with $N > 1$.

(a) Show that

$$\frac{\partial u(a,t)}{\partial x} = \frac{1}{2\Delta}[-3u(a,t) + 4u(x_1,t) - u(x_2,t)] + O(\Delta^2).$$

(b) Use (a) to show that

$$u(a,t) \approx [2\Delta c_a - A_1(4u(x_1,t) - u(x_2,t))]/(2\Delta A_0 - 3A_1).$$

(c) Show the analog

$$u(b,t) \approx [2\Delta c_b + B_1(4u(x_N,t) - u(x_{N-1},t))]/(2\Delta B_0 + 3B_1).$$

(d) How can these expressions be used to generate the differential equation $\dot{\mathbf{Y}} = \mathbf{F}(\mathbf{Y},\lambda)$ that belongs to equation (6.48)?

Exercise 6.18
Specialize the method outlined in Section 6.8 to the computation of a homoclinic orbit.

7 Stability of Periodic Solutions

Periodic solutions of a differential equation are time-dependent or space-dependent orbits, cycles, or oscillations. In this chapter a major concern is time-dependent periodicity ("time periodicity" for short) of solutions of autonomous systems of ODEs

$$\dot{\mathbf{y}} = \mathbf{f}(\mathbf{y}, \lambda). \tag{7.1}$$

For time-periodic solutions, there is a minimum time interval $T > 0$ (the "period") after which the system returns to its original state:

$$\mathbf{y}(t + T) = \mathbf{y}(t)$$

for all t. In Section 7.7 we turn our attention to the non-autonomous case.

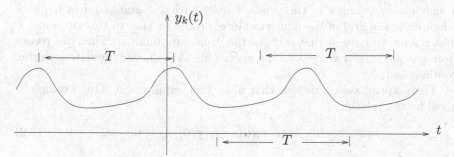

Fig. 7.1. The period T of a periodic solution

Before discussing time periodicity, let us briefly comment on space-dependent periodicity. Space-periodic phenomena are abundant in nature, ranging from the stripes on a zebra to various rock formations to sand dunes on a beach. PDEs are appropriate means for describing spatial patterns, see Section 3.5. The simple example of the PDE

$$\nabla \times \mathbf{y} + \mathbf{y} = \mathbf{0},$$

where $\mathbf{y}$ consists of three components $y_i(x_1, x_2, x_3)$, $i = 1, 2, 3$, depending on three spatial variables x_k may serve as motivation for spatial periodicity

R. Seydel, *Practical Bifurcation and Stability Analysis*,
Interdisciplinary Applied Mathematics 5, DOI 10.1007/978-1-4419-1740-9_7,
© Springer Science+Business Media, LLC 2010

(Exercise 7.1). The main part of this Chapter 7 is confined to time-periodic solutions of (7.1).

7.1 Periodic Solutions of Autonomous Systems

Because the right-hand side $\mathbf{f}(\mathbf{y}, \lambda)$ of equation (7.1) is autonomous, $\mathbf{y}(t + \zeta)$ is a solution of equation (7.1) whenever $\mathbf{y}(t)$ is a solution. This holds for all constant phase shifts ζ. Therefore one is free to start measuring the period T of a periodic orbit at any point $\mathbf{y}(t_0)$ along the profile (see Figure 7.1). We can fix an "initial" moment $t_0 = 0$ wherever we like; this anchors the profile. Such a normalization is frequently called a *phase condition*.

A simple phase condition is provided by

$$p(\mathbf{y}(0), \lambda) := y_k(0) - \eta = 0 \,.$$

Here, $\eta = \eta(k, \lambda)$ is a prescribed value in the range of $y_k(t)$ [AlC84], [HoK84b]. Choosing k and η requires an extra device to determine the current range of $y_k(t)$. This choice of a phase condition has practical disadvantages. Varying λ, the profile of a periodic solution changes. Fixing $y_k(0) = \eta$ does not prevent that peaks and other maxima and minima drift across the time interval. That is, a change in the profile goes along with a shift in time direction. Such a shift makes changes in the profile harder to judge and requires frequent adaption of the grid of the numerical integration. In view of this situation, it makes sense to request that shifts of the profile are minimal when one passes from one parameter value λ to the next. This also allows larger steps during continuation.

There are phase conditions that meet this requirement. One example is given by the relation

$$p(\mathbf{y}(0), \lambda) := \dot{y}_j(0) = f_j(\mathbf{y}(0), \lambda) = 0 \,, \tag{7.2}$$

which demands that $t_0 = 0$ be a critical point of y_j. This normalization, proposed in [Sey81a], does not require any adjustment of j or of other quantities. As pointed out in Section 2.9, the corresponding initial value $y_j(0)$ is well suited to be the ordinate of bifurcation diagrams. The index j is arbitrary ($1' \leq j \leq n$); choose $j = 1$ or pick the component of $\mathbf{f}$ that has the simplest structure. Equation (7.2) forces a maximum or minimum of y_j to stick at $t = 0$.

Another phase condition is the orthogonality

$$p(\mathbf{y}(0), \lambda) := (\mathbf{y}(0) - \hat{\mathbf{y}}(0))^{tr} \mathbf{f}(\hat{\mathbf{y}}(0), \lambda) = 0 \tag{7.3}$$

see [Doe81], [Beyn90a]. In (7.3), $\hat{\mathbf{y}}$ is a known nearby solution, usually the approximation at the previous parameter value in a continuation. Doedel has suggested the integral phase condition

$$p(\mathbf{y}, \lambda) := \int_0^T (\mathbf{y}(t) - \hat{\mathbf{y}}(t))^{tr} \, \mathbf{f}(\mathbf{y}(t), \lambda) \, dt = 0 \qquad (7.4a)$$

which has become popular through the code AUTO [Doe81], [DoH83], [DoKK91]. It minimizes changes in the solution profile, and leads to the alternate phase condition

$$p(\mathbf{y}, \lambda) := \int_0^T \mathbf{y}(t)^{tr} \, \dot{\hat{\mathbf{y}}}(t) \, dt = 0 \qquad (7.4b)$$

(see Exercise 7.2).

The period T of a particular periodic solution $\mathbf{y}(t)$ to equation (7.1) is usually not known beforehand. It must be calculated together with $\mathbf{y}$. Once the phase is set by a phase condition such as (7.2), (7.3), or (7.4), it makes sense to impose the periodicity condition

$$\mathbf{y}(0) = \mathbf{y}(T);$$

this fixes the period T. In such equations defining periodicity it is always assumed that T is the minimum period, $T > 0$. Periodic solutions $\mathbf{y}$ with periods T can be calculated by solving the augmented boundary-value problem

$$\begin{pmatrix} \dot{\mathbf{y}} \\ \dot{T} \end{pmatrix} = \begin{pmatrix} \mathbf{f}(\mathbf{y}, \lambda) \\ 0 \end{pmatrix}, \quad \begin{pmatrix} \mathbf{y}(0) - \mathbf{y}(T) \\ p(\mathbf{y}, \lambda) \end{pmatrix} = \mathbf{0}.$$

For technical reasons (see the treatment of the Duffing equation in Section 6.1), it is advisable to normalize the interval to have unit length. Hence, we rewrite the above boundary-value problem as

$$\begin{pmatrix} \mathbf{y} \\ T \end{pmatrix}' = \begin{pmatrix} T\mathbf{f}(\mathbf{y}, \lambda) \\ 0 \end{pmatrix}, \quad \begin{pmatrix} \mathbf{y}(0) - \mathbf{y}(1) \\ p(\mathbf{y}, \lambda) \end{pmatrix} = \mathbf{0}. \qquad (7.5)$$

The prime in equation (7.5) refers to a normalized time $0 \le t \le 1$. The integral phase conditions (7.4) then stretch over this interval. The phase conditions (7.2) and (7.3) are (nonlinear) initial conditions $p(\mathbf{y}(0), \lambda) = 0$ and closer to the formulation of a standard boundary-value problem. Then, defining $\tilde{n} := n + 1$ and

$$\tilde{\mathbf{y}} := \begin{pmatrix} \mathbf{y} \\ T \end{pmatrix}, \quad \tilde{\mathbf{f}}(\tilde{\mathbf{y}}, \lambda) := \begin{pmatrix} T\mathbf{f}(\mathbf{y}, \lambda) \\ 0 \end{pmatrix}, \quad \tilde{\mathbf{r}} := \begin{pmatrix} \mathbf{y}(0) - \mathbf{y}(1) \\ p(\mathbf{y}(0), \lambda) \end{pmatrix}, \qquad (7.6a)$$

the boundary-value problem (7.5) is equivalent to

$$\tilde{\mathbf{y}}' = \tilde{\mathbf{f}}(\tilde{\mathbf{y}}, \lambda), \quad \tilde{\mathbf{r}}(\tilde{\mathbf{y}}(0), \tilde{\mathbf{y}}(1)) = \mathbf{0}. \qquad (7.6b)$$

This is of the general form of equations (6.1) and (6.2). All the tools available for boundary-value problems can be applied to periodic solutions of autonomous ODEs. Periodic solutions can be calculated by calling standard software for two-point boundary-value problems, and the branching techniques of the previous chapter can be applied to investigate the branching of periodic solutions (see also Section 7.6). Analogously this can be done with the integral phase conditions (7.4), which have been implemented successfully based on collocation, see the code AUTO [Doe81], [Doe97].

To illustrate, we show how previous methods can be adapted to tracing branches of periodic solutions. A local parameterization strategy for solutions of equation (7.6) imposes an additional boundary condition together with the trivial differential equation $\lambda' = 0$ for the parameter. With phase condition (7.2), a boundary-value problem of dimension $n + 2$ results:

$$\begin{pmatrix} \mathbf{y} \\ T \\ \lambda \end{pmatrix}' = \begin{pmatrix} T\mathbf{f}(\mathbf{y}, \lambda) \\ 0 \\ 0 \end{pmatrix}, \quad \begin{pmatrix} \mathbf{y}(0) - \mathbf{y}(1) \\ f_j(\mathbf{y}(0), \lambda) \\ y_k(0) - \eta \end{pmatrix} = \mathbf{0}. \qquad (7.6c)$$

The indices k and j are independent of each other; $j = 1$ works.

Example 7.1   Nerve Impulses
The equations of FitzHugh's nerve model

$$\dot{y}_1 = 3(y_1 + y_2 - \tfrac{1}{3}y_1^3 + \lambda),$$
$$\dot{y}_2 = -(y_1 - 0.7 + 0.8y_2)/3$$

were investigated in Exercises 1.11 and 2.8, and in Example 3.5. In order to calculate periodic solutions, choose, for instance, the phase condition (7.2) with $j = 2$. This choice of j leads to simpler boundary conditions than $j = 1$; the choice $j = 1$ works too. We set $y_3 = T$ and obtain the boundary-value problem of the type in equation (6.1),

$$y_1' = 3y_3(y_1 + y_2 - \tfrac{1}{3}y_1^3 + \lambda),$$
$$y_2' = -\tfrac{1}{3}y_3(y_1 - 0.7 + 0.8\,y_2),$$
$$y_3' = 0,$$

$$y_1(0) - y_1(1) = 0,$$
$$y_2(0) - y_2(1) = 0,$$
$$y_1(0) + 0.8\,y_2(0) - 0.7 = 0.$$

Solving this system for fixed λ is a method for calculating a periodic solution. In case of a continuation with local parameterization, replace λ by y_4 and attach $y_4' = 0$ and $y_k(0) - \eta = 0$. Then, with appropriate values of k and η, a branch of periodic orbits can be traced by calling standard software. Figure 7.2 shows spike behavior of periodic solutions for $\lambda = -1.3$. $\qquad \square$

We note in passing that equation (7.6) does not exclude stationary solutions. Hence, especially in cases when an unstable orbit close to a stable equilibrium is to be calculated, solving equation (7.6) requires an appropriate initial guess. In the context of branch tracing or branch switching, such a guess is usually available. In critical cases, equation (7.6) can be

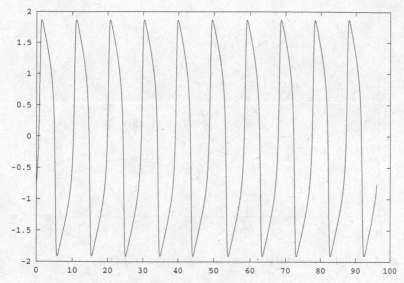

Fig. 7.2. Example 7.1, $y_1(t)$ for $\lambda = -1.3$

modified so that it parameterizes with respect to amplitude [Sey81a]. Fixing a certain generalized amplitude leads to a three-point boundary-value problem that excludes stationary solutions.

7.2 The Monodromy Matrix

When tracing a branch of periodic solutions, the question arises whether the periodic solutions are stable and where and in which way stability is lost. To analyze stability of periodic solutions, one needs the basic tools: the monodromy matrix and the Poincaré map. In what follows, we investigate stability of one particular periodic solution $\mathbf{y}^*$ with period T and defer the dependence on λ for later study (in Section 7.4).

Stability of $\mathbf{y}^*$ manifests itself in the way neighboring trajectories behave (Figure 7.3). Trajectories of the initial-value problem of differential equations are denoted by φ,

$$\varphi(t; \mathbf{z}) \text{ solves } \dot{\mathbf{y}} = \mathbf{f}(\mathbf{y}, \lambda) \text{ with } \mathbf{y}(0) = \mathbf{z}. \tag{7.7}$$

A trajectory that starts from the perturbed initial vector $\mathbf{z}^* + \mathbf{d}_0$ progresses with the distance

$$\mathbf{d}(t) = \varphi(t; \mathbf{z}^* + \mathbf{d}_0) - \varphi(t; \mathbf{z}^*)$$

to the periodic orbit $\mathbf{y}^*$. Here $\mathbf{z}^* = \mathbf{y}^*(0)$ is taken. Measuring the distance after one period T gives

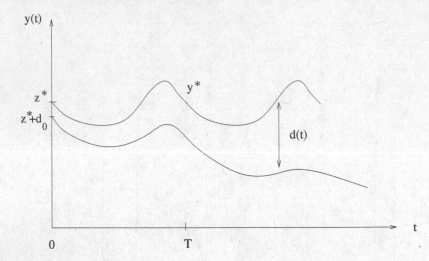

Fig. 7.3. A periodic trajectory $\mathbf{y}^*$ and a neighboring trajectory

$$\mathbf{d}(T) = \varphi(T; \mathbf{z}^* + \mathbf{d}_0) - \varphi(T; \mathbf{z}^*) \,.$$

Taylor expansion yields

$$\mathbf{d}(T) = \frac{\partial \varphi(T; \mathbf{z}^*)}{\partial \mathbf{z}} \mathbf{d}_0 + \text{ terms of higher order} \,.$$

Apparently, the matrix

$$\frac{\partial \varphi(T; \mathbf{z}^*)}{\partial \mathbf{z}} \tag{7.8}$$

plays a role in deciding whether the initial perturbation $\mathbf{d}_0$ decays or grows. The matrix (7.8) is called the *monodromy matrix*.

Some properties of φ help to find another representation of the monodromy matrix. By definition, φ from (7.7) obeys the differential equation (7.1),

$$\frac{d\varphi(t; \mathbf{z})}{dt} = \mathbf{f}(\varphi(t; \mathbf{z}), \lambda) \quad \text{for all } t \,.$$

Differentiating this identity with respect to $\mathbf{z}$ yields

$$\frac{d}{dt} \frac{\partial \varphi(t; \mathbf{z})}{\partial \mathbf{z}} = \frac{\partial \mathbf{f}(\varphi, \lambda)}{\partial \varphi} \frac{\partial \varphi(t; \mathbf{z})}{\partial \mathbf{z}} \,.$$

From $\varphi(0; \mathbf{z}) = \mathbf{z}$, we infer

$$\frac{\partial \varphi(0; \mathbf{z})}{\partial \mathbf{z}} = \mathbf{I} \,.$$

Consequently, the monodromy matrix of equation (7.8) is identical to $\Phi(T)$, when $\Phi(t)$ solves the matrix initial-value problem

$$\dot{\Phi} = \mathbf{f}_{\mathbf{y}}(\mathbf{y}^*, \lambda) \, \Phi, \quad \Phi(0) = \mathbf{I} \,. \tag{7.9}$$

Note that Φ also depends on $\mathbf{y}^*$, $\Phi = \Phi(t; \mathbf{y}^*)$. When we have only one particular trajectory $\mathbf{y}^*(t)$ in mind, we suppress the argument $\mathbf{y}^*$ for convenience. This specific *fundamental solution matrix* Φ gives rise to the following definition:

Definition 7.2   *The n^2* **monodromy matrix** *$\mathbf{M}$ of the periodic solution $\mathbf{y}^*(t)$ with period T and initial vector $\mathbf{z}^*$ is defined by*

$$\mathbf{M} := \Phi(T) = \frac{\partial \varphi(T; \mathbf{z}^*)}{\partial \mathbf{z}} ;$$

φ and Φ are defined in equations (7.7) and (7.9).

The monodromy matrix $\mathbf{M}$ is also called circuit matrix. Local stability of $\mathbf{y}^*$ can be established by means of Floquet theory, based on the monodromy matrix [CoL55], [Hahn67], [Hale69], [Har64], [IoJ81], [Rei91], [Chi99]. We defer results to the next section and close this section with two further properties of $\mathbf{M}$.

Lemma 7.3
(a) $\Phi(jT) = \mathbf{M}^j$, and
(b) $\mathbf{M}$ has $+1$ as eigenvalue with eigenvector $\mathbf{f}(\mathbf{y}(0), \lambda)$.

This can be seen as follows: The Jacobian $\mathbf{J}(t) = \mathbf{f_y}(\mathbf{y}^*, \lambda)$ is periodic, $\mathbf{J}(t + T) = \mathbf{J}(t)$. Hence, any matrix $\mathbf{Z}(t)$ that solves the matrix differential equation $\dot{\mathbf{Z}} = \mathbf{JZ}$ meets the property

$$\mathbf{Z}(t) \text{ is solution implies } \mathbf{Z}(t + T) \text{ is solution.}$$

Thus, there is a constant nonsingular matrix $\mathbf{Q}$ such that

$$\mathbf{Z}(t + T) = \mathbf{Z}(t)\mathbf{Q}.$$

For the special solution Φ from equation (7.9), we have

$$\Phi(0 + T) = \Phi(0)\mathbf{Q} = \mathbf{Q},$$

which shows $\mathbf{Q} = \mathbf{M}$ and

$$\Phi(2T) = \Phi(T + T) = \Phi(T)\mathbf{M} = \mathbf{M}^2.$$

This extends to $\Phi(jT) = \mathbf{M}^j$, which implies

$$\mathbf{d}(jT) = \mathbf{M}^j \mathbf{d}_0 + \text{ terms of higher order.}$$

Because solutions of the autonomous differential equation (7.1) can be shifted in t direction, there are periodic perturbations

$$\mathbf{d}(t) = \mathbf{y}^*(t + \zeta) - \mathbf{y}^*(t)$$

satisfying $\mathbf{d}(t + T) = \mathbf{d}(t)$. This leads to the conclusion that the linearized problem, the *variational equation*

$$\dot{\mathbf{h}} = \mathbf{J}(t)\mathbf{h},$$

has a periodic solution $\mathbf{h}(t)$. In fact, if $\mathbf{y}(t)$ is a period T solution of equation (7.1), then the period T function $\dot{\mathbf{y}}(t)$ solves $\dot{\mathbf{h}} = \mathbf{Jh}$. Every solution of this linearized problem satisfies $\mathbf{h}(t) = \Phi(t)\mathbf{h}(0)$. The relation

$$\mathbf{h}(0) = \mathbf{h}(T) = \Phi(T)\mathbf{h}(0) = \mathbf{Mh}(0)$$

shows that $\mathbf{M}$ has unity as an eigenvalue. The eigenvector is $\mathbf{h}(0) = \dot{\mathbf{y}}(0) = \mathbf{f}(\mathbf{y}(0), \lambda)$.

7.3 The Poincaré Map

The Poincaré map is extremely useful for describing dynamics of different types of oscillation. In particular, arguments based on the Poincaré map establish stability results for periodic orbits.

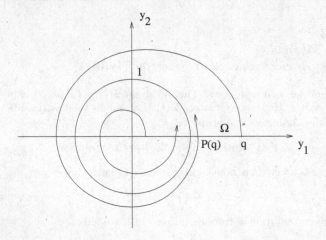

Fig. 7.4. Example 7.4, phase plane

Example 7.4
[Cf. Exercise 1.14 and Equation (2.22) in Section 2.6.] The system

$$\dot{y}_1 = y_1 - y_2 - y_1(y_1^2 + y_2^2),$$
$$\dot{y}_2 = y_1 + y_2 - y_2(y_1^2 + y_2^2)$$

can be written in polar coordinates as

$$\dot{\rho} = \rho(1 - \rho^2),$$
$$\dot{\vartheta} = 1.$$

A stable limit cycle is $\rho = 1$, $\vartheta = t$ (see the illustration of a phase plane in Figure 7.4). Neighboring trajectories approach the limit cycle. Measuring the distance to the limit cycle is difficult when the observer travels along the trajectories. As in other sports, it is easier by far to wait at some finish line Ω and measure how the racetracks have varied after each lap. In this example we choose as the start and finish line the positive y_1-axis,

$$\Omega = \{(\rho, \vartheta) \mid \rho > 0, \ \vartheta = 0\}.$$

The general solution to the differential equation is given by

$$\rho = [1 + (\rho_0^{-2} - 1)e^{-2t}]^{-\frac{1}{2}}, \quad \vartheta = t + \vartheta_0,$$

for initial values ρ_0, ϑ_0. Consequently, a trajectory that starts from position $q \in \Omega$ requires the time 2π to complete one orbit and passes Ω at the radius

$$P(q) = [1 + (q^{-2} - 1)e^{-4\pi}]^{-\frac{1}{2}}.$$

This function $P(q)$ is a simple example of a Poincaré map. The value $q^* = 1$ at which the periodic orbit passes Ω is a fixed point of the Poincaré map, $P(1) = 1$. A trajectory that starts, for instance, at $q = 0.0001$ shows the intermediate results

$$P(q) = 0.05347\ldots,$$
$$P^2(q) := P(P(q)) = 0.99939\ldots$$

during its rapid approach to the stable limit cycle. $\square$

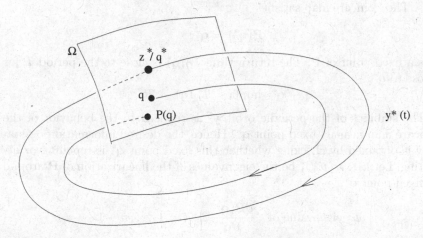

Fig. 7.5. Two trajectories intersecting a Poincaré section Ω

This concept of a Poincaré map can be generalized to the differential equation (7.1) with n components. The set Ω must be an $(n-1)$-dimensional hypersurface (see Figure 7.5). The hypersurface Ω must be chosen such that all trajectories that cross Ω in a neighborhood of a $\mathbf{q}^* \in \Omega$ meet two requirements:

(a) The trajectories intersect Ω transversally, and
(b) the trajectories cross Ω in the same direction.

These requirements characterize Ω as a local set; choosing another $\mathbf{q}^*$ leads to another Ω. The hypersurface Ω is also called the *Poincaré section*. The most important class of Ω are subsets of planes.

The periodic orbit $\mathbf{y}^*$ with period T intersects Ω in $\mathbf{z}^* \in \mathbb{R}^n$. Represented in a coordinate system on Ω, we write $\mathbf{z}^*$ as $\mathbf{q}^*$, where $\mathbf{q}^*$ is $(n-1)$-dimensional. If we restrict φ to Ω, we may summarize this as

$$\mathbf{q}^* = \varphi(T; \mathbf{q}^*).$$

Let $T_\Omega(\mathbf{q})$, with $\mathbf{q} \in \Omega$, be the time taken for a trajectory $\varphi(t; \mathbf{q})$ to first return to Ω,

$$\varphi(T_\Omega(\mathbf{q}); \mathbf{q}) \in \Omega, \quad \varphi(t; \mathbf{q}) \notin \Omega \text{ for } 0 < t < T_\Omega(\mathbf{q}).$$

The *Poincaré map* or *return map* $\mathbf{P}(\mathbf{q})$ is defined by

$$\mathbf{P}(\mathbf{q}) := \mathbf{P}_\Omega(\mathbf{q}) = \varphi(T_\Omega(\mathbf{q}); \mathbf{q}), \quad \text{for } \mathbf{q} \in \Omega. \tag{7.10}$$

The geometrical meaning of $\mathbf{P}(\mathbf{q})$ is illustrated in Figure 7.5. $\mathbf{P}$ takes values in Ω, which is $(n-1)$-dimensional. Because $\mathbf{q}$ is allowed to vary only in Ω, we consider $\mathbf{P}$ as an $(n-1)$-dimensional map—that is, both $\mathbf{q}$ and $\mathbf{P}(\mathbf{q})$ can be interpreted to have $(n-1)$ components with respect to a suitably chosen basis. The Poincaré map satisfies

$$\mathbf{P}(\mathbf{q}^*) = \mathbf{q}^*;$$

$\mathbf{q}^*$ is a fixed point of $\mathbf{P}$. The return time $T_\Omega(\mathbf{q})$ is close to the period T for $\mathbf{q}$ close to $\mathbf{q}^*$,

$$\mathbf{q} \to \mathbf{q}^* \text{ implies } T_\Omega(\mathbf{q}) \to T.$$

The stability of the periodic orbit $\mathbf{y}^*$ is reduced to the behavior of the Poincaré map near its fixed point $\mathbf{q}^*$. Hence, the desired information on stability is obtained by checking whether this fixed point $\mathbf{q}^*$ is repelling or attracting. Let $\mu_1, \ldots, \mu_{n-1}$ be the eigenvalues of the linearization of $\mathbf{P}$ around the fixed point $\mathbf{q}^*$,

$$\mu_j \text{ eigenvalue of } \frac{\partial \mathbf{P}(\mathbf{q}^*)}{\partial \mathbf{q}}, \quad j = 1, \ldots, n-1.$$

The eigenvalues of this matrix are independent of the choice of Ω [Ura67], see Exercise 7.4. Hence, to each periodic orbit there corresponds one set of eigenvalues μ_j. Stability of fixed points is established by Theorem 1.13 in Section 1.4: If the moduli of all $n-1$ eigenvalues are smaller than 1, then $\mathbf{q}^*$ is stable (attracting); if the modulus of at least one eigenvalue is larger than 1, then $\mathbf{q}^*$ is unstable (repelling). The dynamic behavior of the sequence $\mathbf{P}^\nu(\mathbf{q}) := \mathbf{P}(\mathbf{P}^{\nu-1}(\mathbf{q}))$ for $\nu > 1$, $\mathbf{P}^1 := \mathbf{P}$,

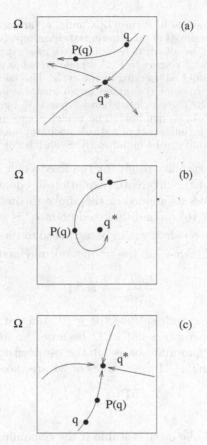

Fig. 7.6. Three Poincaré sections with discrete return points filled to curves. $\mathbf{q}^*$ is the return point of a periodic cycle. (a) Saddle cycle, (b) spiral cycle, and (c) nodal cycle

$$\mathbf{q} \to \mathbf{P}(\mathbf{q}) \to \mathbf{P}^2(\mathbf{q}) \to \dots$$

on Ω depends on the eigenvalues μ_j and the corresponding eigenvectors in an analogous manner, as noted in Section 1.2. Possible paths of the sequence are illustrated in Figure 7.6 by means of two-dimensional hypersurfaces Ω. Trajectories intersect these parts of a plane. The sequences of points $\mathbf{P}^\nu(\mathbf{q})$ of one trajectory do not generate continuous curves, like those depicted in Figure 7.6, but hit them at discrete return points (Exercise 7.3). That is, the continuous curves shown in Figure 7.6 are not trajectories; they indicate the union of an infinite number of possible intersection points of distinct orbits; compare Figure 7.5. Depending on the dynamic behavior on Ω, the periodic orbit related to $\mathbf{q}^*$ is called the *saddle cycle* (Figure 7.6a), the *spiral cycle* (Figure 7.6b), or the *nodal cycle* (Figure 7.6c). By reversing the arrows in (b) and (c), one obtains illustrations of repelling cycles.

The insets and outsets in Figure 7.6a and 7.6c are cross sections of Ω with the invariant manifolds of the differential equations. (Once a trajectory is inside an invariant manifold, it remains inside, see Appendix A.5.) The curves are invariant with respect to $\mathbf{P}$. A saddle cycle has an unstable and a stable invariant manifold intersecting in the cycle. This can be visualized by means of Figure 7.5. Imagine $\mathbf{q}$ situated on an unstable invariant manifold of a saddle-type orbit. Then a closed band "spanned" by both trajectories of Figure 7.5 is part of this manifold. The reader is encouraged to draw an illustration that also includes part of a stable invariant manifold. Instructive illustrations of this kind can be found in [AbS88], [Osi03].

We now apply the stability result for the fixed-point equation $\mathbf{P}(\mathbf{q}) = \mathbf{q}$ to periodic solutions of the differential equation in equation (7.1). In order to obtain the eigenvalues mentioned in the above theorem, the linearization of the Poincaré map (7.10) around the fixed point $\mathbf{q}^*$ is needed. This matrix $\frac{\partial \mathbf{P}(\mathbf{q}^*)}{\partial \mathbf{q}}$ is given by the monodromy matrix restricted to the $(n-1)$-dimensional Ω. As pointed out in Lemma 7.3, the n^2-monodromy matrix

$$\frac{\partial \varphi(T; \mathbf{z}^*)}{\partial \mathbf{z}} = \mathbf{M}$$

has $+1$ as an eigenvalue with eigenvector $\dot{\mathbf{y}}^*(0)$ tangent to the intersecting curve $\mathbf{y}^*(t)$. This eigenvector is not in Ω because the intersection is transversal. Choosing an appropriate basis for the n-dimensional space, one can show that the remaining $n-1$ eigenvalues of $\mathbf{M}$ are those of

$$\frac{\partial \mathbf{P}(\mathbf{q}^*)}{\partial \mathbf{q}} .$$

Consequently, the periodic orbit is stable if the remaining $n-1$ eigenvalues of $\mathbf{M}$ are smaller than unity in modulus. The orbit $\mathbf{y}^*$ is unstable if $\mathbf{M}$ has an eigenvalue μ with $|\mu| > 1$. The eigenvalue 1 corresponds to a perturbation along $\mathbf{y}^*(t)$ leading out of Ω; the other $n-1$ eigenvalues of $\mathbf{M}$ determine what happens to small perturbations within Ω. The eigenvalues of the monodromy matrix are called the *Floquet multipliers* or *(characteristic) multipliers*. Note that calculating $\mathbf{q}^*$ amounts to carrying out a shooting method; compare Section 1.5.4 (Exercise 7.5).

7.4 Mechanisms of Losing Stability

So far we have discussed how (local) stability of a particular periodic solution manifests itself through Floquet multipliers. In general, the multipliers and, hence, the stability vary with λ. First we summarize the previous results on stability. Then we shall discuss mechanisms of losing stability. Classical references are, for example, [Arn83], and [GuH83].

Summary 7.5 *For a value of* λ*, let* $\mathbf{y}(t)$ *be a periodic solution to* $\dot{\mathbf{y}} = \mathbf{f}(\mathbf{y}, \lambda)$ *with period* T. *The monodromy matrix is defined by* $\mathbf{M}(\lambda) = \Phi(T)$, *where* $\Phi(t)$ *solves the matrix initial-value problem*

$$\dot{\Phi} = \mathbf{f}_{\mathbf{y}}(\mathbf{y}, \lambda)\,\Phi, \quad \Phi(0) = \mathbf{I}.$$

The matrix $\mathbf{M}(\lambda)$ *has* n *eigenvalues* $\mu_1(\lambda), \ldots \mu_n(\lambda)$. *One of them is equal to* $+1$, *say* $\mu_n = 1$. *The other* $n - 1$ *eigenvalues determine (local) stability by the following rule:*

$\mathbf{y}(t)$ *is stable if* $|\mu_j| < 1$ *for all* $j = 1, \ldots, n - 1$;
$\mathbf{y}(t)$ *is unstable if* $|\mu_j| > 1$ *for some* j.

Sometimes stability is described by means of *characteristic* or *Floquet exponents* σ, defined by $\mu = \exp(\sigma T)$. Because the exponential function maps the half-plane $\mathrm{Re}(\sigma) < 0$ onto the interior of the unit circle, stability is also signaled by all characteristic exponents having a negative real part.

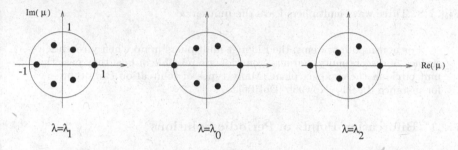

Fig. 7.7. Multipliers (eigenvalues of the monodromy matrix) for three values of λ

Figure 7.7 shows for three values of λ the multipliers μ of fictive periodic solutions that belong to the same branch. The circle is the unit circle, and one eigenvalue is unity for all λ. The left sketch ($\lambda = \lambda_1$) represents a stable solution, because all eigenvalues μ_j lie inside the unit circle ("all" in the sense $j = 1, \ldots, n-1$). The right sketch shows one multiplier outside the unit circle. Consequently, the periodic orbit for $\lambda = \lambda_2$ is unstable. Obviously, for some value λ_0 between λ_1 and λ_2 one multiplier crosses the unit circle and the stability is lost (or gained if we proceed from λ_2 to λ_1). Figure 7.7 assumes that the critical multiplier crosses the unit circle at -1. Depending on where the critical multiplier or pair of complex conjugate multipliers crosses the unit circle, different types of bifurcation occur. One distinguishes three ways of crossing the unit circle, with three associated types of bifurcation. Figure 7.8 shows the path of the critical multiplier only—that is, the eigenvalue with $|\mu(\lambda_0)| = 1$. In Figure 7.8a, the eigenvalue gets unity in addition to μ_n, $\mu(\lambda_0) = 1$. In Figure 7.8b the multiplier crosses the unit circle at the negative real axis, $\mu(\lambda_0) = -1$. In Figure 7.8c the crossing is with nonzero imaginary part—that is, a pair of complex conjugate eigenvalues crosses the unit circle. All three sketches refer to a loss of stability when λ passes λ_0; for an illustration of gain of stability, change the arrows to point in the opposite direction. In what follows, we discuss the three kinds of losing stability,

(a) $\mu_j(\lambda_0) = 1$ for a $j \in \{1, \ldots, n-1\}$,
(b) $\mu_j(\lambda_0) = -1$ for a j,
(c) $\mu_j(\lambda_0)\,\mu_k(\lambda_0) = 1$ for $1 \le k < j < n$

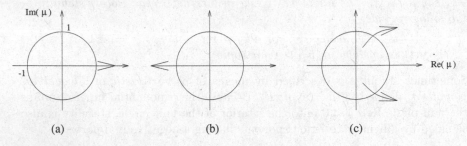

Fig. 7.8. Three ways multipliers leave the unit circle

      For nonsmooth systems, the Floquet multipliers jump when $\mathbf{y}$ reaches a surface of discontinuity. Various scenarios are possible of how they pass the unit circle in the complex plane. Many types of bifurcation can occur, see for instance [Lei00], [Kow06], [DoH06].

7.4.1 Bifurcation Points of Periodic Solutions

Let $\mathbf{P}$ be the Poincaré map associated with the differential equation in equation (7.1). Because we are interested in periodic solutions, we study fixed points of the Poincaré map,

$$\mathbf{P}(\mathbf{q}, \lambda) = \mathbf{q},$$

for $\mathbf{q} \in \Omega$. Assume situation (a) of the above three cases—that is, $\frac{\partial \mathbf{P}}{\partial \mathbf{q}}$ has unity as an eigenvalue for $\lambda = \lambda_0$. The fixed-point equation is equivalent to

$$\tilde{\mathbf{f}}(\mathbf{q}, \lambda) := \mathbf{P}(\mathbf{q}, \lambda) - \mathbf{q} = \mathbf{0}, \qquad (7.11a)$$

which forms a system of $n-1$ scalar equations. For this system of equations, the bifurcation results of Chapter 2 apply. The Jacobian of $\tilde{\mathbf{f}}$ satisfies

$$\frac{\partial \tilde{\mathbf{f}}}{\partial \mathbf{q}} = \frac{\partial \mathbf{P}}{\partial \mathbf{q}} - \mathbf{I}. \qquad (7.11b)$$

Hence, $\mu(\lambda_0) = 1$ implies that for λ_0 the Jacobian of $\tilde{\mathbf{f}}$ has an eigenvalue zero. That is to say, for $\mu(\lambda_0) = 1$ we encounter the stationary bifurcation scenario we discussed in Chapter 2. In particular, assuming the standard situation of a simple eigenvalue [cf. Definitions 2.7 and 2.8], one has simple bifurcation points and turning points of the Poincaré map. This multiplicity of fixed points of **P** on Ω is illustrated by Figure 7.9. This figure shows schematically fixed points of a Poincaré map **P** for varying λ. For λ_1 we have one fixed point only, for λ_2 there are three fixed points. The splitting of fixed points at λ_0 (Figure 7.9 assumes a pitchfork bifurcation) relates to a bifurcation of the corresponding periodic orbits (cf. Figure 7.10). In this illustration, we assume the situation λ_2 of Figure 7.9 with two attracting fixed points and one repelling fixed point. The stable orbits that correspond to the attracting fixed points are emphasized by heavy curves; some approaching trajectories that start from neighboring points on Ω are also indicated.

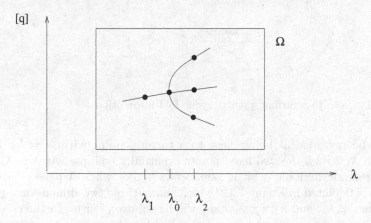

Fig. 7.9. Illustration of a pitchfork bifurcation on a Poincaré section

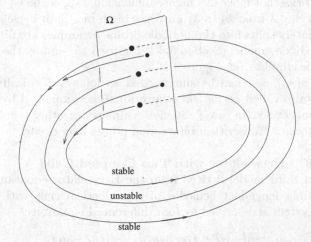

Fig. 7.10. Close to a pitchfork bifurcation; corresponds to λ_2 in Figure 7.9.

Turning points can be illustrated analogously (see Figure 7.11). Figure 7.11 depicts one stable periodic orbit (heavy curve), one unstable periodic orbit, and one neighboring trajectory. The situation of Figure 7.11 (for some

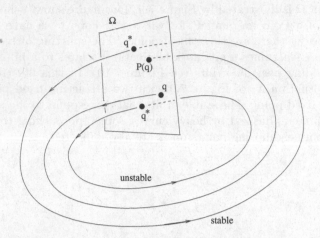

Fig. 7.11. Close to a turning point / cyclic fold bifurcation

λ_1) can be regarded as being close to a turning point with $\lambda = \lambda_0$. If we approach λ_1 to λ_0, the two fixed points eventually collapse. At $\lambda = \lambda_0$, the resulting single periodic orbit is *semistable*; it dies when λ passes λ_0. This scenario is depicted in Figure 7.12, which shows three two-dimensional phase planes for λ_1, λ_0, and λ_2; for λ_2, no periodic solution exists. In the case of a planar situation $(n = 2)$, inside any closed orbit there must be at least one equilibrium. Figure 7.12 is supposed to represent a two-dimensional projection of trajectories that move in a higher-dimensional space; no equilibrium is indicated. Varying λ from λ_2 to λ_1 provokes birth of a limit cycle at $\lambda = \lambda_0$, which immediately splits into two periodic orbits. Examples are furnished by the Hodgkin–Huxley model (Section 3.4.1, Figure 3.15) and by the FitzHugh model of nerve impulses (cf. Figure 3.16).

The case $\mu(\lambda_0) = 1$ can be summarized as follows: Typically, turning points occur (also called *cyclic fold bifurcations*), accompanied by the birth or death of limit cycles. In case **f** satisfies symmetry or other regularity properties, pitchfork or transcritical bifurcation points may occur.

Example 7.6   Brusselator with Two Coupled Cells

We consider a third model derived from the Brusselator equation (5.53)—namely, the time-dependent behavior of two identical cells with mass exchange. The system is described by four differential equations:

$$
\begin{aligned}
\dot{y}_1 &= A - (B + 1)y_1 + y_1^2 y_2 + \delta_1(y_3 - y_1), \\
\dot{y}_2 &= B y_1 - y_1^2 y_2 + \delta_2(y_4 - y_2), \\
\dot{y}_3 &= A - (B + 1)y_3 + y_3^2 y_4 - \delta_1(y_3 - y_1), \\
\dot{y}_4 &= B y_3 - y_3^2 y_4 - \delta_2(y_4 - y_2).
\end{aligned}
\tag{7.12}
$$

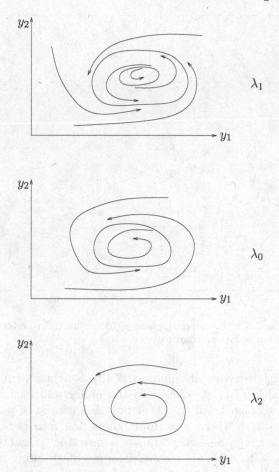

Fig. 7.12. Phase planes illustrating how a stable orbit collapses at a turning point (cyclic fold point) for λ_0

The variables $y_1(t)$ and $y_2(t)$ represent the substances in the first cell, and $y_3(t)$ and $y_4(t)$ represent those of the second cell. The coupling constants δ_1 and δ_2 control the mass exchange between the cells. The equations reflect the Brusselator kinetics in each cell. We choose the constants

$$A = 2, \quad \delta_1 = 1, \quad \delta_2 = 4$$

and take $\lambda = B$ as the bifurcation parameter. There are two kinds of reactions. For a homogeneous reaction, both cells exhibit identical behavior,

$$y_1(t) = y_3(t), \quad y_2(t) = y_4(t) \quad \text{for all } t.$$

The reaction is said to be inhomogeneous if this identity does not hold.

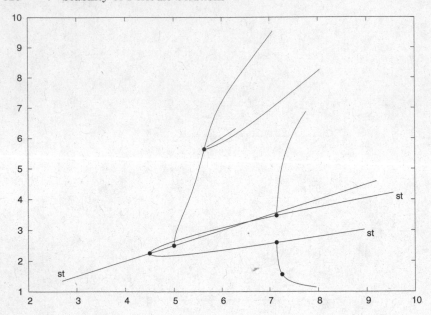

Fig. 7.13. Example 7.6, bifurcation diagram $y_2(0)$ versus λ; "st" refers to stationary solutions; bifurcation points are marked.

The branching behavior of equation (7.12) is depicted in Figure 7.13. The branches marked by "st" consist of stationary solutions. At $\lambda_0 = 4.5$, the stability is exchanged. All the other branches consist of periodic orbits. For the value $\lambda_0 = 5$, there is a Hopf bifurcation into periodic homogeneous orbits. First, the emerging branch is unstable; it gains stability at $\lambda_0 = 5.6369$ ($T_0 = 4.42624$) via a subcritical pitchfork bifurcation. That is, for $\lambda > 5.6369$ there are stable homogeneous oscillations, which are "encircled" by unstable inhomogeneous orbits. The two branches of stationary inhomogeneous solutions, which exist for $\lambda > 4.5$, lose their stability at the two Hopf bifurcation points with $\lambda_0 = 7.14547$, $T_0 = 2.88882$.

The remarkable part of the solutions of equation (7.12) is the Hopf bifurcation at $\lambda_0 = 5$. Because all of the solutions merging in the bifurcation point

$$y_1 = y_3 = 2\,,\quad y_2 = y_4 = 2.5\,,\quad \lambda = 5$$

are homogeneous, the coupling terms in equation (7.12) vanish; the equations of the two cells decouple. Consequently, the same solutions can be calculated by solving only the subsystem of the first two equations ($n = 2$). In this way, both the homogeneous equilibria and the homogeneous oscillations are obtained. But the stability behavior is different. The bifurcations into inhomogeneous solutions do not exist for the reduced system, and there is an exchange of stability at $\lambda_0 = 5$ (Exercise 7.4). This shows that by the process of coupling two identical cells the stability behavior changes drastically. In particular, the homogeneous orbits in the range

$$5 < \lambda < 5.6369$$

lose stability, caused by the onset of additional bifurcations. This serves as another example for the phenomenon that relaxing restrictions can enrich the bifurcation behavior (see Example 6.4). Concerning the underlying "spaces" of admissible functions, distinguish between the space of smooth vector functions with independent components and the small subspace of homogeneous functions satisfying $y_1 = y_3$, $y_2 = y_4$. □

7.4.2 Period Doubling

We now discuss the second case of losing stability, $\mu(\lambda_0) = -1$. From equation (7.11) we expect that the Jacobian of

$$\tilde{\mathbf{f}}(\mathbf{q}, \lambda_0) = \mathbf{P}(\mathbf{q}, \lambda_0) - \mathbf{q}$$

is nonsingular (eigenvalue $-2 \neq 0$). Consequently, at λ_0 there is no bifurcation point of the kind discussed above. Rather, a smooth branch $\mathbf{q}(\lambda)$ passes through $\mathbf{q}(\lambda_0)$ without bifurcating itself in the familiar manner.

Example 7.7   Logistic Map
We study the fixed points of the real-valued function of equation (1.9)

$$P(q, \lambda) = \lambda(1 - q)q \quad \text{for } q > 0. \tag{7.13}$$

From the fixed-point equation, we obtain the two fixed points $q(\lambda)$,

$$q(\lambda) = 0,$$
$$q(\lambda) = \frac{\lambda - 1}{\lambda}.$$

Consequently, there is a transcritical bifurcation at

$$(q_0, \lambda_0) = (0, 1).$$

Stability is determined by the eigenvalues of the Jacobian of $P(q)$, which reduces here to the single value

$$\mu = \frac{dP}{dq} = \lambda(1 - 2q).$$

The bifurcation with exchange of stability at $\lambda_0 = 1$ is confirmed by $\mu(1) = +1$. Evaluated at the nontrivial branch, the expression for μ is

$$\mu = 2 - \lambda.$$

The range of stability is determined by $|\mu| = |2 - \lambda| < 1$, which leads to

$$1 < \lambda < 3.$$

Consequently, the nontrivial branch loses stability at $\lambda = 3$. This point is no bifurcation of fixed points of $P(q, \lambda) = q$ because $\mu = -1$. In order to check where the fixed-point iteration converges for $\lambda > 3$, we choose as an example $\lambda = 3.1$ and $q = 0.1$ and iterate

$$P(q), \ P(P(q)), \ \ldots.$$

After about 30 iterations the values settle down to

$$0.7644\ldots$$
$$0.5583$$
$$0.7645$$
$$0.5582$$
$$0.7645.$$

It turns out that every other iterate converges to a limit—that is, the iteration exhibits a periodicity with period two. For the composed *double-period map*

$$P^2(q, \lambda) = P(P(q, \lambda), \lambda),$$

both the values that occur as limits of the above sequence for $\lambda = 3.1$ are fixed points. For $\lambda > 3$, $\lambda \approx 3$, there is no attracting fixed point of P, but there are two attracting fixed points of P^2. Consequently, at $\lambda_0 = 3$ there is an exchange of stability of period-one fixed points to period-two fixed points. This phenomenon is called *period doubling* or *flip bifurcation*. Period doubling is a bifurcation of the P^2 fixed-point equation, because all fixed points of P are also fixed points of P^2. The eigenvalues of the branch

$$q(\lambda) = \frac{\lambda - 1}{\lambda}$$

of solutions to

$$P^2(q, \lambda) - q = 0$$

indicate the bifurcation at $\lambda_0 = 3$ by $\mu(\lambda_0) = +1$ (Exercise 7.8). We shall meet this example again in another context (Section 9.2).   □

We are now prepared to understand what happens to periodic oscillations when for $\lambda = \lambda_0$ the monodromy matrix has an eigenvalue $\mu = -1$. Because for the related Poincaré map the chain rule implies

$$\frac{\partial}{\partial q}[\mathbf{P}(\mathbf{P}(\mathbf{q}))] = \left(\frac{\partial \mathbf{P}(\mathbf{q})}{\partial \mathbf{q}} \right)^2,$$

the fixed point $\mathbf{q}$ of $\mathbf{P}(\mathbf{q}) = \mathbf{q}$ leads to the eigenvalue $\mu = +1$ for $\mathbf{P}^2(\mathbf{q}) = \mathbf{P}(\mathbf{P}(\mathbf{q}))$. This bifurcation of the double period is illustrated by Figure 7.14.

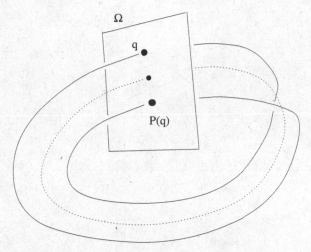

Fig. 7.14. Close to period doubling / flip bifurcation (dashed: unstable periodic orbit of the simple period)

The dashed curve represents an unstable periodic orbit with the simple period; the heavy curve is a stable orbit of the double period. The depicted situation can be seen as representing solutions for a λ close to the value λ_0 of a period doubling. A band spanned by the trajectories of Figure 7.14 forms a Möbius strip. For λ approaching λ_0, the curve that winds twice shrinks to the curve that winds once. Shifting λ in the other direction leads to the phenomenon that at λ_0 the stable single-period oscillation splits into stable double-period oscillations (supercritical case). This splitting requires at least a three-dimensional phase space of the autonomous system equation (7.1), $n \geq 3$. For a calculated example of the schematic Figure 7.14 see Figure 7.18, which shows the results of two successive period doublings.

The term "period" has a slightly different meaning for the Poincaré map **P** and the periodic oscillation $\mathbf{y}(t)$. In the case of a map, the period is the integer that reflects how many iterations of **P** are required to reproduce the point from which the iteration starts. For varying λ, this integer remains constant if no further period doubling occurs. In contrast, the periods T of the oscillations vary with λ. In general, this variation is different for the single-period oscillations and the double-period oscillations. Consequently, for a given $\lambda \neq \lambda_0$ the periods do not differ exactly by a factor of 2. For λ tending to λ_0, this factor is attained as a limit. This will be illustrated in Figure 7.19.

Period doubling is also called *subharmonic bifurcation*. This phenomenon is observed in many applications—for example, in chemical reactions [JoR83], [KaR81], [SmKLB83] (see Example 7.8 below), nerve models [RiM80], or Navier-Stokes equations [FrT79]. Period doubling also occurs in the Brusselator [KuH84b] and Lorenz equations [Spa82]. For the Lorenz equation, see Figure 2.37, which shows a period doubling for $\lambda_0 = 357$; a nearby periodic

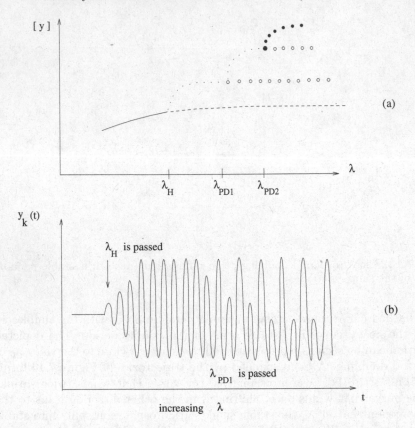

Fig. 7.15. Scenario of a dynamical behavior caused by a subtle Hopf bifurcation followed by period doubling, schematically

orbit "after" the doubling is shown in Figure 2.36. An example of period doubling in electric power generators has been presented in Example 3.8, see the bifurcation diagram of Figure 3.26.

In experiments, for varying λ, often a sequence of period doublings occurs. As pointed out in [CrO84], after a first period doubling a second period doubling is more likely than other bifurcations. Accordingly, sequences of period doublings are typical for a class of bifurcation problems. Figure 7.15a visualizes this bifurcation phenomenon schematically; for the Lorenz equation see Figure 2.37, and for the power generator, see Figure 3.26. After a Hopf bifurcation at λ_H is passed, a series of period doublings at $\lambda_{PD\nu}$ occurs. In a corresponding experiment with increasing λ (Figure 7.15b) a period doubling manifests itself by a lowering of every other maximum (or minimum). Compare also Figure 1.5. Figure 7.15 shows a soft loss of stability. Period doubling can also occur subcritically (see Figure 7.16 and Section 2.9).

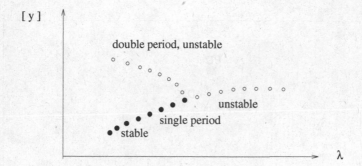

Fig. 7.16. Subcritical period doubling, schematically

A normal form of period doubling can be derived [GuH83], which is of pitchfork shape. This explains why the graphical illustrations of period doublings have a square-root-like shape (Exercise 7.9).

In many real-valued mappings, the sequence of repeated period doubling obeys a certain law concerning the distribution of the period-doubling parameters λ_{PD_ν} [Fei80], [GrT77]. In the limit, these values satisfy

$$\lim_{\nu \to \infty} (\lambda_{\nu+1} - \lambda_\nu)/(\lambda_\nu - \lambda_{\nu-1}) = 0.214169\dots. \qquad (7.14)$$

Frequently, the inverse value is quoted, 4.6692016.... This limit does not only occur in iterations of simple maps; it has also been observed with periodic solutions to ODEs. The scaling law *(Feigenbaum's law)* in equation (7.14) appears to have a universal character. The ratios $\frac{\lambda_{\nu+1}-\lambda_\nu}{\lambda_\nu-\lambda_{\nu-1}}$ have been called *Feigenbaum ratios*.

Example 7.8   Isothermal Reaction
The ODE system from [SmKLB83]

$$\dot{y}_1 = y_1(30 - 0.25y_1 - y_2 - y_3) + 0.001y_2^2 + 0.1\,,$$
$$\dot{y}_2 = y_2(y_1 - 0.001y_2 - \lambda) + 0.1\,, \qquad (7.15)$$
$$\dot{y}_3 = y_3(16.5 - y_1 - 0.5y_3) + 0.1$$

describes an isothermal chemical reaction. A Hopf point ($\lambda_0 = 12.1435$) and a sequence of period-doubling points has been revealed, with values of the parameter

$$\lambda_1 = 10.5710\,,$$
$$\lambda_2 = 10.1465\,,$$
$$\lambda_3 = 10.0912\,,$$
$$\lambda_4 = 10.0808\,.$$

These four values allow calculation of two of the Feigenbaum ratios in equation (7.14). The values are

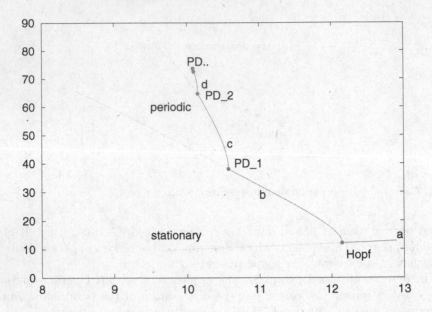

Fig. 7.17. Example 7.8; bifurcation diagram, $y_1(0)$ over λ; cascade of period doublings, solid lines represent stable solutions, thin lines represent UPOs; a,b,c,d refer to the solutions depicted in Figure 1.5 and in Figure 7.18.

$$0.13 ,$$

$$0.188 ,$$

coming closer to the limit in equation (7.14). The bifurcation diagram of Figure 7.17 summarizes the results. Figure 7.18 shows three stable orbits of the sequence of period doublings.   □

Example 7.9   Lorenz's Fourth-Order System
In [Lor84], Lorenz introduced the system of four ordinary differential equations

$$\begin{aligned}
\dot{y}_1 &= -y_1 + 2\lambda - y_2^2 + (y_3^2 + y_4^2)/2 , \\
\dot{y}_2 &= -y_2 + (y_1 y_2 - y_3 y_4) + (y_4^2 - y_3^2)/2 , \\
\dot{y}_3 &= -y_3 + (y_2 - y_1)(y_3 + y_4)/2 , \\
\dot{y}_4 &= -y_4 + (y_2 + y_1)(y_3 - y_4)/2 .
\end{aligned} \tag{7.16}$$

A rigorous bifurcation analysis is given in [KaS86]. There are two stationary bifurcations, one Hopf bifurcation (for $\lambda_0 = 3.8531$), and a sequence of period doublings. Five period-doubling points have been calculated, with the values of λ

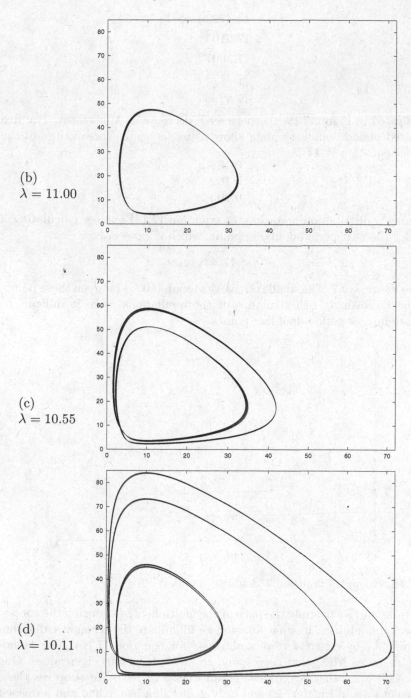

Fig. 7.18. Example 7.8; stable periodic orbits for three values of λ; two period doublings are straddled. (y_1, y_2)-phase plots (projections); b,c,d as in Figure 1.5 and Figure 7.17

$$14.4722\,,$$
$$17.0105\,,$$
$$17.3607\,,$$
$$17.4175\,,$$
$$17.4290\,.$$

As indicated in Figure 7.19, the periods T decrease as λ increases. The five calculated period doubling points allow us to determine three values of the sequence equation (7.14),

$$0.138\,,$$
$$0.162\,,$$
$$0.203\,.$$

Assuming a distribution according to equation (7.14) enables calculating a guess for the next period-doubling point, which is expected for

$$\lambda \approx 17.4314\,;$$

see also Exercise 9.7. The small relative deviation (10^{-4}) between these points and the increasingly rich structure of the oscillations make it difficult to calculate further period-doubling points.

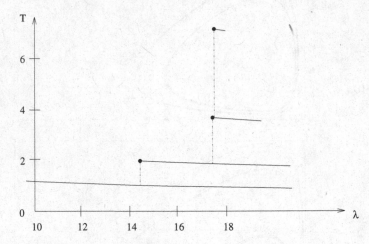

Fig. 7.19. Example 7.9; period T doubles

It is instructive to study the path of the multipliers $\mu(\lambda)$ during the successive period doublings. In what follows, we illuminate the schematic diagram of Figure 7.7 by concrete results obtained for equation (7.16). The monodromy matrices $\mathbf{M}(\lambda)$ that correspond to (7.16) have four eigenvalues. One is always $+1$, another eigenvalue remains close to zero. We suppress these two eigenvalues in Figure 7.20 and only depict in a heavy line and a dashed line the paths $\mu(\lambda)$ of the two remaining multipliers. Figure 7.20 shows the multipliers for the part of the periodic branch between the Hopf point and the first period doubling. The numbers report on values of the bifurcation parameter λ. Inspecting Figure 7.20, we realize that in addition to the multiplier that is unity for all λ, one multiplier is unity at the Hopf bifurcation ($\lambda = 3.8531$). For increasing λ, this multiplier moves inside the unit circle along the real axis. For $\lambda \approx 4$, the multiplier meets the second multiplier and both become a complex-conjugate pair. For increasing λ, the pair of multipliers moves along curved paths (roughly circular) toward the negative real axis. Here, for $\lambda \approx 14.3$, the two multipliers divorce and move in opposite directions. We follow the multiplier with the maximal modulus because this one determines stability. For $\lambda = 14.4722$, this critical multiplier leaves the unit circle at -1, indicating period doubling.

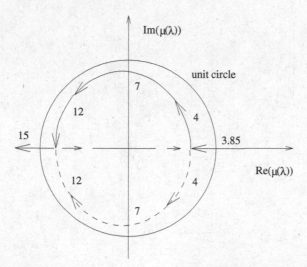

Fig. 7.20. Example 7.9, path of multipliers from Hopf bifurcation to period doubling

Following the multipliers for any stable branch between two period doublings, one obtains an analogous picture. In Example 7.9, the curved paths of conjugate-complex multipliers decrease in diameter after each period doubling. For instance, for the stable periodic orbits in the range $17.4175 < \lambda < 17.429$, the diameter of the curved part of the path is smaller than 0.05. —A reference for similar curves of eigenvalues is [KnE96].   □

7.4.3 Bifurcation into Torus

The third type of losing stability is characterized by a pair of complex-conjugate multipliers crossing the unit circle at

$$\mu(\lambda_0) = e^{\pm i\vartheta} \quad \text{for } \vartheta \neq 0, \ \vartheta \neq \pi$$

(i: imaginary unit). For simplicity, we assume that the angle ϑ is an irrational multiple of 2π (Exercise 7.10). Then the Poincaré map has an *invariant curve* C in Ω (cf. Figure 7.21). That is to say, take any point $\mathbf{q}$ on C and all iterates of the Poincaré map stay on that "drift ring." Consequently, trajectories spiral around a torus-like object. The invariant curve C is the cross section of the torus with the hypersurface Ω. The periodic orbit that has lost its stability to the torus can be seen as the central axis of the torus. For λ tending to λ_0, the diameter of the torus shrinks to zero and eventually reduces to the periodic orbit—that is, at λ_0 a bifurcation from a periodic orbit to a torus takes place. Because this scenario requires complex multipliers in addition to the eigenvalue that is unity, a bifurcation into a torus can take place only for $n \geq 3$.

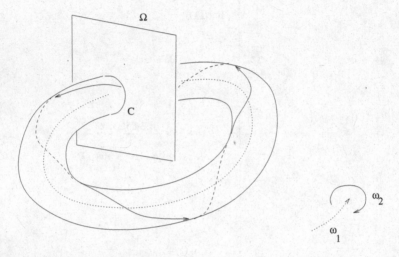

Fig. 7.21. "Torus," a trajectory encircles an unstable periodic orbit

Tori can be attractive or repelling, as with periodic orbits. For $n = 3$, trajectories approach an attractive torus from the inside domain and the outside domain. Trajectories projected to the hypersurface Ω resemble the dynamics close to a Hopf bifurcation, where a stable limit cycle encircles an unstable equilibrium. Accordingly, the phenomenon of bifurcation into a torus is sometimes called *Hopf bifurcation of periodic orbits, secondary Hopf bifurcation*, or *generalized Hopf bifurcation*. Relating to a theoretical result due to Naimark and Sacker [Nai59], [Sac65], bifurcation into tori is frequently called *Naimark–Sacker bifurcation* (Exercise 7.11). A repelling torus can be visualized as a tube surrounding a stable periodic orbit ($n = 3$). For λ tending to λ_0, the domain of attraction of the limit cycle vanishes and the torus dies. This is the scenario of a subcritical torus bifurcation.

For ϑ being an irrational multiple of 2π, a trajectory that starts on the invariant curve C will never arrive in the same position. After the bifurcation from periodic orbit (with frequency $\omega_1(\lambda)$) into torus, there are two frequencies $\omega_1(\lambda), \omega_2(\lambda)$. One frequency describes the component of motion along the axis within the torus (longitudinal motion), the other frequency is measured along the cross section (latitudinal motion, see the inset in Figure 7.21). The related flow is called *quasi-periodic* if the ratio ω_1/ω_2 is irrational. The frequencies ω_1 and ω_2 are said to be *incommensurate* if the equation

$$p\omega_1 + q\omega_2 = 0$$

has no solutions with nonzero integer values of p and q. For some λ, this ratio ω_1/ω_2 may be rational ("locked state"). Then the trajectory on the torus is closed, and the trajectory is periodic. In [ThS86] this phenomenon has been explained as a global bifurcation caused by a saddle node entering the drift ring. For results on quasi-periodic flow on tori, see [Mos66], [Hale69]. Higher-dimensional tori ("m-torus" for m-dimensional torus, 1-torus is a periodic orbit) have more frequencies; for bifurcation from 2-torus to 3-torus, see Section 9.3.1. An analysis of bifurcation into tori is a major mathematical task beyond the scope of this text. Some more insight may be obtained from Section 7.7. For general reference on dynamical systems we refer to [KaH95], [HaK03].

The assumption that $\vartheta/2\pi$ is irrational seems reasonable, because in "nature" almost all numbers are irrational. (On a computer, however, no irrational numbers exist.) For rational fractions, additional resonant terms exist. In particular, the "strongly resonant" cases $\mu^j(\lambda_0) = 1$ for $j = 1, 2, 3, 4$ require specific analysis. We discussed the cases $j = 1$ and $j = 2$ earlier in this chapter; for $j = 3$ and $j = 4$, consult for example [Wan78], [IoJ81], [Flo85a], [Flo85b]. Weak resonances can manifest themselves in that trajectories on the torus concentrate in bands. Graphical illustrations of a related example are shown in [Lan84]; other plots of tori can be found in [AlC84]. Numerical experiments were also conducted in [BoDI78], [KeASP85]. Interactions of Hopf and stationary bifurcations are investigated in [Lan79], [Lan81], [Spi83], [ScM84], [Sch87], [Flo87]. Applications for a Toda chain are discussed in [GeL90].

A *quasi-periodic* function $y(t)$ is one that is periodic in more than one argument,
$$y(\omega_1 t, \dots, \omega_k t),$$
for $k > 1$, and y is periodic with period 2π in each argument. For example, y may be the sum of k periodic functions $u_1(t), \dots, u_k(t)$, where $u_j(t)$ has a minimum period T_j. A simple example is

$$y(t) = \cos(2\pi t) + \cos(2\pi\sqrt{2}t);$$

$y(t)$ is not periodic because $\sqrt{2}$ is irrational. The example $y(t) = \cos(2\pi t) + \cos(7\pi t)$ is periodic, with period $T = 14$. In case the function $y(t)$ has an infinite number of frequencies, it is called *almost periodic*.

Example 7.10   A Model Related to Nerve Impulses

The ODE system

$$\dot{y}_1 = y_1 - \tfrac{1}{3}y_1^3 - y_2 + y_3 + \lambda,$$
$$\dot{y}_2 = 0.08(y_1 + 0.7 - 0.8y_2),$$
$$\dot{y}_3 = 0.0006(-y_1 - 0.775 - y_3)$$

(7.17)

from [RiF84] is an extended version of the FitzHugh model of nerve impulses [Fit61]. There is a Hopf bifurcation at the value of the electric current $\lambda_0 = 0.13825$ and initial period $T_0 = 22.7115$. The emerging branch of periodic orbits loses its stability at $\lambda_0 = 0.1401$ via bifurcation into torus. The system (7.17) exhibits further phenomena—for example, a period doubling has been found at $\lambda_0 = 0.31415$. Characteristic solutions of (7.17) show a double-spike pattern.   □

Example 7.11   Torus

The system

$$\dot{y}_1 = (\lambda - 3)y_1 - 0.25y_2 + y_1[y_3 + 0.2(1 - y_3^2)],$$
$$\dot{y}_2 = 0.25y_1 + (\lambda - 3)y_2 + y_2[y_3 + 0.2(1 - y_3^2)],$$
$$\dot{y}_3 = \lambda y_3 - (y_1^2 + y_2^2 + y_3^2)$$

goes back to Langford [HaK91]. There is a Hopf bifurcation for $\lambda_0 = 1.68$ and a bifurcation into a stable torus for $\lambda_0 = 2$. For the parameter value $\lambda = 2.002$, we show in Figure 7.22 a characteristic time dependence of a trajectory on the torus. Almost visible seems the unstable periodic orbit, with frequency $\omega_1 \approx 25$, which forms the core of the motion. The small-amplitude wiggles are due to the second frequency.   □

7.5   Calculating the Monodromy Matrix

We learned about the outstanding role of the monodromy matrix $\mathbf{M}$ from Definition 7.2. In this section, we discuss methods for calculating $\mathbf{M}$. In the first subsection, we assume that a periodic solution $\mathbf{y}^*(t)$ to equation (7.1) with period T has been calculated and that $\mathbf{M}$ is calculated subsequently.

7.5.1   A Posteriori Calculation

Integrating the n^2 scalar differential equations in equation (7.9) for $0 \leq t \leq T$ yields the monodromy matrix $\mathbf{M}$ from Definition 7.2. Because the right-hand side

$$\mathbf{f_y}(\mathbf{y}^*, \lambda)\Phi$$

varies with $\mathbf{y}^*(t)$ and hence with t, the Jacobian matrix must be made available for each t. A convenient way of providing the periodic oscillation $\mathbf{y}^*(t)$ for

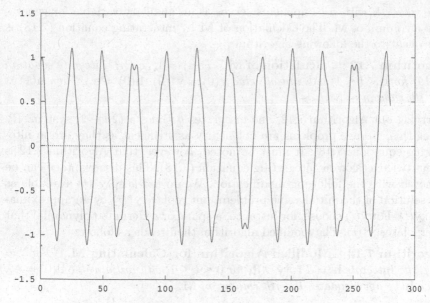

Fig. 7.22. Example 7.11, trajectory on a torus, $y_1(t)$, $\lambda = 2.002$

all t is to attach the original ODE system equation (7.1) as an initial-value problem,

$$\mathbf{y}^*(t) = \varphi(t; \mathbf{y}^*(0)) \, ;$$

see equation (7.7) for the definition of φ. A first method for calculating $\mathbf{M}$ is as follows: Integrate the initial-value problem

$$\begin{pmatrix} \dot{\mathbf{y}} \\ \dot{\Phi} \end{pmatrix} = \begin{pmatrix} \mathbf{f}(\mathbf{y}, \lambda) \\ \mathbf{f_y}(\mathbf{y}, \lambda)\Phi \end{pmatrix}, \quad \begin{pmatrix} \mathbf{y}(0) \\ \Phi(0) \end{pmatrix} = \begin{pmatrix} \mathbf{y}^*(0) \\ \mathbf{I} \end{pmatrix} \tag{7.18}$$

until $t = T$, then $\mathbf{M}$ is given by $\Phi(T)$. This method requires that $n + n^2$ scalar differential equations be integrated simultaneously. The enormous storage requirements usually restrict the use of equation (7.18) to differential equations with small n. The differential equation (7.18) was incorporated in [HoK84b] as part of the solution procedure to calculate $\mathbf{y}^*(t)$.

As an alternative, the monodromy matrix can be calculated columnwise [Sey85a]. For each column of $\mathbf{M}$, a separate system is integrated, consisting of essentially $2n$ scalar differential equations. For practical reasons that will soon become apparent, one may include trivial differential equations for λ and T and integrate

$$\begin{pmatrix} \mathbf{y} \\ T \\ \lambda \\ \mathbf{h} \end{pmatrix}' = \begin{pmatrix} T\mathbf{f}(\mathbf{y}, \lambda) \\ 0 \\ 0 \\ T\mathbf{f_y}(\mathbf{y}, \lambda)\mathbf{h} \end{pmatrix}. \tag{7.19}$$

The vector $\mathbf{h}$ with n components serves as an auxiliary variable for generating a column of $\mathbf{M}$. The calculation of $\mathbf{M}$ by integrating equation (7.18) is equivalent to the following algorithm:

Algorithm 7.12   Calculation of M. *For $j = 1, \ldots, n$: Integrate equation (7.19) for $0 \le t \le 1$ with initial value $\mathbf{y}(0) = \mathbf{y}^*(0)$, $\mathbf{h}(0) = \mathbf{e}_j$. Then $\mathbf{h}(1)$ is the jth column of $\mathbf{M}$.*

Carrying out Algorithm 7.12, one integrates n times a $(2n + 2)$ system. By doing this, storage problems are not nearly as stringent as they are in integrating equation (7.18). It must be noted, however, that Algorithm 7.12 is about twice as slow as integrating equation (7.18). This disadvantage can be removed with the following modification: We do not keep $\mathbf{y}^*(t)$ available as the solution of the initial-value problem, but replace $\mathbf{y}^*(t)$ by an approximation $\bar{\mathbf{y}}(t)$. For $\bar{\mathbf{y}}(t)$ choose, for instance, a spline or a Hermite polynomial that interpolates $\mathbf{y}^*(t)$. The modified algorithm then reads as follows:

Algorithm 7.13   Modified Algorithm for Calculating M. *For $j = 1, \ldots, n$: Integrate $\mathbf{h}' = T\mathbf{f}_\mathbf{y}(\bar{\mathbf{y}}, \lambda)\mathbf{h}$ for $0 \le t \le 1$ with initial value $\mathbf{h}(0) = \mathbf{e}_j$. Then $\mathbf{h}(1)$ approximates the jth column of $\mathbf{M}$.*

Algorithm 7.13 is efficient with regard to both storage and computing time. The drawbacks of Algorithm 7.13 are technical: Storing $\bar{\mathbf{y}}$ is more tedious than saving the vector $\mathbf{y}^*(0)$; in addition, an error control takes more effort. Hence, Algorithm 7.13 is much more complex than carrying out Algorithm 7.12 or integrating equation (7.18). Note that $\mathbf{M}$ is still an n^2-matrix, the storage efficiency of Algorithms 7.12 and 7.13 refers to the calculation only.

7.5.2 Monodromy Matrix as a By-Product of Shooting

The amount of work required for an a posteriori calculation of the monodromy matrix $\mathbf{M}$ is significant. So it is worthwhile to think about alternatives that provide $\mathbf{M}$ as a side result during the calculation of a periodic orbit. In what follows, we show as an example how to obtain $\mathbf{M}$ without any cost as a by-product of the (multiple) shooting approach. As in Section 6.3, the Jacobian $\mathbf{f}_\mathbf{y}$ is not required explicitly.

The iteration matrix of a shooting algorithm in condensed form was introduced in equations (6.21) and (6.22). The definitions of $\mathbf{M} = \Phi(T)$ and $\mathbf{G}_j$ imply that

$$\mathbf{M} = \mathbf{G}_{m-1} \cdot \ldots \cdot \mathbf{G}_1 .$$

From the definitions of $\mathbf{A}$ and $\mathbf{B}$ [cf. equation (6.12)], we infer that for periodic boundary conditions

$$\mathbf{r}(\mathbf{y}(0), \mathbf{y}(1)) = \mathbf{y}(0) - \mathbf{y}(1) = \mathbf{0} ,$$

the relations $\mathbf{A} = \mathbf{I}$ and $\mathbf{B} = -\mathbf{I}$ hold. Consequently,

$$\mathbf{E} = \mathbf{A} + \mathbf{B}\mathbf{G}_{m-1} \cdot \ldots \cdot \mathbf{G}_1 = \mathbf{I} - \mathbf{M} . \tag{7.20}$$

Because shooting is applied to the extended system equation (7.5) or (7.6) rather than equation (7.1), the approximation of the $(n+1)$-vector $\tilde{\mathbf{y}} = (\mathbf{y}^*, T)$ terminates with an iteration matrix $\tilde{\mathbf{E}}$, which is an $(n+1)^2$-matrix or $(n+2)^2$-matrix. By investigating the structure of equations (7.5) and (7.6), one realizes that $\mathbf{E}$ is the leading n^2-submatrix of $\tilde{\mathbf{E}}$.

> This gives an explanation of why standard shooting cannot be applied directly to the n system in equation (7.1) with boundary conditions $\mathbf{y}(0) = \mathbf{y}(T)$: Equation (7.20) implies that the iteration matrix $\mathbf{E}$ is singular because $\mathbf{M}$ has unity as eigenvalue. However, the extended iteration matrix $\tilde{\mathbf{E}}$ is mostly nonsingular. $\tilde{\mathbf{E}}$ may get singular at bifurcations of the type at which $\mathbf{M}$ has unity as an additional eigenvalue (turning point, stationary bifurcation point). The eigenvalue $+1$ then has algebraic multiplicity two. The singularity of $\tilde{\mathbf{E}}$ may explain difficulties that have occurred when applying equation (7.6) for the calculation of periodic orbits close to turning points.

We summarize the above strategy as follows [Sey85a]:

Algorithm 7.14   Simultaneous Calculation of $\mathbf{y}^*$ and $\mathbf{M}$. *Solve equations (7.5) or (7.6) by (multiple) shooting. $\mathbf{E}$ is the leading n^2-submatrix of the final iteration matrix. Then $\mathbf{M} = \mathbf{I} - \mathbf{E}$.*

Because $\tilde{\mathbf{E}}$ is obtained by numerical differentiation, the accuracy of the monodromy matrix calculated by Algorithm 7.14 is strongly affected by the approximation strategies in a particular implementation of shooting. Of course, the essence of Algorithm 7.14 for calculating $\mathbf{M}$ can be carried out independently of the shooting algorithm—that is, the version of $\mathbf{M}$ in equation (7.8) is approximated by numerical differentiation.

7.5.3 Numerical Aspects

The calculated monodromy matrix $\mathbf{M}$ has a feature that must be appraised highly—namely, a measure of its accuracy is accessible. Let $\mathbf{M}'$ be a numerically approximated monodromy matrix with eigenvalues μ_j' $(j = 1, \ldots, n)$. If $\mathbf{M}'$ and its eigenvalues are obtained accurately and the underlying solution is periodic, then one eigenvalue (say, μ_n') is close to unity. Consequently, the distance

$$\delta = |1 - \mu_n'| \qquad (7.21)$$

serves as a measure of the quality of approximation of $\mathbf{M}'$. If δ is small (say, $\delta < 0.005$), we accept the other $n - 1$ multipliers $\mu_1', \ldots, \mu_{n-1}'$ as reasonable approximations to the exact multipliers. Because the methods of equation (7.18) and of Algorithms 7.12 and 7.14 are essentially based on numerically integrating initial-value problems, the measure δ reflects how accurately this "shooting" is carried out.

Numerical experience suggests the following conclusions: Shooting methods work well for stable orbits and weakly unstable orbits. Here we call a periodic orbit *weakly unstable* if the multipliers are not large in modulus, say,

$$\max |\mu_j| < 10^3 .$$

Shooting encounters difficulties in providing enough accuracy when there are large multipliers.

> The above bound of 10^3 is artificial; it depends on the chosen numerical integrator and the prescribed error tolerance. Hence, the conclusions above do not necessarily apply to all problems and all integrators in that extent. In a small number of problems with strongly unstable orbits, backward shooting may help. That is, the differential equations are *integrated in reverse time*. Because all changes in stability occur in the stable and weakly unstable range, shooting can be expected to behave well in a bifurcation and stability analysis. For example, all results shown in the figures of this chapter have been calculated by shooting methods. With inaccurate values of $\mathbf{M}'$ (δ large), there can be difficulties deciding safely which of the eigenvalues approximates unity; this frequently happens when $\mathbf{M}$ has a multiplier close to unity.

The number δ from equation (7.21) allows one to compare the algorithms for approximating $\mathbf{M}'$. Clearly, Algorithm 7.12 and the integration of equation (7.18) yield the same results, because both methods differ only in their organization. Algorithm 7.14, which exploits the iteration matrix of multiple shooting, is strongly affected by the strategies of both continuation and approximating the shooting matrix (compare Section 5.1). As a rule, it is worthwhile to calculate the eigenvalues of $\mathbf{M}'$ from Algorithm 7.14. In the following cases, $\mathbf{M}'$ must be recalculated by a more accurate method, such as Algorithm 7.12:

> when δ based on Algorithm 7.14 is too large,
> after every five (say) continuation steps, in order to check the accuracy of Algorithm 7.14 by comparing it with Algorithm 7.12,
> when close to a loss or gain of stability—that is, when a multiplier approaches unity in modulus.

Such a combination of the cheap Algorithm 7.14 with the accurate Algorithm 7.12 appears to be a practical compromise.

In a similar way, as Algorithm 7.14 exploits results of shooting, other solution methods can be exploited. For example, [DoKK91], [Doe97] calculate periodic orbits by means of collocation, from which a monodromy matrix is derived. For critical cases of calculating Floquet multipliers an alternative approach has been presented in [FaJ92]. For references on calculating Floquet multipliers see also [Lust01].

If the underlying system $\dot{\mathbf{y}} = \mathbf{f}(\mathbf{y}, \lambda)$ results from a discretization of PDEs—that is, if the dimension n is large—a calculation of $\mathbf{M}$ and its eigenvalues may be too costly. In several of these cases, it is advisable to test stability by means of a coarser discretization, thereby assuming that the stability result is not severely affected by discretization errors. That is, smaller monodromy matrices that reflect the stability behavior of the large-scale system are to be generated. This has been done for finite-element equations modeling a wind energy converter [MaN85]. Restricting the use of the above algorithms to problems with small n makes the use of the QR method for calculating the eigenvalues tolerable. For periodic solutions of large systems, consult [LuRSC98].

7.6 Calculating Bifurcation Behavior

In the two preceding chapters, we discussed methods for handling bifurcation points, with emphasis on systems of nonlinear equations and boundary-value problems of ODEs. The main ideas underlying these methods also apply to periodic solutions. Recall the correspondence between the problem of calculating periodic solutions and the boundary-value problem in equations (7.6a/b). Instead of closing this section with that remark, it is worthwhile to discuss some specific aspects of periodic orbits.

7.6.1 Test Functions

The standard approach for locating bifurcation points is to monitor an appropriate test function. Because bifurcation is related to a multiplier μ_j that escapes or enters the unit disk of the complex plane, we can choose the modulus for a general type of test function: The function $\tau(\lambda)$ defined by

$$\tau := |\mu_j| - 1, \text{ with } |\mu_j| = \max\{|\mu_1|, \ldots, |\mu_{n-1}|\} \tag{7.22}$$

signals a change in stability by a change of sign. This assumes that the multipliers are ordered such that $\mu_n = 1$. In order to locate bifurcation points within an unstable range, choose the j of the term that minimizes the sequence

$$||\mu_1| - 1|, \ldots, ||\mu_{n-1}| - 1|.$$

These test functions correspond to equations (5.7) and (5.8).

Specific test functions can be defined as products of the eigenvalues

$$\tau_{\text{fold}} := \prod_{j=1}^{n-1} (\mu_j - 1)$$

$$\tau_{\text{flip}} := \prod_{j=1}^{n-1} (\mu_j + 1)$$

$$\tau_{\text{NS}} := \prod_{j<k<n} (\mu_k \mu_j - 1)$$

Obviously, these specific test functions refer to the three cases of losing stability that are discussed in Section 7.4. For the evaluation of these test functions, see for instance [Kuz98], [DoGK03]. Indirect methods based on bifurcation test functions have proved successful in the context of periodic solutions; strategies like that outlined in Section 5.4 work.

7.6.2   Applying Branching Systems

Apart from indirect methods, direct methods have also been applied. For this purpose, extended systems were formulated. In this introduction, we stay with "fully extended" systems. For *minimally extended* systems, we refer to [DoGK03]. For nonsmooth systems, defining systems are given in [KuRG03].

The branching system for periodic solutions, which results from equation (6.14), consists of $2n + 3$ components [Sey88]. Because numerical results obtained in applying this system have shown poor convergence, it cannot be recommended. In the special case of Hopf bifurcation, however, the system can be reduced favorably. Observing $\mathbf{f}(\mathbf{y}, \lambda) = \mathbf{0}$, the specific branching system

$$
\begin{pmatrix} \mathbf{y} \\ T \\ \lambda \\ \mathbf{h} \end{pmatrix}' = \begin{pmatrix} T\mathbf{f}(\mathbf{y}, \lambda) \\ 0 \\ 0 \\ T\mathbf{f_y}(\mathbf{y}, \lambda)\,\mathbf{h} \end{pmatrix}, \quad \begin{pmatrix} \mathbf{y}(0) - \mathbf{y}(1) \\ \mathbf{h}(0) - \mathbf{h}(1) \\ \sum_i h_i\, \partial f_1/\partial y_i \\ h_1(0) - 1 \end{pmatrix} = \mathbf{0} \qquad (7.23)
$$

results [Sey81a], [Sey85a]. The summation term in the boundary condition of equation (7.23) is evaluated at $t = 0$,

$$
\sum_{i=1}^{n} h_i(0)\, \frac{\partial f_1(\mathbf{y}(0), \lambda)}{\partial y_i} = 0\,.
$$

The convergence of an iteration procedure (for example, shooting combined with quasi-Newton) that solves equation (7.23) has shown satisfactory results in a wide range of examples. Concerning singularity and convergence speed, the remarks in Section 5.4.1 apply. Notice that the right-hand side of (7.23) is the same one used in Algorithm 7.12 for calculating the monodromy matrix. Solving equation (7.23) could be used as alternative to the direct method of solving equation (5.70). The two methods approach the Hopf point from different sides: Solving equation (5.70) is a "stationary method," whereas solving equation (7.23) is a "dynamic method." A further reduction of equation (7.23) toward a differential-algebraic system is possible, making use of $\mathbf{y}' = \mathbf{0}$. In general, solving the ODEs (7.23) is more expensive than solving the nonlinear equations (5.70). In our introductory context, equation (7.23) sets the stage for the calculation of period doubling points in Section 7.6.4.

Discontinuous systems are an ideal setting for shooting approaches. For a discontinuity of type (3.15), the solution $\mathbf{h}(t)$ of the linearized system is subjected to a similar jump,

$$\mathbf{h}(t_{\text{jump}}^+) = \frac{\partial \mathbf{g}}{\partial \mathbf{y}} \, \mathbf{h}(t_{\text{jump}}^-)$$

(shown in [Sey77]). For example, oscillations subjected to outer impacts and their bifurcation behavior were calculated this way in [Bec79], and an optimal-control problem (where such discontinuities arise naturally) in [Sey77]. The good features of shooting in the context of discontinuities were also stressed in [DoH06].

7.6.3   Calculating a Periodic Orbit Close to a Hopf Bifurcation

Branch switching at a Hopf bifurcation point amounts to calculate a periodic orbit close to the Hopf point. After a Hopf point $(\mathbf{y}_0, T_0, \lambda_0, \mathbf{h}_0)$ is calculated, an initial guess for small-amplitude periodic solutions is given by

$$\bar{\mathbf{y}} = \mathbf{y}_0 + \delta \mathbf{h}_0, \quad \bar{T} = T_0, \quad \bar{\lambda} = \lambda_0 . \tag{7.24}$$

These data (for small $|\delta|$) serve as an initial approximation for equation (6.3), which is based on the boundary-value problem in equations (7.6a) and (7.6b). Consequently, the $(n + 2)$-dimensional boundary-value problem in equation (7.6c) is solved, starting from equation (7.24). The boundary conditions equation (7.6c) require the value

$$\eta = \bar{y}_k(0) .$$

Solving equation (7.6) allows one to trace the branch of periodic solutions. Depending on the solution method, difficulties can arise at turning points (cyclic fold points). Note that the method of parallel computation can also be applied. It is straightforward to modify equation (6.28) for the case where $\mathbf{y}$ is the stationary solution and $\mathbf{z}$ is the periodic solution.

In case the Hopf point is calculated by the steady-state method of solving equation (5.70), the vector function $\mathbf{h}_0(t)$ is not immediately available. Instead, the complex-conjugate eigenvector $\mathbf{w}$ is part of the solution. Hence, in order to use equation (7.24), $\mathbf{h}_0(t)$ must be calculated based on the real part $\mathbf{h}$ and the imaginary part $\mathbf{g}$ of $\mathbf{w}$. [Here the constant vector $\mathbf{h}$ in equation (5.70) is not to be confused with $\mathbf{h}_0(t)$.] The derivation of $\mathbf{h}_0(t)$ is a standard analysis in ODE textbooks; see also Appendix A.3. Let the eigenvalues of $\mathbf{f}_\mathbf{y}(\mathbf{y}_0, \lambda_0)$ be $\mu_1 = \beta$, $\mu_2 = -\beta$, $\mu_3, \mu_4, \ldots, \mu_n$. Then solutions $\mathbf{h}_0(t)$ to

$$\mathbf{h}' = T \mathbf{f}_\mathbf{y}(\mathbf{y}_0, \lambda_0) \mathbf{h}$$

are given by

$$\begin{aligned} \mathbf{h}_0(t) =& (c_1 \mathbf{h} + c_2 \mathbf{g}) \cos T\beta t + (c_2 \mathbf{h} - c_1 \mathbf{g}) \sin T\beta t \\ &+ \sum_{i=3}^{n} c_i \mathbf{z}^i \exp(T\mu_i t) . \end{aligned} \tag{7.25}$$

In equation (7.25) we assume that the Jacobian $\mathbf{f}_\mathbf{y}(\mathbf{y}_0, \lambda_0)$ has n linear independent eigenvectors $\mathbf{z}^i$ corresponding to μ_i. The constants c_i are determined by the boundary conditions imposed on $\mathbf{h}_0(t)$ [see equation (7.23)]. The resulting vector function $\mathbf{h}_0(t)$ is

$$\mathbf{h}_0(t) = \mathbf{h} \cdot \cos 2\pi t - \mathbf{g} \cdot \sin 2\pi t \,. \qquad (7.26)$$

This function is based on the normalization $w_1 = 1$ and matches the normalizations chosen in equation (7.23) (Exercise 7.13). Other normalizations in equation (5.70) or (7.23) lead to expressions similar to that in equation (7.26). The solutions to equations (7.23) and (5.70) are related via equation (7.26).

Note that, with the chosen normalization, the first component of $\mathbf{h}_0(t)$ is known without solving equation (7.23) or (5.70),

$$h_{01}(t) = \cos 2\pi t \,.$$

In the special case $n = 2$, the other component can also be precalculated (Exercise 7.14). Thus, in the case of $n = 2$ the initial guess (7.24) can be calculated without requiring any eigenvectors. The approach equation (7.23) has been modified to parabolic PDEs [Sey85a].

7.6.4   Handling of Period Doubling

Direct methods were also proposed for calculating period-doubling points. For instance, [KuH84b] use the equation

$$(\mathbf{M} + \mathbf{I})\mathbf{z} = \mathbf{0} \,,$$

which exploits the fact that -1 is the eigenvalue of the monodromy matrix $\mathbf{M}$. One component of $\mathbf{z}$ is prescribed as normalization. Hence, this system constitutes n equations for $(n-1)$ unknown components of $\mathbf{z}$ and the parameter value λ of a period doubling. In addition, equations that constitute $\mathbf{M}$ must be attached.

Our branching systems work too. To this end, consider the boundary-value problem for the $2T$ solution. In the bifurcation point the solution $\mathbf{y}(t)$ consists of two identical halves; the symmetry of type (6.30d) applies with $b = 2T$, $a = 0$. In this sense, the $2T$-periodic solution is considered as "asymmetric." Then, the asymmetric $2T$-periodic branch is parameterized via

$$\gamma = z_k(0) - z_k(T),$$

compare (6.32) in Section 6.4, with z_k from (6.27) [or (7.24) if the double period is observed]. For λ_0 the monodromy matrix of the $2T$-periodic solution is $\mathbf{M}^2$. According to Lemma 7.3 the $2T$-periodic solution $\mathbf{h}(t)$ of the variational equation satisfies $\mathbf{h}(0) = \mathbf{M}^2\mathbf{h}(0)$. Since $\mathbf{h}(T) \neq \mathbf{h}(0)$ we can identify $\mathbf{h}(0)$ with the eigenvector of $\mathbf{M}$ that corresponds to -1. Hence, $\mathbf{h}(T) = -\mathbf{h}(0)$. This allows to reduce the double period $2T$ to the simple period T. The corresponding branching system for the period-doubling case is

$$\begin{pmatrix} \mathbf{y} \\ T \\ \lambda \\ \mathbf{h} \end{pmatrix}' = \begin{pmatrix} T\mathbf{f}(\mathbf{y},\lambda) \\ 0 \\ 0 \\ T\mathbf{f_y}(\mathbf{y},\lambda)\mathbf{h} \end{pmatrix}, \quad \begin{pmatrix} \mathbf{y}(0) - \mathbf{y}(1) \\ p(\mathbf{y}(0),\lambda) \\ \mathbf{h}(0) + \mathbf{h}(1) \\ h_k(0) - 1 \end{pmatrix} = \mathbf{0}. \qquad (7.27)$$

Concerning branch switching, the methods of Section 6.4 also apply in the case of periodic solutions. The kind of regularity breaking that is appropriate for period doubling was introduced by equations (6.34) and (6.30d)/(6.31d).

   The above handling of period doubling is part of the methods of [Sey79a], [Sey83a], see Section 6.4 or [Sey88]. Numerical methods for handling bifurcation into tori have been established in [KeASP85], [Vel87], [DiLR91], [Lor92], [Kuz98], [DaT06]. A computation of bifurcations that is based on Fourier expansions has been advocated in [Del92]. For delay equations, consult [LuER01].

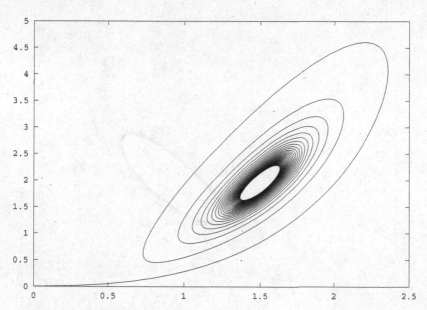

Fig. 7.23a. Example 7.15, (y_1, y_2)-plane, $\lambda = 1.3$

7.6.5   Applying Simulation

Simulation (Algorithm 1.8) is successfully applied in a practical bifurcation and stability analysis of periodic solutions. Sometimes, because it often requires long integration times until the dynamic behavior becomes clear, simulation is discredited. The uncertainty about where to choose the starting vector and how to choose the termination time t_f gives simulation a random character. On the other hand, compared with solving boundary-value problems like equation (7.5), simulation has strong advantages:

The computational overhead is small (only an integrator is used).
Results are more global and include information on stability.

In many instances, a simulation yields a limit cycle even faster than a solution of the boundary-value problem in equation (7.5) does—not counting the effort of checking the stability, which is also required when equation (7.5) is applied. Simulation makes possible an easy branch switching from repellors to attractors; this includes equilibria, periodic orbits, and tori. But simulation can be tedious near Hopf bifurcations; here a solution is only weakly attracting or repelling, because the real part of the critical eigenvalue is close to zero. In case of a hard loss of stability, however, simulation can generate a periodic orbit rapidly. If simulation is avoided in instances where it works slowly, it is highly recommended. The advantages of simulation seem to grow with the complexity of the underlying model.

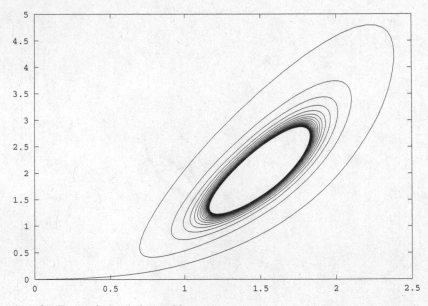

Fig. 7.23b. Example 7.15, $\lambda = 1.33$

Example 7.15   Chemical Reaction with Autocatalytic Step

This example is attributed to [KrK85]. Dimensionless variables are

$y_i(t)$: concentrations of chemicals;
ψ: flow ($\psi = 3$); and
λ: velocity coefficient of the autocatalytic step.

The reaction is defined by the differential equations

$$
\begin{aligned}
\dot{y}_1 &= \psi - y_1 - \lambda y_1 y_3 \,, \\
\dot{y}_2 &= y_1 - y_2 y_3 \,, \\
\dot{y}_3 &= y_2 y_3 - \lambda y_1 y_3 \,.
\end{aligned}
\tag{7.28}
$$

There is a Hopf bifurcation with

$$\lambda_0 = 1.30176\,, \quad T_0 = 6.03555\,,$$

calculated by means of equation (5.70). A branch of stable periodic orbits branches off for $\lambda \geq \lambda_0$. Figures 7.23a,b,c show three simulations starting from $\mathbf{z} = (0, 0, 1)$ for three parameter values; the phase space is projected to the (y_1, y_2)-plane. In Figure 7.23a, the trajectory is approaching the stable equilibrium close to the Hopf point so slowly that the simulation had to be terminated before the underlying dynamics became evident. Figure 7.23b shows a trajectory approaching a stable limit cycle, still at a dilatory rate. This illustrates that a simulation too close to a Hopf bifurcation with soft loss of stability can be a waste of time. In such situations, simulation is no alternative to Newton-like procedures for calculating equilibria and period orbits. If we try simulation again for $\lambda = 1.4$ (Figure 7.23c), it works well: The trajectory encircles the limit cycle only a few times before being absorbed. Generating the limit cycle in this way compares favorably with calculating the limit cycle by solving the boundary-value problem in equation (7.5). Note that each numerical differentiation involved in solving (7.5) requires calculation of n trajectories. In equation (7.28), with $n = 3$, one numerical differentiation corresponds to three rotations of the spiraling trajectory. Hence, establishing the limit cycle and its stability in Figure 7.23c by simulation requires less effort than solving a boundary-value problem such as equation (7.5).

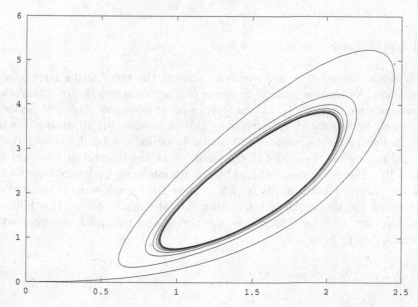

Fig. 7.23c. Example 7.15, $\lambda = 1.4$

The periodic oscillations of equation (7.28) become a relaxation oscillation for $\lambda \approx 5$. Such an oscillation is depicted in Figure 7.24 for $\lambda = 6$. Close to a Hopf bifurcation, oscillations have a sinusoidal shape. This sinusoidal function is approximated by equation (7.24), which is the function in equation (7.26) added to a constant vector. In equation (7.28), the initial smoothness diminishes gradually with increasing λ and gives way to the saw-toothed shape of Figure 7.24.   □

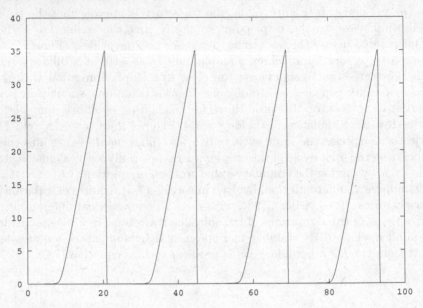

Fig. 7.24. Example 7.15, $y_2(t)$ for $\lambda = 6$.

It is not necessary to use positive values of the terminating time t_f of a simulation. A negative value of t_f means *backward integration*, or integration in reverse time, $0 \geq t \geq t_f$. In this way, in some examples, unstable periodic orbits can be generated. In particular, this is possible for all examples with $n = 2$. For instance, consider the unstable periodic orbit for $\lambda = -0.337$ in the FitzHugh nerve model (Example 3.5.; see the bifurcation diagram Figure 3.16). This orbit was obtained by simulation using backward integration (see the phase diagram in Figure 7.25). Here the convergence is again slow, because of the nearby Hopf bifurcation. On the other side of this bifurcation, however, there is a hard loss of stability and the quick success that a simulation can produce.

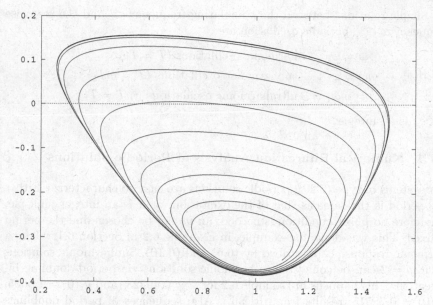

Fig. 7.25 Example 3.5, phase plane, $\lambda = -0.337$

7.7 Nonautonomous Systems

So far, this chapter has dwelt on periodic solutions of autonomous systems. We now focus our attention to nonautonomous systems

$$\dot{\mathbf{y}} = \mathbf{f}(t, \mathbf{y}, \lambda), \tag{7.29a}$$

where $\mathbf{f}$ is periodic in t,

$$\mathbf{f}(t + T_1, \mathbf{y}, \lambda) = \mathbf{f}(t, \mathbf{y}, \lambda) \tag{7.29b}$$

for some minimal $T_1 > 0$. Solutions to equation (7.29a) can be periodic, quasi-periodic, irregular, or "chaotic." The period T of a periodic solution of equation (7.29a,b) may be closely related to T_1. In Section 7.7.2 we shall study possible relations between T_1 and T for the important class of forced oscillators of the second order,

$$\ddot{u} + g(u, \dot{u}) = \gamma \cos(\omega_1 t). \tag{7.30}$$

The forcing term $\gamma \cos(\omega_1 t)$ is a harmonic function with amplitude γ and driving frequency ω_1. Clearly, $T_1 = 2\pi/\omega_1$, and equation (7.30) is of the type of equation (7.29a,b). The Duffing equation (2.9) is of that type, see Example 6.2. If the stimulation is absent ($\gamma = 0$), the system may have a free oscillation with frequency ω_0. For forced oscillations ($\gamma \neq 0$), the frequency of a periodic oscillation is denoted ω with period $T = \frac{2\pi}{\omega}$.

Simple relations between the stimulating frequency ω_1 and the response frequency ω of a periodic oscillation are

$$\omega = \omega_1 : \quad \text{harmonic oscillations } (T = T_1);$$
$$m\omega = \omega_1 : \quad \text{subharmonic oscillations } (T = mT_1);$$
$$\omega = m\omega_1 : \quad \text{ultraharmonic oscillations } (mT = T_1);$$

$m > 1$ is an integer.

7.7.1 Numerical Bifurcation Analysis of Periodic Solutions

For systems of type (7.29), periodic solutions are easy to characterize: Either the period is the same as that of the excitation, or it is an integer multiple. Therefore no phase condition is needed; an m can be chosen and the period is fixed. This was done, for example, in Example 6.2 in Section 6.1, where a harmonic response is prescribed by (6.6) and (6.10). Subharmonic solutions with $m = 2$ can be connected to harmonic solutions via period-doubling bifurcation. This happens in Example 6.2 for $\omega = 0.49228$ (and $y_1(0) = 1.0055$, $y_2(0) = 0.78515$; results from [Rie96]). Also sequences of period doublings can occur.

The numerical bifurcation analysis is analogous as in Chapter 6 and Sections 7.2 through 7.6. For a periodic solution $\mathbf{y}^*$ the monodromy matrix is defined as $\Phi(T)$, where $\Phi(t)$ solves

$$\dot{\Phi} = \mathbf{f}_\mathbf{y}(t, \mathbf{y}^*, \lambda)\,\Phi, \quad \Phi(0) = \mathbf{I}.$$

The branching system (6.14) applies; its boundary conditions specialize to

$$\begin{pmatrix} \mathbf{y}(0) - \mathbf{y}(T) \\ h_k(0) - 1 \\ \mathbf{h}(0) - \mathbf{h}(T) \end{pmatrix} = \mathbf{0},$$

or to a version for the normalized interval as in (7.23). The matrix $\mathbf{E} = \mathbf{I} - \mathbf{M}$ from (7.20) is important here too, with $\mathbf{E}\,\mathbf{h}(0) = \mathbf{0}$, and $\mathbf{h}(0)$ can be approximated by (6.24). In this way, for example, cascades with eight subsequent doublings were calculated for the Duffing equation of Example 6.2. Thereby, the Feigenbaum limit (7.14) was verified with three correct digits [Rie96].

7.7.2 Phase Locking

The transformation

$$\theta(t) = \omega_1 t \mod 2\pi \tag{7.31}$$

assigns to each t an angle θ. Hence, $\theta(t)$ maps $\mathbb{R}$ onto the unit circle S^1. Substituting $t = \theta/\omega_1$ into equation (7.29a) produces the autonomous system of size $n + 1$

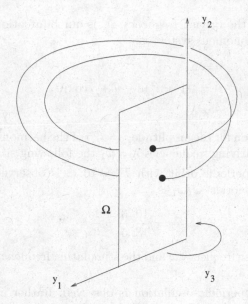

Fig. 7.26. Phase space with subharmonic orbit

$$\dot{\mathbf{y}} = \tilde{\mathbf{f}}(\theta, \mathbf{y}, \lambda) := \mathbf{f}(\frac{\theta}{\omega_1}, \mathbf{y}, \lambda),$$
$$\dot{\theta} = \omega_1$$

(7.32)

with $(\mathbf{y}, \theta) \in \mathbb{R}^n \times S^1$. The angles $\theta = 0$ and $\theta = 2\pi$ are identified, and the phase space belonging to equation (7.32) can be visualized as in Figure 7.26.

Because θ is monotonically increasing it is easy to set up a Poincaré section Ω that is necessarily intersected by any trajectory,

$$\Omega = \{(\mathbf{y}, \theta) \in \mathbb{R}^n \times S^1 \mid \theta = \theta_0\},$$

(7.33)

with $\theta_0 = 0$. Hence, the Poincaré map amounts to observing the state of the system at discrete time instances

$$t_k := k\frac{2\pi}{\omega_1} = kT_1,$$

sampled at constant time intervals T_1. This means to locate the trajectory at $(u(t_k), \dot{u}(t_k))$ by "flashing the stroboscope" once per period T_1. Accordingly, this Poincaré map has been called a *stroboscopic map* [Min74]. Figure 7.26 depicts a subharmonic orbit of order two, which has the double period compared with the reference period T_1 of the excitation.

Example 7.16   Forced van der Pol Equation
A forced van der Pol oscillator is given by

$$\ddot{u} - \delta(1 - u^2)\dot{u} + u = \gamma \cos(\omega_1 t),$$

(7.34)

with $\delta = 4$, where the driving frequency ω_1 is our bifurcation parameter λ. The equivalent autonomous system is

$$\dot{u} = v,$$
$$\dot{v} = 4(1 - u^2)v - u + \gamma \cos \lambda t,$$
$$\dot{\theta} = \lambda.$$

Numerical simulation for the amplitude $\gamma = 0.5$ of the harmonic forcing yields for three selected driving frequencies $\lambda = \omega_1$ the following situation:

$\omega_1 = 1.8$: A periodic orbit with $T = 10.48$ is observed. Hence, $\omega = 2\pi/T = 3/5$. The quotient ω/ω_1 is

$$\frac{\omega}{\omega_1} = \frac{10}{18} \frac{3}{5} = \frac{1}{3}.$$

That is, the response frequency ω and the stimulating frequency ω_1 are related by the quotient $1 : 3$.

$\omega_1 = 1.0$: No periodic oscillation is observed. Rather a quasi-periodic oscillation results, which is shown in Figure 7.27 for $0 \le t \le 100$.

$\omega_1 = 0.5$: Simulation reveals a periodic oscillation with period $T = 12.54$, which amounts to a harmonic oscillation with relation $\omega : \omega_1 = 1 : 1$. $\square$

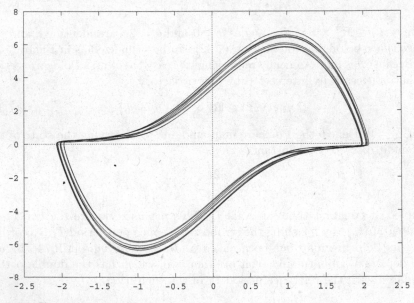

Fig. 7.27. Example 7.16, phase portrait, $\omega_1 = 1$

Example 7.16 shows that the response frequency ω can be in resonance with the driving frequency ω_1; the forced van der Pol equation synchronizes. The simulation for $\omega_1 = 1$ in Figure 7.27 illustrates that a torus need not have a circular cross section as symbolically drawn in Figure 7.21.

Definition 7.17 *A trajectory is called **phase locked** (or **frequency locked** or **entrained**) if ω/ω_1 is rational. That is, there are integers p and q such that $\omega : \omega_1 = p : q$. In this case, the frequencies ω_1 and ω are **commensurate**.*

Investigating Example 7.16 further by choosing stimulating frequencies λ traversing some interval reveals that there are subintervals of λ-values with constant phase-locking ratios $\omega : \lambda$. The width of these λ-subintervals varies with the second parameter γ. The bifurcation structure of driven van der Pol oscillators is discussed in [MePL93]. The frequency-locked regions in a (λ, γ)-parameter plane extend to tongue-like objects called *Arnold's tongues*. Schematically, these tongues look as in the parameter chart of Figure 7.28. For all combinations $(\omega_1, \gamma) = (\lambda, \gamma)$ within a "tongue" the same phase-locking ratio $p : q$ holds. The borders of the tongues are bifurcation curves; for their computation see Section 7.7.1. Three tongues of (7.34) with $\delta = 1$ are shown in Figure 7.29.

Between each tongue there are more tongues, most of them so thin that they are not noticed. Upon plotting the ratio $\omega : \omega_1$ versus an underlying parameter λ (which may be ω_1), locking manifests itself by horizontal "steps." The ratio $\omega : \omega_1$ is a simple definition of a *rotation number* or *winding number*. For general concepts of rotation numbers see, for instance, [Arn83], [GuH83]. For the forced van der Pol oscillator cascades of period doublings are also observed. The reader with an integrator at hand is encouraged to study the dynamics of the forced van der Pol oscillator (7.34) for various parameter values in the (λ, γ)-plane.

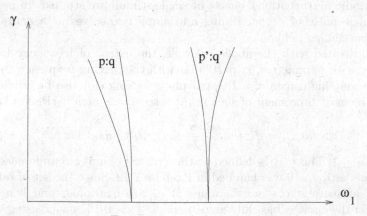

Fig. 7.28. Parameter chart with Arnold tongues, schematically

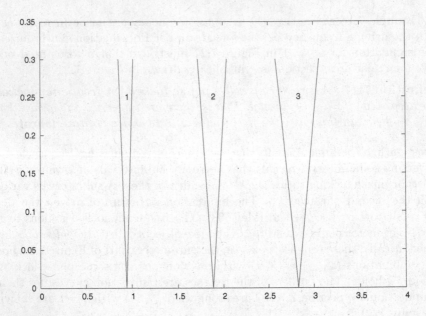

Fig. 7.29. Parameter chart with Arnold tongues, for (7.34), $\delta = 1$. Horizontal axis: ω_1, vertical axis: γ. For $\gamma = 0$, $T_0 = 6.663287$, $\omega_0 = 0.942956$.

Synchronization has two aspects: resonance occurs (amplitude increases), or the phases of the excitation and the system become locked [Min74]. Phase locking takes place in all fields—for example, biology [PeG84], [GlM88], electrical engineering [UeA81], and chemistry [Sch85], [Aro86]. In [TaK91] coupled chemical oscillators are investigated, showing that by coupling of two autonomous systems a structure typical for resonance occurs naturally. Apart from periodic perturbations, effects of single stimuli are studied. In particular, a single pulse of proper timing and amplitude serves as a neural and cardiac pacemaker [ChL85].

As illustrated with the above example, the extent of frequency locking may vary with parameters, in particular with the driving frequency ω_1, and with the coupling constant γ. Related phase locking can also be studied for maps. The most prominent of such maps is the *circle map* [ArP82], [Arn83], [Schu84]

$$\vartheta_{\nu+1} = \vartheta_\nu + \omega_1 + \frac{\gamma}{2\pi}\sin(2\pi\vartheta_\nu) \quad \text{mod } 1 \qquad (7.35)$$

for $0 \le \omega_1 \le 1$. This map is defined on the circle S^1 of unit circumference. The linear case with $\gamma = 0$ was handled in Exercise 7.12. Since the set of rational ω_1 is of measure zero, "most" values of ω_1 are irrational, and hence the motion on the circle is basically quasi-periodic ($\gamma = 0$). For increasing values of the coupling γ, phase locking occurs for *intervals* of ω_1-values. That is, the phase-locking intervals along the ω_1-axis have positive measure (length) for $\gamma \ne 0$ (Exercise 7.15). For weak coupling (γ small), the phase-locked intervals are narrow and the motion is mostly quasi-periodic (ratio irrational). For increasing coupling (γ larger), the phase-locked portions increase. The resulting step function, called the *devil's staircase* [Man83], has been observed in many applications (see [Har61], [Bak86], [MaM86], [MaS87], [IsBR88]).

7.8 Further Examples and Phenomena

Oscillatory phenomena are richer and more widespread than indicated in this chapter. We complete the preceding material with hints about further examples and investigations, including phenomena that are nonoscillatory. This section extends the applications discussed in Chapter 3. Faced with the great variety of nonlinear phenomena that have been reported, we confine our-selves to relatively few hints. The reader will know which journals to consult for recent reports on nonlinear phenomena in her or his own field. General references are, for example, the journals *Physica D*, *International Journal of Bifurcation and Chaos*, *Chaos Solitions & Fractals*, *Nonlinear Dynamics*, *Nonlinear Science*, and related journals.

In *chemistry* and chemical engineering, much attention has been given to flames and combustion (see, for example, [MaS78], [HeOR79], [PeOH81], [MaM83], [MaM85], [SmMK85], [BaM90], [Mar90], [Mar91]. Some papers on chemical kinetics have already been quoted; more hints can be found in [Ray77], [Eps83]. For pattern formation and waves, see Section 3.5. Feed-back is related to bifurcation [Gla75]; it was even introduced artificially to discriminate among different dynamic behavior [KüK84], [LyKB84]. For bio-chemical phenomena, see [HeM85], for example.

As examples of bifurcation in *physics*, we mention Hopf bifurcation in a model describing glaciation cycles [GhT83] and in ocean circulation [TiKF02], equations describing meteorological effects more accurately than the Lorenz system in equation (2.25) does [Cur78], and the study of the earth's magnetic field [Bus77]. Other discussions of climate modeling include [Het89], [HeS91], [Sch92]; Hopf bifurcation in a model of cloud physics has been discussed in [Wac92]. Impressive results on the 3-body problem can be found in [Doe03], [Doe07]. Bifurcation of lasers has been studied in [Ren79], [HaFM89].

A great variety of bifurcation problems occur in *fluid dynamics*. Reviews and handbooks are [Kir75], [Jos76], [Stu77], [SwG81], [ZiO82]. In fluid dy-namics, much depends on the underlying geometry. Accordingly, different studies for different geometries have been conducted. Flows in channels or tu-bes [Vel64], [BrA81], and the flow between two rotating concentric cylinders (the Taylor problem) [Tay23], [Vel66], [KiS69], [Ben78], [MeK80] have found specific interest. For numerical aspects see [ClST00]. Convection driven by buoyancy or surface tension is mentioned in Section 2.8; for further references, see [Pea58], [SwG81], [ZiO82]. Specifically for the Marangoni effect (convec-tion driven by surface tension), which is related to pattern formation and oscillations, consult, for instance [RoDH82], [PrSS83], [SmD83], [GoNP02]. Acoustic cavitation noise has revealed interesting bifurcations [PaESL90].

In *biology*, many of the current investigations related to bifurcation are devoted to nerve models, see Section 3.4. Some of these models are designed to model heartbeats, consult also [Chay95]. Oscillatory problems are ubiquitous in biology. For circadian rhythm of two proteins in cells see [SiV06]. Interesting nonlinear effects occur in *predator–prey models*, which describe oscillations of populations of certain species that struggle for existence [Bra78], [KaD78]. Related phenomena occur, for example, with fish [Vol31] and starfish [AnFK87], in immune response [Bell73], with the Canadian lynx (a cycle with a period of 10 years) [ScK85], or with plant-herbivore evolution [ThS86]. Many kinds of predator-prey models have been discussed in the literature. We give one example of a three-species food-chain model, where y_1 represents plankton, y_2 fish, and y_3 shark,

$$
\begin{aligned}
\dot{y}_1 &= y_1(1 - y_1) - 3y_1y_2, \\
\dot{y}_2 &= -\tfrac{1}{4}y_2 + 3y_1y_2 - 3y_2y_3 - \lambda(1 - e^{-5y_2}), \\
\dot{y}_3 &= -\tfrac{1}{2}y_3 + 3y_2y_3.
\end{aligned}
\tag{7.36}
$$

The term $\lambda(1 - e^{-5y_2})$ represents fishing [DoKK91]. Hopf bifurcation also occurs in animal orientation [HaMS80]. Phase locking in magnetoencephalography is studied in [Tass98].

Classical applications for bifurcation occur in mechanics. In particular, the bending and buckling of structures have found much attention. We mention mode jumping of rectangular plates, which are modeled by the von Kármán equations [HoS84], [List78], [SuTW85], the bending of tubes of elliptic cross sections [Wei70], and the buckling of spherical shells (see [GrSTW85] and the references therein). General references for nonlinear phenomena in mechanical engineering are [ThH83], [Tro84]. Specific experiments are a double pendulum (motivated by vibrations of the Trans-Arabian pipeline in 1950) [Hol77], [ScTZ83], snap-through of imperfect shallow arches [Hua68] and the tractor–semitrailer, the straight motion of which loses stability for certain critical speeds [TrZ84]. The Hopf bifurcation of the Watt steam engine is analyzed in [AlK85]. Nonlinear phenomena of railway cars have been investigated, for instance, in [XuST92] or [Rei91], [Sim88]. Bifurcations in systems with dry friction have been reported in [Ste90]; flow-induced vibrations are discussed in [Duc92]. Loss of stability is an issue in axial flow compressors [AbHH93] and flight dynamics [AbL90]. Bifurcations were created artificially by adding appropriate control [ChWC01]. A classical application of bifurcation is the onset of flutter [ChPL95]. Magnetic bearings have interesting bifurcation effects [KnE96]. Vibrations of suspended cables have been investigated by [Rega04].

In electrical engineering many models involve bifurcation as an integral mechanism. For the control of electric power systems see Section 3.6. For example, the test functions of Section 5.4.3 and the methods outlined by

Exercises 5.7, 5.8, 5.9 have been applied to predict voltage stability [WaG96]. A review on bifurcations in switching converters is [ElA05].

This book is devoted to deterministic systems. In recent years, stochastic systems have been investigated repeatedly, and the knowledge on *stochastic bifurcation* has increased significantly. For a review see [Nam90]; for applications and theoretical results see [NaT89], [LeN91], [Nam91], [Arn98].

Exercises

Exercise 7.1
Consider the PDE $\nabla \times \mathbf{y} + \mathbf{y} = \mathbf{0}$ for $\mathbf{y}$ defined on part of $\mathbb{R}^3$. Imagine the solution

$$\mathbf{y} = \begin{pmatrix} c_2 \sin(x_2) + c_3 \cos(x_3) \\ c_3 \sin(x_3) + c_1 \cos(x_1) \\ c_1 \sin(x_1) + c_2 \cos(x_2) \end{pmatrix}$$

to be a velocity field; draw streamlines in the (x_1, x_2)-plane for

$$0 \le x_1 \le 3\pi, \quad 0 \le x_2 \le 2\pi$$

and $c_1 = c_2 = 1$, $c_3 = 0$. In which parts of the plane is the flow upward? In which parts is it downward?

Exercise 7.2
(a) Illustrate the geometrical meaning of the phase conditions (7.2) and (7.3) by means of sketches in a phase plane.
(b) Let $\mathbf{y}$ be a periodic solution. We look for a shift ζ^* that minimizes the translation from a reference solution $\hat{\mathbf{y}}$. That is, minimize

$$g(\zeta) := \int_0^T \|\mathbf{y}(t + \zeta) - \hat{\mathbf{y}}(t)\|_2^2 \, dt.$$

Show

$$0 = \int_0^T (\mathbf{y}(t + \zeta^*) - \hat{\mathbf{y}}(t))^{tr} \, \dot{\mathbf{y}}(t + \zeta^*) \, dt$$

Argue why the phase condition (7.4a) follows.
(c) Assume T-periodicity of $\mathbf{y}$ and $\hat{\mathbf{y}}$, apply integration by parts and derive from (7.4a) the alternative phase condition (7.4b)

$$0 = \int_0^T \mathbf{y}(t)^{tr} \dot{\hat{\mathbf{y}}}(t) \, dt.$$

Exercise 7.3

A linear two-dimensional map is defined by

$$\mathbf{P q} = \mathbf{P} \begin{pmatrix} q_1 \\ q_2 \end{pmatrix} = \begin{pmatrix} 1 & 1 \\ 1 & 2 \end{pmatrix} \begin{pmatrix} q_1 \\ q_2 \end{pmatrix}.$$

Calculate eigenvalues and eigenvectors and decide whether the fixed point is attractive or not. Draw the eigenvectors and indicate motion along these two lines by arrows. For

$$\mathbf{q} - \begin{pmatrix} 3 \\ -1.85 \end{pmatrix}$$

calculate the sequence $\mathbf{P(q)}, \dots, \mathbf{P}^8(\mathbf{q})$ and enter the iterates in your sketch. (Remark: In [GuH83] this map is discussed on a torus rather than on the plane.)

Exercise 7.4

Let Ω and $\hat{\Omega}$ be two Poincaré sections for a periodic orbit $\mathbf{y}$ of $\dot{\mathbf{y}} = \mathbf{f(y)}$, and let $\mathbf{q}$ and $\hat{\mathbf{q}}$ be the intersection points, and $\mathbf{P}$, $\hat{\mathbf{P}}$ the Poincaré maps. Show that the matrices

$$\frac{\partial \mathbf{P(q)}}{\partial \mathbf{q}} \quad \text{and} \quad \frac{\partial \hat{\mathbf{P}}(\hat{\mathbf{q}})}{\partial \mathbf{q}}$$

are similar. (That is, they have the same eigenvalues.)

Exercise 7.5

Design and implement an algorithm that approximates a hitting point where a trajectory intersects a given planar hypersurface Ω. Start with a first approximation obtained by linear interpolation of results calculated on both sides of Ω.

Exercise 7.6

In a manner similar to that in Figure 7.12, draw two-dimensional phase planes illustrating the dynamic behavior indicated in Figures 7.9 and 7.10.

Exercise 7.7

Consider the bifurcation diagram in Figure 7.13.

(a) Draw a diagram depicting only those solutions that can be observed in biochemical reality. [Assume that equation (7.12) represents reality.]
(b) Draw a bifurcation diagram of the solutions to equation (7.12) in the decoupled case ($\delta_1 = \delta_2 = 0$).

Exercise 7.8

Consider the fixed-point equation that corresponds to equation (7.13).

(a) Establish explicitly the fixed-point equation that belongs to P^2.
(b) To which values does the fixed-point iteration $q^{(j)} = P(q^{(j-1)}, \lambda)$ settle for $\lambda = 3.5$, with $q^{(0)} = 0.1$?
(c) Sketch a bifurcation diagram. (Hint: Something "happens" at $\lambda = 3.45$.)

Exercise 7.9
Consider the mapping $P(y) = -(1 + \lambda)y \pm y^3$ which is the normal form for period doubling. Truncate the mapping P^2 after the y^3 term, and investigate its fixed points for $\lambda \approx 0$.

Exercise 7.10
Consider the linear map in the plane,

$$\mathbf{P}(\vartheta)\mathbf{q} := \begin{pmatrix} \cos \vartheta & -\sin \vartheta \\ \sin \vartheta & \cos \vartheta \end{pmatrix} \mathbf{q}.$$

(a) Show that the eigenvalues of $\mathbf{P}$ lie on the unit circle of the complex plane.
(b) Show that the map establishes a rotation.
(c) Show that $(\mathbf{P}(\vartheta))^m = \mathbf{P}(m\vartheta)$.
(d) For which values of ϑ is $(\mathbf{P}(\vartheta))^m = \mathbf{I}$ (identity matrix)?

Exercise 7.11
A truncated normal form of Naimark–Sacker bifurcation is given in polar coordinates by the map

$$\rho_{\nu+1} = \rho_\nu + (\delta\rho_\nu + a\rho_\nu^2)\rho_\nu \,,$$
$$\vartheta_{\nu+1}' = \vartheta_\nu + \gamma + \zeta\lambda + b\rho_\nu^2 \,, \tag{7.37}$$

with constants $\delta, a, b, \zeta, \gamma$ [Wig90]. Show that the circle $\{(\rho, \vartheta) \mid \rho = (-\lambda\delta/a)^{1/2}\}$ is invariant under the mapping.

Exercise 7.12
By Exercise 7.11, the dynamics of equation (7.37) can be described for fixed λ by the mapping
$$\vartheta_{\nu+1} = \vartheta_\nu + \omega_1 \,.$$
Show that the mapping has a periodic solution for ω_1 rational and no periodic solution for ω_1 irrational.

Exercise 7.13
Confirm equation (7.25). Determine the constants c_i such that $\mathbf{h}_0(t)$ obeys the boundary conditions in equation (7.23). Assume hereby that $\mathbf{h}$ and $\mathbf{g}$ are normalized by $h_1 = 1$, $g_1 = 0$.

Exercise 7.14
Assume that $n = 2$ and $\partial f_1 / \partial y_2 \neq 0$ in the Hopf bifurcation point. Establish a formula for the second component of $\mathbf{h}_0(t)$.

Exercise 7.15
(Requires the use of a computer.) Consider the circle map

$$\vartheta_{j+1} = \vartheta_j + \omega_1 + \frac{\gamma}{2\pi}\sin(2\pi\vartheta_j), \qquad 0 \le \omega_1 \le 1$$

and the winding number

$$\lim_{j\to\infty} \frac{1}{j}(\vartheta_{j+1} - \vartheta_1).$$

For at least three values of the coupling γ ($\gamma = 0$, $\gamma = 0.8$, $\gamma = 1.0$), calculate and draw the winding number versus ω_1. (For nonzero γ, you must resort to approximations.) For increasing γ, increasing ω_1-intervals of phase locking become visible, leading to a devil's staircase. For continuous γ, in a (ω_1, γ)-plane, these intervals extend to Arnold tongues.

8 Qualitative Instruments

8.1 Significance

Numerical calculations are primarily quantitative in nature. Appropriate interpretation transforms quantitative information into meaningful results. The diversity of nonlinear phenomena requires more than just the ability to understand numerical output. The preceding chapters have concentrated on one-parameter problems, with only brief excursions (in Sections 2.10 and 5.9) to two-parameter models, mainly to show how one-parameter studies can be applied to investigate certain aspects of two-parameter models. Reducing a parameter space to lower-dimensional subsets will remain a standard approach for obtaining quantitative insight. However, a full qualitative interpretation of multiparameter models requires instruments that have not yet been introduced in the previous chapters. These instruments are provided by singularity theory and catastrophe theory. Both these fields are qualitative in nature; in no way do they replace numerical parameter studies. Knowledge of singularity theory and catastrophe theory helps organize a series of partial results into a global picture.

Before numerical methods became widely used, analytical methods, such as the Liapunov–Schmidt method [Sta71], [VaT74] and the center manifold theory (to be discussed briefly in Section 8.7), were the standard means for carrying out bifurcation analysis. Today, and presumably to an even larger extent in the future, the analysis is mainly accomplished by numerical methods. This is due to the high level of sophistication that numerical methods have reached, as well as to limitations of analytical methods. These limitations are twofold. First, numerical methods are required for evaluating analytical expressions anyway. Second, analytical results are generally local— that is, they consist of assertions of the type "holds for a sufficiently small distance," where "sufficiently small" is left unclarified. Although remarkable results have been obtained by analytical methods (for example, [RoDH82], [MaM83], [ScTZ83]), and although analytical methods have been supported by the availability of symbolic manipulators [RaA87], the extensive application of numerical methods is indispensable for practical bifurcation and stability analysis. With a black box of numerical methods at hand, it is easy—sometimes too easy—to take advantage of the skills of others. Unfortunately, the use of black boxes occasionally results in a shallower understan-

R. Seydel, *Practical Bifurcation and Stability Analysis*,
Interdisciplinary Applied Mathematics 5, DOI 10.1007/978-1-4419-1740-9_8,
© Springer Science+Business Media, LLC 2010

ding. Nevertheless, the former rule, "Analysis must come ahead of numerical analysis," has lost its validity. But analytical methods are needed to classify and understand bifurcation phenomena.

8.2   Construction of Normal Forms

A first step toward a classification of bifurcation phenomena is to set up equations in *normal form*. A normal form is the simplest representative of a class of equations featuring a specific phenomenon. We met normal forms in Chapter 2; see, for example, equation (2.13)

$$\dot{y} = \lambda y - y^3$$

as the simplest equation exhibiting supercritical pitchfork bifurcation. In principle, normal forms can be constructed. The procedure is described in this section. Assume the original system of equations is

$$\dot{\mathbf{y}} = \mathbf{f}(\mathbf{y}),$$

with a known stationary solution $\mathbf{y}^s$. By a series of transformations this system will be reduced to normal form; intermediate vectors will include

$$\mathbf{y}(t) \to \mathbf{d}(t) \to \mathbf{x}(t) \to \mathbf{z}(t).$$

Note that the final transformation(s) will require "small" vectors in the sense $\mathbf{z} \approx \mathbf{0}$. That is, the results will be valid only in a neighborhood of $\mathbf{y}^s$. The final normal form is *not* equivalent to the original system $\dot{\mathbf{y}} = \mathbf{f}(\mathbf{y})$; the normal form only reflects *local* dynamical behavior. This is sufficient information when $\mathbf{y}^s$ is chosen to be a bifurcation solution and the goal is to analyze the type of bifurcation.

The first transformation is the coordinate translation

$$\mathbf{d} := \mathbf{y} - \mathbf{y}^s,$$

which leads to

$$\dot{\mathbf{d}} = \mathbf{f}(\mathbf{d} + \mathbf{y}^s) =: \mathbf{H}(\mathbf{d}) \tag{8.1}$$

and shifts the equilibrium to the origin, $\mathbf{H}(\mathbf{0}) = \mathbf{0}$. The second transformation splits off the linear part. Assuming $\mathbf{f}$ to be sufficiently often continuously differentiable, a Taylor series is set up,

$$\mathbf{H}(\mathbf{d}) = \mathbf{H}(\mathbf{0}) + \mathbf{DH}(\mathbf{0})\,\mathbf{d} + \bar{\mathbf{H}}(\mathbf{d}),$$

where $\bar{\mathbf{H}}(\mathbf{d}) = O(\|\mathbf{d}\|^2)$ contains the nonlinear terms of order 2 and higher, and $\mathbf{DH}(\mathbf{0})$ is the matrix of the first-order partial derivatives evaluated at $\mathbf{0}$. This leads to

$$\dot{\mathbf{d}} = \mathbf{DH}(\mathbf{0})\,\mathbf{d} + \bar{\mathbf{H}}(\mathbf{d}).$$

The third transformation is to simplify the linear part as much as possible. Let $\mathbf{J}$ be the Jordan canonical form of the matrix $\mathbf{DH}(0)$, and $\mathbf{B}$ the matrix that transforms $\mathbf{DH}(0)$ into $\mathbf{J}$ using the similarity transformation

$$\mathbf{J} = \mathbf{B}^{-1}\,\mathbf{DH}(0)\,\mathbf{B}\,.$$

The transformation $\mathbf{Bx} := \mathbf{d}$ defines $\mathbf{x}$, and leads to

$$\dot{\mathbf{x}} = \mathbf{Jx} + \mathbf{B}^{-1}\bar{\mathbf{H}}(\mathbf{Bx})\,.$$

Denoting the nonlinear terms as $\mathbf{N}(\mathbf{x})$ we have $\mathbf{N}(\mathbf{x}) = O(\|\mathbf{x}\|^2)$, and summarize the result of the transformations as

$$\dot{\mathbf{x}} = \mathbf{Jx} + \mathbf{N}(\mathbf{x})\,. \tag{8.2}$$

The remaining transformations try to simplify the nonlinear term $\mathbf{N}(\mathbf{x})$. To this end we rewrite $\mathbf{N}(\mathbf{x})$ as the sum of Taylor terms $\mathbf{N}_i$ of order i, and obtain

$$\dot{\mathbf{x}} = \mathbf{Jx} + \mathbf{N}_2(\mathbf{x}) + \mathbf{N}_3(\mathbf{x}) + ... + \mathbf{N}_{\nu-1}(\mathbf{x}) + O(\|\mathbf{x}\|^\nu)\,.$$

This assumes $\mathbf{f} \in C^\nu$. For $n = 2$, for example, $\mathbf{N}_2(\mathbf{x})$ consists of two components that consist of the quadratic terms,

$$\mathbf{N}_2(\mathbf{x}) = \begin{pmatrix} a_1 x_1^2 + a_2 x_1 x_2 + a_3 x_2^2 \\ b_1 x_1^2 + b_2 x_1 x_2 + b_3 x_2^2 \end{pmatrix}\,.$$

A *nonlinear* coordinate transformation,

$$\mathbf{x} = \mathbf{z} + \mathbf{n}_2(\mathbf{z}), \qquad \mathbf{n}_2(\mathbf{z}) = O(\|\mathbf{z}\|^2)\,,$$

is the next step. The goal is to choose a second-order function $\mathbf{n}_2(\mathbf{z})$ such that as many terms of $\mathbf{N}_2(\mathbf{x})$ as possible will be eliminated. To study related options, the nonlinear coordinate transformation are substituted. Differentiating the transformation yields

$$\dot{\mathbf{x}} = \dot{\mathbf{z}} + \mathbf{Dn}_2(\mathbf{z})\dot{\mathbf{z}} = (\mathbf{I} + \mathbf{Dn}_2(\mathbf{z}))\dot{\mathbf{z}}\,.$$

The transformed differential equation (8.2) is

$$(\mathbf{I} + \mathbf{Dn}_2(\mathbf{z}))\dot{\mathbf{z}} = \mathbf{Jz} + \mathbf{Jn}_2(\mathbf{z}) + \mathbf{N}_2(\mathbf{z} + \mathbf{n}_2(\mathbf{z})) + O(\|\mathbf{z}\|^3)\,. \tag{8.3}$$

Since $\mathbf{n}_2(\mathbf{z})$ is of second order, its derivative is of first order, $\mathbf{Dn}_2(\mathbf{z}) = O(\|\mathbf{z}\|)$, and hence small for small $\mathbf{z}$. Sufficient smallness of $\mathbf{z}$ (closeness of $\mathbf{y}$ to $\mathbf{y}^s$) is now assumed. This justifies the relation

$$(\mathbf{I} + \mathbf{Dn}_2(\mathbf{z}))^{-1} = \mathbf{I} - \mathbf{Dn}_2(\mathbf{z}) + O(\|\mathbf{z}\|^2)\,.$$

Applying this to equation (8.3), and ordering terms according to their order leads to

$$\dot{\mathbf{z}} = \mathbf{Jz} + \mathbf{Jn}_2(\mathbf{z}) + \mathbf{N}_2(\mathbf{z}) - \mathbf{Dn}_2(\mathbf{z})\mathbf{Jz} + O(\|\mathbf{z}\|^3)\,.$$

Ideally, the goal might be to choose $\mathbf{n}_2(\mathbf{z})$ such that all second-order terms vanish,

$$\mathbf{Dn}_2(\mathbf{z})\mathbf{Jz} - \mathbf{Jn}_2(\mathbf{z}) = \mathbf{N}_2(\mathbf{z}).$$

But in general it is not possible to fully eliminate $\mathbf{N}_2(\mathbf{z})$. Note that the nonlinear coordinate transformation also changes the higher-order terms such as $\mathbf{N}_3$. Taking $\mathbf{n}_2(\mathbf{z})$ as a polynomial of the same type as illustrated above for $\mathbf{N}_2(\mathbf{x})$ in the case $n = 2$ leads to linear equations in the coefficients of $\mathbf{n}_2(\mathbf{z})$; this facilitates the calculation. At the end of the transformation the differential equation reads

$$\dot{\mathbf{z}} = \mathbf{Jz} + \tilde{\mathbf{N}}_2(\mathbf{z}) + O(\|\mathbf{z}\|^3), \tag{8.4}$$

where $\tilde{\mathbf{N}}_2(\mathbf{z})$ in general contains fewer terms than $\mathbf{N}_2(\mathbf{x})$. Truncating the higher-order terms leads to the truncated normal form

$$\dot{\mathbf{z}} = \mathbf{Jz} + \tilde{\mathbf{N}}_2(\mathbf{z}). \tag{8.5}$$

Depending on the choice of the function $\mathbf{n}_2(\mathbf{z})$ different normal forms may result.

Example 8.1   Bogdanov–Takens

A prominent example of a class of differential equations with two normal forms of the same phenomenon are those equations with $n = 2$ that lead to the Jordan matrix

$$\mathbf{J} = \begin{pmatrix} 0 & 1 \\ 0 & 0 \end{pmatrix}.$$

This matrix has zero as an eigenvalue with algebraic multiplicity two and geometric multiplicity one. Takens [Tak74] proposed the normal form

$$\begin{aligned} \dot{y}_1 &= y_2 + a_1 y_1^2, \\ \dot{y}_2 &= b_1 y_1^2, \end{aligned} \tag{8.6}$$

in which we rewrite $\mathbf{z}$ as $\mathbf{y}$. For the same class of differential equations, Bogdanov [Bog75] came up with the normal form

$$\begin{aligned} \dot{y}_1 &= y_2, \\ \dot{y}_2 &= b_1 y_1^2 + b_2 y_1 y_2. \end{aligned} \tag{8.7}$$

The latter system arises from Exercise 2.11 for $\lambda_0 = \delta_0 = 0$. It serves as a normal form for the *Bogdanov–Takens bifurcation*, which is defined as a stationary solution with a Jacobian having an algebraically double and geometrically simple eigenvalue zero. For Exercise 2.11 the parameter chart is presented in Figure 8.1. Bifurcation curves are indicated by heavy lines. One curve of turning points (fold bifurcations) is indicated. To the left of that fold line ($\delta < 0$), there are stationary solutions. A curve of Hopf bifurcation points merges with the fold line in the Bogdanov–Takens point (BT).

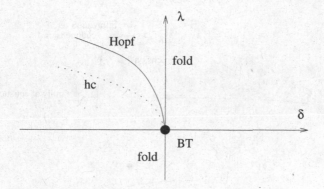

Fig. 8.1. Parameter chart of a Bogdanov–Takens scenario, Example 8.1

In addition, this example has a curve of homoclinic bifurcations (hc). In the Bogdanov–Takens point, a curve of Hopf points ends.

The above exposition of how to construct normal forms follows [Wig90]. The nonlinear transformation can be similarly applied in further stages to simplify $\tilde{N}_3(z)$ or other higher-order terms. A discussion of equations with Jacobians of the Bogdanov–Takens situation can be found in [Arn83], [GuH83], [Wig90]. Numerical aspects have also been studied; see [FiK87], [DeRR90], [WeJ91], [Spe93].

8.3   A Program Toward a Classification

In this section, we heuristically describe the general idea for classifying bifurcation phenomena. In Section 8.4, this plan will be filled with some details.

Different bifurcation phenomena can be characterized by their extent in being singular. Recall that the Jacobian of a simple bifurcation point (Definition 2.7) has a rank-deficiency one. Higher rank-deficiencies are possible and lead to other phenomena, which are observed less frequently and under specific circumstances. Below, the degree of a bifurcation to be degenerate will be characterized by its *codimension*. (A k-dimensional set $\mathcal{M}$ in an l-dimensional space $\mathcal{N}$ has codimension $l - k$. In finite-dimensional space, the codimension is the number of scalar equations needed to define $\mathcal{M}$.) In a three-dimensional space, a plane is of codimension one, and a curve is of codimension two. For a point arbitrarily picked in $\mathbb{R}^3$, it is a rare event if the point happens to be within a fixed set of codimension one. The higher the codimension of $\mathcal{M}$, the more likely it is to avoid $\mathcal{M}$ when moving about the ambient space $\mathcal{N}$.

To visualize the program toward a classification of bifurcation, see Figure 8.2. Assume that the space underlying the depicted scenario is the space of all bifurcation problems $f(y, \lambda) = 0$. This is the full set of codimension zero.

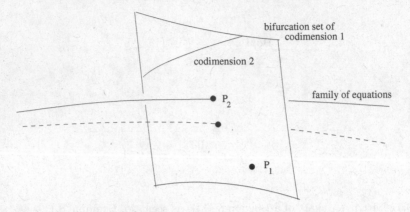

Fig. 8.2. Problems/equations P_1 and P_2 in the space of all bifurcation problems

The hypersurface illustrates a subset of those bifurcation problems that have a certain bifurcation phenomenon, say a simple bifurcation point (codimension one). The points P_1 and P_2 symbolize two related bifurcation problems. An *arbitrary* perturbation of the equations causes the point to leave the surface of codimension one. This reflects the fact that a simple bifurcation is structurally unstable, being dissolved by perturbations (Section 2.10). But, if a *family* of bifurcation problems is investigated, the situation changes. Assume that the family of equations is parameterized by γ, $f(y, \lambda, \gamma) = 0$. This is considered as a *one*-parameter problem within the space of problems of type $f(y, \lambda) = 0$. Hence, in Figure 8.2, this family corresponds to a curve. It may be the curve that connects the bifurcation problems shown in Figures 2.48 to 2.50. Whereas it is unlikely for an *arbitrary problem* $f(y, \lambda)$ to lie on the codimension-one set (point P_1), there is a good chance that the (one-dimensional) curve intersects a surface (point P_2). In this sense, codimension-one bifurcations are important for problems $f(y, \lambda, \gamma) = 0$, and not essential for $f(y, \lambda) = 0$. As an example of a family of bifurcation problems take equation (2.28), $0 = \lambda y + y^3 + \gamma$. As outlined in Section 2.10, this family of equations includes for $\gamma_0 = 0$ an equation that exhibits bifurcation. The hypersurface may be defined by $\gamma = 0$. Not every curve necessarily intersects the bifurcation surface; the family $0 = \lambda y + y^3 + 1 + \gamma^2$ does not. The family $0 = \lambda y + y^3 + \gamma^2$ touches the bifurcation surface. If the intersection at P_2 is transversal, then small perturbations of the family of equations deform the curve in such a way (dashed curve) that it still contains a codimension-one bifurcation problem for some γ_0.

Assume now that a bifurcation phenomenon is more degenerate or "bizarre" than a simple bifurcation, say it is of codimension two. Then the one-parameter family $f(y, \lambda, \gamma) = 0$, in general, does not hit the set of codimension two. An additional parameter δ leads to a two-dimensional surface representing the two-parameter family (parameters γ, δ). This family now has a good chance to hit the codimension-two phenomenon. Generally, phenomena of codimension m need not be expected in equations with less than m parameters. Here, λ is not counted; we distinguish between the bifurcation parameter λ and the other parameters, which we shall later call *unfolding parameters*. Although *codimension* is not yet defined, we already take the stand that the higher the codimension the stranger the bifurcation is.

8.4   Singularity Theory for One Scalar Equation

The program outlined in the previous section is now filled with some details. A framework for classifying bifurcation phenomena or normal forms is provided by singularity theory. Many of the related results are attributable to Golubitsky and Schaeffer [GoS79], [GoS85] (see also [Arn83], [GoKS81], [GoL81]). In this section we give a brief account of parts of singularity theory. A review of Exercise 2.14 (finite-element analog of a beam) will help in understanding the present section. We restrict ourselves to one scalar equation

$$0 = f(y, \lambda)$$

with a scalar variable y. A singularity is a solution (y, λ) with $f_y(y, \lambda) = 0$. Both simple bifurcation points and turning points are singular points. For convenience, assume that the singularity under investigation is situated in the origin—that is, in particular, $f(0,0) = 0$.

A first step toward a classification of singularities and bifurcation phenomena is to ignore differences that are not essential. For example, a bifurcation problem does not change its quality when the coordinates are transformed by some translation. The attempt to designate two similar problems as "equivalent" leads to the following definition:

Definition 8.2 $f(y, \lambda) = 0$ *and* $\tilde{f}(\tilde{y}, \tilde{\lambda}) = 0$ *are said to be* **contact equivalent** *when the qualitative features close to* $(0,0)$ *are the same. Or, more formally, there is a scaling factor* $\zeta(\tilde{y}, \tilde{\lambda})$ *such that for the transformation*

$$\tilde{y}, \tilde{\lambda} \quad \rightarrow \quad y(\tilde{y}, \tilde{\lambda}), \lambda(\tilde{\lambda})$$

the following holds:

$$\tilde{f}(\tilde{y}, \tilde{\lambda}) = \zeta(\tilde{y}, \tilde{\lambda})\, f(y(\tilde{y}, \tilde{\lambda}), \lambda(\tilde{\lambda}))\,,$$
$$y(0,0) = 0, \;\; \lambda(0) = 0, \;\; \zeta(0,0) \neq 0\,,$$
$$\partial y(0,0)/\partial \tilde{y} > 0, \;\; \partial \lambda(0)/\partial \tilde{\lambda} > 0\,.$$

Definition 8.2 is local in the sense that it is tied to the specific singularity $(0,0)$. The definition carries over to the n-dimensional vector case [GoS85].

Example 8.3

The two bifurcation problems

$$\tilde{f}(\tilde{y}, \tilde{\lambda}) = 1 - \tilde{\lambda} - \cos(\tilde{y}) = 0\,,$$
$$f(y, \lambda) = y^2 - \lambda = 0$$

are contact equivalent in a neighborhood of $y = 0$. This is prompted by the power series of $\cos(y)$,

$$1 - \tilde{\lambda} - \cos(\tilde{y}) = 1 - \tilde{\lambda} - (1 - \tfrac{1}{2}\tilde{y}^2 + \ldots) = \tfrac{1}{2}\tilde{y}^2 - \tilde{\lambda} \pm \ldots\,.$$

The dots refer to terms of order $\tilde{y}^4$. The formal transformation requested in Definition 8.2 is easily found (Exercise 8.1). □

The definition of contact equivalence allows one to identify bifurcation problems that differ only by a change of coordinates. The second step toward a classification is to investigate how many perturbations are required to "unfold" a bifurcation phenomenon.

Definition 8.4   *An **unfolding** or **deformation** of $f(y, \lambda)$ is an m-parameter family*

$$U(y, \lambda, \Lambda_1, \ldots, \Lambda_m)$$

with

$$U(y, \lambda, 0, \ldots, 0) = f(y, \lambda)\,.$$

Here $\Lambda_1, \ldots, \Lambda_m$ are m additional parameters. A simple example is furnished by equation (2.28): The expression

$$y^3 + \lambda y + \Lambda_1$$

unfolds

$$y^3 + \lambda y$$

(m=1). This unfolding is illustrated in Figure 2.44. Attaching an additional term with parameter Λ_2 is also an unfolding; hence, the above meaning of unfolding is still too general to be useful for a classification. This leads to a more distinguishing definition, which gets rid of redundant parameters.

Definition 8.5   *A **universal unfolding** is an unfolding with the following two features:*

(1) It includes all possible small perturbations of $f(y, \lambda)$ (up to contact equivalence).

(2) It uses the minimum number of parameters $\Lambda_1, \ldots, \Lambda_m$.

*This number m is called the **codimension** of $f(y, \lambda)$.*

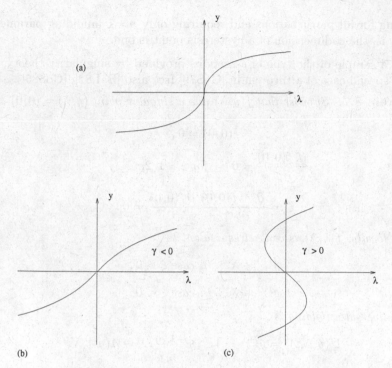

Fig. 8.3. Example 8.6, unperturbed situation (a) and perturbations (b), (c)

Example 8.6   Codimension of a Hysteresis Point

The equation

$$f(y,\lambda) = y^3 - \lambda = 0,\qquad (8.8)$$

the normal form of a hysteresis point, has a singularity at $(y,\lambda) = (0,0)$ (see Figure 8.3a). Up to contact equivalence, there are two possible qualitative changes (see Figures 8.3b and c). Both situations (b) and (c) are characterized by one equation

$$y^3 - \gamma y - \lambda = 0$$

with two signs of γ. The unfolding of f

$$U(y,\lambda,\Lambda_1) = y^3 + \Lambda_1^2 y - \lambda$$

is not universal because it does not allow for both signs of γ; perturbation (c) is ruled out, and hence feature (1) in Definition 8.5 is not satisfied. Also, the unfolding

$$U(y,\lambda,\Lambda_1,\Lambda_2) = y^3 - \Lambda_1 y^2 - \Lambda_2 y - \lambda$$

is not universal, because the two parameters Λ_1,Λ_2 are not the minimum; feature (2) is violated. Apparently, the universal unfolding is given by

$$U(y,\lambda,\Lambda_1) = y^3 - \Lambda_1 y - \lambda,$$

allowing for all perturbations and requiring only $m=1$ unfolding parameter —that is, the codimension of a hysteresis point is one.   □

As a sample of the type of assertions provided by singularity theory, we give a special case of a theorem in [GoS79] (see also [BaL82], [GoS85]).

Theorem 8.7 *Suppose that $f(y, \lambda)$ has a singular point $(y, \lambda) = (0,0)$ that obeys*

$$f(0,0) = 0 \, ,$$

$$\frac{\partial^i f(0,0)}{\partial y^i} = 0 \quad \text{for } i = 1, 2, \ldots, j \, ,$$

$$\xi := \frac{\partial^{j+1} f(0,0)}{\partial y^{j+1}} \frac{\partial f(0,0)}{\partial \lambda} \neq 0 \, .$$

Then, locally, $f(y, \lambda)$ is contact equivalent to

$$y^{j+1} + \lambda \quad \text{in case } \xi > 0 \, ,$$

$$y^{j+1} - \lambda \quad \text{in case } \xi < 0 \, .$$

The universal unfolding is

$$U(y, \lambda, \Lambda) = y^{j+1} - \Lambda_{j-1} y^{j-1} - \ldots - \Lambda_1 y \mp \lambda \, .$$

The theorem states that bifurcation problems obeying the assumptions of the theorem can be represented by polynomials, which can be seen as normal forms. Their codimension is $j - 1$. The application is as follows: Calculate the order j of a singularity. The higher the integer j, the more degenerate a singularity is. The singularity of highest order is the solution (y, λ) at which the maximum number of derivatives of f with respect to y vanish. After j is determined, check the sign of ξ. In case the singularity is nondegenerate, $\xi \neq 0$, $j + 1$ different solutions can be found locally. In Example 8.6 [equation (8.8)], $(y, \lambda) = (0,0)$ is the only singular point; we find $j = 2$, $\xi = -6$. As illustrated in Figure 8.3, there are parameter combinations for which $j + 1 = 3$ different solutions y exist.

Generalizations of the above theorem are established (see [GoS85]). For instance, in case $f = f_y = f_{yy} = f_\lambda = 0$, $f_{yyy} f_{y\lambda} \neq 0$, the normal form is that of a pitchfork,

$$y^3 \pm \lambda y \, .$$

The unfolding of a pitchfork (no symmetry assumed) requires two parameters,

$$y^3 \pm \lambda y + \Lambda_1 + \Lambda_2 y^2 \, .$$

Theorems of this kind focus on the "worst case"—that is, the solution with a singularity of maximal order. Turning points are the least degenerate singular points ($f = f_y = 0$, $f_{yy} \neq 0$, $f_\lambda \neq 0$). Here $j = 1$ and the codimension of turning points is zero; a turning point cannot be unfolded. Simple bifurcation

TABLE 8.1.   Elementary normal forms.

Defining conditions	Normal Form	Codimension	Nomenclature
$f = f_y = 0$	$y^2 + \lambda$	0	Fold or turning point
$f = f_y = f_\lambda = 0$	$y^2 - \lambda^2$	1	Simple bifurcation
$f = f_y = f_\lambda = 0$	$y^2 + \lambda^2$	1	Isola center
$f = f_y = f_{yy} = 0$	$y^3 + \lambda$	1	Hysteresis point
$f = f_y = f_{yy} = f_\lambda = 0$	$y^3 + \lambda y$	$2^{*)}$	Pitchfork bifurcation

$^{*)}$ 0 in case of $\mathbf{Z}_2$ symmetry

points and isolas ($f_\lambda = 0$), and hysteresis points ($f_{yy} = 0$) are singularities of higher order, their codimension is one. Tables of normal forms, codimension, and unfoldings are presented in [GoS85].

Table 8.1 lists the cases of codimension ≤ 1 completely. The left column gives information to recognize and classify the bifurcation; nondegeneracy conditions are omitted. Of the three possible normal forms of codimension two we only include the pitchfork case because it has codimension zero in case $f(y, \lambda)$ satisfies a $\mathbf{Z}_2$ symmetry, $f(-y, \lambda) = -f(y, \lambda)$. That is, for the restricted class of $\mathbf{Z}_2$-symmetric problems, the pitchfork bifurcation is as generic as turning points are. This is the pitchfork we have met in several examples of this book. Equivalent normal forms, which can be identified by sign changes $\pm f(y, \pm \lambda)$, are considered as identified and are not listed in Table 8.1.

In what follows, the parameter vector Λ stands for the "given" problem-inherent control parameters instead of unfolding parameters. Suppose the singularity of highest order takes place for the specific parameter vector Λ^0. Then for some parameters Λ close to Λ^0, lower-order singularities arise. The lower-order singularities are organized by the singularity of highest order, which prompts the name "organizing center." It is tempting to calculate organizing centers beforehand and then to fill the parameter space with other solutions. This works in a few scalar examples, but in practice it is usually illusive. First, because higher-order derivatives of f are evaluated in only the simplest examples, a direct calculation of high-order organizing centers is too laborious. Second, even if an organizing center is calculated, its actual structure becomes apparent only *after* global results are calculated. Because the entire formation, not only the location of the organizing center is of interest, the approximation of the manifolds defined by $f(y, \lambda, \Lambda) = 0$ is most important (see Section 5.9).

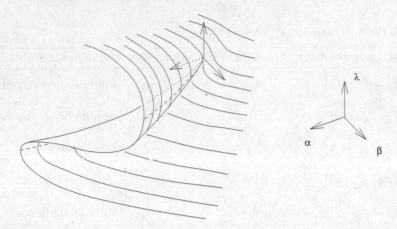

Fig. 8.4. Equation (8.10), singular set

In order to obtain a global picture, the parameter space can be sub-divided in a systematic way, and samples can be taken. This is best described for the scalar equation

$$f(y, \lambda, \Lambda_1, \ldots, \Lambda_m) = 0 \,. \tag{8.9}$$

As mentioned earlier, singularity theory sees λ as a distinguished bifurcation parameter that is not mixed with the other m parameters $\Lambda_1, \ldots, \Lambda_m$. Accordingly, we interpret the parameter space here as being m-dimensional. (The advantage of this will become apparent later.) Singularities of equation (8.9) are given by

$$\frac{\partial f(y, \lambda, \Lambda)}{\partial y} = 0 \,.$$

This equation and equation (8.9) define an m-dimensional surface in the (y, λ, Λ)-space, called the *singular set*. The projection of the singular set onto the (λ, Λ)-space is the bifurcation set or catastrophe set. For instance, in the equation

$$f(y, \lambda, \alpha, \beta) = y - \alpha - 2\lambda \sin(y) + \beta \cos(y) = 0 \tag{8.10}$$

of Exercise 2.14 we have $m = 2$ $(\Lambda_1 = \alpha, \ \Lambda_2 = \beta)$, and the parameters for which there are singular points form a two-dimensional surface in the (λ, α, β)-space. This bifurcation set is illustrated schematically in Figure 8.4.

For some purposes, the bifurcation set is too large to distinguish between different bifurcation phenomena. The bifurcation set includes all parameters (λ, Λ) for which turning points arise, and these least-degenerate singular points of codimension zero are abundant. It is a question on a "higher" level to ask for higher-order singularities, which separate different kinds of bifurcation diagrams. For example, a hysteresis point (Figure 8.3a) stands between a region without a turning point (Figure 8.3b) and a region with two turning points (Figure 8.3c). Similarly, bifurcation points separate different bifurcation diagrams (see Figure 2.47). Higher-order singularities of the scalar equation in equation (8.9) are defined in what follows.

Definition 8.8 *A **hysteresis variety*** V_{HY} *is defined by the three equations*

$$
\begin{aligned}
f(y, \lambda, \Lambda) &= 0\,, \\
f_y(y, \lambda, \Lambda) &= 0\,, \\
f_{yy}(y, \lambda, \Lambda) &= 0\,.
\end{aligned}
\tag{8.11}
$$

After eliminating y and λ (not always possible explicitly), there remains one equation in Λ, defining an $(m-1)$-dimensional hypersurface in the Λ-space. Obviously, every hysteresis point satisfies equation (8.11). Accordingly, the hypersurface defined by equation (8.11) separates parameter combinations Λ with bifurcation diagrams like that in Figure 8.3b from parameter combinations Λ' with bifurcation diagrams like that in Figure 8.3c. In the example of equation (8.10), the two additional equations defining a hysteresis variety are

$$
\begin{aligned}
f_y &= 1 - 2\lambda \cos(y) - \beta \sin(y) = 0\,, \\
f_{yy} &= 2\lambda \sin(y) - \beta \cos(y) = 0\,.
\end{aligned}
$$

A straightforward analysis (Exercise 8.2) reveals that the hysteresis variety forms a curve (one-dimensional manifold) in the (α, β)-plane. This curve is given by

$$
\beta = \sin \alpha\,.
$$

Another hypersurface of higher-order singularities is characterized by the following definition:

Definition 8.9 *A **bifurcation and isola variety*** V_{BI} *is defined by the three equations*

$$
\begin{aligned}
f(y, \lambda, \Lambda) &= 0\,, \\
f_y(y, \lambda, \Lambda) &= 0\,, \\
f_\lambda(y, \lambda, \Lambda) &= 0\,.
\end{aligned}
\tag{8.12}
$$

Both simple bifurcation points and isola centers solve equation (8.12). To see this, consider $u = f(y, \lambda)$ to describe a surface in the three-dimensional (y, λ, u)-space. Solutions of $f(y, \lambda) = 0$ are given as the intersection of the surface with the plane $u = 0$, which in view of equation (8.12) are critical points of $u = f(y, \lambda)$—that is, the gradient vanishes. Bifurcation points correspond to saddles of u, and isola centers correspond to maxima or minima. (The discriminant of the second-order derivatives decides which case occurs; see calculus textbooks. This can be seen as a simple special case related to the more general analysis outlined in Section 5.6.1.) The varieties V_{BI} and V_{HY} are hypersurfaces separating the bifurcation diagrams of the two right columns in Figure 2.47; in addition, they separate a closed branch from a situation with no local solution.

A further exception occurs when two turning points exist for the same parameter value. This hypersurface in the Λ-space is defined as follows:

Definition 8.10   *A **double limit variety** V_{DL} satisfies for $y_1 \neq y_2$ the four equations*

$$f(y_1, \lambda, \Lambda) = 0\,,$$
$$f(y_2, \lambda, \Lambda) = 0\,,$$
$$f_y(y_1, \lambda, \Lambda) = 0\,,$$ (8.13)
$$f_y(y_2, \lambda, \Lambda) = 0\,.$$

In the example of equation (8.10), the curves of V_{HY} and V_{BI} separate different regions in the (α, β)-parameter plane (see Figure 8.5). Five insets in Figure 8.5 illustrate sample bifurcation diagrams that are characteristic of the three regions and the two separating curves. Each bifurcation diagram depicts y versus λ. For example, for any parameter combination in the shaded region, the bifurcation diagram has three turning points. If we move a sample parameter combination $\Lambda = (\alpha, \beta)$ toward the bifurcation and isola variety V_{BI}, two of the turning points approach each other. Moving the sample Λ along a curve corresponds to a family of bifurcation problems as indicated in Figure 8.2. When Λ reaches and crosses the codimension-one set V_{BI}, the two turning points collapse into a simple bifurcation point and detach immediately, as indicated by two further sample bifurcation diagrams in Figure 8.5. On the other hand, if we start with Λ again in the shaded region and move toward the hysteresis variety V_{HY}, the two turning points that form a hysteresis approach each other and join. After V_{HY} is crossed, the hysteresis has disappeared. This is illustrated by the two remaining sample bifurcation diagrams in Figure 8.5. For any point (α, β) in Figure 8.5, there is at least one singularity. Figure 8.4 illustrates the two-dimensional bifurcation set in the (λ, α, β)-space that represents turning points. The dashed lines in Figure 8.4 correspond to V_{HY} and V_{BI}. The collapsing of turning points at the dashed lines of higher-order singularities becomes apparent.

Diagrams like that in Figure 8.5 are frequently used to summarize the variety of nonlinear phenomena. These multidiagrams are especially useful in case $m = 2$, making possible a two-dimensional representation of the bifurcation behavior in the higher-dimensional (y, λ, Λ)-space. Modifications are used frequently. For instance, with $m = 1$ and $n = 2$, it is instructive to draw phase diagrams in insets surrounding a (λ, Λ_1)-parameter chart.

The hypersurfaces V_{HY}, V_{BI}, V_{DL} separate the Λ-parameter space into connected components (Figure 8.5). Let us denote the interior of any such connected component by C (for example, the shaded area in Figure 8.5, excluding V_{BI} and V_{HY}). For each of these subsets C of the parameter space Λ, the following holds:

Theorem 8.11   *Suppose the two parameter vectors Λ and Λ' are in the same region C. Then $f(y, \lambda, \Lambda)$ and $f(y, \lambda, \Lambda')$ are contact equivalent.*

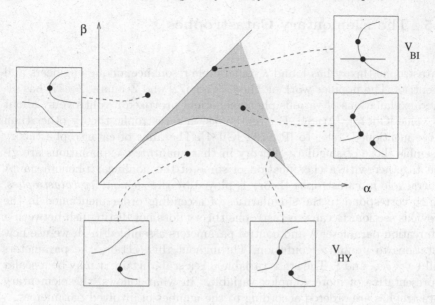

Fig. 8.5. Example of Exercise 2.14, equation (8.10), parameter chart

This statement justifies choosing only one sample Λ in each C for the purpose of illustrating characteristic bifurcation behavior. The resulting bifurcation diagram is representative of the entire class C and stable under variations of Λ that remain inside C.

The above remarks illuminate what was introduced in Section 2.10 under the name "generic bifurcation." Inside any component C, there are only turning points, which are stable and ubiquitous. Hopf bifurcations cannot occur in our scalar setting; for $n \geq 2$ they are also generic. As seen in the above discussion, generic bifurcation has the meaning of codimension-zero bifurcation. Higher-order singularities are restricted to the boundaries of the components C; these hypersurfaces (curves in the core part of Figure 8.5) may intersect or touch when Λ is varied. The higher the order of a singularity, the more unlikely it is to pass the singularity. There is a hierarchy of singularities. Singularities of a higher order are escorted on their flanks by singularities of a lower order. The least-degenerate turning points may coalesce in singularities of the next higher order—namely, bifurcation, isola, and hysteresis points. On their part, the hypersurfaces V_{HY} or V_{BI} may collapse in singularities of an even higher order. In Figure 8.5 this happens at $\alpha = \beta = 0$, where the codimension-one sets V_{HY} and V_{BI} meet in a pitchfork, which comprises the defining conditions of V_{HY} and V_{BI}; compare Table 8.1. In the following section, we shall find out how high-order singularities look.

8.5   The Elementary Catastrophes

Catastrophe theory has found a remarkable resonance among engineers and scientists. The popular work of Thom [Thom72] and Zeeman [Zee77] has offered explanations of various phenomena and provoked a controversy about its value [Guc78], [Arn84]. For textbooks on catastrophe theory other than those mentioned, refer to [PoS78], [Gil81]. The aims of catastrophe theory resemble those of singularity theory in that qualitative explanations are given together with a classification of some of the nonlinear phenomena. A central role in catastrophe theory is played by the *elementary catastrophes*, which correspond to the singularities of ascending order mentioned in the previous section. Because catastrophe theory does not distinguish between a bifurcation parameter λ and control parameters assembled in Λ, we use new notations to minimize confusion. Throughout this section, the parameters will be γ, δ, ζ, and ξ. The state variable y is a scalar again; it may be a scalar representative of more complex variables. In what follows, the elementary catastrophes are ordered according to the number of involved parameters.

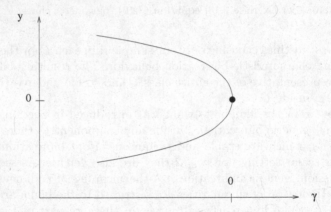

Fig. 8.6. Equation (8.14), a fold

8.5.1   The Fold

The *fold* is the name of a manifold, the simplest representative of which is defined by

$$0 = y^2 + \gamma. \tag{8.14}$$

This equation is a parabola in the two-dimensional (y, γ)-space (see Figure 8.6). The singularity at $(y_0, \gamma_0) = (0, 0)$ is a turning point. Because there are two solutions for $\gamma < \gamma_0$ and no solution for $\gamma > \gamma_0$, the multiplicity changes at γ_0. This point γ_0 establishes the bifurcation set of the parameter "space."

For later reference we point out that the bifurcation set of a fold is zero-dimensional in the one-dimensional parameter space. The least degenerate singularity represented by a fold is well-known to us from the examples that exhibit turning points. The geometry shown in Figure 8.6 illustrates why turning points are also called fold bifurcations.

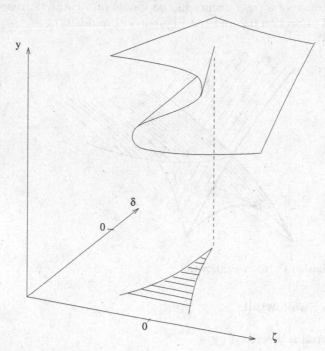

Fig. 8.7. Equation (8.15), a cusp

8.5.2   The Cusp

The model equation that represents a cusp is

$$0 = y^3 + \zeta y + \delta. \tag{8.15}$$

This example was discussed in detail in Section 2.10 and Example 8.6. The manifold of a cusp is illustrated in Figure 8.7. The bifurcation set given by

$$4\zeta^3 + 27\delta^2 = 0$$

[cf. equation (2.29)] bounds a region in the parameter space (shaded in Figure 8.7). Equation (8.15) has three solutions for parameter values inside the hatched area and one solution for outside parameter values. The bifurcation set is one-dimensional in the two-dimensional parameter space; the cusp manifold is a two-dimensional surface in the (y, ζ, δ)-space. The singularity of highest order occurs at $(\zeta, \delta) = (0,0)$. Cutting one-dimensional slices into the parameter plane (vertical planes in Figure 8.7), one obtains folds. These folds escort the cusp center—that is, singularities of low order form the flanks of the singularity of higher order. Planar cross sections of the cusp are depicted in Figures 2.44 and 2.46. On one "side" of the cusp there is hysteresis. Another example of a cusp occurs in the model of a catalytic reaction (see Example 6.3 in Section 6.2 and the Figures 5.21 and 5.22).

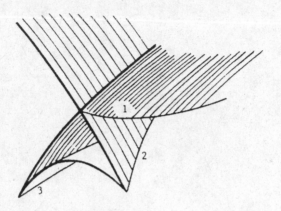

Fig. 8.8. Equation (8.16), a swallowtail

8.5.3 The Swallowtail

The swallowtail is described by

$$0 = y^4 + \gamma y^2 + \delta y + \zeta. \tag{8.16}$$

This equation defines a three-dimensional manifold in the four-dimensional $(y, \gamma, \delta, \zeta)$-space. In order to obtain the bifurcation set, differentiate equation (8.16) with respect to y,

$$0 = 4y^3 + 2\gamma y + \delta.$$

This equation and equation (8.16) define a surface in the (γ, δ, ζ)-space. For a parameterization of the bifurcation set, two parameters ρ and σ are needed; we choose $\rho = y$ and $\sigma = \gamma$ and obtain

$$\begin{aligned} \gamma &= \sigma, \\ \delta &= -4\rho^3 - 2\sigma\rho, \\ \zeta &= 3\rho^4 + \sigma\rho^2. \end{aligned} \tag{8.17}$$

To visualize the bifurcation set given by equation (8.17), we take cross sections. For instance, discussing the relation between γ and ζ for the three cases

$\sigma = 0$, $\sigma > 0$, and $\sigma < 0$ provides enough information to draw an illustration of the bifurcation set (Figure 8.8, see Exercise 8.4). The heavy curve is a cross section of the bifurcation set with a plane defined by $\sigma = \gamma = $ constant, $\gamma <' 0$. The bifurcation set of Figure 8.8 separates the (γ, δ, ζ)-space into three parts. A core part for $\gamma < 0$ extends to the curves 1, 2, and 3, reminding one of a tent-like structure. The two other parts are "above" and "below" the manifold sketched in Figure 8.8. In the center of the swallowtail, the three curves 1, 2, and 3 collapse. This center is the singularity of highest order, organizing the hierarchy of lower-order singularities. Curve 1 is a curve of self-intersection, curves 2 and 3 are formed by cusp ridges, which are accompanied by the lowest-order singularities (Exercise 8.4).

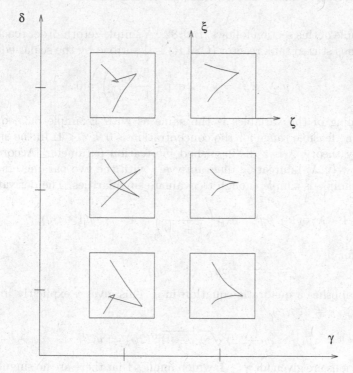

Fig. 8.9. Equation (8.18), illustration of a butterfly

Other Catastrophes

The normal form of a *butterfly* is given by

$$0 = y^5 + \gamma y^3 + \delta y^2 + \zeta y + \xi. \qquad (8.18)$$

Because its bifurcation set extends in the four-dimensional $(\gamma, \delta, \zeta, \xi)$-space, an illustration of a butterfly is complicated. We show typical slices, holding two parameters fixed (γ and δ). Then the cross sections are (ζ, ξ)-planes, drawn in the insets in Figure 8.9—that is, Figure 8.9 represents a two-dimensional family of two-dimensional cross sections [PoS78].

There is another group of elementary catastrophes—the *umbilics*. These are described by two scalar equations involving either three or four parameters. For details and illustrations, see the special literature.

We close this section with a disillusioning remark: The higher the order of a singularity, the more unlikely to discover a singularity in a practical parameter study, and the more costly any attempt to calculate it.

8.6 Zeroth-Order Reaction in a CSTR

The example of this section follows [BaL82]. A simple zeroth-order reaction in a continuous stirred tank reactor (CSTR) is described by the scalar equation

$$f(y, \lambda, \gamma, \beta) = y - \lambda \exp\left(\frac{\gamma \beta y}{1 + \beta y}\right) = 0. \qquad (8.19)$$

The meaning of the variables is the same as with Example 6.3 [equation (6.15)]. The feasible range for the concentration is $0 < y < 1$. In the spirit of singularity theory, λ is a distinguished bifurcation parameter. Accordingly, we ask how (y, λ)-bifurcation diagrams vary with the two parameters β and γ. The example is simple enough to evaluate singularities. The derivative

$$f_y = 1 - \lambda \gamma \beta (1 + \beta y)^{-2} \exp\left(\frac{\gamma \beta y}{1 + \beta y}\right) = 1 - \gamma \beta y (1 + \beta y)^{-2} \qquad (8.20)$$

vanishes for

$$\gamma \beta y = (1 + \beta y)^2 \,,$$

which establishes a quadratic equation in y. This gives y explicitly in terms of β and γ, $y = u(\beta, \gamma)$,

$$y_{1,2} = [\gamma - 2 \pm \sqrt{\gamma(\gamma - 4)}]/(2\beta) =: u(\beta, \gamma) \,.$$

Real solutions are given for $\gamma \geq 4$, which implies that there are no singularities and hence no bifurcation phenomena for $0 < \gamma < 4$. Substituting $y_{1,2}$ for $\gamma \geq 4$ into the equation $f = 0$ yields the bifurcation set in implicit form,

$$0 = u(\beta, \gamma) - \lambda \exp\left(\frac{\gamma \beta u(\beta, \gamma)}{1 + \beta u(\beta, \gamma)}\right) . \qquad (8.21)$$

Equation (8.21) involves only the three parameters; it defines a two-dimensional surface in the (λ, β, γ)-parameter space.

Next we investigate the subset formed by singularities of higher order. To this end, we shall discuss the varieties introduced in Section 8.4. Clearly, $f_\lambda \neq 0$, because the exponential term in equation (8.19) cannot vanish. Hence,

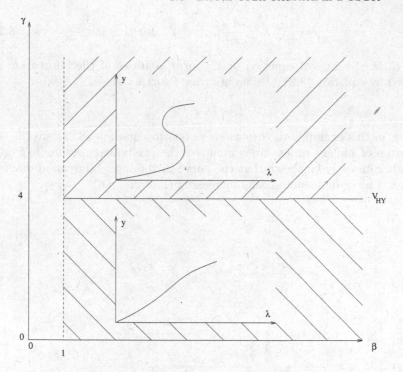

Fig. 8.10. CSTR (8.19), first results

there is no bifurcation and isola variety V_{BI} in this example [cf. equation (8.12)]. In order to find a hysteresis variety V_{HY} [cf. equation (8.11)], we must evaluate the second-order derivative with respect to y,

$$f_{yy} = -\gamma\beta(1 - \beta y)(1 + \beta y)^{-3},$$

which vanishes for

$$y = \frac{1}{\beta}.$$

Inserting this expression into the equation $f_y = 0$ gives

$$\gamma = 4.$$

This establishes V_{HY} as a straight line in the (β, γ)-plane. Because $y < 1$, V_{HY} is defined only for $\beta > 1$. Corresponding values of λ are given by equation (8.12),

$$\lambda = y \cdot \exp\left(-\frac{\gamma\beta y}{1 + \beta y}\right) = \frac{1}{\beta}\exp(-2).$$

The loci of hysteresis points are now known; parameterized by β, the hysteresis points are given by

$$y = \frac{1}{\beta}, \quad \lambda = \frac{1}{\beta e^2}, \quad \gamma = 4 \quad \text{for } \beta > 1 . \tag{8.22}$$

The (λ, β, γ)-values of equation (8.22) form a subset of the bifurcation set defined by equation (8.21). Differentiating f further shows

$$f_{yyy} \, f_\lambda \neq 0 .$$

Hence, in this example, the hysteresis points in equation (8.22) are the singularities of highest order. We summarize the preliminary results in Figure 8.10. It can easily be shown that the curve in the inset bifurcation diagram for $\gamma < 4$ must be monotonically increasing (Exercise 8.6).

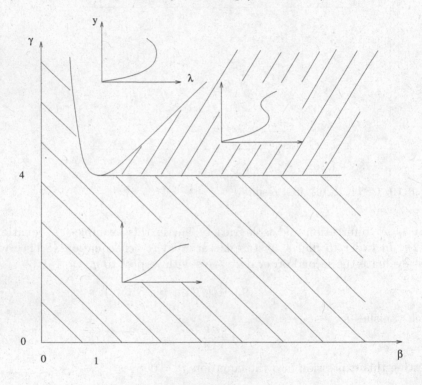

Fig. 8.11. CSTR (8.19), parameter chart

The restriction $\beta > 1$ is essential only for the existence of the hysteresis points with $y < 1$. It is still unclear what happens for $\beta \leq 1$. The role of β in equation (8.19) is artificial; β can be removed by introducing new variables, but this would conceal the feasible range of y. What we can do is establish the curve in the (β, γ)-plane, for which the y value of a singularity reaches unity. A straightforward calculation establishes the curve

$$\gamma = (1 + \beta)^2 / \beta .$$

This curve is shown in Figure 8.11. Above that curve, only one of the turning points satisfies $y < 1$, for both $\beta > 1$ and $\beta \leq 1$. It remains to discuss the region between the curve and the straight line $\gamma = 4$. Here, for $\beta > 1$ both turning points satisfy $y < 1$, whereas for $\beta \leq 1$ none of the turning points satisfies $y < 1$. A characteristic bifurcation behavior of this latter part resembles the behavior for $\gamma < 4$; accordingly, we join these two components of the (β, γ)-plane. Hence, the (β, γ)-quadrant is subdivided into three regions, with bifurcation diagrams $(y < 1)$ as indicated in Figure 8.11.

8.7 Center Manifolds

In Section 1.2 we investigated trajectories close to stationary solutions by means of linearizations. In nondegenerate cases, the linearized problem describes the flow correctly in the neighborhood of the equilibrium. This hyperbolic situation is described by the theorems of Liapunov, and Grobman and Hartman. Because most of the nonlinear phenomena have their origin in solutions that are characterized by a zero real part of an eigenvalue of the Jacobian, it is natural to ask whether there are analytical tools for analyzing the flow in the degenerate situation. The answer is yes, and the proper framework is provided by center manifold theory. A detailed study of how to apply center manifold theory is given in [Carr81].

Consider an autonomous system of ODEs, with a stationary solution assumed in the coordinate origin,

$$\dot{\mathbf{y}} = \mathbf{f}(\mathbf{y}), \quad \mathbf{f}(0) = \mathbf{0}.$$

After carrying out a Taylor expansion and after a suitable transformation to a normal form, the system can be written as

$$\begin{aligned}
\dot{\mathbf{y}}^- &= \mathbf{A}^- \mathbf{y}^- + \mathbf{f}^- (\mathbf{y}^-, \mathbf{y}^+, \mathbf{y}^0), \\
\dot{\mathbf{y}}^+ &= \mathbf{A}^+ \mathbf{y}^+ + \mathbf{f}^+ (\mathbf{y}^-, \mathbf{y}^+, \mathbf{y}^0), \\
\dot{\mathbf{y}}^0 &= \mathbf{A}^0 \mathbf{y}^0 + \mathbf{f}^0 (\mathbf{y}^-, \mathbf{y}^+, \mathbf{y}^0)
\end{aligned} \tag{8.23}$$

with square matrices $\mathbf{A}^-, \mathbf{A}^+$, and $\mathbf{A}^0$ of sizes i, j, and k, whose eigenvalues have negative, positive, and zero real parts, respectively. In equation (8.23) the n-vector $\mathbf{y}$ is partitioned into three subvectors of lengths i, j, and k such that $i + j + k = n$. Not all the dimensions i, j, or k need to be nonzero. The nonlinear functions $\mathbf{f}^-, \mathbf{f}^+$, and $\mathbf{f}^0$ and their first-order derivatives vanish at the equilibrium $(\mathbf{y}^-, \mathbf{y}^+, \mathbf{y}^0) = \mathbf{0}$. Decomposing the original system $\dot{\mathbf{y}} = \mathbf{f}(\mathbf{y})$ as in equation (8.23) enables splitting off stable and unstable parts; in this way a stability analysis can be reduced to a subsystem for the subvector $\mathbf{y}^0$.

To gain more insight into the structure of the flow of equation (8.23), one can use the concept of invariant manifolds. Recall that an invariant manifold $\mathcal{M}$ is a manifold with the additional property that a trajectory that starts

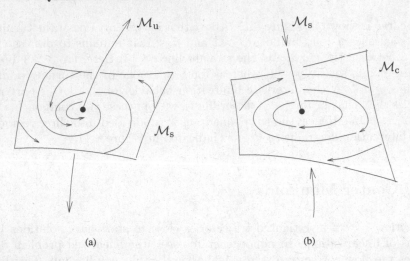

Fig. 8.12. Invariant manifolds in three-dimensional space

on $\mathcal{M}$, stays on it. Center manifold theory says that for equation (8.23) there are five types of invariant manifolds (locally) that pass $(\mathbf{y}^-, \mathbf{y}^+, \mathbf{y}^0) = \mathbf{0}$ —namely, the stable manifold $\mathcal{M}_s$, the unstable manifold $\mathcal{M}_u$, the center manifold $\mathcal{M}_c$, the center-stable manifold $\mathcal{M}_{cs}$, and the center-unstable manifold $\mathcal{M}_{cu}$ [Kel67]. The manifolds $\mathcal{M}_s$, $\mathcal{M}_u$, $\mathcal{M}_c$ are known from the theory of ordinary differential equations; see Appendix A.5. Not all these manifolds need to exist simultaneously. This is pictured by two sketches in Figure 8.12, both illustrating the three-dimensional situation $n = 3$. In Figure 8.12a we depict the situation $i = 2$, $j = 1$, $k = 0$ —that is, there is a two-dimensional stable manifold $\mathcal{M}_s$ (here with a focus), a one-dimensional unstable manifold $\mathcal{M}_u$ and no center manifold. The two manifolds $\mathcal{M}_s$ and $\mathcal{M}_u$ intersect in the stationary solution, tangentially to the span of the corresponding eigenvectors. Any flow line starting on the invariant manifold $\mathcal{M}_s$ remains there and spirals toward the equilibrium, although it is unstable. Each trajectory not starting exactly on $\mathcal{M}_s$ spirals toward $\mathcal{M}_u$, leaving the neighborhood of the stationary solution. Figure 8.12b shows a center on a two-dimensional center manifold $\mathcal{M}_c$ ($k = 2$), intersected by a stable manifold $\mathcal{M}_s$ ($i = 1$). This is the situation depicted in Figure 1.16 for the Lorenz equations. Along $\mathcal{M}_s$ the flow tends to $\mathcal{M}_c$. In Figure 8.13 we illustrate an example $i = j = k = 1$. Here also a center-stable manifold $\mathcal{M}_{cs}$ (spanned by $\mathcal{M}_c$ and $\mathcal{M}_s$) and a center-unstable manifold $\mathcal{M}_{cu}$ (spanned by $\mathcal{M}_c$ and $\mathcal{M}_u$) are indicated.

For $j > 0$, the stationary solution is unstable; this situation is of little physical interest. Therefore, we restrict the discussion to the case $j = 0$,

$$\begin{aligned}
\dot{\mathbf{y}}^- &= \mathbf{A}^- \mathbf{y}^- + \mathbf{f}^-(\mathbf{y}^-, \mathbf{y}^0), \\
\dot{\mathbf{y}}^0 &= \mathbf{A}^0 \mathbf{y}^0 + \mathbf{f}^0(\mathbf{y}^-, \mathbf{y}^0).
\end{aligned} \tag{8.24}$$

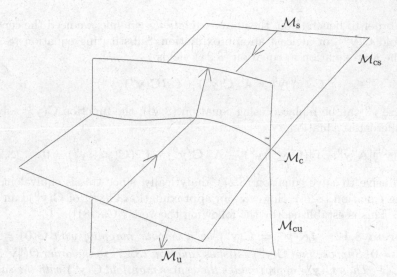

Fig. 8.13. One-dimensional invariant manifolds

Such a scenario is depicted in Figure 8.12b, compare also Figure 1.16. Assume temporarily that $\mathbf{f}^-$ and $\mathbf{f}^0$ are zero. Then the problem is linear and has two invariant manifolds—the stable manifold $\mathbf{y}^0 = \mathbf{0}$ and the center manifold $\mathbf{y}^- = \mathbf{0}$. On the center manifold, the linear system is reduced to

$$\dot{\mathbf{y}}^0 = \mathbf{A}^0 \mathbf{y}^0 .$$

This k-dimensional equation on the center manifold determines whether $(\mathbf{y}^-, \mathbf{y}^0) = \mathbf{0}$ is attractive or not. The reduction also works in the nonlinear case equation (8.24). Let the center manifold be described by a function $\mathbf{C}$ with

$$\mathbf{y}^- = \mathbf{C}(\mathbf{y}^0), \tag{8.25a}$$

$$\mathbf{C}(\mathbf{0}) = \mathbf{0}, \quad \mathbf{C}'(\mathbf{0}) = \mathbf{0} . \tag{8.25b}$$

The existence of such a function $\mathbf{C}(\mathbf{y}^0)$ is established by the invariant manifold theorem; see Appendix A.5. In equation (8.25b), $\mathbf{C}'(\mathbf{y}^0)$ denotes the rectangular matrix of the first-order partial derivatives of $\mathbf{C}$. The center manifold $\mathcal{M}_c$ satisfies equation (8.25b) because it passes the equilibrium tangentially to the center manifold $\mathbf{y}^- = \mathbf{0}$ of the linearized problem. Substituting equation (8.25a) into the second equation of equation (8.24) leads to the k-dimensional system

$$\dot{\mathbf{y}}^0 = \mathbf{A}^0 \mathbf{y}^0 + \mathbf{f}^0(\mathbf{C}(\mathbf{y}^0), \mathbf{y}^0) . \tag{8.26}$$

One of the main conclusions of center manifold theory states that the stability/instability of the equilibrium $\mathbf{y} = \mathbf{0}$ of equation (8.24) is reflected by the reduced system equation (8.26).

In order to benefit from the above *reduction principle*, we need the center manifold $\mathbf{C}(\mathbf{y}^0)$, or at least an approximation. Substituting equation (8.25) into the first equation in equation (8.24) yields

$$\mathbf{C}'(\mathbf{y}^0)\dot{\mathbf{y}}^0 = \mathbf{A}^- \mathbf{C}(\mathbf{y}^0) + \mathbf{f}^-(\mathbf{C}(\mathbf{y}^0), \mathbf{y}^0).$$

Because $\dot{\mathbf{y}}^0$ can be replaced using equation (8.26), the function $\mathbf{C}(\mathbf{y}^0)$ solves the differential equation

$$\mathbf{C}'(\mathbf{y}^0)[\mathbf{A}^0\mathbf{y}^0 + \mathbf{f}^0(\mathbf{C}(\mathbf{y}^0), \mathbf{y}^0)] - \mathbf{A}^-\mathbf{C}(\mathbf{y}^0) - \mathbf{f}^-(\mathbf{C}(\mathbf{y}^0), \mathbf{y}^0) = 0. \quad (8.27)$$

It is illusive to solve equation (8.27) analytically, since this is equivalent to solving equation (8.24). However, an approximation $\bar{\mathbf{C}}(\mathbf{y}^0)$ of $\mathbf{C}(\mathbf{y}^0)$ can be found. This is established by the following theorem ([Carr81]):

Theorem 8.12   *Let* $\mathbf{y}^- = \bar{\mathbf{C}}(\mathbf{y}^0)$ *be a smooth mapping with* $\bar{\mathbf{C}}(\mathbf{0}) = \mathbf{0}$, $\bar{\mathbf{C}}'(\mathbf{0}) = \mathbf{0}$. *Suppose that* $\bar{\mathbf{C}}(\mathbf{y}^0)$ *satisfies equation (8.27) to the order* $O(\|\mathbf{y}^0\|)^q$ *for* $q > 1$. *Then* $\bar{\mathbf{C}}(\mathbf{y}^0)$ *approximates the center manifold* $\mathbf{C}(\mathbf{y}^0)$ *with the same order q.*

Analog reduction and approximation principles hold for equation (8.23).

Example 8.13
We apply the above ideas to the simple example

$$\begin{aligned}
\dot{y}_1 &= -y_1 + y_2^2 + y_1 y_2^2, \\
\dot{y}_2 &= y_1 y_2 + y_2^3 + y_1^2 y_2.
\end{aligned} \quad (8.28)$$

Because the Jacobian at the only equilibrium $\mathbf{y} = \mathbf{0}$ has zero as an eigenvalue, the principle of linearized stability does not apply. We try to obtain stability results by means of center manifold theory. The system in equation (8.28) is already written in the form of equation (8.24). Only the first equation has a linear term; hence, equation (8.28) corresponds to equation (8.24) as follows:

$$\begin{aligned}
A^- &= -1, \quad y_1 = y^-, \quad A^0 = 0, \quad y_2 = y^0, \\
f^-(y^-, y^0) &= f_1(y_1, y_2) = (y_1 + 1)y_2^2, \\
f^0(y^-, y^0) &= f_2(y_1, y_2) = y_2(y_1 + y_1^2 + y_2^2).
\end{aligned}$$

To approximate the center manifold $y_1 = C(y_2)$, we form equation (8.27),

$$C'(y_2)[y_2(C(y_2) + (C(y_2))^2 + y_2^2)] + C(y_2) - (C(y_2) + 1)y_2^2 = 0. \quad (8.29)$$

The simplest trial that satisfies equation (8.25b) is

$$\bar{C}(y_2) = \gamma y_2^2.$$

Substituting this expression into equation (8.29) yields

$$\gamma y_2^2 - y_2^2 + y_2^4(2\gamma^2 - \gamma) + 2\gamma^3 y_2^6 = 0.$$

For $\gamma = 1$, all terms of order less than four drop out, and the differential equation (8.29) is satisfied to the order $O(|y_2|^4)$. Accordingly, by Theorem 8.12, $\bar{C}(y_2) = y_2^2$ approximates the center manifold $C(y_2)$ with the same order $q = 4$ —that is,

$$C(y_2) = y_2^2 + O(y_2^4).$$

This version of C is substituted into the second equation in (8.28), leading to the scalar differential equation

$$\dot{u} = u^2 u + u^3 + u^4 u + O(u^5)$$
$$= 2u^3 + O(u^5) = u^3(2 + O(u^2)).$$

Locally, for small $|u|$, $u > 0 \Leftrightarrow \dot{u} > 0$, and $u = 0$ is an unstable equilibrium of this differential equation. Hence, the equilibrium $\mathbf{y} = \mathbf{0}$ of equation (8.28) is unstable.

Troubles with linearization techniques and center manifold procedures are reported in [Aul84]. Note that the setup of $\dot{\mathbf{y}} = \mathbf{f}(\mathbf{y})$ into the decomposition in equation (8.23) is not guaranteed to be practical. Specific applications of center manifold theory can be found, for instance, in [Hol77], [ScTZ83], [TrS91]. The computation of invariant manifolds is discussed in [Kra05], [EnKO07].

Exercises

Exercise 8.1
Verify the criterion of Definition 8.2 for contact equivalence for Example 8.3.

Exercise 8.2
Calculate the hysteresis variety and the bifurcation and isola variety of equation (8.10). Is there a double limit variety? Compare your results with Figure 8.5 and Exercise 2.14.

Exercise 8.3
Consider equation (2.29). What are V_{HY} and V_{BI}? Draw a sketch that corresponds to Figure 8.5 ($\Lambda_1 = \gamma$).

Exercise 8.4
In equation (8.17), assume three fixed values of σ ($\sigma > 0$, $\sigma = 0$, $\sigma < 0$) and discuss the curves in the (δ, ζ)-plane. Draw a diagram for each case. Try to find your curve for $\sigma < 0$ in Figure 8.8. Which parts of the surface in Figure 8.8 are the loci of the lowest-order singularities?

Exercise 8.5
Extend Figure 8.9 (that is, vary γ and δ further) to illustrate the birth and death of swallowtails.

Exercise 8.6
Consider equation (8.19). Show that for $\gamma < 4$ the solution increases monotonically with λ.

9 Chaos

The oscillations we have encountered in previous chapters have been periodic. Periodicity reflects a high degree of regularity and order. Frequently, however, one encounters *irregular* oscillations, like those illustrated in Figure 9.1. Functions or dynamical behavior that is not stationary, periodic or quasi-periodic may be called *chaotic*, sometimes *aperiodic* or *erratic*. We shall use "chaos" in this broad meaning; note that chaos can be defined in a restricted way, characterizing orbits with a positive Liapunov exponent (to be explained later). Such dynamic behavior can be seen as the utmost flexibility a dynamical system may show. It is expected that systems with chaotic behavior can be easily modulated or stabilized. Note that the "irregularity" of chaos is completely deterministic and not stochastic.

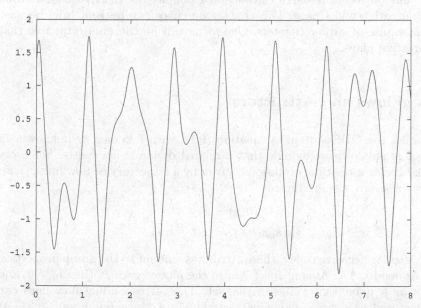

Fig. 9.1. Duffing equation (9.4); $u(t)$ for $\omega = 0.48$

Irregularity does not imply that the oscillation is completely arbitrary. Notice that the amplitude in Figure 9.1 is bounded. Moreover, the "frequency" seems to vary only slightly. These observations nurture the impression that

R. Seydel, *Practical Bifurcation and Stability Analysis*,
Interdisciplinary Applied Mathematics 5, DOI 10.1007/978-1-4419-1740-9_9,
© Springer Science+Business Media, LLC 2010

erratic oscillation, or "chaos," may conceal an inherent structure. In this chapter, we briefly discuss how chaos arises and characterize the structure in chaos. Such questions are of immediate interest in many fields, form acoustics to fluid dynamics to orbital motion [RuT71], [LaH91], [SpW91]. Some aspects of chaos have stimulated deep philosophical questions about determinism in nature, including the question of whether the outcome of an experiment is predictable.

Throughout this book, we have assumed that the equations modeling a physical situation are given. Calculations were based, for instance, on a known system of ODEs

$$\dot{\mathbf{y}} = \mathbf{f}(\mathbf{y}, \lambda) . \tag{9.1}$$

For the most part of this chapter, the dependence on λ is not essential, and we may drop λ. Aperiodic oscillations can arise not only from numerical integrations but also in measurements of experiments. Frequently, only one scalar variable $y_1(t)$ is observable (Figure 9.1), and the equation of the underlying dynamical system is unknown. In the present chapter, we address this situation.

This is no book about chaos. On that field there is ample literature. Classical surveys on chaos include [Rue80], [Eck81], [Ott81], [Shaw81], [Schu84], [Dev86], [ThS86], [Hao87], [Moon87], [PaC89], [ElN90], [PePRS92], [Str94]. For hints on recent research consult, for example, the International Journal of Bifurcation and Chaos. This chapter on chaos can be kept brief because it makes use of earlier chapters. One focus will be the important role that bifurcation plays.

9.1   Flows and Attractors

Consider the ODE system in equation (9.1), with $\mathbf{f}$ known or unknown. In case $\mathbf{f}$ is unknown we assume that a related dynamic law exists. For every initial vector $\mathbf{z}$ the time evolution is given by a trajectory or flow line $\varphi(t; \mathbf{z})$. Flow lines have the attributes

$$\varphi(0; \mathbf{z}) = \mathbf{z} ,$$
$$\varphi(s; \varphi(t; \mathbf{z})) = \varphi(t + s; \mathbf{z}) .$$

Allowing for negative time t, these attributes amount to the group properties; see Appendix A.7. Assume a set $\mathcal{M}_0$ in the phase space at time $t_0 = 0$, and let $\mathbf{z}$ be a point that varies in this set. The corresponding flow lines can be pictured as densely filling part of the state space (cf. Figure 9.2). At time instant t, the original set $\mathcal{M}_0$ has been moved or deformed to a set $\mathcal{M}_t = \varphi(t; \mathcal{M}_0)$. The question is, what structure the final "ω-limit set"

$$\mathcal{M}_\infty := \lim_{t \to \infty} \varphi(t; \mathcal{M}_0)$$

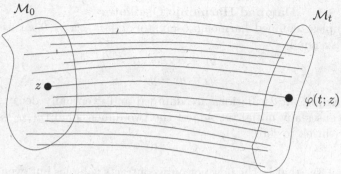

Fig. 9.2. A set $\mathcal{M}$ deformed by a flow

may have. In what follows, we discuss how, for a fixed value of λ, a set $\mathcal{M}_0$ is deformed by the flow for $t \to \infty$.

First, let the set $\mathcal{M}_0$ consist of a single stationary point $\mathbf{y}^s$ of equation (9.1): $\mathbf{f}(\mathbf{y}^s, \lambda) = \mathbf{0}$. Then, theoretically, the flow line remains at this point,

$$\varphi(t; \mathbf{y}^s) = \mathbf{y}^s \quad \text{for all } t \,.$$

In other words, the stationary solution is a fixed point of the flow; the set $\mathcal{M}_0$ is *invariant under the flow*. As a second example, consider the set $\mathcal{M}_0$ consisting of a limit cycle of equation (9.1). Then, for any $\mathbf{z}$ in $\mathcal{M}_0$, the flow line follows the limit cycle, and $\mathcal{M}_0$ remains invariant under the flow:

$$\varphi(t; \mathcal{M}_0) = \mathcal{M}_0 \quad \text{for all } t \,.$$

Apart from these two exceptional cases, no flow line can intersect with itself. General sets $\mathcal{M}_0$ are deformed by the flow.

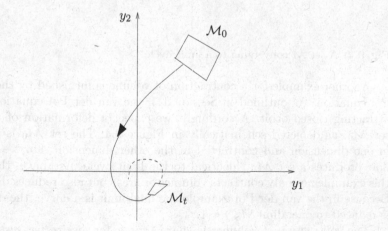

Fig. 9.3. A set $\mathcal{M}$ shrinking to 0

Example 9.1   Damped Harmonic Oscillator

To the simplest damped harmonic oscillator $\ddot{u} + \dot{u} + u = 0$ an associated first-order system is

$$\dot{y}_1 = -y_2\,,$$
$$\dot{y}_2 = y_1 - y_2\,.$$

In this example, all oscillations are damped and eventually decay to zero. The flow causes any initial set $\mathcal{M}_0$ of the two-dimensional (y_1, y_2)-space to eventually shrink to zero,

$$\mathcal{M}_\infty = \{0\}$$

(cf. Figure 9.3)—that is, the flow not only contracts volumes but even reduces the dimension, $\dim(\mathcal{M}_\infty) = 0$.   □

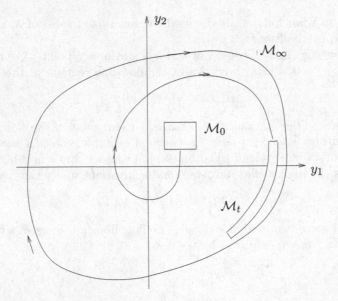

Fig. 9.4. A set $\mathcal{M}$ converging to a limit cycle

Another example for a contraction of volume is furnished by the van der Pol equation. As outlined in Section 1.3, the van der Pol equation has an attracting closed orbit. Accordingly, we expect a deformation of an initial set $\mathcal{M}_0$, as depicted schematically in Figure 9.4. The set $\mathcal{M}_0$ is stretched in one dimension and contracted in the other dimension. For $t \to \infty$, the flow produces a set $\mathcal{M}_\infty$ identical to the limit cycle. As above, the flow of this example not only contracts volumes or areas but also reduces dimension. Because in the van der Pol example the final limit is a curve, the dimension is reduced to one, $\dim(\mathcal{M}_\infty) = 1$.

The contraction of volumes is characteristic for *dissipative systems* (systems with friction). Let $V(t)$ be the volume of the set $\mathcal{M}_t$. Then changes of the volume are given by

$$\frac{dV}{dt} = \int_{\mathcal{M}_t} \nabla \cdot \mathbf{f} \, dV \, . \tag{9.2}$$

Hence, div $\mathbf{f} = \nabla \cdot \mathbf{f} = 0$ implies that the volume remains invariant under the flow (Liouville's theorem). The system must be dissipative if $\nabla \cdot \mathbf{f} < 0$ holds along $\mathcal{M}_t$ (Exercise 9.1).

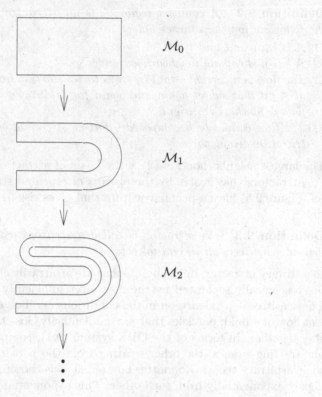

Fig. 9.5. Illustration of a horseshoe map

Next we imagine a flow that not only stretches into one dimension and contracts in another dimension, but also folds (see Figure 9.5). The question arises, Where does $\mathcal{M}$ finally arrive after an infinite number of stretchings and foldings? A related example, the "horseshoe map," was constructed by Smale [Sma67]. Imagine in Figure 9.5 a map that deforms the rectangle $\mathcal{M}_0$ into the curved object $\mathcal{M}_1$. The same transformation is applied again to the rectangle in which $\mathcal{M}_1$ can be embedded. The result is $\mathcal{M}_2$. After a sequence of these transformations a "strange" attractor arises that seems to consist of a bunch of lines. Attractors of such flows cannot easily be compared to the elementary attractors formed by stable equilibria and limit cycles.

Attractors are formed by the sets to which trajectories tend for $t \to \infty$. Hence, attractors are closely related to stability; they describe the long-term behavior of physical systems. The invariant sets of attractors are generalizations of stable equilibria and limit cycles. In spite of the significance of attractors, there has not been much agreement on which definition of an attractor is most useful. Here we give a definition following [Eck81]. (The reader may like to draw sketches of situations in which hypotheses (3) or (4) are not satisfied.)

Definition 9.2 *A compact region* $\mathcal{A}$ *is an* **attractor** *of a flow* $\varphi(t; \mathbf{z})$*, if the following four hypotheses hold:*

(1) $\mathcal{A}$ *is invariant under* φ*.*
(2) $\mathcal{A}$ *has a shrinking neighborhood under* φ*.*
(3) The flow is recurrent—that is, trajectories starting from any (open) subset of $\mathcal{A}$ *hit that subset again and again for arbitrarily large values of* t*; the flow is nowhere transient.*
(4) The flow cannot be decomposed—that is, $\mathcal{A}$ *cannot be split into two nontrivial invariant parts.*

The largest neighborhood of (2) is the basin of attraction.

Attractors that evolve by the process of repeated stretching and folding (cf. Figure 9.5) have a peculiar feature that gives rise to the following definition:

Definition 9.3 *An attractor is called a* **strange attractor** *if flow lines depend sensitively on the initial values.*

In a strange attractor, initial points that are arbitrarily close to each other are macroscopically separated by the flow after sufficiently long time intervals. The sensitive dependence on initial conditions can be visualized by a turbulent flow of a fluid; particles that are momentarily close to each other will not stay together. In terms of the ODE system (9.1), compare two trajectories, one starting from $\mathbf{z}$, the other starting from the perturbed point $\mathbf{z} + \epsilon\mathbf{d}$, $\mathbf{d}$ some arbitrary vector. No matter how small $|\epsilon|$ is chosen, the two trajectories depart exponentially from each other. This exponential divergence is a local property in that the distance between the trajectories remains bounded by the diameter of the attractor. The sensitive dependence on initial conditions can be formally defined; this is illustrated in Exercise 9.2. A famous example for sensitive dependence on initial data is the Lorenz equation (2.25). Apparently this was the first example for which the sensitivity was investigated. Figure 9.6 shows two trajectories with close initial vectors: There is only a one-percent deviation in $y_1(0)$; $y_2(0)$ and $y_3(0)$ are identical. We notice at around $t = 5$ a separation of the two trajectories. Already, for $t = 7$ the two trajectories are completely different.

The sensitive dependence on initial conditions has practical consequences. Because small deviations of initial conditions are always present, the position of a trajectory inside a strange attractor is not accurately predictable. The longer the time interval of a prediction, the less that can be said about the future state. This affects, for example, a long-term weather forecast. Inte-

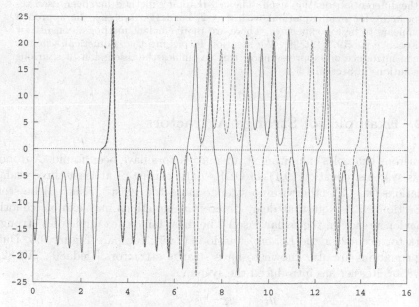

Fig. 9.6. Lorenz equation (2.25), $y_1(t)$ for $\lambda = R = 35$; two trajectories that are very close initially

grating an ODE system that exhibits a strange attractor is, to some extent, a nondeterministic problem; limits of computability are reached. Note again that the chaos represented by strange attractors is governed by deterministic laws, not by stochastic perturbations. Sensitive dependence on initial conditions can also be studied for maps; see Exercise 9.3.

Conservative systems, which include the dynamic systems of classical mechanics [AbM67], do not exhibit strange attractors. In conservative systems other kinds of irregular or chaotic motions may appear [Hel80], [LiL82]. Here we focus on dissipative systems. A general reference on the behavior of flows is [ArP82], see also [KaH95], [HaK03].

Simplifying one might say that strange attractors and chaotic attractors are more or less the same; we do not try to distinguish between them but use these two terms loosely and interchangeably. But the terms "strange" and "chaotic" are not equivalent. There are chaotic attractors that are not strange, and it is possible to have attractors which are strange but not chaotic [GrOPY84], [PrNR01]. Strange rather refers to the shape of the attractor, while the word chaos refers to the dynamics of the attractor's orbits.

Computational methods have been developed for constructing attractors. An early approach is the *cell-mapping method* [Hsu87], which subdivides a part of the phase space into cells and studies where each cell is mapped. This information is compiled into a probabilistic description of the mapping, setting up a matrix of transition probabilities. Then the analysis is based on the probabilistic representation of the flow rather than on the differential equation itself. The cell-mapping method has been used repeatedly for small equations with $n = 2$. For larger systems the approach appears to be too expensive. There are more efficient methods, consult for instance [DeH97], [Stä99]. We mention in passing that numerical schemes can introduce or suppress chaos; this resembles the discussion of spurious solutions in Section 1.5.2.

9.2 Examples of Strange Attractors

Numerous examples exhibiting strange attractors have been found. Autonomous systems, equation (9.1), of ODEs must consist of at least three scalar equations ($n \geq 3$), otherwise no strange attractors exist. (This will be seen in Section 9.7.3. Note that delay differential equations may exhibit chaotic behavior already in the scalar case.) The most famous example of a strange attractor occurs for the Lorenz equations (2.25); compare Figure 9.6. Duffing equations are also known to have strange attractors [Ueda80], [Tro82], [Sey85b]. Rössler has introduced the system

$$\dot{y}_1 = -y_2 - y_3\,,$$
$$\dot{y}_2 = y_1 + ay_2\,,$$
$$\dot{y}_3 = b + y_1 y_3 - cy_3\,,$$

which exhibits strange attractors, for instance, for $a = 0.55$, $b = 2$, $c = 4$ [Rös76]. Chua's attractor is defined by

$$\dot{y}_1 = \alpha[y_2 - g(y_1)]\,,$$
$$\dot{y}_2 = y_1 - y_2 + y_3\,,$$
$$\dot{y}_3 = -\beta y_2\,,$$

for a piecewise linear function

$$g(x) := \begin{cases} a_1 x - (a_0 - a_1) & \text{for } x < -1 \\ a_0 x & \text{for } |x| \leq 1 \\ a_1 x + (a_0 - a_1) & \text{for } x > 1\,. \end{cases}$$

with $\alpha = 9$, $\beta = \frac{100}{7}$, $a_0 = -\frac{1}{7}$, $a_1 = \frac{2}{7}$. Its "double scroll" has become popular [ChDK87], [PaC89]. Other equations that show chaos are forced van der Pol oscillators [GuH83], [MePL93]. In recent years strange attractors have been analyzed for so many ODE problems that it appears impossible to give a review.

Strange attractors have also been discussed for maps $\mathbf{y}^{(\nu+1)} = \mathbf{P}(\mathbf{y}^{(\nu)})$. Recall the important class of Poincaré maps, which are generated by flows (Section 7.3). For maps, the restriction to dimensions greater than two does not hold; simple examples of dimension one or two can be studied. We mention the logistic map ($n = 1$), $y^{(\nu+1)} = \lambda y^{(\nu)}(1 - y^{(\nu)})$. This equation serves

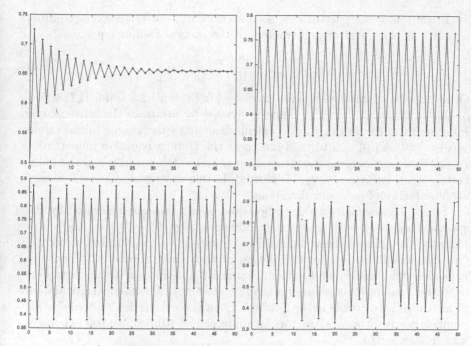

Fig. 9.7. Logistic map (1.9), 50 iterates starting with 0.5 for four parameter values: $\lambda = 0.29$ (top left), $\lambda = 0.31$ (top right), $\lambda = 0.35$ (bottom left), $\lambda = 0.36$ (bottom right)

as an elementary example to illustrate period doubling; compare equation (7.13) and Exercise 7.8. Recall that we encounter period doublings at $\lambda = 3$ and $\lambda \approx 3.45$. For example, for $\lambda = 3.5$, there is an attractor that consists of four points (Exercise 9.4). After further period doublings, chaos sets in for $\lambda \approx 3.57$. Figure 9.7 depicts iterations for λ-values of the four ranges stationary, simple period, double period, chaotic. Embedded in the chaotic range of λ-values there are several "windows" with periodic behavior. For example, period three sets in for $\lambda = 3.83$. The logistic map has been treated extensively in the literature [May76], [GrT77], [Fei80], [Ott81], [Schu84].

Another popular map is the Hénon map,

$$
\begin{aligned}
y_1^{(\nu+1)} &= 1 - a\big(y_1^{(\nu)}\big)^2 + y_2^{(\nu)}\,,\\
y_2^{(\nu+1)} &= b y_1^{(\nu)}\,,
\end{aligned}
\tag{9.3}
$$

with, for instance, $a = 1.4$, $b = 0.3$ [Hén76]. It is highly instructive to iterate this map (Exercises 9.5 and 9.6). Plots of the Hénon attractor and other strange attractors are included in [Rue80]. — The circle map equation (7.35) exhibits chaos for $K \geq 1$. — We remark in passing that orbits can be constructed that connect fixed points [BeHK04].

In the following part of this section we study an ODE example more closely, namely, the strange attractors of the specific Duffing equation

$$\ddot{u} + \frac{1}{25}\dot{u} - \frac{1}{5}u + \frac{8}{15}u^3 = \frac{2}{5}\cos\omega t. \qquad (9.4)$$

Other aspects of this equation are discussed in Sections 2.3 and 6.1. The flow of this equation in the $(u, \dot{u})$-plane is studied by means of the stroboscopic Poincaré map; see Section 7.7. To calculate an attractor, choose initial values $y_1(0) = u(0)$, $y_2(0) = \dot{u}(0)$, and integrate the Duffing equation numerically. At the time instances t_k, the integration is halted temporarily and the point $(u(t_k), \dot{u}(t_k)) = (y_1(t_k), y_2(t_k))$ is plotted. In this way we obtain the *Poincaré set*, a sequence of points in the plane.

Figure 9.8 assembles 8000 stroboscopic points that are generated by the specific trajectory, with $\omega = 0.2$ starting from $(y_1(0), y_2(0)) = (0, 0)$. Apparently, the points indicate lines that are nested in a complicated structure. One can imagine that after calculation of an infinite number of points the gaps along the lines would be filled. Figure 9.8 illustrates a strange attractor or, rather, an approximation of a strange attractor. A numerical calculation of a strange attractor is never finished. Even the calculation of "only" 8000 points is costly. Strange attractors for maps can be studied in a less expensive way (Exercise 9.6).

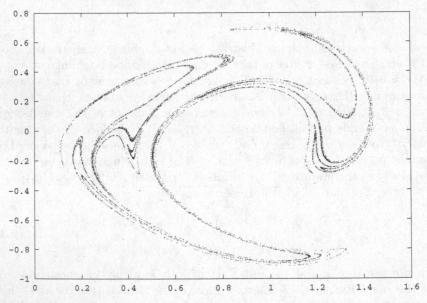

Fig. 9.8. Duffing equation (9.4), $(u, \dot{u})$-plane, Poincaré set of the stroboscopic map, $\omega = 0.2$, fractal dimension $D = 2.381$

The impression that the "lines" in Figure 9.8 give is misleading. A strong enlargement of any of the "lines" reveals that they consist of a bunch of "sublines." Each magnification of sublines leads to the same observation— lines are resolved into more lines. Theoretically, there is an infinite number of lines, comparable to a *Cantor set*. Notice that the strange attractor illustrated by its Poincaré set in Figure 9.8 is actually a surface-type set in the three-dimensional space (cf. Figure 7.26). From the line structure indicated in Figure 9.8 we infer that this strange attractor has a leaved structure, with leaves densely folded and nested in a highly complex way.

A *Cantor set* can be constructed as follows: Take a straight line that fills the unit interval and remove the inner third. Both the remaining short lines are treated in the same way, resulting in a total of four short intervals, each of length $\frac{1}{9}$ (cf. Figure 9.9). The process of removing thirds is repeated infinitely many times; the remaining structure converges to the Cantor set. This way of constructing the Cantor set can be generalized to higher dimensions. The removed parts need not be just a third; similar structures are obtained when other portions are removed. Looking at Figure 9.9 (second line, say) from a large enough distance, the small gaps are not visible and the structure resembles a collection of (two) dots. Now imagine that a cross section of one of the "lines" in Figure 9.8 is the fictitious "dot" of a Cantor set. This explains how, with increasing magnification, points are resolved into more points, lines into more lines, leaves into more leaves. Only the outer leaves are visible. The structure repeats itself on finer and finer scales. This *self-similarity* becomes apparent with the Cantor set. For instance, each half of the second line in Figure 9.9 is similar to the first stage of the construction. "Self-similar" patterns are frequently found in the context of chaos (Exercise 9.6b). Theoretically, the fineness of the self-similar pattern reaches zero, but in practice uncertainties present in any system (noise, discretization errors) truncate Cantor-like sets, whole families of trajectories become identified, and the number of points (lines, leaves) is finite.

Fig. 9.9. Constructing the middle-third Cantor set

The order in which stroboscopic points are generated by trajectories is unpredictable. Recalculating Figure 9.8 with another accuracy or another integrator leads to a different arrangement of points along the same "lines" of the attractor. This reflects the attribute of strange attractors that trajectories, once inside the attractor, diverge from each other.

Observations on chaotic phenomena have been reported repeatedly. Some of the numerous examples are investigations on nerve membranes [ChR85], brain activity [BaD86], biophysics (glycolysis) [MaKH85], Rayleigh–Bénard cells [BeDMP80], [GiMP81], driven semiconductor oscillations [TePJ82], cardiac cells [GuGS81], and ecological systems [ScK85]. Poincaré maps obtained experimentally are illustrated in [Moon80]. Classical examples are reported in [Eck81], [Ott81], [Schu84] and in the reprint selection [Cvi84]. Chaotic dynamics of suspended cables were studied in [BeR90]. For a review on chaos in mechanical systems see [SeN95]. Spatiotemporal patterns have been investigated by means of cellular dynamical systems [AbCD91]. Chaotic trajectories frequently have been observed for periodically driven oscillators such as the Duffing equation (9.4) and the Van der Pol equation (7.34). Existence of chaotic solutions of systems with almost periodic forcing has been proved by [Sch86].

9.3   Routes to Chaos

Because our focus is on parameter-dependent equations, it is natural to ask how chaos depends on λ. In particular, we like to know for which values of λ one may expect chaotic behavior. As motivation, imagine that one wants to predict when a hitherto laminar flow becomes turbulent. Both theoretical and experimental investigations have revealed that there is no unique way in which chaos arises. Some of the different possible scenarios are discussed in this section. As we shall see, bifurcations play a crucial role in initiating chaos.

9.3.1   Route via Torus Bifurcation

An early conjecture attributed to Landau (1944, [LaL59]) postulated that chaos is caused by an infinite sequence of secondary Hopf bifurcations. That is, after a first Hopf bifurcation from equilibrium to limit cycle, a sequence of bifurcations into tori arises, each bifurcation adding another fundamental frequency. It suggests that, as more and more frequencies occur, the motion gets more and more turbulent. However, the Landau scenario has not been backed by experiments. Theoretically it was shown that the infinite sequence of Hopf bifurcations is not generic.

Another route to chaos based on torus bifurcation was proposed by Newhouse, Ruelle, and Takens [NeRT78]; this scenario is backed by both theoretical investigations and experimental evidence. According to this conjecture, only two bifurcations are precursive to chaos—one standard Hopf bifurcation and one subsequent bifurcation to a two-frequency torus. Then chaos is more likely than bifurcation to a 3-torus. When the third frequency is about to appear, simultaneously a strange attractor arises, since a three-frequency flow is destroyed by certain small perturbations. Bifurcation into a 3-torus is not generic [IoL80]. The Ruelle–Takens–Newhouse route to chaos has been observed in Bénard-type experiments (see Section 2.8) and in the Taylor experiment [SwG78]. The scenario is stable with respect to small external noise [CrH80], [Eck81].

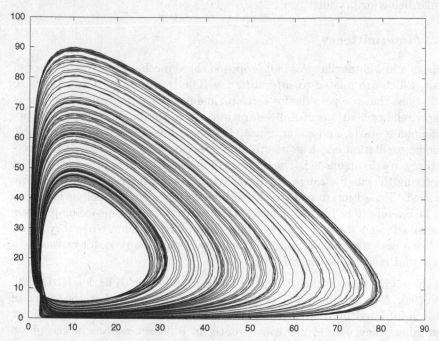

Fig. 9.10. Example 7.8, chaotic orbit projected onto the (y_1, y_2)-plane, for $\lambda = 10$. This figure extends Figure 7.18

9.3.2 Period-Doubling Route

As noted in Section 7.4.2, there are models exhibiting a sequence of period-doubling bifurcations. In case the sequence is infinite, the periods quickly approach infinity. The "speed" of how rapidly chaos is reached (in terms of λ) is given by the scaling law in equation (7.14). For instance, in the logistic map in equation (1.9), the entire cascade of period doublings takes place for $3 \leq \lambda < 3.5699$. The latter value is the accumulation point of the period-doubling sequence [Fei80], [GrT77] (Exercise 9.7). Compare also Figure 9.7. An ODE example with the period-doubling route to chaos is Example 7.8. The sequence of period doublings that starts in Figure 7.18 ends in a chaotic regime, see Figure 9.10.

Noise affects the sequence of period doublings in that higher-order subharmonics are truncated. The period-doubling route to chaos has been supported by many experiments, including chemical reactions [JoR83], [KaRV81],

[SmKLB83], driven nonlinear oscillations [TePJ82], and Rayleigh–Bénard cells [GiMP81]. We note in passing that inside the chaotic regime, sequences of strange attractors were discovered that resemble a sequence of period doublings [Cru80], [Far81]. Frequency divisions other than doubling were reported in [LiM82]. Chaos may be interrupted by "windows" of λ-values, for which regular behavior prevails.

9.3.3 Intermittency

Pomeau and Manneville [PoM80] proposed three mechanisms for the onset of chaos, which are related to *intermittency*. The term intermittency refers to oscillations that are periodic for certain time intervals, interrupted by bursts of aperiodic oscillations of finite duration. After the burst stops, a new periodic phase starts, and so on. Since there are three ways in which a stable periodic oscillation can lose its stability (Section 7.4), three kinds of intermittency are distinguished. The number of aperiodic bursts varies with λ; for λ entering the chaotic range, the duration of chaotic intervals increases. The so-called "Type-I intermittency" is explained by a scenario such as is described in Exercise 9.8. In this scenario, the iteration of a Poincaré map enters a "channel" and stays in it for a while. The corresponding trajectory behaves like a periodic solution. Intermittency has been observed, for instance, in Rayleigh–Bénard cells [BeDMP80].

The above three routes to chaos are not the only scenarios [GrOY83], [KaY79a]. All the different hypotheses have in common that the onset of chaos is initiated by bifurcations. The different scenarios do not contradict each other; they may evolve concurrently in different regions of the phase space [Eck81]. For the circle map, chaos is initiated by overlapping phase-locked intervals. Other mechanisms of how chaos arises exist for dry-friction models [diBKN03].

9.4 Phase Space Construction

In Figure 9.8 a set of many points $(u(t_j), \dot{u}(t_j))$ in phase space is plotted. As reported in Section 9.2, these points were calculated by integrating a known differential equation. The two-dimensional illustration is a natural consequence of the order two of the underlying second-order differential equation. Now we address the frequent case when no equation is known. Instead, a time series of observations is given, which may consist of scalar measurements. In this situation a phase space with illustrations similar to Figure 9.8 can be constructed, but finding a proper "dimension" is more difficult.

Suppose we are given a sequence of scalar measurements

$$y(t_1),\ y(t_1 + \Delta t),\ y(t_1 + 2\Delta t)\,,\ y(t_1 + 3\Delta t)\,,\ \ldots,$$

which may be obtained by sampling a continuous output at time instances $t_j = t_1 + (j - 1)\Delta t$. (To make sure to be inside the attractor, it may be necessary to discard initial transients.) In case the recorded measurements are vectors, pick an arbitrary component. The time increment or time lag Δt is often inherently given by the time series; otherwise, it can be chosen rather freely (see below) but must be kept constant. A two-dimensional representation of an underlying attractor is obtained by plotting the pairs of consecutive numbers

$$(y(t_j), y(t_{j+1})), \quad j = 1, 2, \dots ,$$

which can be motivated by the relation between $y(t_{j+1})$ and $\dot{y}(t_j)$ via the difference quotient. This is a special case of a general result due to [PaCFS80], [Tak81], which states that m-vectors $\mathbf{z}(t)$, formed by m consecutive values

$$y(t), \ y(t + \Delta t), \ y(t + 2\Delta t), \dots, y(t + (m - 1)\Delta t),$$

"embed" the underlying flow. Here the values of t are the discrete times t_j of the time measurements. That is, a set $\mathcal{P}_m$ of m-vectors $\mathbf{z}(t_1), \mathbf{z}(t_2), \dots$ is constructed with

$$\mathbf{z}(t_j) := (y(t_j), y(t_{j+1}), \dots, y(t_{j+m-1})), \tag{9.5}$$

$j = 1, 2, \dots$, which forms points in an m-dimensional phase space. These points can be used in the same way for further analysis as points calculated from a known set of equations.

The proper dimension m is called the *embedding dimension*. It is related to the dimension of the flow, which is usually unknown when the time series is derived from measurements. The determination of the embedding dimension m proceeds in an experimental way: One starts with a small integer m and calculates the corresponding points $\mathbf{z}(t_j)$, $j = 1, 2, \dots$ and the set $\mathcal{P}_m$ in m-dimensional space. Then one proceeds with the next larger integer m and compares the new set of $\mathbf{z}$ points with the previous set. For "small" m, the transition $m \to m+1$ leads to changes in the set $\mathcal{P}_m$ (distribution of $\mathbf{z}$ points); for "large" m this transition leaves $\mathcal{P}_m$ qualitatively unchanged. The "true" embedding dimension m is the smallest integer such that the set $\mathcal{P}_m$ of $\mathbf{z}$ points "does not change" when m is further increased. Then one concludes that the attractor is correctly represented by those points.

There remains the difficulty of how to judge that the set of $\mathbf{z}$ points "does not change." A visual inspection is somewhat vague. The set $\mathcal{P}_m$ needs to be characterized by a quantitative instrument. Fractal dimensions D do the job (next section). One calculates these characteristic numbers and increases m as long as D still changes.

The above method of embedding observed data into an m-dimensional space has been applied even in the case when equations are given. In this situation the time lag Δt can be varied freely, and one can study the effect of a proper choice of Δt. If Δt is chosen too small, then $z_1 \approx z_2 \approx \dots \approx z_m$

and noise will hide the attractor. (In case of time series, remove this diffi-culty by taking a multiple of Δt.) If Δt is chosen too large, then $\mathbf{z}(t_{j+1})$ will not be correlated to $\mathbf{z}(t_j)$. The time scale Δt does affect the choice of coor-dinates. Procedures for choosing Δt and m have been discussed frequently; see [BrK86], [Fra89]. As a rule of thumb, choose Δt close to the shortest period of oscillations that can be recognized in the data.

9.5   Fractal Dimensions

Frequently, one has calculated or measured some oscillation $y(t)$ and wants to decide whether the oscillation is periodic or aperiodic. A seemingly irregular oscillation might be periodic, with a period so large that it can hardly be assessed. For example, the Duffing equation in equation (9.4) possesses for $\omega = 0.38$ a subharmonic orbit of order 11, which has the period $T = 181.88$ and looks irregular. By merely inspecting a graphical output of a time history, it is often difficult to decide whether an oscillation is chaotic. Hence, we require instruments that facilitate a decision. Among the tools that are helpful in this context are fractal dimensions (this section) and Liapunov exponents (Section 9.7).

Throughout this book, we so far have encountered only integer dimensi-ons; these integer dimensions are also called *topological dimension*. In Eucli-dian space, the dimension is the minimum number of parameters or coordi-nates needed to specify a point. The hosting space $\mathbb{R}^n$ in which flows and attractors of (9.1) live, clearly has the integer dimension n. The attractors typically are subsets with dimensions smaller than n. The structure of Eucli-dian space with its topological dimension can be assigned to standard regular attractors of ODEs, and the attractors can be distinguished accordingly:

> stationary point: dimension zero;
> limit cycle: dimension one; and
> two-frequency torus $(n = 3)$:   dimension two.

Hence, following the route to chaos that is based on primary and secondary Hopf bifurcation, we encounter an increasing of the attractor's dimension,

$$0 \to 1 \to 2 \to ?$$

The question arises, What dimension has a resulting strange attractor? Since the n-dimensional volume of the strange attractor is zero, its dimension must be smaller than the dimension n of the state space. Consider, for instance, the Duffing equation (9.4) or the Lorenz equation (2.25), both of which live in the three-dimensional state space. Here a zero volume of the strange attrac-tor means that its dimension must be smaller than three. Hence, we expect attractors with noninteger dimension. Sets with noninteger dimension are called *fractals*.

Noninteger or *fractal dimensions* require refined concepts of dimension [FaOY83], [Man83], [Ott81]. The aim is to measure the dimension of a set $\mathcal{P}$ in $\mathbb{R}^n$. $\mathcal{P}$ may be an attractor of a flow, or any other set. A general idea is to introduce a scale ϵ and to count the points in $\mathcal{P}$ that meet a specified requirement of size ϵ, where properties of smaller size than ϵ are ignored. Let $N(\epsilon)$ denote this number of points. Then one investigates whether there is a power law

$$N(\epsilon) \propto \epsilon^{-D}.$$

That means, $N(\epsilon)$ is inversely proportional to ϵ^D, and D is taken as fractal dimension. In other words, for small $\epsilon > 0$, $N(\epsilon)$ behaves like

$$N(\epsilon) = c \left(\frac{1}{\epsilon}\right)^D$$

for a constant c.

A simple example of a fractal dimension is defined as follows: Suppose that $N(\epsilon)$ is the minimum number of small n-dimensional cubes of side length ϵ that are needed to cover $\mathcal{P}$. That is, all the points $\mathbf{z} \in \mathcal{P}$ are inside the union of the $N(\epsilon)$ cubes. Then a fractal dimension of this region is defined by

$$D_{\text{cap}} := \lim_{\epsilon \to 0} \frac{\ln(N(\epsilon))}{-\ln \epsilon}, \qquad (9.6)$$

provided the limit exists. (ln denotes the natural logarithm.) This dimension is called *capacity dimension* or *box dimension*. The computation of the capacity is cumbersome, especially if $n > 2$, and convergence may be slow. If the region is covered by more general n-dimensional volumes (including cubes) then we arrive at the *Hausdorff dimension*. We omit a formal presentation of the Hausdorff dimension because it mostly takes the same values as the capacity dimension and is even harder to calculate. To illustrate the concept of fractal dimension, we discuss two simple examples; a third example is given in Exercise 9.9.

Example 9.4   Unit Square
To cover the unit square, $n = 2$, we take small squares with side length ϵ. Then exactly

$$N(\epsilon) = \frac{1}{\epsilon^2}$$

of the small squares are required to cover the unit square. The dimension is now readily calculated,

$$D_{\text{cap}} = \lim_{\epsilon \to 0} \frac{-2\ln \epsilon}{-\ln \epsilon} = 2.$$

Here the capacity dimension specializes to the standard integer dimension, which corresponds to our intuition.   □

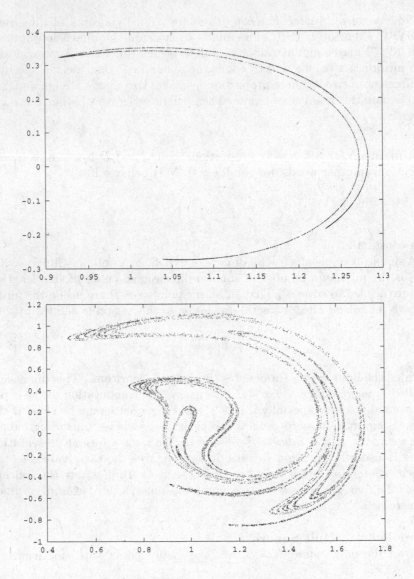

Fig. 9.11. Poincaré sets of the Duffing equation (9.4). Top: $\omega = 0.04$, $D = 2.167$; bottom: $\omega = 0.32$, $D = 2.526$. Compare with Figure 9.8.

Example 9.5   Middle-Third Cantor Set

In Figure 9.9, second line, $N(\epsilon) = 4$ short lines are required to cover the structure in $\mathbb{R}^1$, each of length $\epsilon = \frac{1}{9}$. In the jth stage of the set that converges to the Cantor set,

$$N(\epsilon) = 2^j \text{ with } \epsilon = 3^{-j}$$

of these short lines are needed. Hence,

$$D_{\text{cap}} = \lim_{j \to \infty} \frac{\ln 2^j}{\ln 3^j} = \frac{\ln 2}{\ln 3} = 0.6309 \ldots .$$

Since the length of the jth stage of the construction is

$$\epsilon N(\epsilon) = \left(\frac{2}{3}\right)^j$$

and each subsequent stage has smaller length, the entire length of the Cantor set is $\lim_{j \to \infty} \epsilon N(\epsilon) = 0$. Hence the topological dimension of the Cantor set is 0. The noninteger dimension D_{cap} with $0 < D_{\text{cap}} < 1$ reflects that the Cantor set is "thicker" than a point, but its one-dimensional volume (length) is zero. □

Fractal dimensions are characteristic for strange attractors and basin boundaries. Some examples are listed in Table 9.1 below (values from the literature; for example, from [TrKS85]). The first column of Table 9.1 displays the integer dimension of the hosting state space; this number n is to be related to the fractal dimension D of the strange attractor. For instance, the Hénon attractor (Exercise 9.6) has a line structure, with lines densely folded in a Cantor set structure. The richness of the line structure leads to a dimension $D > 1$ ($D = 1$ for an "ordinary" line). On the other hand, the lines are sparse in the plane, thus $D < 2$; a nonzero volume (here nonzero area) would mean $D = 2$. Similarly, the Duffing strange attractor with its leaved structure appears "thicker" than an ordinary surface, but still has volume zero, hence, $2 < D < 3$. In higher-dimensional spaces, strange attractors are possible with higher dimensionality [Ott81]. Table 9.1 shows for the Duffing equation (9.4) how the complexity of a strange attractor can be measured by fractal dimensions. The related Poincaré sections are shown in Figures 9.8 and 9.11. Further figures that illustrate the growing complexity of the Duffing attractor for increasing ω can be found in [Sey85b]. Bear in mind that the actual attractor has a leaved structure; imagine Figure 9.11 illustrating cuts through growing lettuce plants.

TABLE 9.1. Fractal dimensions of three examples.

n	Example	D
1	Cantor set	$D_{\text{cap}} \approx 0.6309$
2	Hénon map, eq. (9.3)	$D_{\text{cap}} \approx 1.28$
		$D_{\text{I}} \approx 1.23$
		$D_{\text{corr}} \approx 1.21$
3	Duffing eq. (9.4)	
	$\omega = 0.04$ (Fig. 9.11)	$D_{\text{L}} \approx 2.17$
	$\omega = 0.2$ (Fig. 9.8)	$D_{\text{L}} \approx 2.38$
	$\omega = 0.32$ (Fig. 9.11)	$D_{\text{L}} \approx 2.52$

The capacity dimension and the Hausdorff dimension are just two examples of fractal dimensions. There are further concepts of fractal dimensions. An infinite sequence of dimensions can be defined, all of them being less or equal to D_{cap} [GrP83], [Hal86], [HeP83]. For instance, the *information dimension* is defined by

$$D_{\text{I}} := \lim_{\epsilon \to 0} \left[\sum_{i=0}^{N(\epsilon)} p_i \frac{\ln p_i}{\ln \epsilon} \right]. \tag{9.7}$$

The notion of ϵ and $N(\epsilon)$ is the same as for the capacity dimension in (9.6). p_i is the probability that a point of the attractor lies within the ith cell. If K points of an attractor are calculated and if for a partition of the phase space into cells K_i points are in the ith cell, then

$$p_i = \frac{K_i}{K}.$$

The information dimension D_{I} accounts how often a cell is visited by a trajectory.

The *correlation dimension* D_{corr} takes into account the density of the points on $\mathcal{P}$. Take K points $\mathbf{z}^{(1)}, \ldots, \mathbf{z}^{(K)}$ in $\mathcal{P}$ and study their density via their distances. The computation is based on the correlation integral [GrP83]

$$C(\epsilon) := \lim_{K \to \infty} \frac{1}{K^2} \sum_{i,j=1}^{K} H(\epsilon - \|\mathbf{z}^{(i)} - \mathbf{z}^{(j)}\|), \tag{9.8a}$$

where H is the Heaviside step function [$H(x) = 1$ for $x > 0$, and zero elsewhere]. That is, the sum counts the number of pairs $(\mathbf{z}^{(i)}, \mathbf{z}^{(j)})$ with Euclidian distance smaller than ϵ. In contrast to $N(\epsilon)$ above, $C(\epsilon)$ is proportional to ϵ, and hence

$$D_{\text{corr}} = \lim_{\epsilon \to 0} \frac{\ln C(\epsilon)}{\ln \epsilon}. \tag{9.8b}$$

A log-log plot of $\ln C(\epsilon)$ versus $\ln \epsilon$, or a least-squares line help disclosing the slope D_{corr}. In practice, very large and very small values of ϵ will not work and should not be used for the fit. The correlation dimension is widely used since it is relatively easy to calculate. — Another fractal dimension that can be calculated with relative ease is the *Liapunov dimension* D_{L} of (9.20); this will be introduced in Section 9.7.

A general family of dimensions is defined by

$$D_q := \frac{1}{q-1} \cdot \lim_{\epsilon \to 0} \frac{\ln \sum_{i=0}^{N(\epsilon)} p_i^q}{\ln \epsilon}$$

for any $q \geq 0$. It specializes for $q = 0$ to the capacity, $D_0 = D_{\text{cap}}$. And taking the limit $q \to 1$, the information dimension results, $D_1 = D_{\text{I}}$. (The reader may check!) The value D_2 is the correlation dimension. It has been shown that this family of dimensions is ordered in the sense

$$D_q \leq D_p \text{ for } q > p$$

(see [HeP83]). That is, D_{cap} is upper bound for the entire family of dimensions D_q, and

$$D_{\text{corr}} \leq D_I \leq D_{\text{cap}}.$$

Differences between different fractal dimensions are often small; for a large class of attractors, even $D_q = D_p$ holds.

Algorithms for calculating fractal dimensions, and practical hints can be found in [PaC89]. Note that for a data set $\mathcal{P}$ consisting of a finite number of M discrete points, there is an $\epsilon_0 > 0$ such that $N(\epsilon) = M$ for all $\epsilon < \epsilon_0$. Then a limit $\epsilon \to 0$ can be meaningless, and (9.6), for example, should be replaced by $-\ln N(\epsilon_0)/\ln \epsilon_0$. For a noisy and truncated data set one should not expect good accuracy.

Exciting applications of fractal dimensions include time series obtained from objects from an electroencephalogram. This reveals that brain activities are characterized by strange attractors [BaD86]; jumps in the dimensionality are observed when an epileptic seizure occurs. For correlation dimension in brain electric activity, see also [OsL07]. Correlation dimension of heartbeat is estimated in [NiMA06].

9.6 Control of Chaos

Is it possible to stabilize unstable states? This can be achieved easily as long as the unstable states are embedded in a chaotic regime. Then the perspective of chaos as a state of utmost flexibility suggests that the stabilization may be possible even with little impact. Such a stabilization endeavor is called *control of chaos*. One of the most popular approaches of chaos control is the "OGY-method," named after the authors of [OttGY90]. Here we explain the OGY-method for a representative special case, in which a saddle is stabilized.

Consider a parameter-dependent mapping

$$\mathbf{q}^{(\nu+1)} = \mathbf{P}(\mathbf{q}^{(\nu)}, \lambda), \tag{9.9}$$

for a specific parameter value $\lambda = \lambda^*$. This mapping may result from a Poincaré map of an ODE system, and is supposed to represent chaotic behavior. Assume a fixed point $\mathbf{q}^*$,

$$\mathbf{q}^* = \mathbf{P}(\mathbf{q}^*, \lambda^*),$$

which is unstable but possesses a stable manifold $\mathcal{M}_s$. Accordingly the matrix

$$\frac{\partial \mathbf{P}(\mathbf{q}^*, \lambda^*)}{\partial \mathbf{q}}$$

has eigenvalues μ with both $|\mu| > 1$ and $|\mu| < 1$, compare Section 1.4.2. In a two-dimensional illustration this amounts to a saddle at $\mathbf{q}^*$ for the specific parameter value λ^*, see Figure 9.12(a).

If the parameter is varied from the value λ^* to another value λ^{**}, then in general the fixed point will drift. Figure 9.12(b) illustrates a possible path of

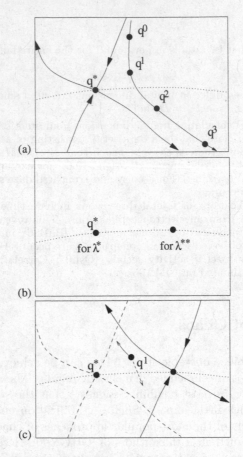

Fig. 9.12. Scenario of the OGY-method

the fixed point as parameterized by λ. Assume further that for a parameter interval $\lambda_{min} \leq \lambda \leq \lambda_{max}$ we move in an area of the space where the dynamics is qualitatively the same, namely, chaos embedding a saddle-type fixed point (Figure 9.12(c)). The chaotic dynamics (9.9) maps any starting vector $\mathbf{q}^{(0)}$ eventually into a neighborhood of $\mathbf{q}^*$ (for λ^*). In practice this might require a long initial precursive iteration. Here we discard this initial phase and, without loss of generality, think of sitting on a point $\mathbf{q}^{(1)}$ close to $\mathbf{q}^*$ (for λ^*). This sets the stage for the OGY-method.

The idea of OGY is to shift the parameter λ in such a way that the next iterate $\mathbf{q}^{(2)}$ will not drift away from $\mathbf{q}^*$ (as it would do for the value λ^*) but rather is "kicked" back towards the stable manifold $\mathcal{M}_s$ of $\mathbf{q}^*$. This strategy is illustrated in Figure 9.12(c). The artificial perturbation by changing λ amounts to a switch from the "dashed" dynamics belonging to λ^* to the "solid" dynamics belonging to another suitably chosen value of λ. Under the new dynamics, $\mathbf{q}^{(2)}$ will be pushed to stay close to $\mathbf{q}^*$. Of course, after each iteration $\nu \longrightarrow \nu + 1$, a new perturbed parameter λ must be chosen. The procedure reminds of a ping-pong game where the paddle is moved in the right position to bounce the iterate back to the target regime $\mathbf{q}^*$. It suffices to keep the current iterate close to the stable manifold. Then the iterates $\mathbf{q}^{(\nu)}$ are trapped in a neighborhood of $\mathbf{q}^*$.

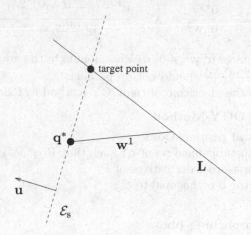

Fig. 9.13. Geometry of the OGY-method

How is the perturbed λ adjusted properly? Let us enter the analysis of how λ is calculated. Note that $\mathbf{P}(\mathbf{q}^{(1)}, \lambda)$ parameterized by λ is a curve. Start from the Taylor expansion of $\mathbf{P}(\mathbf{q}^{(1)}, \lambda)$ around $(\mathbf{q}^*, \lambda^*)$ for arbitrary λ,

$$\mathbf{q}^* + \underbrace{\frac{\partial \mathbf{P}(\mathbf{q}^*, \lambda^*)}{\partial \mathbf{q}}(\mathbf{q}^{(1)} - \mathbf{q}^*)}_{=:\mathbf{w}^1} + \underbrace{\frac{\partial \mathbf{P}(\mathbf{q}^*, \lambda^*)}{\partial \lambda}(\lambda - \lambda^*)}_{=:\mathbf{w}^2} + t.h.o. \qquad (9.10)$$

For λ^* this would give the unperturbed $\mathbf{q}^{(2)}$. The linearized version of (9.10) is parameterized by λ,

$$\mathbf{L}(\lambda) := \mathbf{q}^* + \mathbf{w}^1 + \mathbf{w}^2(\lambda - \lambda^*).$$

This defines a straight line $\mathbf{L}$; $\mathbf{w}^1$ and $\mathbf{w}^2$ are vectors. This line is illustrated in Figure 9.13; it approximates the curve $\mathbf{P}(\mathbf{q}^{(1)}, \lambda)$. If all assumptions hold, then the line $\mathbf{L}(\lambda)$ intersects the linearized stable manifold $\mathcal{E}_\mathrm{s}$, which is tangent to $\mathcal{M}_\mathrm{s}$ in $\mathbf{q}^*$. Let $\mathbf{u}$ be a normal vector to $\mathcal{E}_\mathrm{s}$. In this linearization, λ is optimal when $\mathcal{E}_\mathrm{s}$ and $\mathbf{L}(\lambda)$ intersect. So require orthogonality

$$\mathbf{L}(\lambda) - \mathbf{q}^* \perp \mathbf{u}. \qquad (9.11)$$

$\mathbf{L}(\lambda) - \mathbf{q}^*$ is the ray from $\mathbf{q}^*$ that follows a point on the straight line $\mathbf{L}(\lambda)$; the special ray for λ^* is $\mathbf{w}^1$, shown in the figure. The requirement $\mathbf{L}(\lambda) - \mathbf{q}^* \perp \mathbf{u}$ is the linearized version of the aim that the curve $\mathbf{P}(\mathbf{q}^{(1)}, \lambda)$ intersects $\mathcal{M}_\mathrm{s}$, compare Figure 9.12(c). From the requirement (9.11) the proper λ is given by

$$\mathbf{u}^{tr}(\mathbf{w}^1 + \mathbf{w}^2(\lambda - \lambda^*)) = \mathbf{0},$$

which leads to

$$\lambda = \lambda^* - \frac{\mathbf{u}^{tr}\mathbf{w}^1}{\mathbf{u}^{tr}\mathbf{w}^2} = \lambda^* - \frac{\mathbf{u}^{tr}\frac{\partial \mathbf{P}(\mathbf{q}^*,\lambda^*)}{\partial \mathbf{q}}(\mathbf{q}^{(1)} - \mathbf{q}^*)}{\mathbf{u}^{tr}\frac{\partial \mathbf{P}(\mathbf{q}^*,\lambda^*)}{\partial \lambda}}.$$

Of course, this requires $\mathbf{u}^{tr}\mathbf{w}^2 \neq 0$, which amounts to the additional assumption that $\mathcal{E}_s$ and $\frac{\partial \mathbf{P}(\mathbf{q}^*,\lambda^*)}{\partial \lambda}$ should not be parallel.

We summarize the algorithm of the OGY-method as follows:

Algorithm 9.6   OGY-Method

phase 1: initial preparations
calculate the unstable fixed point $\mathbf{q}^*$ such that $\mathbf{P}(\mathbf{q}^*, \lambda^*) = \mathbf{q}^*$;
calculate the first-order derivatives of $\mathbf{P}$;
calculate a vector $\mathbf{u}$ orthogonal to $\mathcal{E}_s$;
choose $\mathbf{q}^{(0)}$.

phase 2: approaching phase
iterate the unperturbed system $\mathbf{q}^{(\nu+1)} = \mathbf{P}(\mathbf{q}^{(\nu)}, \lambda^*)$ for $\nu = 0, 1, 2, \ldots$
until a $\mathbf{q}^{(\nu)}$ close to $\mathbf{q}^*$ is reached.

phase 3: stabilization
iterate $\mathbf{q}^{(\nu+1)} = \mathbf{P}(\mathbf{q}^{(\nu)}, \lambda)$ where λ is adapted by

$$\lambda := \lambda^* - \frac{\mathbf{u}^{tr}\frac{\partial \mathbf{P}(\mathbf{q}^*,\lambda^*)}{\partial \mathbf{q}}(\mathbf{q}^{(\nu)} - \mathbf{q}^*)}{\mathbf{u}^{tr}\frac{\partial \mathbf{P}(\mathbf{q}^*,\lambda^*)}{\partial \lambda}} \tag{9.12}$$

Practical remarks

Phase 2 can last long. An efficient variant based on interpolation is presented in [Stä94]. In this method, a data set of $\mathbf{q}$-points is created, and the information is used in a learning phase to reach the target $\mathbf{q}$ quickly. The procedure enables an easy transition to another unstable fixed point.

In case $\mathbf{P}$ is a Poincaré map and a UPO is to be stabilized, it may not be sufficient to apply only one correction per cycle. This happens when the instability leads to a rapid drift-off of the trajectory; then $\mathbf{q}^{(\nu+1)}$ is not close enough to $\mathbf{q}^*$, or even fails to exist.

The flexibility based on many underlying unstable orbits, has made chaotic systems ideal candidates for control. These *applications of chaos* are most promising. There are also other methods for chaos control in addition to the OGY-method. To stabilize an UPO, Pyragas' method corrects continuously [Pyr92]. Instead of integrating $\dot{\mathbf{y}} = \mathbf{f}(t, \mathbf{y})$, the system

$$\dot{\mathbf{y}} = \mathbf{f}(t, \mathbf{y}) - \mathbf{G}(\mathbf{y} - \overline{\mathbf{y}})$$

is integrated, where $\bar{\mathbf{y}}$ is the UPO that is to be stabilized, and $\mathbf{G}$ is the "gain-matrix." It is sufficient to use a diagonal matrix for $\mathbf{G}$, as long as its elements are large enough. For specific methods of chaos control, see for instance [DrN92], [GaSDW92], [Shi93], [MaCV03], [ChZXW04], [WaJ04].

Chaos control is closely related to *anticontrol* of chaos. This means the artificial chaotification of systems to create beneficial flexibility. For these kinds of methods, see for instance [GeCC06], [MoBC05], [ShC05], [WaC08]. The flexibility of chaotic systems facilitates *synchronization* of two systems [PiRK01], [PeC90], [PaC92]. This has many applications [Kur00], and has also been used for encoding and decoding purposes.

9.7 Liapunov Exponents

An important tool for analyzing attractors is provided by the concept of Liapunov exponents. For motivation we begin with scalar maps.

9.7.1 Liapunov Exponents for Maps

As pointed out in Section 9.1, a strange attractor can be seen as the result of an infinite number of stretchings in one direction and contractions in another direction, combined with foldings. Let us look at how to measure a stretching or contraction in case of a one-dimensional map

$$y^{(\nu+1)} = P(y^{(\nu)}).$$

y is mapped to $P(y)$, and a neighboring $y+\delta$ is mapped to the value $P(y+\delta)$, with some distance $\bar{\delta}$ to $P(y)$. The factor $\exp(L)$ measures an exponential stretching of the distance δ to the distance $\bar{\delta}$,

$$|\bar{\delta}| = |\delta| e^L.$$

Because, in general, the stretching varies with y, the exponent L depends on where the distance δ is measured on the y-axis, $L = L(y)$. We generalize the statement

$$|P(y + \delta) - P(y)| = |\delta| e^L$$

to the situation for the map P^N, which is the composed map of P iterated N times. This gives L a slightly changed meaning:

$$|P^N(y + \delta) - P^N(y)| = |\delta| e^{NL},$$

which can be written

$$L = \frac{1}{N} \ln |(P^N(y + \delta) - P^N(y))/\delta|.$$

This attempt to define a stretching measure L still suffers from being dependent on N and δ. Hence, we consider the stretching of an infinitesimally small distance $\delta \to 0$ after an infinite number of iterations $N \to \infty$. This leads to the definition

$$L(y) = \lim_{N \to \infty} \lim_{\delta \to 0} \frac{1}{N} \ln |(P^N(y + \delta) - P^N(y))/\delta|, \qquad (9.13)$$

which, by definition of the derivative via the limit of difference quotients, is equivalent to

$$L(y) = \lim_{N \to \infty} \frac{1}{N} \ln \left| \frac{\mathrm{d}P^N(y)}{\mathrm{d}y} \right|.$$

Applying the chain rule

$$\frac{\mathrm{d}P^N(y)}{\mathrm{d}y} = \frac{\mathrm{d}P(y^{(N-1)})}{\mathrm{d}y} \cdot \ldots \cdot \frac{\mathrm{d}P(y)}{\mathrm{d}y},$$

and taking the logarithm of this product, yields a sum and the expression

$$L(y) = \lim_{N \to \infty} \frac{1}{N} \sum_{\nu=0}^{N-1} \ln |P'(y^{(\nu)})|. \qquad (9.14)$$

This number is called the *Liapunov exponent* of a scalar map; it serves as a measure for exponential divergence ($L > 0$) or contraction ($L < 0$). The exponential divergence is local. Note that some authors prefer to define Liapunov exponents using $\log_2$ rather than $\ln$.

9.7.2 Liapunov Exponents for ODEs

For systems of ODEs (n scalar equations), the derivation of a Liapunov exponent is similar. Imagine a small test volume around the starting point of a flow line—for instance, take a ball with the radius ρ. The flow deforms the ball toward an ellipsoid-like object; for dissipative systems the volume eventually shrinks to zero. We want to know how the principal axes $\rho_k(t)$ ($k = 1, \ldots, n$) evolve with time. In analogy to equation (9.13) ($t \leftrightarrow N$, $\rho \leftrightarrow \delta$), a Liapunov exponent can be defined by

$$L_k = \lim_{t \to \infty} \lim_{\rho \to 0} \frac{1}{t} \ln |\rho_k(t)/\rho|.$$

For $\rho \to 0$ the infinitesimal deformation of the ball is described by the linearization of the flow. For computational purposes we next develop an equivalent definition for L_k. For the further discussion we repeat analytical background from Section 7.2.

The linearization of equation (9.1) is the variational equation

$$\dot{\mathbf{h}} = \mathbf{f_y}(\mathbf{y}, \lambda)\,\mathbf{h}.$$

Here the Jacobian $\mathbf{f_y}$ is evaluated along the particular flow line $\mathbf{y}$ that is investigated. For any initial vector $\mathbf{h}(0)$, the evolution of the solution $\mathbf{h}(t)$

can be expressed by means of the fundamental solution matrix $\Phi(t)$ or $\Phi(t; \mathbf{y})$ from equation (7.9),

$$\mathbf{h}(t) = \Phi(t)\,\mathbf{h}(0)\,.$$

Accordingly, the matrix function $\Phi(t)$ describes the linearized flow. Choosing $\mathbf{h}(0)$ as a suitable axis of the initial ball, the matrix $\Phi(t)$ will tell how the relevant principal axis of the ellipsoid will evolve. Let the eigenvalues of $\Phi(t)$ be $\mu_1(t), \ldots, \mu_n(t)$. To first order, these eigenvalues represent the stretching of the principal axes; a special case is handled in Exercise 9.10. This leads to a definition of Liapunov exponents L_k.

Definition 9.7 *Let $\mu_k(t)$, $k = 1, \ldots, n$, be the eigenvalues of $\Phi(t; \mathbf{y})$ from equation (7.9).* **Liapunov exponents** L_k *of the trajectory $\mathbf{y}(t)$ are defined by*

$$L_k = \lim_{t \to \infty} \frac{1}{t} \ln |\mu_k(t)|\,. \tag{9.15}$$

These Liapunov coefficients $L_k(\mathbf{y})$, $k = 1, \ldots, n$, are real numbers. The existence of this limit was established in [Ose68]. For stationary and periodic solutions the L_k are closely related to the characteristic numbers known from earlier chapters; see Exercise 9.11.

For computational purposes, the expression of (9.15) is not yet satisfactory, and we analyze this definition further. Assume that $\mathbf{z}(t)$ is eigenvector of the eigenvalue $\mu_k(t)$,

$$\Phi(t)\,\mathbf{z} = \mu_k\,\mathbf{z}\,.$$

Then, applying vector norms $\|\ \|$, $|\mu_k|$ can be expressed as

$$|\mu_k| = \frac{\|\Phi(t)\,\mathbf{z}\|}{\|\mathbf{z}\|}\,.$$

Assuming further that $\mathbf{h}(t)$ solves the variational equation with suitable initial vector $\mathbf{h}(0)$, we are motivated to replace $|\mu_k|$ by the quotient $\|\mathbf{h}(t)\|/\|\mathbf{h}(0)\|$. In fact, the expression

$$\lim_{t \to \infty} \frac{1}{t} \ln \frac{\|\Phi(t)\,\mathbf{h}(0)\|}{\|\mathbf{h}(0)\|} \tag{9.16}$$

turns out to be helpful for computational purposes. What happens when we pick an *arbitrary* vector $\mathbf{h}(0)$ in equation (9.16)? This question is settled by the following theorem:

Theorem 9.8 *Assume that L_1 is strictly the largest Liapunov exponent, and the exponents are ordered in descending order,*

$$L_1 > L_2 \geq \ldots \geq L_n\,.$$

Then for almost any choice of $\mathbf{h}(0)$ the limit in equation (9.16) yields L_1,

$$\lim_{t \to \infty} \frac{1}{t} \ln \frac{\|\Phi(t)\mathbf{h}(0)\|}{\|\mathbf{h}(0)\|} = L_1\,. \tag{9.17}$$

We sketch the derivation of this result. Assume $\mathbf{z}^k$ are eigenvectors of $\Phi(t)$ with eigenvalues μ_k, $k = 1, \ldots, n$. Then any vector $\mathbf{h}(0)$ can be expressed as a linear combination

$$\mathbf{h}(0) = \sum_{k=1}^{n} c_k \mathbf{z}^k,$$

which implies

$$\mathbf{h}(t) = \Phi(t)\mathbf{h}(0) = \sum_{k=1}^{n} c_k \Phi \mathbf{z}^k = \sum_{k=1}^{n} c_k \mu_k \mathbf{z}^k.$$

By (9.15), the value

$$\epsilon_k(t) := L_k - \frac{1}{t} \ln |\mu_k(t)|$$

is close to zero for large t,

$$\lim_{t \to \infty} \epsilon_k(t) = 0.$$

This leads to

$$\mu_k = \exp(L_k t - \epsilon_k t + i\vartheta_k).$$

The angles ϑ_k take care of the imaginary part or the sign. Note that $c_k, \mathbf{z}^k, \epsilon_k$, and ϑ_k vary with t; i is the imaginary unit. This expression for μ_k leads to

$$\mathbf{h}(t) = e^{L_1 t}\left(c_1 e^{i\vartheta_1 - \epsilon_1 t} \mathbf{z}^1 + \sum_{k=2}^{n} c_k e^{(L_k - L_1 - \epsilon_k)t} e^{i\vartheta_k} \mathbf{z}^k \right),$$

which is meaningful for $c_1 \neq 0$. The next step is to realize

$$\ln \frac{\|\mathbf{h}(t)\|}{\|\mathbf{h}(0)\|} = L_1 t + \ln \frac{\zeta(t)}{\|\mathbf{h}(0)\|},$$

where $\zeta(t)$ is the norm of the term in parentheses. Dividing by t and taking limits, we arrive at

$$\lim_{t \to \infty} \frac{1}{t} \ln \frac{\|\mathbf{h}(t)\|}{\|\mathbf{h}(0)\|} = L_1.$$

This assumes boundedness of $|c_1(t)|$ and $\|\mathbf{z}^1\|$, and $c_1 \neq 0$. Because is it most likely that any arbitrary $\mathbf{h}(0)$ has a component in the direction of $\mathbf{z}^1$, $c_1 \neq 0$, the assertion of the theorem holds for almost all $\mathbf{h}(0)$.

For large t, equation (9.17) implies

$$\|\mathbf{h}(t)\| \approx \|\mathbf{h}(0)\| \exp(L_1 t).$$

This result shows that for $L_1 > 0$ the vector $\mathbf{h}(t)$ grows like $e^{L_1 t}$; for $t \to \infty$ each initial vector $\mathbf{h}(0)$ under the flow tends to fall along the direction of most rapid growth. This can be generalized. Rather than studying how a single vector $\mathbf{h}(0)$ evolves under the flow, we follow the dynamical behavior of a two-dimensional area element. To this end we take instead of $\mathbf{h}(0)$ *two* vectors h_1 and h_2 as edges of a two-dimensional parallelepiped and ask how its volume $V_2(t)$ changes under the flow. An analysis analogous to the above shows that $V_2(t)$ grows like $e^{(L_1+L_2)t}$,

$$\lim_{t\to\infty} \frac{1}{t}\ln\frac{|V_2(t)|}{|V_2(0)|} = L_1 + L_2. \tag{9.18}$$

The resulting number $L^{(2)} := L_1 + L_2$ is the mean rate of exponential growth of a two-dimensional parallelepiped. This concept of *Liapunov exponents of second order* is generalized to parallelepipeds of any dimension $k \le n$. Consequently the sum of k Liapunov exponents

$$L^{(k)} := L_1 + L_2 + \ldots + L_k \tag{9.19}$$

describes the mean rate of exponential growth of a k-dimensional volume. $L^{(k)}$ is called the Liapunov exponent of order k, or a k-dimensional Liapunov exponent. The existence of the limits in equations (9.15) through (9.18) was established in [Ose68].

9.7.3   Characterization of Attractors

As seen above, the sum $L^{(n)}$ of all Liapunov exponents is the average volume contraction rate. Hence, for dissipative systems, the sum of all Liapunov exponents must be negative,

$$L_1 + L_2 + \ldots + L_n < 0.$$

Chaos is characterized by at least one positive Liapunov exponent, which reflects a stretching into one or more directions while the overall volume is shrinking. That is, after ordering the Liapunov exponents

$$L_1 \ge L_2 \ge \ldots \ge L_n,$$

a criterion for chaos is given by $L_1 > 0$. In fact, the criterion $L_1 > 0$ has been used as a definition of chaos. A system that is dissipative must have at least one negative Liapunov coefficient, $L_n < 0$, in order to satisfy $L^{(n)} < 0$. Any nonstationary attractor has at least one Liapunov coefficient that vanishes, because there is no exponential growth or shrinking along the trajectory. The dimensions of the attractors stationary solution, periodic orbit, and torus are reflected by the number of vanishing Liapunov coefficients. For a periodic attractor, $L_1 = 0$; this corresponds to the eigenvalue 1 of the monodromy matrix; compare Exercise 9.11(b). Hence the sign pattern of the Liapunov exponents of limit cycles is $(0, -, -, \ldots)$.

Requiring for dissipative chaos at least one positive, one negative, and one zero Liapunov coefficient shows that chaos is possible only for those autonomous ODE systems that have size $n \ge 3$. Recall that this minimal size corresponds to $n = 2$ for a nonautonomous differential equation such as Duffing's equation.

Kaplan and Yorke [KaY79b] have defined a fractal dimension D_L via the Liapunov exponents L_k by

$$D_L := l - \left(\sum_{k=1}^{l} L_k \right) / L_{l+1} \qquad (9.20)$$

For this "Liapunov dimension," l is the maximum index such that

$$L_1 + L_2 + \dots + L_l \geq 0 \, ;$$

the Liapunov exponents are again assumed to be ordered. For chaotic flows such an l exists, because $L_1 > 0$. The l is smaller than n because the system is dissipative. It has been shown that D_L is close to the information dimension D_I (see equation (9.7)), and $D_L \geq D_I$ holds [GrP83].

In the special case of a strange attractor that arises for an autonomous dissipative ODE system with $n = 3$, L_3 must be negative, since $L_1 > 0$, $L_2 = 0$, and the average volume contraction rate is negative. Consequently, $l = 2$, and the Liapunov dimension (9.20) specializes to

$$D_L = 2 - L_1 / L_3 \, .$$

The sign pattern of the three Liapunov exponents in this case can be characterized by $(+, 0, -)$; see also Exercise 9.12.

The equation (9.20) offers a way to calculate a fractal dimension of a strange attractor. In ODE problems, it has been shown to be more convenient to calculate the Liapunov exponents than to count, for example, small boxes as required by equation (9.6). Accordingly, the fractal dimensions of the orbits shown in Figure 9.8 and in Figure 9.11 (from [TrKS85], see Table 9.1) were calculated via (9.20). For further results on Liapunov exponents see [EcKRC86], [SaS85], [StM88]; a FORTRAN program can be found in [WoSSV85].

9.7.4   Computation of Liapunov Exponents

The evaluation of Liapunov exponents is based on the matrix $\mathbf{f_y}$, which is evaluated along the trajectory $\mathbf{y}(t)$. First one integrates the differential equation $\dot{\mathbf{y}} = \mathbf{f}(\mathbf{y}, \lambda)$ for some time until it is most likely that transients have died out, and $\mathbf{y}(t)$ has entered the attractor. Here we assume for simplicity that the numerical integrator reveals the chaotic structure of a trajectory in a trustworthy way. This is not obvious because the extreme sensitivity of chaos makes it sensitive to truncation errors; see, for example, [Cor92].

A numerical computation of equation (9.17) by integrating equation (7.9) may suffer from overflow because the linearization has at least one exponentially diverging solution, $\|\mathbf{h}(t)\| \approx \|\mathbf{h}(0)\| e^{L_1 t}$ (chaotic trajectory, $L_1 > 0$).

The question arises what happens earlier: the divergence of $\mathbf{h}(t)$ or the convergence of (9.17). In general the divergence occurs first. Therefore, the integration must be halted repeatedly to rescale $\mathbf{h}(t)$. The calculation of higher-dimensional Liapunov coefficients such as $L^{(2)}$ from equation (9.18) suffers from a further complication: The initially chosen vectors $\mathbf{h}_1$ and $\mathbf{h}_2$ tend to fall along the direction of most rapid growth. Eventually, $\mathbf{h}_1(t)$ and $\mathbf{h}_2(t)$ can hardly be distinguished, and discretization errors and noise will dominate. Therefore, the integration must be halted repeatedly to reorthogonalize the vectors $\mathbf{h}_1$ and $\mathbf{h}_2$. This is done by applying the Gram–Schmidt orthonormalization procedure; see, for instance, [WoSSV85].

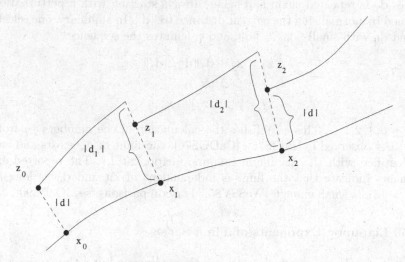

Fig. 9.14. Rescaling a neighboring trajectory

Here we restrict our attention to the calculation of L_1 because of its decisive role in characterizing chaos. Using the chain rule similar to the one in the derivation of equation (9.14), we can replace the single logarithm in equation (9.17) by a sum. That is, for the study of divergence of nearby trajectories, the quantities

$$q_N = \frac{1}{N\Delta t} \sum_{j=1}^{N} \ln \frac{\|\mathbf{d}_j\|}{\|\mathbf{d}\|} \tag{9.21}$$

are calculated. Here Δt is a sampling interval for rescaling and $\|\mathbf{d}\|$ and $\|\mathbf{d}_j\|$ are norms of distances that are defined as follows (cf. Figure 9.14): Denote by

$$\mathbf{x}_j = \mathbf{y}(j\Delta t)$$

points calculated along the particular trajectory $\mathbf{y}$ that is being investigated. These points are evaluated at integral multiples of the sampling rate Δt (chosen rather arbitrarily, but fixed). At these regularly occurring time instances, the behavior of nearby trajectories is measured. To this end, select a small perturbation $\mathbf{d}_0$ with norm $\|\mathbf{d}_0\| = \|\mathbf{d}\|$ (with, for instance, the Euclidian vector norm $\|\mathbf{d}\|$; see Appendix A.1), and calculate a part of the trajectory that emanates at

$$\mathbf{z}_0 = \mathbf{x}_0 + \mathbf{d}_0 \,.$$

After the time increment Δt is elapsed, the difference

$$\mathbf{d}_1 = \varphi(\Delta t; \mathbf{z}_0) - \mathbf{x}_1$$

gives a first hint on the divergence behavior. The process of calculating distances $\mathbf{d}_j$ is repeated again and again, always starting with a perturbation obtained by normalizing the current distance to $\|\mathbf{d}\|$. In summary, one selects Δt and $\mathbf{d}_0$ with small $\|\mathbf{d}_0\| = \|\mathbf{d}\|$, and calculates the sequence

$$
\begin{aligned}
\mathbf{z}_j &= \mathbf{x}_j + \mathbf{d}_j \|\mathbf{d}\| / \|\mathbf{d}_j\| \,, \\
\mathbf{x}_{j+1} &= \varphi(\Delta t; \mathbf{x}_j) \,, \\
\mathbf{d}_{j+1} &= \varphi(\Delta t; \mathbf{z}_j) - \mathbf{x}_{j+1}
\end{aligned}
\tag{9.22}
$$

for $j = 0, 1, 2, \ldots$. This establishes the calculation of the numbers q_N from (9.21). As observed in [BeGS76], [CaDGS76], the limit of q_N exists and can be identified with the maximum Liapunov exponent L_1. The reported experiments indicate that the limit is independent of Δt and $\mathbf{d}_0$ as long as $\|\mathbf{d}_0\| = \|\mathbf{d}\|$ is small enough [WoSSV85]. For comparisons, see [GePL90].

9.7.5 Liapunov Exponents of Time Series

We return to the situation when no equation is known, but a time series is at hand. Assume that the time series has been embedded into an m-dimensional phase space (see Section 9.4), and a large set of points $\mathcal{P} = \{\mathbf{z}(t_1), \mathbf{z}(t_2), \ldots\}$ is available. A too small data set, and a bad quality of the data, will severely affect the quality of the approach. Without accurate data, one should not expect more than an accuracy of, say, 10%. Here we ignore this practical problem, and describe a procedure for calculating Liapunov coefficients of time series. This method was developed in [WoSSV85]; this paper also includes a FORTRAN program.

The idea is as follows: Search the point set $\mathcal{P}$ for the point $\mathbf{z}(t_k) \in \mathcal{P}$ that is closest to $\mathbf{x}(t_1)$ in the Euclidian norm. Here we use the notation $\mathbf{x}(t_k)$ for the points in $\mathcal{P}$ in their natural numbering; $\mathbf{z}(t_\nu)$ are also in $\mathcal{P}$ but are only ordered in a piecewise fashion. The distance is $\delta_0 := \|\mathbf{x}(t_1) - \mathbf{z}(t_k)\|$. This corresponds to the initial perturbation $\|\mathbf{h}(0)\|$ of the Liapunov exponents of ODEs. We next study how this length element is propagated. To this end the evolution of the *pair* of nearby points is monitored, assuming that they lie on different trajectories. The length element that is propagated through the attractor has the length $\delta_j := \|\mathbf{x}(t_{1+j}) - \mathbf{z}(t_{k+j})\|$; it corresponds to $\|\mathbf{h}(t)\|$. When the separation of the pair has become too large, a *replacement procedure* must be performed, at some $j = j_r$. Let the distance of δ obtained at the replacement step be denoted by δ'. The procedure corresponds to the rescaling in the continuous case, which is illustrated in Figure 9.14. A replacement point $\mathbf{z} \in \mathcal{P}$ is searched close to $\mathbf{x}(t_{j_r+1})$ in order to find a new starting point for a neighboring trajectory. Because there is no hope to find a point $\mathbf{z} \in \mathcal{P}$ exactly on the line $\mathcal{L}$ that connects $\mathbf{x}(t_{j_r+1})$ with $\mathbf{z}(t_{k+j_r})$, a replacement point $\mathbf{z} \in \mathcal{P}$ is required that is not only close to $\mathbf{x}(t_{j_r+1})$ but also close to that line $\mathcal{L}$. Now the process of monitoring adjacent trajectories is restarted, with replacement steps performed when necessary. In this way, the $\mathbf{x}(t_j)$ traverse the entire data set. At each replacement step the value

$$\ln \frac{\delta'}{\delta}$$

is calculated analogously as in equation (9.21); their sum gives an estimate for the largest Liapunov coefficient L_1,

$$\frac{1}{t_M - t_1} \sum \ln \frac{\delta'}{\delta} \approx L_1 \, ;$$

the summation is over all replacement steps. In a similar way the growth rate of a k-volume element can be estimated; for details see [WoSSV85]. The replacement steps bound the size of the k-volume elements in an attempt to simulate infinitesimally small length scales. The reader is encouraged to draw a sketch similar to Figure 9.14 to illustrate a replacement step. The identification of true and spurious Liapunov exponents is discussed in [Par92]. A review on how to detect nonlinearities in time series is [Tak93].

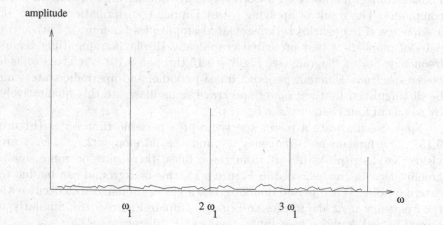

amplitude

Fig. 9.15. Frequencies of a periodic signal, schematically

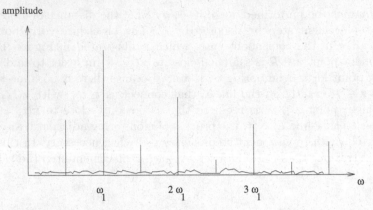

Fig. 9.16. After period doubling, schematically

9.8   Power Spectra

Consider a time series

$$y(0),\ y(1),\ y(2),\ldots$$

of a scalar variable. Upon integrating systems of ODEs, a time series can be obtained by sampling one component $y_k(t)$ at sampling intervals Δt,

$$y_k(0),\ y_k(\Delta t),\ y_k(2\Delta t),\ldots.$$

The power spectrum of such a time series is usually calculated by the *fast Fourier transformation* (FFT). Corresponding routines are available on most computers. The result of applying a fast Fourier transformation consists of a sequence of frequencies with associated amplitudes (energies). This set of pairs of numbers is best presented graphically, displaying amplitude versus frequency. Such a diagram (see Figures 9.15 through 9.19) is referred to as a *power spectrum*. Different periodic, quasi-periodic, and aperiodic states can be distinguished by their power spectra. Let us illustrate this qualitatively by means of four characteristic figures.

First, we illustrate a power spectrum of a periodic time series (Figure 9.15). The fundamental frequency ω_1 and the harmonics $2\omega_1, 3\omega_1,\ldots$ are visible as sharp peaks. Apart from these lines, there may be some small-amplitude noise, as indicated in Figure 9.15; this background can be due to external noise. When period doubling takes place, subharmonics occur with the frequency $\omega_1/2$ and integer multiples (compare Figure 9.16). Similarly, a second period doubling contributes peaks with "distance" $\omega_1/4$.

A bifurcation from periodic motion to quasi-periodic motion on a 2-torus adds a fundamental frequency ω_2. This new frequency is generally not commensurable with ω_1. With two fundamental frequencies, a power spectrum may resemble Figure 9.17. Both periodic motions (Figures 9.15 and 9.16) and

amplitude

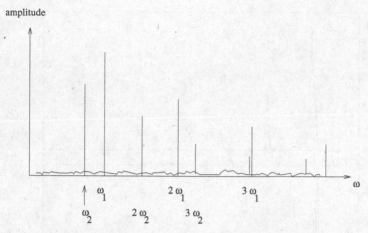

Fig. 9.17. Frequencies of a trajectory on a torus, schematically

amplitude

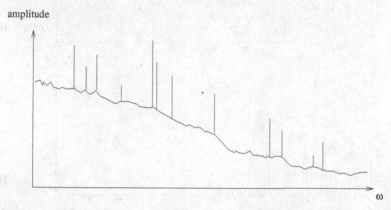

Fig. 9.18. Power spectrum of a chaotic signal, schematically

the quasi-periodic motion (Figure 9.17) are characterized by power spectra with small-amplitude noise and more or less discrete peaks.

This is in contrast to the situation of chaos, which is characterized by a broadband noise (Figure 9.18). There may be peaks sitting on top of the broadband noise, which are more or less accentuated. The construction of power spectra has been shown to be a standard means for characterizing chaos. Most references quoted in the context of chaos illustrate their results with power spectra. Thereby, the onset of broad-band noise serves as confirmation of chaotic phenomena.

We conclude the tour into chaos land with the power spectra of the isothermal reaction of Example 7.8, see Figure 9.19. This shows how the schematic diagrams of Figures 9.15 through 9.18 may look in a real experiment. In practice it can be hard to see a clear difference in the power spectra of a

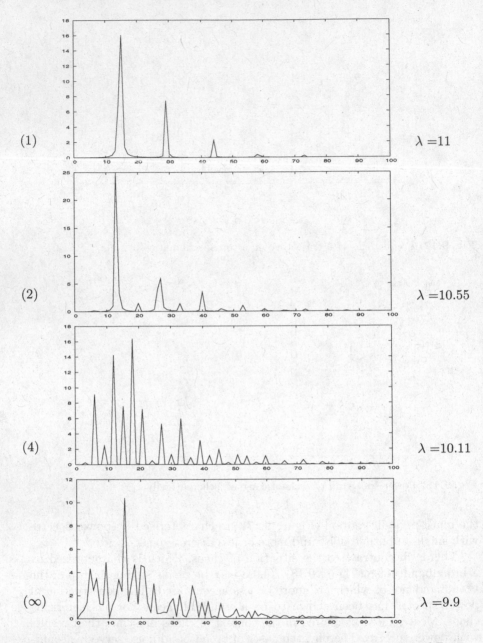

Fig. 9.19. Power spectra of four trajectories of the isothermal reaction of Example 7.8. Compare to Figure 1.5 and to Figure 7.18. Horizontal axis: index of the frequency, vertical axis: artificially scaled amplitudes; the first peak is removed in all of the four spectra. The positions of the peaks are not immediately comparable, because the frequencies vary with the parameter λ. (1) simple period of Figure 1.5(b); (2) double period of Figure 1.5(c), (4) fourfold period of Figure 1.5(d), (∞) chaotic orbit of Figure 1.5(e).

chaotic and an eventful periodic signal. Often several different tools must be applied to analyze a signal, namely, fractal dimensions, Liapunov exponents, and power spectra.

Exercises

Exercise 9.1

(a) Derive equation (9.2) from the convection theorem [Mey71]

$$\frac{\mathrm{d}}{\mathrm{d}t} \int_{\mathcal{M}_t} u \, \mathrm{d}V = \int_{\mathcal{M}_t} \left(\frac{\mathrm{d}u}{\mathrm{d}t} + u \nabla \cdot \mathbf{f} \right) \mathrm{d}V,$$

which holds for arbitrary C^1-smooth functions $u(\mathbf{x}, t)$, and for $\frac{\mathrm{d}}{\mathrm{d}t}$ being the material derivative.
(b) Assume $\gamma := \nabla \cdot \mathbf{f} = $ constant. Show $V(t) = V(0)e^{\gamma t}$ and apply this to the Lorenz equation (2.25).
(c) Check the sign of $\nabla \cdot \mathbf{f}$ for the van der Pol equation (1.6).

Exercise 9.2
Consider the differential equation $\dot{y} = ay$ for $a > 0$ with initial value $y(0) = z_1$. Let U be a neighborhood (interval) that includes z_1 and the solution $\varphi(t, z_1)$. Show sensitive dependence on initial conditions by proving the following statement: There is a $\epsilon > 0$ such that for all U (no matter how small) there is a $z_2 \in U$ and a $t > 0$ such that $|\varphi(t, z_1) - \varphi(t, z_2)| > \epsilon$.

Exercise 9.3
Let y be binary numbers in the interval $0 \leq y \leq 1$,

$$y = \sum_{i=1}^{\infty} \gamma_i 2^{-i} = 0.\gamma_1 \gamma_2 \gamma_3 \dots$$

with γ_i either 0 or 1. The *Bernoulli map* is defined as

$$y_{\nu+1} = B(y_\nu) = \begin{cases} 2y_\nu & \text{if } y_\nu < 0.5, \\ 2y_\nu - 1 & \text{if } y_\nu \geq 0.5. \end{cases}$$

Carrying out $B(y)$ amounts to drop γ_1 and shift the remaining part of the mantissa one place to the left; this explains the alternate name *shift map* for $B(y)$.

(a) Show that $0 \leq y_0 \leq 1$ implies $0 \leq y_\nu \leq 1$ for $\nu > 0$, and draw a sketch of the function $B(y)$.
(b) Assume that y has a periodic symbol sequence with $\gamma_{i+m} = \gamma_i$ for all i and some integer $m \geq 1$. Show that the Bernoulli map has periodic orbits with period m.
(c) Show for this map sensitive dependence on initial conditions.

Exercise 9.4
Draw parallel y-axes and illustrate the attractors of the logistic map in equation (7.9) at least for the parameter values $\lambda = 2.7$, 3.1, and 3.5 (see Section 7.4.2 and Exercise 7.8).

Exercise 9.5
Let $\mathbf{J}$ be the Jacobian of a map. A map is dissipative if $|\det \mathbf{J}| < 1$. For which values of the constants a and b is the Hénon map in equation (9.3) dissipative?

Exercise 9.6
(programming assignment)
(a) Iterate equation (9.3) as often as possible (according to how patient you are) and plot all iterates $\mathbf{y}^{(\nu)}$ for $\nu > N$. The first N iterates are suppressed; they represent the transient initial phase until the iteration is trapped by the strange attractor. N depends on your choice of a starting vector $\mathbf{y}^{(0)}$.
(b) If you have even more patience, it is instructive to enlarge a "line" of the attractor in order to see that the "line" consists of more lines. (The figures in [Ott81] are based on more than 10,000 points.)

Exercise 9.7
From equation (7.14) derive the estimate for the accumulation point of a sequence λ_ν of period-doubling points

$$\lambda_\infty \approx (\delta \lambda_{\nu+1} - \lambda_\nu)/(\delta - 1),$$

where $\delta = 4.6692016\ldots$.

Exercise 9.8
Consider the scalar map $y^{(\nu+1)} = P(y^{(\nu)})$ with

$$P(y) := \lambda + y + y^2.$$

(a) Discuss fixed points and their stability in dependence of the parameter λ.
(b) Draw diagrams depicting $y^{(\nu+1)}$ versus $y^{(\nu)}$ for the specific parameter values $\lambda = +0.01$ and $\lambda = -0.01$; the sketches should include the graph of $P(y)$ and some typical iteration.
(c) Imagine $P(y)$ being the Poincaré map of a continuous dynamical system. What is the effect on the iterates when λ is varied from -0.01 to $+0.01$?

Exercise 9.9
A widely known process for obtaining snowflake-like objects starts with an equilateral triangle and constructs iteratively a two-dimensional object following the rule: Glue to the middle third of each boundary straight line an equilateral triangle of the proper size. The first iterate is

In the limit the boundary "converges" to a curve of infinite length, called *Koch curve*. Show that the fractal dimension equation (9.6) of the Koch curve is $\ln 4/\ln 3$.

Exercise 9.10

The transformation $\mathbf{z} = \mathbf{A}\mathbf{y}$ transforms the sphere $\mathbf{y}^{tr}\mathbf{y} - \rho^2 = 0$ in $\mathbf{y}$-space into an ellipsoid-like surface in $\mathbf{z}$-space. Here $\mathbf{A}$ is an n^2-matrix with eigenvalues $\mu_1, \ldots, \mu_n$. Show that the principal axes of the ellipsoid are given by $\rho_k = \rho\mu_k$.

Exercise 9.11

Consider Liapunov exponents L_k from equation (9.15). The matrices $\Phi(t)$ and $\mathbf{f_y}$ as well as their eigenvalues depend on the specific trajectory $\mathbf{y}_0$ around which they are evaluated.

(a) Assume $\mathbf{y}_0$ is stationary solution and let μ_k be eigenvalue of $\mathbf{f_y}$. Show that $L_k = \mathrm{Re}(\mu_k)$.

(b) Assume $\mathbf{y}_0(t)$ is a periodic orbit with period T and monodromy matrix $\mathbf{M}$. Let μ_k be an eigenvalue of $\mathbf{M}$. Show that $L_k = \frac{1}{T}\ln|\mu_k|$.

Exercise 9.12

Assume a dissipative and autonomous ODE system with $n = 4$ having a chaotic attractor. Show that exactly three patterns of signs of the Liapunov exponents are possible.

Appendices

A.1 Some Basic Glossary

The set of real numbers is denoted by $\mathbb{R}$ or $\mathbb{R}^1$. Geometrically, $\mathbb{R}$ represents a line. $\mathbb{R}^2$ denotes our "standard" (two-dimensional) plane and $\mathbb{R}^3$ the three-dimensional Euclidian space; $\mathbb{R}^n$ is the generalization to n-dimensional space. Writing $\mathbf{y} \in \mathbb{R}^n$ means that $\mathbf{y}$ is a column vector consisting of n real components $y_1, \ldots, y_n$. (Bold face notation refers to vectors and matrices.) A *norm* of a vector $\mathbf{y}$ is denoted by $\|\mathbf{y}\|$. Important examples are the *Euclidian norm* ("length")

$$\|\mathbf{y}\| = \|\mathbf{y}\|_2 := \sqrt{y_1^2 + \ldots + y_n^2}\,,$$

and the *maximum norm* or *infinity norm*,

$$\|\mathbf{y}\| = \|\mathbf{y}\|_\infty := \max_{i=1}^{n} |y_i|\,.$$

Functions $y(t)$ can be distinguished according to their degree of smoothness. Writing $y \in C^0$ is shorthand describing continuity of a function. A function $y \in C^k$ (k a positive integer) is k times differentiable: The derivatives $y^{(j)}$ of y for $j = 1, \ldots, k$ exist and are continuous. In case the smoothness is restricted to some domain $\mathcal{D}$, one writes $y \in C^k(\mathcal{D})$.

Functions can be vectors. For example, for $t \in \mathbb{R}$, $\mathbf{y}(t) \in \mathbb{R}^3$,

$$\mathbf{y}(t) = \begin{pmatrix} y_1(t) \\ y_2(t) \\ y_3(t) \end{pmatrix}$$

is a parametric representation of a *curve* in $\mathbb{R}^3$. Here the curve parameter is t, which may be time. The tangent to the curve is

$$\dot{\mathbf{y}}(t) = (\dot{y}_1(t), \dot{y}_2(t), \dot{y}_3(t))^{tr}\,,$$

where ^{tr} indicates transpose. Two curves intersect *transversally* if their tangents in the intersection point form a nonzero angle—that is, the curves are not tangential.

A curve is a one-dimensional set. A curve in an ambient space $\mathbb{R}^{n+1}$ can be defined by n scalar equations. If all scalar equations are linear, then the

R. Seydel, *Practical Bifurcation and Stability Analysis*,
Interdisciplinary Applied Mathematics 5, DOI 10.1007/978-1-4419-1740-9,
© Springer Science+Business Media, LLC 2010

curve specializes to a straight line. A curve that has the structure of $\mathbb{R}^1$ is called a one-dimensional manifold. Generally, an m-dimensional *manifold* $\mathcal{M} \subset \mathbb{R}^n$ ($m \leq n$) is a set such that each $\mathbf{y} \in \mathcal{M}$ has a neighborhood $\mathcal{U}$ for which there is a smooth invertible transformation ("diffeomorphism") that maps $\mathbb{R}^n$ to $\mathcal{U}$. For example, the unit circle S^1 is a one-dimensional manifold, the unit sphere S^2 is a two-dimensional manifold, $\mathbb{R}^m$ is an m-dimensional manifold. In this book, it is sufficient to think of an m-dimensional manifold as an m-dimensional surface embedded in $\mathbb{R}^n$. In case $m = n - 1$ we speak of a *hyper*surface.

A.2 Some Basic Facts from Linear Algebra

Assume $\mathbf{A}$ is a square matrix with n rows and n columns. The n^2 entries of $\mathbf{A}$ are mostly real numbers. $\mathbf{I}$ denotes the identity matrix. All its entries are zero, except for the diagonal elements, which are unity. A useful vector is the unit vector $\mathbf{e}_k$; all its components are zero except for the kth component, which is unity. The identity matrix $\mathbf{I}$ can be seen as consisting of the columns $\mathbf{e}_1, ..., \mathbf{e}_n$. If there is a vector $\mathbf{w} \neq \mathbf{0}$ and a scalar μ such that

$$\mathbf{A}\mathbf{w} = \mu\mathbf{w},$$

then $\mathbf{w}$ is called *(right) eigenvector*, and μ is called *eigenvalue*. A left eigenvector $\mathbf{g}$ is defined by

$$\mathbf{g}^{tr}\mathbf{A} = \mu\mathbf{g}^{tr},$$

which is equivalent to $\mathbf{A}^{tr}\mathbf{g} = \mu\mathbf{g}$. The vectors $\mathbf{w}$ and $\mathbf{g}$, and the scalar μ can be complex. A *generalized eigenvector* $\mathbf{w}$ satisfies $(\mathbf{A} - \lambda\mathbf{I})^j\mathbf{w} = \mathbf{0}$ for an integer j.

Assume that $\mathbf{z}^1, \mathbf{z}^2, \ldots, \mathbf{z}^m$ are vectors with n components. A *linear combination* of m vectors $\mathbf{z}^i$ is a sum

$$\sum_{i=1}^{m} c_i \mathbf{z}^i \, ;$$

the c_i are arbitrary numbers. Assembling the columns $\mathbf{z}_1, ..., \mathbf{z}_m$ into a matrix $\mathbf{A}$ (n rows, m columns), and the scalars c_i into a vector $\mathbf{c}$, the linear combination can be written $\mathbf{A}\mathbf{c}$. The set of all possible linear combinations is called the *span* of $\mathbf{z}^1, ..., \mathbf{z}^m$. A vector $\mathbf{z} \in \mathbb{R}^n$ is called *linear dependent* of $\mathbf{z}^1, \mathbf{z}^2, \ldots, \mathbf{z}^m$ if $\mathbf{z}$ can be represented as a linear combination of these vectors $\mathbf{z}^i$; otherwise, $\mathbf{z}$ is linear independent of $\mathbf{z}^1, \ldots, \mathbf{z}^m$. The *rank* of a matrix is the maximum number of linear independent columns (equivalently, rows). A square matrix $\mathbf{A}$ ($m = n$) is called *nonsingular* if rank$(\mathbf{A}) = n$. The following statements are equivalent:

$\mathbf{A}$ is singular
$\Leftrightarrow \operatorname{rank}(\mathbf{A}) < n$
$\Leftrightarrow \mathbf{A}$ has 0 as eigenvalue
$\Leftrightarrow$ there exists a nonzero vector $\mathbf{h}$ with $\mathbf{Ah} = \mathbf{0}$
$\Leftrightarrow \det(\mathbf{A}) = 0$.

The *range* or *column space* of a matrix $\mathbf{A}$ with column vectors $\mathbf{z}^1, \ldots, \mathbf{z}^n$ is the set of all linear combinations,

$$\operatorname{range}(\mathbf{A}) = \{\mathbf{Ac}, \text{ for all vectors } \mathbf{c} \in \mathbb{R}^n\}$$
$$= \left\{ \sum_{i=1}^n c_i \mathbf{z}^i, \text{ for arbitrary } c_i \in \mathbb{R} \right\}.$$

That is, the range of a matrix is the span of its columns. The system of linear equations represented by the equation $\mathbf{Az} = \mathbf{b}$ for $\mathbf{b} \in \mathbb{R}^n$ is solvable by definition if and only if

$$\mathbf{b} \in \operatorname{range}(\mathbf{A}).$$

This is equivalent to

$$\operatorname{rank}(\mathbf{A}) = \operatorname{rank}(\mathbf{A}|\mathbf{b}),$$

where $(\mathbf{A}|\mathbf{b})$ is the rectangular matrix that consists of the n columns of $\mathbf{A}$ and the column vector $\mathbf{b}$. The *null space* or *kernel* of a matrix $\mathbf{A}$ is the solution set for $\mathbf{b} = \mathbf{0}$; the null space has the dimension

$$n - \operatorname{rank}(\mathbf{A}).$$

This implies that the null space of a "full-rank" matrix has dimension 0 and consists only of the zero vector. In the case of $\operatorname{rank}(\mathbf{A}) = n - 1$, there is one nonzero vector $\mathbf{h} \in \mathbb{R}^n$ that *spans* the null space,

$$\operatorname{null}(\mathbf{A}) = \{\gamma \mathbf{h} \text{ for all } \gamma \in \mathbb{R}\}.$$

The null space is the linear combination of $(n - \operatorname{rank}(\mathbf{A}))$ vectors.
The range of a matrix is orthogonal to the null space of its transposure,

$$\operatorname{range}(\mathbf{A}) = (\operatorname{null}(\mathbf{A}^{tr}))^{\perp}$$

Proof: For $\mathbf{x} \in \operatorname{null}(\mathbf{A}^{tr}), \mathbf{v} \in \operatorname{range}(\mathbf{A})$ conclude
$\mathbf{A}^{tr}\mathbf{x} = \mathbf{0}, \mathbf{v} = \mathbf{Az}$ for a $\mathbf{z}$,
and consequently $\mathbf{v}^{tr}\mathbf{x} = (\mathbf{Az})^{tr}\mathbf{x} = \mathbf{z}^{tr}\mathbf{A}^{tr}\mathbf{x} = 0$.

General references for linear algebra include [Str88]; see also [GoL96].

428   Appendices

A.3   Some Elementary Facts from ODEs

Here we report on a few elementary facts from ordinary differential equations. For further background, consult textbooks on ODEs, such as [CoL55], [Har64], [Hale69], [Arn73], [HiS74], [JoS77], [Bra78], [ArP90], [Chi99].

A scalar ODE of the order m for a function $u(t)$ is given by an equation of the type

$$F(t, u(t), \dot{u}(t), ..., u^{(m)}(t)) = 0.$$

The dot and the superscript represent differentiation with respect to the independent variable t. Solutions $u(t)$ are implicitly defined by this equation. We are mainly interested in explicit differential equations of the first order,

$$\dot{\mathbf{y}}(t) = \mathbf{f}(t, \mathbf{y}(t)).$$

This equation stands for a system consisting of n scalar components,

$$\begin{pmatrix} \dot{y}_1 \\ \dot{y}_2 \\ \vdots \\ \dot{y}_n \end{pmatrix} = \begin{pmatrix} f_1(t, y_1, \ldots, y_n) \\ f_2(t, y_1, \ldots, y_n) \\ \vdots \\ f_n(t, y_1, \ldots, y_n) \end{pmatrix}.$$

An equation (system) of the type

$$\dot{\mathbf{y}} = \mathbf{f}(\mathbf{y}),$$

in which the independent variable t does not occur explicitly, is called *autonomous*. Autonomous systems have the feature that whenever $\mathbf{y}(t)$ is a solution then $\mathbf{y}(t + c)$ is also a solution for all c. Consequently, initial values

$$\mathbf{y}(t_0) = \mathbf{y}_0$$

can be assumed for $t_0 = 0$. A nonautonomous equation can be transformed into an extended autonomous system by means of

$$y_{n+1} := t.$$

This leads to a system with $n + 1$ components,

$$\begin{pmatrix} \dot{y}_1 \\ \vdots \\ \dot{y}_n \\ \dot{y}_{n+1} \end{pmatrix} = \begin{pmatrix} f_1(y_{n+1}, y_1, \ldots, y_n) \\ \vdots \\ f_n(y_{n+1}, y_1, \ldots, y_n) \\ 1 \end{pmatrix}.$$

Existence and Uniqueness

An existence and uniqueness theorem for the intial-value problem $\dot{\mathbf{y}} = \mathbf{f}(t, \mathbf{y})$, $\mathbf{y}(a) = \mathbf{y}_0$, is as follows: Suppose $\mathbf{f}$ is defined on the strip

$$\mathcal{S} := \{(t,\mathbf{y}) \mid a \le t \le b,\ \mathbf{y} \in \mathbb{R}^n\},$$

and $\mathbf{f}$ is finite and continuous on $\mathcal{S}$, and *Lipschitz-continuous* in $\mathbf{y}$,

$$\|\mathbf{f}(t,\mathbf{y}) - \mathbf{f}(t,\mathbf{z})\| \le L\|\mathbf{y} - \mathbf{z}\|$$

for some constant L. Then for all initial vectors $\mathbf{y}_0$ there is exactly one solution $\mathbf{y}(t)$ of the initial-value problem for $a \le t \le b$. The solution is C^1-smooth.

Linear ODE System with Constant Coefficients

We review the general solution of a system of n homogeneous linear differential equations

$$\dot{\mathbf{y}} = \mathbf{A}\mathbf{y}.$$

Suppose the real n^2-matrix $\mathbf{A}$ is constant and consists of n rows and n columns. A general solution is obtained by linear superposition of n linearly independent solutions $\mathbf{y}^k(t)$,

$$\mathbf{y}(t) = \sum_{k=1}^{n} c_k \mathbf{y}^k(t)$$

(the superscript is no exponent). The constants c_k can be determined by initial conditions or by boundary conditions. The solution structure is most simple when $\mathbf{A}$ has n linear independent eigenvectors $\mathbf{w}^k$ ($k = 1,\dots,n$) associated with eigenvalues μ_k. Then the *basis functions* $\mathbf{y}^k(t)$ take the form

$$\mathbf{y}^k(t) = \exp(\mu_k t)\,\mathbf{w}^k.$$

Recall that since $\mathbf{A}$ is real, the eigenvalues are pairwise complex-conjugate. The basis functions $\mathbf{y}^k(t)$ can be taken as real even when μ_k is complex: Let one pair of eigenvalues be

$$\mu_k = \alpha_k + \mathrm{i}\beta_k, \quad \mu_{k+1} = \alpha_k - \mathrm{i}\beta_k$$

(i: imaginary unit) with eigenvectors

$$\mathrm{Re}(\mathbf{w}^k) \pm \mathrm{i}\,\mathrm{Im}(\mathbf{w}^k)$$

(Re: real part, Im: imaginary part). Then two linearly independent basis functions are

$$\mathbf{y}^k(t) = \exp(\alpha_k t)\,\{\mathrm{Re}(\mathbf{w}^k)\cos\beta_k t - \mathrm{Im}(\mathbf{w}^k)\sin\beta_k t\},$$
$$\mathbf{y}^{k+1}(t) = \exp(\alpha_k t)\,\{\mathrm{Re}(\mathbf{w}^k)\sin\beta_k t + \mathrm{Im}(\mathbf{w}^k)\cos\beta_k t\}.$$

Bernoulli Differential Equation

In Chapter 1 we mention the Bernoulli differential equation, which is of the form

$$\dot{y} = a(t)y + b(t)y^m.$$

This equation is nonlinear for $m \neq 1$, $m \neq 0$. The transformation

$$y = u^{1/(1-m)}$$

leads to the differential equation

$$\dot{u} = (1-m)au + (1-m)b\,.$$

This version is linear in u, and the calculation of a closed-form solution is elementary.

A.4 Implicit Function Theorem

The equation

$$0 = \mathbf{f}(\mathbf{y}, \lambda)$$

with $\mathbf{y} \in \mathbb{R}^n$, $\lambda \in \mathbb{R}$, $\mathbf{f}(\mathbf{y}, \lambda) \in \mathbb{R}^n$ consists of n components. The function $\mathbf{f}$ is often defined only on a subset of the $(n+1)$-dimensional space. We ignore equations with empty solution set. In general, the equation $\mathbf{0} = \mathbf{f}(\mathbf{y}, \lambda)$ defines implicitly one or more curves in the $(n+1)$-dimensional $(\mathbf{y}, \lambda)$-space. A simple example is furnished by the scalar equation

$$0 = -1 + y^2 + \lambda^2\,,$$

which defines the unit circle in the (y, λ)-plane. The question is whether an equation $0 = f(y, \lambda)$ implicitly defines a function

$$y = F(\lambda)\,.$$

The above equation defining the unit circle has two implicitly defined functions. In this simple example, we encounter the rare situation that the functions are known explicitly,

$$y = F(\lambda) = +\sqrt{1 - \lambda^2} \quad \text{and} \quad y = F(\lambda) = -\sqrt{1 - \lambda^2}\,.$$

Both functions are defined only for $-1 \leq \lambda \leq 1$.

The general assertion is as follows:

Implicit Function Theorem
Assume that
(1) $\mathbf{f}(\mathbf{y}^\star, \lambda^\star) = \mathbf{0}$,
(2) $\mathbf{f}$ *is continuously differentiable on its domain, and*
(3) $\mathbf{f}_\mathbf{y}(\mathbf{y}^\star, \lambda^\star)$ *is nonsingular.*
Then there is an interval $\lambda_1 < \lambda^\star < \lambda_2$ *about* $\lambda^\star$, *in which a vector function* $\mathbf{y} = F(\lambda)$ *is defined by* $\mathbf{0} = \mathbf{f}(\mathbf{y}, \lambda)$ *with the following properties holding for all* λ *with* $\lambda_1 < \lambda < \lambda_2$:
(a) $\mathbf{f}(F(\lambda), \lambda) = \mathbf{0}$,
(b) $F(\lambda)$ *is unique with* $\mathbf{y}^\star = F(\lambda^\star)$,
(c) $F(\lambda)$ *is continuously differentiable, and*
(d) $\mathbf{f}_\mathbf{y}(\mathbf{y}, \lambda)\, d\mathbf{y}/d\lambda + \mathbf{f}_\lambda(\mathbf{y}, \lambda) = \mathbf{0}$.

Remarks.

(1) If in assumption (3) C^k smoothness is assumed, the same smoothness is guaranteed in statement (c).
(2) An implicit function theorem was proved in [HiG27]; other proofs can be found in any analysis textbook.
(3) Analog theorems apply for more general spaces than $\mathbb{R}^n$ (function spaces, Banach spaces).

For smooth functions $\mathbf{f}$, the implicitly defined function $F(\lambda)$ can be extended until $\mathbf{f_y}$ gets singular. That is, for λ_1 and λ_2 one has the property

$$\mathbf{f_y}(F(\lambda_1), \lambda_1) \text{ singular}, \quad \mathbf{f_y}(F(\lambda_2), \lambda_2) \text{ singular},$$

here is $\lambda_1 < \lambda < \lambda_2$ the maximum interval such that the hypotheses of the theorem are valid. In the simple example of the unit circle, one has $\lambda_1 = -1$ and $\lambda_2 = +1$.

A.5 Special Invariant Manifolds

Let the vector function $\varphi(t; \mathbf{z})$ denote the solution (trajectory) of the initial-value problem $\dot{\mathbf{y}} = \mathbf{f}(\mathbf{y})$, $\mathbf{y}(0) = \mathbf{z}$. A set $\mathcal{M}$ is called *invariant* under $\dot{\mathbf{y}} = \mathbf{f}(\mathbf{y})$ when the trajectory $\varphi(t; \mathbf{z}) \in \mathcal{M}$ for all $\mathbf{z} \in \mathcal{M}$ and $t \in \mathbb{R}$. A C^r-invariant manifold is a manifold that is invariant and C^r-smooth in that it can be locally represented as a C^r-graph. Any subset of the $\mathbf{y}$-space made up entirely of trajectories is an invariant manifold. Next we define specific invariant manifolds.

After Taylor expansion and suitable transformations (Chapter 8), the nonlinear problem (NLP)

$$\dot{\mathbf{y}} = \mathbf{f}(\mathbf{y}), \quad \mathbf{f}(\mathbf{y}^s) = 0$$

of calculating solutions close to a stationary solution $\mathbf{y}^s$ can be written as

$$
\begin{aligned}
\dot{\mathbf{y}}^- &= \mathbf{A}^- \mathbf{y}^- + \mathbf{f}^-(\mathbf{y}^-, \mathbf{y}^+, \mathbf{y}^0), \\
\dot{\mathbf{y}}^+ &= \mathbf{A}^+ \mathbf{y}^+ + \mathbf{f}^+(\mathbf{y}^-, \mathbf{y}^+, \mathbf{y}^0), \\
\dot{\mathbf{y}}^0 &= \mathbf{A}^0 \mathbf{y}^0 + \mathbf{f}^0(\mathbf{y}^-, \mathbf{y}^+, \mathbf{y}^0).
\end{aligned}
\qquad \text{(NLP)}
$$

This assumes that $\mathbf{A}^-$, $\mathbf{A}^+$, and $\mathbf{A}^0$ have eigenvalues with negative, positive, and zero real parts, respectively. The vector $\mathbf{y}$ is decomposed into $\mathbf{y} = (\mathbf{y}^-, \mathbf{y}^+, \mathbf{y}^0)$ according to the sizes of the square matrices $\mathbf{A}^-$, $\mathbf{A}^+$, and $\mathbf{A}^0$. The functions $\mathbf{f}^-$, $\mathbf{f}^+$, and $\mathbf{f}^0$ contain all nonlinear terms. Assume further that $\mathbf{w}^k$ are the (generalized) eigenvectors, and

$$\mathcal{E}_s := \text{span}\{\mathbf{w}^1, ..., \mathbf{w}^i\} \quad \text{(eigenvalues have negative real part)},$$

$$\mathcal{E}_u := \text{span}\{\mathbf{w}^{i+1}, ..., \mathbf{w}^{i+j}\} \quad \text{(eigenvalues have positive real part)},$$

$$\mathcal{E}_c := \text{span}\{\mathbf{w}^{i+j+1}, ..., \mathbf{w}^{i+j+k}\} \quad \text{(eigenvalues have zero real part)}$$

with $i + j + k = n$. The spaces $\mathcal{E}_s$, $\mathcal{E}_u$, $\mathcal{E}_c$ are called the *stable, unstable,* and *center manifolds* of the linearized problem, which decouples into

$$\dot{\mathbf{y}}^- = \mathbf{A}^-\mathbf{y}^-, \quad \dot{\mathbf{y}}^+ = \mathbf{A}^+\mathbf{y}^+, \quad \dot{\mathbf{y}}^0 = \mathbf{A}^0\mathbf{y}^0. \tag{LP}$$

The manifolds $\mathcal{E}$ are invariant under (LP).

The nonlinear counterparts of $\mathcal{E}_s$, $\mathcal{E}_u$, and $\mathcal{E}_c$ are the so-called *local stable manifold* $\mathcal{M}_s$, *local unstable manifold* $\mathcal{M}_u$, and *local center manifold* $\mathcal{M}_c$ of the stationary solution. The manifold $\mathcal{M}_s$ is the set of initial vectors $\mathbf{z}$ such that $\varphi(t; \mathbf{z})$ approach the stationary solution as $t \to \infty$. This feature is also called "$\mathbf{y}^s$ is the ω-limit set of $\mathcal{M}_s$," and $\mathcal{M}_s$ is called *inset*. The same characterization holds for the *outset* $\mathcal{M}_u$ when $t \to \infty$ is replaced by $t \to -\infty$. The manifolds $\mathcal{M}_u$, $\mathcal{M}_s$, and $\mathcal{M}_c$ are invariant under (NLP). The stationary solution is an element of all these manifolds; here the manifolds intersect. In case (NLP) is of class C^r, $r \geq 2$, the manifolds $\mathcal{M}$ in the stationary solution are *tangent* to their linear counterparts $\mathcal{E}$. The dimensions of the $\mathcal{M}$ manifolds match the dimensions of their linear counterparts $\mathcal{E}$. Locally, the manifolds can be represented as graphs. For example, functions $\mathbf{g}$ and $\mathbf{h}$ exist such that

$$\mathcal{M}_s = \{(\mathbf{y}^-, \mathbf{y}^+, \mathbf{y}^0) \in \mathbb{R}^n \mid \mathbf{y}^+ = \mathbf{g}(\mathbf{y}^-), \ \mathbf{y}^0 = \mathbf{h}(\mathbf{y}^-),$$
$$\mathbf{g}'(\mathbf{0}) = \mathbf{0}, \ \mathbf{h}'(\mathbf{0}) = \mathbf{0}, \ \text{for } \|\mathbf{y}^-\| \text{ sufficiently small}\}.$$

The manifolds $\mathcal{M}_u$ and $\mathcal{M}_s$ have the asymptotic properties of their counterparts $\mathcal{E}_u$ and $\mathcal{E}_s$.

In the simplest case, for $n = 2$, an illustration can be found in Figure 1.11. In this figure, there are one-dimensional manifolds $\mathcal{M}_s$ and $\mathcal{M}_u$ (solid curves). The corresponding lines $\mathcal{E}_s$ and $\mathcal{E}_u$ are dotted. The illustration does not show the local character of the manifolds. For example, in Figure 1.10, $\mathcal{M}_s$ is two-dimensional but need not extend over the entire plane. Illustrations for $n = 3$ are shown in Figures 1.15, 1.16, and 8.12.

Related theorems can be found, for example, in [HiS74], [GuH83], or [Wig90].

A.6   Numerical Integration of ODEs

Among the many successful codes designed to integrate initial-value problems of ODEs, Runge–Kutta-type methods are most popular [Kut01], [Run1895]. Combined with error formulas of Fehlberg [Feh69], [Feh70], these methods have proved to be robust and widely applicable. Runge–Kutta-type methods have been adopted to integrate stiff differential equations as well (cf., for instance, [Ren85]). Here we focus on the nonstiff situation and present the formulas on which an explicit Runge–Kutta–Fehlberg method of order four is based. For more background consult Section 1.5.2.

Suppose that the initial-value problem

$$\mathbf{y}' = \mathbf{f}(t, \mathbf{y}), \quad \mathbf{y}(t_0) = \mathbf{y}_0$$

is to be integrated. A typical integration step approximates $\mathbf{y}$ at $t = t_0 + \Delta$; Δ is the step length. The formulas are

$$\mathbf{y} = \mathbf{y}_0 + \Delta \sum_{k=0}^{4} c_k \mathbf{f}^{(k)},$$

$$\bar{\mathbf{y}} = \mathbf{y}_0 + \Delta \sum_{k=0}^{5} \bar{c}_k \mathbf{f}^{(k)}$$

with

$$\mathbf{f}^{(0)} := \mathbf{f}(t_0, \mathbf{y}_0),$$

$$\mathbf{f}^{(k)} := \mathbf{f}\left(t_0 + \alpha_k \Delta, \mathbf{y}_0 + \Delta \sum_{j=0}^{k-1} \beta_{kj} \mathbf{f}^{(j)}\right).$$

Both $\mathbf{y}$ and $\bar{\mathbf{y}}$ approximate the exact solution, $\bar{\mathbf{y}}$ is the approximation of higher order. The difference $\mathbf{y} - \bar{\mathbf{y}}$ serves as an estimate of the error in $\mathbf{y}$. The coefficients from [Feh70] are given in Table A.1.

TABLE A.1.   Coefficients of a RKF method of order four.

k	α_k	β_{k0}	β_{k1}	β_{k2}	β_{k3}	β_{k4}	c_k	$\bar{c}_k$
0	0	0					$\frac{25}{216}$	$\frac{16}{135}$
1	$\frac{1}{4}$	$\frac{1}{4}$					0	0
2	$\frac{3}{8}$	$\frac{3}{32}$	$\frac{9}{32}$				$\frac{1408}{2565}$	$\frac{6656}{12825}$
3	$\frac{12}{13}$	$\frac{1932}{2197}$	$-\frac{7200}{2197}$	$\frac{7296}{2197}$			$\frac{2197}{4104}$	$\frac{28561}{56430}$
4	1	$\frac{439}{216}$	-8	$\frac{3680}{513}$	$-\frac{845}{4104}$		$-\frac{1}{5}$	$-\frac{9}{50}$
5	$\frac{1}{2}$	$-\frac{8}{27}$	2	$-\frac{3544}{2565}$	$\frac{1859}{4104}$	$-\frac{11}{40}$		$\frac{2}{55}$

Apart from Runge–Kutta methods, multistep methods and extrapolation methods have proved successful; see, for example, [StB80], [HaNW87]. Widely used integrators for stiff equations are based on the backward differentiation formula of [Gear71]. Another example of an implicit method is the *trapezoidal method*

$$y(t + \Delta) = y(t) + \frac{\Delta}{2}[f(t + \Delta, \, y(t + \Delta)) + f(t, y(t))],$$

which is of second order. For differential-algebraic systems; see, for instance, [HaLR89], [ReRS89], [HaW91], and the references therein.

A.7   Symmetry Groups

In [Mil72], a *group* is defined as a set $\mathcal{G}$ of objects $\{g, h, k, \ldots\}$ together with a binary operation that associates with any ordered pair of elements g, h in $\mathcal{G}$ a third element gh. The binary operation (*group multiplication*) is subject to the three requirements:

(1) There exists an *identity element* e in $\mathcal{G}$ such that $ge = eg = g$ for all $g \in \mathcal{G}$.
(2) For every $g \in \mathcal{G}$, there exists in $\mathcal{G}$ an *inverse element* g^{-1} such that $gg^{-1} = g^{-1}g = e$.
(3) The *associative law* $(gh)k = g(hk)$ holds for all $g, h, k \in \mathcal{G}$.

The reader will do well to test the above properties for the following examples of groups $\mathcal{G}1$ to $\mathcal{G}6$:

$\mathcal{G}1$ The group containing the two elements $\{1, -1\}$ with the operation "multiplication" is an example of the *cyclic group of order two* $\mathbf{Z}_2 = \{e, g\}$ with $g^2 = e$.

As a realization of $g \in \mathbf{Z}_2$ consider g as corresponding to a reflection about a plane. Clearly the identity e results when carrying out the reflection g twice.

$\mathcal{G}2$ The *dihedral group* $\mathbf{D}_4$ is the group of the four rotations and four reflections that transform a square (equilateral four-gon) onto itself.

The reflections in $\mathbf{D}_4$ are about the two diagonals and about the axes through midpoints of opposite sides of the square. The rotations have angles of multiples of $\pi/2$ radians; the "fourth" rotation is the identity. The dihedral group $\mathbf{D}_m$ is defined analogously; $\mathbf{D}_3$ is used to describe symmetries of an equilateral triangle.

$\mathcal{G}3$ The nonsingular $n \times n$ matrices with matrix multiplication form a group.

In this group, the identity element is the identity matrix $\mathbf{I}$ and the inverse elements are the inverse matrices. To each nonsingular matrix $\mathbf{S}$ corresponds an invertible linear transformation via the mapping $\mathbf{z} = \mathbf{Sy}$.

An n-dimensional *matrix representation* is a transformation $\mathcal{T}$ of a group $\mathcal{G}$ into the group $\mathcal{G}3$ such that

$$\mathcal{T}(g_1)\mathcal{T}(g_2) = \mathcal{T}(g_1 g_2).$$

Representations are not unique. Physically, the matrices $\{\mathbf{S} = \mathcal{T}(g)\}$ represent transformations of the symmetries of the geometry into the symmetry of the equations.

$\mathcal{G}4$ The 6×6 matrix $\mathbf{S}$ in equation (5.55) together with the identity matrix of the same size is a representation of $\mathbf{Z}_2$.

$\mathcal{G}5$ $\mathbf{O}(n)$ is the *orthogonal group* of all orthogonal $n \times n$ matrices.

That is, all matrices $\mathbf{S} \in \mathbf{O}(n)$ satisfy $\mathbf{S}^{tr}\mathbf{S} = \mathbf{I}$. The matrices of $\mathbf{O}(3)$ correspond exactly to those transformations in $\mathbb{R}^3$ that preserve length. As a consequence, angles are also preserved. $\mathbf{O}(3)$ are the rotations and reflections that keep the origin fixed.

A *subgroup* of $\mathcal{G}$ is a subset that is itself a group under the group operation of $\mathcal{G}$. A trivial subgroup is $\{e\}$.

$\mathcal{G}6$ A subgroup of $\mathbf{O}(n)$ is the *special orthogonal group* $\mathbf{SO}(n)$ consisting of those $\mathbf{S} \in \mathbf{O}(n)$ that satisfy $\det \mathbf{S} = +1$.

Transformations by matrices of $\mathbf{SO}(3)$ are rotations. For instance, a rotation about the x_3-axis is carried out by

$$\begin{pmatrix} \cos\vartheta & \sin\vartheta & 0 \\ -\sin\vartheta & \cos\vartheta & 0 \\ 0 & 0 & 1 \end{pmatrix} \begin{pmatrix} x_1 \\ x_2 \\ x_3 \end{pmatrix}.$$

A reflection about the (x_1, x_2)-plane is defined by

$$\begin{pmatrix} 1 & 0 & 0 \\ 0 & 1 & 0 \\ 0 & 0 & -1 \end{pmatrix} \begin{pmatrix} x_1 \\ x_2 \\ x_3 \end{pmatrix}.$$

A.8   Proof of Theorem 5.8

Theorem 5.8. *For a matrix* $\mathbf{J}$ *with* $\operatorname{rank}(\mathbf{J}) = n-1$ *the following statements are equivalent:*

(1) $\mathbf{B}$ *is nonsingular*
(2) $\mathbf{b} \notin \operatorname{range}(\mathbf{J})$ *and* $\mathbf{c} \notin \operatorname{range}(\mathbf{J}^{tr})$
(3) $\mathbf{g}^{tr}\mathbf{b} \neq 0$ *and* $\mathbf{c}^{tr}\mathbf{h} \neq 0$
(4) $(\mathbf{I} - \mathbf{b}\mathbf{b}^{tr})\mathbf{J} + \mathbf{b}\mathbf{c}^{tr}$ *has full rank* n
(for the latter equivalence assume in addition $\mathbf{b}^{tr}\mathbf{b} = 1$.)

Proof:
Note that as in Section 5.5.1

$$\operatorname{range}(\mathbf{J}) = \{\mathbf{y} \in \mathbb{R}^n \mid \mathbf{g}^{tr}\mathbf{y} = 0\},$$
$$\operatorname{range}(\mathbf{J}^{tr}) = \{\mathbf{y} \in \mathbb{R}^n \mid \mathbf{h}^{tr}\mathbf{y} = 0\},$$

and hence

$$\mathbf{b} \notin \text{range}(\mathbf{J}) \iff \mathbf{g}^{tr}\mathbf{b} \neq 0$$
$$\mathbf{c} \notin \text{range}(\mathbf{J}^{tr}) \iff \mathbf{h}^{tr}\mathbf{c} \neq 0.$$

proof of the first equivalence

(2) $\Rightarrow$ (1): assume $\mathbf{B}$ is singular. Then

$$\left.\begin{cases} \mathbf{Jz} + \mathbf{b}\xi = \mathbf{0} \\ \mathbf{c}^{tr}\mathbf{z} + \zeta\xi = 0 \end{cases}\right\} \text{ for } \begin{pmatrix} \mathbf{z} \\ \xi \end{pmatrix} \neq \mathbf{0}$$

case $\xi \neq 0$: the first equation states $\mathbf{b} \in \text{range}(\mathbf{J})$. contradiction.

case $\xi = 0$:

   1st equation $\Rightarrow$ null $(\mathbf{J}) = \text{span}(\mathbf{z})$

   2nd equation $\Rightarrow$ $\mathbf{c}$ orthogonal to span $(\mathbf{z}) = \text{null}(\mathbf{J}) \Rightarrow \mathbf{c} \in \text{range}(\mathbf{J}^{tr})$.

   contradiction.

(1) $\Rightarrow$ (2):

$\mathbf{b} \in \text{range}(\mathbf{J}) \Rightarrow \text{rank}(\mathbf{J} \mid \mathbf{b}) = n - 1 \Rightarrow \mathbf{B}$ singular.

$\mathbf{c} \in \text{range}(\mathbf{J}^{tr}) \Rightarrow \exists \mathbf{v}$ with $\mathbf{J}^{tr}\mathbf{v} = \mathbf{c} \Rightarrow \mathbf{c}$ is linear combination of the rows of $\mathbf{J}$

$$\Rightarrow \text{rank}\left(\frac{\mathbf{J}}{\mathbf{c}^{tr}}\right) = n - 1 \Rightarrow \mathbf{B} \text{ singular.}$$

(2) $\iff$ (3): see above

(3) $\Rightarrow$ (4): abbreviation $\bar{\mathbf{J}} := (\mathbf{I} - \mathbf{bb}^{tr})\mathbf{J} + \mathbf{bc}^{tr}$

show $\bar{\mathbf{J}}\mathbf{x} = \mathbf{0} \Rightarrow \mathbf{x} = \mathbf{0}$.

$$\mathbf{b}^{tr}\bar{\mathbf{J}} = (1 - \mathbf{b}^{tr}\mathbf{b})\mathbf{b}^{tr}\mathbf{J} + \mathbf{b}^{tr}\mathbf{bc}^{tr} = \mathbf{c}^{tr} \quad \text{(extra assumption } \|\mathbf{b}\|_2 = 1)$$

$$\bar{\mathbf{J}}\mathbf{x} = \mathbf{0} \Rightarrow (\mathbf{I} - \mathbf{bb}^{tr})\mathbf{J}\mathbf{x} + \mathbf{bc}^{tr}\mathbf{x} = \mathbf{0}$$

$$\Rightarrow \mathbf{b}^{tr}\bar{\mathbf{J}}\mathbf{x} = \mathbf{c}^{tr}\mathbf{x} = 0 \qquad \Rightarrow (\mathbf{I} - \mathbf{bb}^{tr})\mathbf{J}\mathbf{x} = \mathbf{0}$$

$$\Rightarrow \mathbf{J}\mathbf{x} = \mathbf{bb}^{tr}\mathbf{J}\mathbf{x} = \mathbf{b}\underbrace{(\mathbf{b}^{tr}\mathbf{J}\mathbf{x})}_{\text{real}} \underset{\mathbf{b} \notin \text{range}(\mathbf{J})}{\Rightarrow} \mathbf{J}\mathbf{x} = \mathbf{0}$$

$$\Rightarrow \mathbf{x} = \gamma \cdot \mathbf{h} \Rightarrow \mathbf{c}^{tr}\mathbf{x} = \gamma \underbrace{\mathbf{c}^{tr}\mathbf{h}}_{\neq 0}$$

$$\mathbf{c}^{tr}\mathbf{x} = 0 \Rightarrow \gamma = 0 \Rightarrow \mathbf{x} = \mathbf{0}.$$

(4) $\Rightarrow$ (3): Suppose rank $(\bar{\mathbf{J}}) = n$. Then $\bar{\mathbf{J}}\mathbf{h} \neq \mathbf{0}$.

$$\bar{\mathbf{J}}\mathbf{h} = \mathbf{0} + \mathbf{bc}^{tr}\mathbf{h} \Rightarrow \mathbf{bc}^{tr}\mathbf{h} \neq \mathbf{0} \Rightarrow \mathbf{c}^{tr}\mathbf{h} \neq 0.$$

Assume $\mathbf{g}^{tr}\mathbf{b} = 0$. Then

$$\bar{\mathbf{J}}^{tr}\mathbf{g} = \left[\mathbf{J}^{tr}(\mathbf{I} - \mathbf{bb}^{tr}) + \mathbf{cb}^{tr}\right]\mathbf{g} = \mathbf{J}^{tr}\mathbf{g} = \mathbf{0}$$

$\bar{\mathbf{J}}^{tr}\mathbf{g} = \mathbf{0}$ is a contradiction to $\bar{\mathbf{J}}$ being nonsingular ($\Rightarrow \mathbf{g}^{tr}\mathbf{b} \neq 0$).

A.9 Numerical Software and Packages

General Numerical Methods

There are so many computer programs of numerical software that it is impossible to give a review, so we shall mention only briefly some of the standard algorithms and packages used for this text. Classical numerical linear algebra has been based on the handbook [WiR71]. The most recent versions of these packages are collected in LAPACK, see www.netlib.org/lapack. Most of the boundary-value problems in this text have been solved by shooting, consult, for instance, [Bul71], [StB80], and the references therein. Shooting was used along with integrators such as those from [BuS66], [Feh69].

Algorithms Devoted to Nonlinear Phenomena

Many codes have been written for handling dynamical systems and bifurcation problems. Apparently, the first bifurcation packages were AUTO and BIFPACK [Doe81], [Sey80]. The diagrams and figures in this book have been obtained using programs of the package BIFPACK [Sey83c], which was designed as an interface between standard routines of numerical analysis and the particular problem of the user. BIFPACK served as research FORTRAN platform testing prototype algorithms. AUTO has become a standard package; it has been applied successfully on a wide range of problems [DoKK91], [Doe97]. Other related packages include CONTENT (references in [Kuz98]), DsTool [Back92], CANDYS [FeJ92], LOCBIF [KhKLN93], and XPPAUT [Erm02].

Apart from these packages with general applicability, there are other programs for specific tasks. For an overview on early programs see the appendix in [Sey88]. Other codes are described in [PaC89], [Khi90], [GaH91], [GuK91], [HuW92], [NuY94]. Many of these codes are specifically devoted to simulating dynamical systems and do not necessarily include a bifurcation analysis. For software to calculate invariant manifolds, see [EnKO07]. How to calculate limit-cycle bifurcations in a semi-analytic way is discussed in [RoAM04]. At present, it is impossible to review related software; the above list of codes is certainly not complete.

List of Major Examples

Unforced Duffing equation: Examples 1.3, 1.4; Equation (1.5)
Lorenz equation: Example 1.7, Section 2.8
Unforced van der Pol equation: Example 1.9; Equation (1.6)
Logistic differential equation: Example 1.11; Equation (1.7)
Logistic map: Equation (1.9), Example 7.7
Student migration: Example 1.12, Equation (1.10)
Cells in a ring: Example 1.14
Model of heartbeat: Exercise 1.6
Predator-prey Lotka–Volterra model: Exercise 1.9
Oscillation of a beam, Duffing equation: Section 2.3, Example 6.2
CSTR: Example 2.14; Equation (2.27)
Finite-element analog of Euler's beam: Exercise 2.14
Production of blood cells: Example 3.1, Equation (3.3)
Business cycles: Example 3.2, Equation (3.10)
Dry friction: Section 3.2.1, Example 3.3
Flip-flop circuit: Example 3.4
Hodgkin–Huxley nerve model: Section 3.4.1
FitzHugh's nerve model: Section 3.4.3, Example 7.1
Fisher's model of migration of genes: Example 3.6
FitzHugh–Nagumo nerve model: Section 3.5.2
Gierer–Meinhardt reaction: Example 3.7
Electric power generator: Example 3.8
Electrical circuit: Section 5.4.2
Temperature in a reacting material: Example 5.6, Equation (5.30)
Coupled cell reaction, Brusselator: Section 5.6.5
Brusselator with diffusion: Example 6.1, Equation (6.4)
Catalytic reaction: Example 6.3, Equation (6.15)
Superconductivity in a slab: Example 6.4, Equation (6.35)
Buckling a of rod: Example 6.6
Heat and mass transfer: Example 6.7
Brusselator with two coupled cells: Example 7.6, Equation (7.12)
Isothermal reaction: Example 7.8, Equation (7.15)
Model related to nerve impulses: Equation (7.17)
Chemical reaction with autocatalytic step: Example 7.15, Equation (7.28)

Forced van der Pol equation: Equation (7.34)
Predator-prey model: Equation (7.36)
Zeroth-order reaction in a CSTR: Section 8.6

References

Abbreviations:

ARMA   Archive for Rational Mechanics and Analysis
IJBC   International Journal of Bifurcation and Chaos
CSF   Chaos, Solitons & Fractals
IMA   Institute for Mathematics and its Applications
ISNM   International Series of Numerical Mathematics
SIAM   Society for Industrial and Applied Mathematics
ZAMM   Zeitschrift für Angewandte Mathematik und Mechanik,
          Journal of Applied Mathematics and Mechanics
ZAMP   Zeitschrift für Angewandte Mathematik und Physik,
          Journal of Applied Mathematics and Physics

[Abb77]   Abbott, J.P.: Numerical continuation methods for nonlinear equations and bifurcation problems. Ph.D. Thesis. Australian National University 1977

[Abb78]   Abbott, J.P.: An efficient algorithm for the determination of certain bifurcation points. J. Comput. Appl. Math. **4** (1978) 19–27

[AbHH93]   Abed, E.H., Houpt, P.K., Hosny, W.M.: Bifurcation analysis of surge and rotating stall in axial flow compressors. J. Turbomachinery **115** (1993) 817–824

[AbL90]   Abed, E.H., Lee, H.-C.: Nonlinear stabilization of high angle-of-attack flight dynamics using bifurcation control. Proc. American Control Conference, San Diego 1990

[AbV84]   Abed, E.H., Varaiya, P.P.: Nonlinear oscillations in power systems. Int. J. Electric Power and Energy Systems **6** (1984) 37–43

[AbCD91]   Abraham, R.H., Corliss, J.B., Dorband, J.E.: Order and Chaos in the Toral Logistic Lattice. IJBC **1** (1991) 227-234

[AbM67]   Abraham, R.H., Marsden, J.E.: Foundation of Mechanics. Benjamin, New York 1967

[AbS88]   Abraham, R.H., Shaw, C.D.: Dynamics—The Geometry of Behavior. Aerial Press, Santa Cruz. four parts (1982-1988)

[AgS05]   Ageno, A., Sinopoli, A.: Lyapunov's exponents for nonsmooth dynamics with impacts: stability analysis of the rocking block. IJBC **15** (2005) 2015-2039

[AlCK79]   Albrecht, J., Collatz, L., Kirchgässner, K. (Eds.): Constructive Methods for Nonlinear Boundary Value Problems and Nonlinear Oscillations. Proceedings of a Conference in Oberwolfach 1978. Birkhäuser, Basel 1979, ISNM **48**

[AlK85]   Al-Humadi, A., Kazarinoff, N.D.: Hopf bifurcation in the Watt steam engine. J. Inst. Math. Appl. **21** (1985) 133–136

[AlB88]   Allgower, E.L., Böhmer, K.: Resolving singular nonlinear equations. Rocky Montain J. of Math. **18** (1988) 225–268

[AlBG92]   Allgower, E.L., Böhmer, K., Golubitsky, M.: Bifurcation and Symmetry. Birkhäuser, Basel 1992, ISNM **104**

[AlG80]   Allgower, E.L., Georg, K.: Simplicial and continuation methods for approximating fixed points and solutions to systems of equations. SIAM Review **22** (1980) 28–85

[AlG90]   Allgower, E.L., Georg, K.: Numerical Continuation Methods. Springer, Berlin 1990

[AlGP80]   Allgower, E.L., Glashoff, K., Peitgen, H.-O. (Eds.): Numerical Solution of Nonlinear Equations. Proceedings, Bremen 1980. Lecture Notes in Math. **878**, Springer, Berlin 1981

[AlS85]   Allgower, E.L., Schmidt, P.H.: An algorithm for piecewise-linear approximation of an implicitly defined manifold. SIAM J. Numer. Anal. **22** (1985) 322–346

[AlC84]   Aluko, M., Chang, H.-C.: PEFLOQ: An algorithm for the bifurcation analysis of periodic solutions of autonomous systems. Computers and Chemical Engineering **8** (1984) 355–365

[AnVK87]   Andronov, A.A., Vitt, A.A., Khaikin, S.E.: Theory of Oscillators. Dover Publications, New York 1987

[AnM66]   Anselone, P.M., Moore, R.H.: An extension of the Newton-Kantorovic method for solving nonlinear equations with an application to elasticity. J. Math. Anal. Appl. **13** (1966) 476–501

[AnFK87]   Antonelli, P.L., Fuller, K.D., Kazarinoff, N.D.: A study of large amplitude periodic solutions in a model for starfish predation of coral. IMA J. Math. Appl. in Medicine and Biol. **4** (1987) 207–214

[Arn98]   Arnold, L.: Random Dynamical Systems. Springer, Berlin 1998

[Arn73]   Arnol'd, V.I.: Ordinary Differential Equations. MIT Press, Cambridge 1973

[Arn83]   Arnol'd, V.I.: Geometrical Methods in the Theory of Ordinary Differential Equations. Springer, New York 1983

[Arn84]   Arnol'd, V.I.: Catastrophe Theory. Springer, Berlin 1984

[Aro86]   Aronson, D.G., McGehee, R.P., Kevrekidis, I.G., Aris, R.: Entrainment regions for periodically forced oscillators. Phys. Rev. **A 33** (1986) 2190–2192

[ArW75]   Aronson, D.G., Weinberger, H.F.: Nonlinear Diffusion in Population Genetics, Combustion, and Nerve Propagation. Lecture Notes in Math. **446**. Springer, Berlin 1975

[ArP82]   Arrowsmith, D.K., Place, C.M.: Ordinary Differential Equations. Chapman & Hall, London 1982

[ArP90]   Arrowsmith, D.K., Place, C.M.: An Introduction to Dynamical Systems. Cambridge University Press, Cambridge 1990

[AsMR88]   Ascher, U.M., Mattheij, R.M.M., Russell, R.D.: Numerical Solution of Boundary Value Problems for Ordinary Differential Equations. Prentice Hall, Englewood Cliffs 1988

[AsR81]   Ascher, U.M., Russell, R.D.: Reformulation of boundary value problems into "standard" form. SIAM Review **23** (1981) 238–254

[AsR85]   Ascher, U.M., Russell, R.D. (Eds.): Numerical Boundary Value Problems. Proceedings Vancouver 1984. Progress in Scientific Computing, Vol. 5. Birkhäuser, Boston 1985

[Aul84]   Aulbach, B.: Trouble with linearization. In: Neunzert, H. (Ed.): Mathematics in Industry. Teubner, Stuttgart 1984

[AwL03]   Awrejcewicz, J., Lamarque, C.-H.: Bifurcation and Chaos in Nonsmooth Mechanical Systems. World Scientific, New Jersey 2003

[BaD86]   Babloyantz, A., Destexhe, A.: Low-dimensional chaos in an instance of epilepsy. Proc. Natl. Acad. Sci. USA **83** (1986) 3513

[Back92]   Back, A., Guckenheimer, J., Myers, M., Wilcin, F. and Worfolk, P.: DsTool; Computer assisted exploration of dynamical systems. Notices of the ACM **39** (4) (1992) 303–309

[BaLS96]   Bai, F., Lord, G.J., Spence, A.: Numerical computations of connecting orbits in discrete and continuous dynamical systems. IJBC **6** (1996) 1281–1293

[Bak86]   Bak, P.: The devil's staircase. Physics Today **39** (1986) 38–45

[BaL82]   Balakotaiah, V., Luss, D.: Structure of the steady-state solutions of lumped-parameter chemically reacting systems. Chem. Eng. Sci. **37** (1982) 1611–1623

[Bar09]   Barton, D.A.W.: Stability Calculations for piecewise-smooth delay equations. IJBC **19** (2009) 639-650

[BaM90]   Bayliss, A., Matkowsky, B.J.: Two routes to chaos in condensed phase combustion. SIAM J. Appl. Math. **50** (1990) 437–459

[BaW78]   Bazley, N.W., Wake, G.C.: The disappearance of criticality in the theory of thermal ignition. ZAMP **29** (1978) 971–976

[BaW81]   Bazley, N.W., Wake, G.C.: Criticality in a model for thermal ignition in three or more dimensions. J. Appl. Math. Phys. ZAMP **32** (1981) 594–601

[Bec79]   Becker, K.-H.: Numerische Lösung von Verzweigungsproblemen bei nichtlinearen Schwingungen mit der Hilfe der Mehrzielmethode. Diploma Thesis, Mathematisches Institut, Techn. Univ. München 1979

[BeS80]   Becker, K.-H., Seydel, R.: A Duffing equation with more than 20 branch points. In: [AlGP80]

[BeR77]   Beeler, G.W., Reuter, H.: Reconstruction of the action potential of ventricular myocardial fibres. J. Physiol. **268** (1977) 177–210

[Bell73]   Bell, G.I.: Predator–prey equations simulating an immune response. Math. Biosci. **16** (1973) 291–314

[BeZ03]   Bellen, A., Zennaro, M.: Numerical Methods for Delay Differential Equations. Oxford 2003

[Bén1901]   Bénard, H.: Les tourbillons cellulaires dans une nappe liquide transportant de la chaleur par convection en regime permanent. Ann. Chim. Phys. **7**, Ser.23 (1901) 62

[BeR90]   Benedettini, F., Rega, G: Numerical simulations of chaotic dynamics in a model of an elastic cable. Nonl. Dyn. **1** (1990) 23–38

[BeGS76]   Benettin, G., Galgani, L., Strelcyn, L.-M.: Kolmogorov entropy and numerical experiments. Phys. Rev. **A 14** (1976) 2338–2345

[Ben78]   Benjamin, T.B.: Bifurcation phenomena in steady flows of a viscous fluid. I. Theory. Proc. R. Soc. Lond. **A 359** (1978) 1–26

[BeDMP80]   Bergé, P., Dubois, M., Manneville, P., Pomeau, Y.: Intermittency in Rayleigh-Bénard convection. J. Phys. Lett. **41** (1980) 341

[BeGMT03]   Beuter, A., Glass, L., Mackey, M.C., Titcombe, M.S.: Nonlinear Dynamics in Physiology and Medicine. Springer, New York 2003

[Beyn80]   Beyn, W.-J.: On discretizations of bifurcation problems. ISNM **54** (1980) 46–73

[Beyn84]   Beyn, W.-J. : Defining equations for singular solutions and numerical applications. ISNM **70** (1984) 42–56. In: [KüMW84]

[Beyn85]   Beyn, W.-J.: Zur numerischen Berechnung mehrfacher Verzweigungs-
              punkte. ZAMM **65** (1985) T370–T371
[Beyn87a]   Beyn, W.-J.: On invariant closed curves for one-step methods. Numer.
              Math **51** (1987) 103–122
[Beyn87b]   Beyn, W.-J.: The effect of discretization on homoclinic orbits. ISNM
              79 (1987) 1–8
[Beyn90a]   Beyn, W.-J.: The numerical computation of connecting orbits in dy-
              namical systems. IMA J.Numer.Anal. **9** (1990) 379–405
[Beyn90b]   Beyn, W.-J.: Global bifurcations and their numerical computation.
              In: Roose, D. et al. (Eds.): Continuation and Bifurcations. Kluwer
              Academic Publishers (1990) 169–181
[BeD81]   Beyn, W.-J., Doedel, E.: Stability and multiplicity of solutions to dis-
              cretizations of nonlinear ordinary differential equations. SIAM J. Sci.
              Stat. Comput. **2** (1981) 107–120
[BeHK04]   Beyn, W.-J., Hüls, T., Kleinkauf, J.-M.: Numerical Analysis of dege-
              nerate connecting orbits for maps. IJBC **14** (2004) 3385–3407
[Bog75]   Bogdanov, R.I.: Versal deformations of a singular point on the plane in
              the case of zero eigenvalues. Functional Anal. Appl. **9** (1975) 144–145
[Bohl79]   Bohl, E.: On the bifurcation diagram of discrete analogues of ordinary
              bifurcation problems. Math. Meth. Appl. Sci. **1** (1979) 566–571
[Bor02]   Borckmans, P. (et al.): Diffusive instabilities and chemical reactions.
              IJBC **12** (2002) 2307–2332
[BoDI78]   Bouc, R., Defilippi, M., Iooss, G.: On a problem of forced nonlinear
              oscillations. Numerical example of bifurcation into an invariant torus.
              Nonlinear Analysis **2** (1978) 211–224
[BrA81]   Brady, J.F., Acrivos, A.: Steady flow in a channel or tube with an
              accelerating surface velocity. An exact solution to the Navier-Stokes
              equations with reverse flow. J. Fluid Mech. **112** (1981) 127–150
[Bra78]   Braun, M.: Differential Equations and Their Applications. Springer,
              New York 1978
[Bre73]   Brent, R.P.: Some efficient algorithms for solving systems of nonlinear
              equations. SIAM J. Numer. Anal. **10** (1973) 327–344
[BrRR81]   Brezzi, F., Rappaz, J., Raviart, P.-A.: Finite dimensional approxima-
              tion of nonlinear problems. Part III: Simple bifurcation points. Numer.
              Math. **38** (1981) 1–30
[BrUF84]   Brezzi, F., Ushiki, S., Fujii, H.: "Real" and "ghost" bifurcation dyna-
              mics in difference schemes for ODEs. ISNM **70** (1984) 79–104
[BrK86]   Broomhead, D.S., King, G.P.: Extracting qualitative dynamics from
              experimental data. Physica **20D** (1986) 217–236
[BrG71]   Brown, K.M., Gearhardt, W.B.: Deflation techniques for the calcu-
              lation of further solutions of a nonlinear system. Numer. Math. **16**
              (1971) 334–342
[Bro65]   Broyden, C.G.: A class of methods for solving nonlinear simultaneous
              equations. Math. Comput. **19** (1965) 577–593
[Bul71]   Bulirsch, R.: Die Mehrzielmethode zur numerischen Lösung von nicht-
              linearen Randwertproblemen und Aufgaben der optimalen Steuerung.
              Report der Carl-Cranz-Gesellschaft, 1971
[BuS66]   Bulirsch, R., Stoer, J.: Numerical treatment of ordinary differential
              equations by extrapolation methods. Numer. Math. **8** (1966) 1–13
[Bus62]   Busch, W.: Max und Moritz. (1865) Facsimile W. Busch Gesellschaft,
              Hannover 1962
[Bus77]   Busse, F.H.: Mathematical problems of dynamo theory. In: [Rab77]

[Bus81]   Busse, F.H.: Transition to turbulence in Rayleigh-Bénard convection.
            In: [SwG81]

[Bus82]   Busse, F.H.: Transition to turbulence in thermal convection. In: [ZiO82]

[Carr81]   Carr, J.: Applications of Center Manifold Theory. Springer, New York
            1981

[CaP92]   Cartwright, J.H.E., Piro, O.: The dynamics of Runge-Kutta methods.
            IJBC **2** (1992) 427–449

[CaDGS76]   Casartelli, M., Diana, E., Galgani, L., Scotti, A.: Numerical compu-
            tations on a stochastic parameter related to the Kolmogorov entropy.
            Phys. Rev. **A 13** (1976) 1921–1930

[ChK94]   Champneys, A.R., Kuznetsov, Yu.A.: Numerical detection and con-
            tinuation of codimension-two homoclinic bifurcations. IJBC **4** (1994)
            785–882

[Chan84a]   Chan, T.F.: Newton-like pseudo-arclength methods for computing sim-
            ple turning points. SIAM J. Sci. Stat. Comput. **5** (1984) 135–148

[Chan84b]   Chan, T.F.: Techniques for large sparse systems arising from continua-
            tion methods. ISNM **70** (1984) 116–128

[Chan85]   Chan, T.F.: An approximate Newton method for coupled nonlinear
            systems. SIAM J. Numer. Anal. **22** (1985) 904–913

[ChK82]   Chan, T.F., Keller, H.B.: Arclength continuation and multigrid tech-
            niques for nonlinear elliptic eigenvalue problems. SIAM J. Sci. Stat.
            Comput. **3** (1982) 173–194

[ChPL95]   Chandiramani, N.K., Plaut, R.P., Librescu, L.I.: Nonperiodic flutter
            of a buckled composite panel. Sādhanā **20** (1995) 671–689

[ChCJ05]   Chang, S.-L., Chien, C.-S., Jeng, B.-W.: Tracing the solution surface
            with folds or a two-parameter system. IJBC **15** (2005) 2689–2700

[Chay95]   Chay, T.R.: Bifurcations in heart rhythms. IJBC **5** (1995) 1439–1486

[ChFL95]   Chay, T.R., Fan, Y.S., Lee, Y.S.: Bursting, spiking, chaos, fractals,
            and universality in biological rhythms. IJBC **5** (1995) 595–635

[ChL85]   Chay, T.R., Lee, Y.S.: Phase resetting and bifurcation in the ventricu-
            lar myocardium. Manuscript 1985, Department of Biological Sciences,
            Univ. Pittsburgh

[ChR85]   Chay, T.R., Rinzel, J.: Bursting, beating, and chaos in an excitable
            membrane model. Biophys. J. **47** (1985) 357–366

[ChHW01]   Chen, K., Hussein, A., Wan, H.-B.: An analysis of Seydel's test func-
            tion methods for nonlinear power flow equations. Int. J. Comput.
            Math. **78** (2001) 451–470

[ChWC01]   Chen, D.S., Wang, H.O., Chen, G.: Anti-control of Hopf bifurcation.
            in: Circuits and Systems I: Fundamental Theory and Applications.
            IEEE Transactions **48** (2001) 661–672

[ChZXW04]   Chen, S., Zhang, Q., Xie, J., Wang, C.: A stable-manifold-based me-
            thod for chaos control and synchronization. CSF **20** (2004) 947–954

[Chi99]   Chicone, C.: Ordinary Differential Equations with Applications. Sprin-
            ger, New York 1999

[ChLM90]   Chossat, P., Lauterbach, R., Melbourne, I.: Steady-state bifurcation
            with $O(3)$-symmetry. ARMA **113** (1990) 313–376

[ChH82]   Chow, S.-N., Hale, J.K.: Methods of Bifurcation Theory. Springer, New
            York 1982

[ChMY78]   Chow, S.N., Mallet-Paret, J., Yorke, J.A.: Finding zeros of maps: ho-
            motopy methods that are constructive with probability one. Math.
            Comp. **32** (1978) 887–899

[ChDK87]   Chua, L.O., Desoer, C.A., Kuh, E.S.: Linear and Nonlinear Circuits.
            McGraw-Hill, New York 1987

[ChCX06] Chung, K.W., Chan, C.L., Xu, J.: A perturbation-incremental method for delay differential equations. IJBC **16** (2006) 2529–2544

[ClB74] Clever, R.M., Busse, F.H.: Transition to time-dependent convection. J. Fluid Mech. **65** (1974) 625–645

[ClST00] Cliffe, K.A., Spence, A., Tavener, S.J.: Numerical Analysis of Bifurcation Problems. Acta Numerica **9** (2000) 39–131

[CoL55] Coddington, E.A., Levinson, N.: Theory of Ordinary Differential Equations. McGraw-Hill, New York 1955

[Coo72] Cooperrider, N.K.: The hunting behavior of conventional railway trucks. Transactions of the ASME, Journal of Engineering for Industry **94** (1972) 752–762

[Cor92] Corless, R.M.: Defect-controlled numerical methods and shadowing for chaotic differential equations. Physica **60D** (1992) 323–334

[Cos94] Costa, P.J.: The Hartman-Grobman theorem for nonlinear partial differential evolution equations. Center for Applied Mathematics, University of St. Thomas (1994)

[CrR71] Crandall, M.G., Rabinowitz, P.H.: Bifurcation from simple eigenvalues. J. Functional Anal. **8** (1971) 321–340

[CrR77] Crandall, M.G. Rabinowitz, P.H.: The Hopf bifurcation theorem in infinite dimensions. ARMA **67** (1977/78) 53–72

[CrO84] Crawford, J.D., Omohundro, S.: On the global structure of period doubling flows. Physica **13D** (1984) 161–180

[Cru80] Crutchfield, J., Farmer, D., Packard, N., Shaw, R., Jones, G., Donnelly, R.J.: Power spectral analysis of a dynamical system. Phys. Lett. **76A** (1980) 1–4

[CrH80] Crutchfield, J.P., Huberman, B.A.: Fluctuations and the onset of chaos. Phys. Lett. **77A** (1980) 407

[CrK87] Crutchfield, J.P., Kaneko, K.: Phenomenology of spatio-temporal chaos. In: Directions in Chaos (Ed.: Hao Bai-lin) World Scientific, Singapore 1987

[Cur78] Curry, J.H.: A generalized Lorenz system. Comm. Math. Phys. **60** (1978) 193–204

[Cvi84] Cvitanovic, P. (Ed.): Universality in Chaos. A Reprint Selection. A. Hilger, Bristol 1984

[DaB08] Dahlquist, G., Björck, Å: Numerical Methods in Scientific Computing. Vol. 1, SIAM, Philadelphia 2008

[DaT06] Dankowicz, H., Thakur, G.: A Newton method for locating invariant tori of maps. IJBC **16** (2006) 1491–1503

[Dav53] Davidenko, D.F.: On a new method of numerical solution of systems of nonlinear equations. Dokl. Akad. Nauk. SSSR **88** (1953) 601–602 Mathematical Reviews **14**, p. 906

[DeK81] Decker, D.W., Keller, H.B.: Path following near bifurcation. Comm. Pure Appl. Math. **34** (1981) 149–175

[DeKK83] Decker, D.W., Keller, H.B., Kelley, C.T.: Convergence rates for Newton's method at singular points. SIAM J. Numer. Anal. **20** (1983) 296–314

[DeRR90] De Dier, B., Roose, D., Van Rompay, P.: Interaction between fold and Hopf curves leads to new bifurcation phenomena. ISNM **92** (1990) 171–186

[DeS67] Deist, F.H., Sefor, L.: Solution of systems of non-linear equations by parameter variation. Computer J. **10** (1967) 78–82

[Del92] Dellnitz, M.: Computional bifurcation of periodic solutions in system with symmetry. IMA J. Math. Appl. **12** (1992) 429–455

[DeH97]   Dellnitz, M., Hohmann, A.: A subdivision algorithm for the compu-
            tation of unstable manifolds and global attractors. Numer. Math. **75**
            (1997) 293–317
[DeW89]   Dellnitz, M., Werner, B.: Computational methods for bifurcation pro-
            blems with symmetries—with special attention to steady state and
            Hopf bifurcation points. J. Comput. and Appl. Math. **26** (1989) 97–
            123
[DeR81]   Den Heijer, C., Rheinboldt, W.C.: On steplength algorithms for a class
            of continuation methods. SIAM J. Numer. Anal. **18** (1981) 925–948
[DeM77]   Dennis, J.E., Moré, J.J.: Quasi-Newton methods, motivation and
            theory. SIAM Review **19** (1977) 46–89
[Deu79]   Deuflhard, P.: A stepsize control for continuation methods and its
            special application to multiple shooting techniques. Numer. Math. **33**
            (1979) 115–146
[DeFK87]   Deuflhard, P., Fiedler, B., Kunkel, P.: Efficient numerical path follo-
            wing beyond critical points. SIAM J. Numer. Anal. **24** (1987) 912–927
[Dev86]   Devaney, R.L.: An Introduction to Chaotic Dynamical Systems. Menlo
            Park, Benjamin 1986
[DíF02]   Díaz-Sierra, R., Fairén, V.: New methods for the estimation of domains
            of attraction of fixed points from Lyapunov functions. IJBC **12** (2002)
            2467–2477
[diB08]   di Bernardo, M. et al.: Bifurcations in Nonsmooth Dynamical Systems.
            SIAM Review **50** (2008) 629–701
[diBFHH99] di Bernardo, M., Feigin, M.I., Hogan, S.J., Homer, M.E.: Local ana-
            lysis of C-bifurcations in n-dimensional piecewise-smooth dynamical
            Systems. CSF **10** (1999) 1881–1908
[diBKN03]   di Bernardo, M., Kowalczyk, P., Nordmark, A.: Sliding bifurcations: a
            novel mechanism for the sudden onset of chaos in dry friction oscilla-
            tors. IJBC **13** (2003) 2935–2948
[DiLR91]   Dieci, L., Lorenz, J., Russell, R.D.: Numerical calculation of invariant
            tori. SIAM J. Sci. Stat. Comput. **12** (1991) 607–647
[Dob93]   Dobson, I.: Computing a closest bifurcation instability in multidimen-
            sional parameter space. J. Nonlinear Science **3** (1993) 307–327
[Dob03]   Dobson, I.: Distance to bifurcation in multidimensional parameter
            space: margin sensitivity and closest bifurcations. in: Bifurcation Con-
            trol, Lecture Notes in Control and Information Sciences. Springer, Ber-
            lin 2003
[DoC89]   Dobson, I., Chiang, H.-D.: Towards a theory of voltage collapse in
            electric power systems. Systems and Control Letters **13** (1989) 253–
            262
[Doe81]   Doedel, E.J.: AUTO: A program for the automatic bifurcation analysis
            of autonomous systems. Congressus Numerantium **30** (1981) 265–284
[Doe97]   Doedel, E.J., Champneys, A.R., Fairgrieve, T.F., Kuznetsov, Yu.A.,
            Sandstede, B., Wang, X.J.: AUTO97: Continuation and bifurcation
            software for ordinary diffential equations.
            http://cmvl.cs.concordia.ca
[DoGK03]   Doedel, E.J., Govaerts, W., Kuznetsov, Yu.A.: Computation of peri-
            odic solution bifurcations in ODEs using bordered systems. SIAM J.
            Numer. Anal. **41** (2003) 401–435
[DoH83]   Doedel, E.J., Heinemann, R.F.: Numerical computation of periodic
            solution branches and oscillatory dynamics of the stirred tank reactor
            with $A \rightarrow B \rightarrow C$ reactions. Chem. Eng. Sci. **38** (1983) 1493–1499

[DoKK91]   Doedel, E.J., Keller, H.B., Kernevez, J.P.: Numerical analysis and control of bifurcation problems. IJBC **1** (1991) Part I: 493–520, Part II: 745–772

[DoK86]   Doedel, E.J., Kernevez, J.P.: A numerical analysis of wave phenomena in a reaction diffusion model. In: Othmer, H.G. (Ed.): Nonlinear Oscillations in Biology and Chemistry. Springer Lecture Notes in Biomathem. **66** (1986)

[Doe03]   Doedel, E.J., Paffenroth, R.C., Keller, H.B, Dichmann, D.J., Galán-Vioque, J., Vanderbauwhede, A.: Computation of periodic solutions of conservative systems with application to the 3-body problem. IJBC **13** (2003) 1353–1381

[Doe07]   Doedel, E.J., Romanov, V.A., Paffenroth, R.C., Keller, H.B., Dichmann, D.J., Galán-Vioque, A., Vanderbauwhede, A.: Elemental periodic orbits associated with the libration points in the circular restricted 3-body problem. IJBC **17** (2007) 2625–2677

[DoH06]   Donde, V., Hiskens, I.A.: Shooting methods for locating grazing phenomena in hybrid systems. IJBC **16** (2006) 671–692

[DrN92]   Dressler, U., Nitsche, G.: Controlling chaos using time delay coordinates. Phys. Rev. Lett. **68** (1992) 1–4

[Duc92]   Ducci, A.: Identifikation strömungsselbsterregter Schwingungssysteme. VDI-Verlag Düsseldorf 1992

[Duf18]   Duffing, G.: Erzwungene Schwingungen bei veränderlicher Eigenfrequenz. Vieweg, Braunschweig 1918

[Eck81]   Eckmann, J.P.: Roads to turbulence in dissipative dynamical systems. Reviews of Modern Physics **53** (1981) 643–654

[EcKRC86]   Eckmann, J.P., Kamphorst, S.O., Ruelle, D., Ciliberto, S.: Lyapunov exponents from time series. Phys. Rev. **A 34** (1986) 4971–4979.

[Ede88]   Edelstein-Keshet, L.: Mathematical Models in Biology. Birkhäuser, Boston 1988

[ElA05]   El Aroudi, A., Debbat, M., Giral, R., Olivar, G., Benadero, L., Toribio, E.: Bifurcations in DC-DC switching converters: review of methods and applications. IJBC **15** (2005) 1549–1578

[ElN90]   El Naschie, M.S.: Stress, Stability and Chaos in Structural Engineering: An Energy Approach. McGraw-Hill, London 1990

[EnKO07]   England, J.P., Krauskopf, B., Osinga, H.M.: Computing two-dimensional global invariant manifolds in slow-fast systems. IJBC **17** (2007) 805–822

[EnL91]   Englisch, V., Lauterborn, W.: Regular window structure of a double-well duffing oscillator. Phys. Rev. **A 44** (1991) 916–924

[Eps83]   Epstein, I.R.: Oscillations and chaos in chemical systems. Physica **7D** (1983) 47–56

[Erm02]   Ermentrout, B.: Simulating, Analyzing, and Animating Dynamical Systems. SIAM, Philadelphia 2002

[Eul52]   Euler, L.: De Curvis Elasticis, Methodus Inveniendi Lineas Curvas Maximi Minimive Proprietate Gaudentes. Additamentum I. (1744) In: Opera Omnia I, Vol.24, 231–297, Zürich 1952

[FaJ92]   Fairgrieve, T.F., Jepson, A.D.: O.K. Floquet multipliers. SIAM J. Numer. Anal. **28** (1992) 1446–1462

[Far81]   Farmer, J.D.: Spectral broadening of period-doubling bifurcation sequences. Phys. Rev. Lett. **47** (1981) 179–182

[FaOY83]   Farmer, J.D., Ott, E., Yorke, J.A.: The dimension of chaotic attractors. Physica **7D** (1983) 153–180

[FäS92]   Fässler, A., Stiefel, E.: Group Theoretical Methods and Their Appli-
             cations. Birkhäuser, Boston 1992
[Feh69]   Fehlberg, E.: Klassische Runge–Kutta Formeln fünfter und siebenter
             Ordnung mit Schrittweitenkontrolle. Computing **4** (1969) 93–106
[Feh70]   Fehlberg, E.: Klassische Runge–Kutta Formeln vierter und niedrige-
             rer Ordnung mit Schrittweiten-Kontrolle und ihre Anwendung auf
             Wärmeleitungsprobleme. Computing **6** (1970) 61–71
[Fei92]   Feichtinger, G.: Hopf bifurcation in an advertising diffusion model. J.
             Economic Behavior and Organization **17** (1992) 401-411
[Fei80]   Feigenbaum, M.J.: Universal behavior in nonlinear systems. Los Ala-
             mos Science **1** (1980) 4–27
[FeJ92]   Feudel, U., Jansen, W.: CANDYS/QA—A software system for the
             qualitative analysis of nonlinear dynamical systems. IJBC **2** (1992)
             773–794
[FiMD91]   Fidkowski, Z.T., Malone, M.F., Doherty, M.F.: Nonideal multicompo-
             nent distillation: use of bifurcation theory for design. AIChE Journal
             37 (1991) 1761–1779
[Fie88]   Fiedler, B.: Global Bifurcation of Periodic Solutions with Symmetry.
             Springer, Berlin 1988
[FiK87]   Fiedler, B., Kunkel, B.: A quick multiparameter test for periodic so-
             lutions. ISNM **79** (1987) 61–70
[FiN74]   Field, R.J., Noyes, R.M.: Oscillations in chemical systems. IV. Limit
             cycle behavior in a model of a chemical reaction. J. Chem. Physics **60**
             (1974) 1877–1884
[Fife79]   Fife, P.C.: Mathematical Aspects of Reacting and Diffusing Systems.
             Lecture Notes in Biomathematics **28**. Springer, Berlin 1979
[Fil88]   Filippov, A.F.: Differential Equations with Discontinuous Righthand
             Sides. Kluwer, Dordrecht 1988
[Fis37]   Fisher, R.A.: The wave of advance of advantageous genes. Ann. Euge-
             nics **7** (1937) 355–369
[Fit61]   FitzHugh, R.: Impulses and physiological states in theoretical models
             of nerve membrane. Biophys. J. **1** (1961) 445–466
[Flo85a]   Flockerzi, D.: Resonance and bifurcation of higher-dimensional tori.
             Proceedings of the AMS **94** (1985) 147–157
[Flo85b]   Flockerzi, D.: On the $T^k \to T^{k+1}$ bifurcation problem. In: Pnevmati-
             kos, S.N. (Ed.): Singularities and Dynamical Systems. Elsevier 1985
[Flo87]   Flockerzi, D.: Codimension two bifurcations near an invariant n-torus.
             In: [KüST87]
[Fof03]   Fofana, M.S.: Delay dynamical systems and applications to nonlinear
             machine-tool chatter. CSF **17** (2003) 731–747
[Foi88]   Foias, C., Jolly, M.S., Kevrekidis, I.G., Sell, G.R., Titi, E.S.: On the
             computation of inertial manifolds. Physics Letters **A 131** (1988) 433–
             436
[FrT79]   Franceschini, V., Tebaldi, C.: Sequences of infinite bifurcations and
             turbulence in a five-mode truncation of the Navier-Stokes equations.
             J. Statistical Phys. **21** (1979) 707
[Fra99]   Franke, C.: Periodic motions and nonlinear dynamics of a wheelset
             model. IJBC **9** (1999) 1983-1994
[FrF01]   Franke, C., Führer, C.: Collocation methods for the investigation of
             periodic motions of constrained multibody systems. Multibody System
             Dynamics **5** (2001) 133–158

[Fra89]   Fraser, A.M.: Reconstructing attractors from scalar time series: A comparison of singular system and redundancy criteria. Physica **34D** (1989) 391–404

[FrD91]   Friedman, M.J., Doedel, E.J.: Numerical computation and continuation of invariant manifolds connecting fixed points. SIAM J. Numer. Anal. **28** (1991) 789–808

[Ful68]   Fuller, A.T.: Conditions for a matrix to have only characteristic roots with negative real parts. J. of Mathem. Analysis and Applications **23** (1968) 71–98

[GaBB95]   Galvanetto, U., Bishop, S.R., Briseghella, L.: Mechanical stick-slip vibrations. IJBC **5** (1995) 637–651

[GaSDW92] Garfinkel, A., Spano, M.L., Ditto, W.L., Weiss, J.N.: Controlling cardiac chaos. Science **257** (1992) 1230–1235

[GaMS91]   Garratt, T.J., Moore, G., Spence, A.: Two methods for the numerical detection of Hopf bifurcations. ISNM **97** (1991) 129–133

[GaH91]   Gatermann, K., Hohmann, A.: Symbolic exploitation of symmetry in numerical pathfollowing. Impact of Computing in Science and Engineering **3** (1991) 330–365

[GeCC06]   Ge, Z.-M., Chang, C.-M., Chen, Y.-S.: Anti-control of chaos of single time scale brushless dc motors and chaos synchronization of different order systems. CSF **27** (2006) 1298–1315

[Gear71]   Gear, C.W.: Numerical Initial Value Problems in Ordinary Differential Equations. Englewood Cliffs, Prentice Hall 1971

[GeL90]   Geist, K., Lauterborn, W.: The nonlinear dynamics of the damped and driven toda chain. Physica **41D** (1990) 1–25

[GePL90]   Geist, K., Parlitz, U., Lauterborn, W.: Comparison of different methods for computing Lyapunov exponents. Progress of Theor. Physics **83** (1990) 875–893

[Geo81a]   Georg, K.: Numerical integration of the Davidenko equation. In: [AlGP81]

[Geo81b]   Georg, K.: On tracing an implicitly defined curve by quasi-Newton steps and calculating bifurcation by local perturbations. SIAM J. Sci. Stat. Comput. **2** (1981) 35–50

[GeST90]   Gerhardt, M., Schuster, H., Tyson, J.J.: A cellular automaton model of excitable media including curvature and dispersion. Science **247** (1990) 1563–1566

[GhT83]   Ghil, M., Tavantzis, J.: Global Hopf bifurcation in a simple climate model. SIAM J. Appl. Math. **43** (1983) 1019–1041

[GiM72]   Gierer, A., Meinhardt, H.: Theory of biological pattern formation. Kybernetik **12** (1972) 30–39

[GiMP81]   Giglio, M., Musazzi, S., Perini, U.: Transition to chaotic behavior via reproducible sequence of period-doubling bifurcations. Phys. Rev. Lett. **47** (1981) 243

[Gil81]   Gilmore, R.: Catastrophe Theory for Scientists and Engineers. John Wiley, New York 1981

[GlP71]   Glansdorff, P., Prigogine, I.: Thermodynamic Theory of Structure, Stability and Fluctuations. Wiley-Interscience, London 1971

[Gla75]   Glass, L.: Combinatorial and topological methods in nonlinear chemical kinetics. J. Chem. Phys. **63** (1975) 1325–1335

[GlM88]   Glass, L., Mackey, M. C.: From Clocks to Chaos: The Rhythms of Life. Princeton University Press, 1988

[GoNP02]   Golovin, A.A., Nepomnyashchy, A.A., Pismen, L.M.: Nonpotential effects in nonlinear dynamics of Marangoni convection. IJBC **12** (2002) 2487–2500

[GoL96]   Golub, G.H., van Loan, C.F.: Matrix Computations. Third Edition. J.Hopkins Univ. Press, Baltimore 1996

[GoKS81]   Golubitsky, M., Keyfitz, B.L., Schaeffer, D.G.: A singularity theory analysis of a thermal-chainbranching model for the explosion peninsula. Comm. Pure Appl. Math. **34** (1981) 433–463

[GoL81]   Golubitsky, M., Langford, W.F.: Classification and unfoldings of degenerate Hopf bifurcations. J. Diff. Eq. **41** (1981) 375–415

[GoS79]   Golubitsky, M., Schaeffer, D.: A theory for imperfect bifurcation via singularity theory. Comm. Pure Appl. Math. **32** (1979) 21–98

[GoS85]   Golubitsky, M., Schaeffer, D.G.: Singularities and Groups in Bifurcation Theory. Vol.1. Springer, New York 1985

[GoS86]   Golubitsky, M., Stewart, I.: Symmetry and stability in Taylor-Couette flow. SIAM J. Math. Anal. **17** (1986) 249–288

[GoSS88]   Golubitsky, M., Stewart, I., Schaeffer, D.G.: Singularities and Groups in Bifurcation Theory. Volume II. Springer, New York 1988

[Gov00]   Govaerts, W.: Numerical Methods for Bifurcations of Dynamical Equilibria. SIAM, Philadelphia 2000

[GrSTW85] Graff, M., Scheidl, R., Troger, H., Weinmüller, E.: An investigation of the complete post-buckling behavior of axisymmetric spherical shells. J. Appl. Math. Phys. (ZAMP) **36** (1985) 803–821

[GrP83]   Grassberger, P., Procaccia, I.: Measuring the strangeness of strange attractors. Physica **9D** (1983) 189–208

[GrOPY84] Grebogi, C., Ott, E., Pelikan, S., Yorke, J.A.: Strange attractors that are not chaotic. Physica **13D** (1984) 261–268

[GrOY83]   Grebogi, C., Ott, E., Yorke, J.A.: Crises, sudden changes in chaotic attractors, and transient chaos. Physica **7D** (1983) 181–200

[GrHH78]   Greenberg, J.M., Hassard, B.D., Hastings, S.P.: Pattern formation and periodic structures in systems modeled by reaction-diffusion equations. Bull. Amer. Math. Soc. **84** (1978) 1296

[GrH78]   Greenberg, J.M., Hastings, S.: Spatial patterns for discrete models of diffusion in excitable media. SIAM J. Appl. Math. **34** (1978) 515–523

[Gri85]   Griewank, A.: On solving nonlinear equations with simple singularities or nearly singular solutions. SIAM Review **27** (1985) 537–563

[GrR83]   Griewank, A., Reddien, G.W.: The calculation of Hopf bifurcation points by a direct method. IMA J. Numer. Anal. **3** (1983) 295–303

[GrR84a]   Griewank, A., Reddien, G.W.: Characterization and computation of generalized turning points. SIAM J. Numer. Anal. **21** (1984) 176–185

[GrR84b]   Griewank, A., Reddien, G.W.: Computation of generalized turning points and two-point boundary value problems. ISNM **70** (1984) 385–400. In: [KüMW84]

[GrR90]   Griewank, A., Reddien, G.W.: Computation of cusp singularities for operator equations and their discretizations. ISNM **92** (1990) 133–153

[GrT77]   Großmann, S., Thomae, S.: Invariant distributions and stationary correlation functions of one-dimensional discrete processes. Z. Naturforschung **32a** (1977) 1353–1363

[Guc78]   Guckenheimer, J.: The catastrophe controversy. The Mathem. Intelligencer **1** (1978) 15–20

[GuH83]   Guckenheimer, J., Holmes, Ph.: Nonlinear Oscillations, Dynamical Systems and Bifurcation of Vector Fields. Springer, New York 1983

[GuK91]   Guckenheimer, J., Kim, S.: Computational environments for exploring dynamical systems. IJBC **1** (1991) 269–276

[GuMS97]   Guckenheimer, J., Myers, M., Sturmfels, B.: Computing Hopf bifurcations I. SIAM J. Numer. Anal. **34** (1997) 1–21

[GuGS81]   Guevara, M.R., Glass, L., Shrier, A.: Phase locking, period-doubling bifurcations, and irregular dynamics in periodically stimulated cardiac cells. Science **214** (1981) 1350–1353

[HaHS76]   Hadeler, K.P., an der Heiden, U., Schumacher, K.: Generation of nervous impulse and periodic oscillations. Biol. Cybernetics **23** (1976) 211–218

[HaMS80]   Hadeler, K.P., de Mottoni, P., Schumacher, K.: Dynamic models for animal orientation. J. Mathem. Biol. **10** (1980) 307–332

[HaR75]   Hadeler, K.P., Rothe, F.: Travelling fronts in nonlinear diffusion equations. J. Math. Biol. **2** (1975) 251–263

[Hahn67]   Hahn, W.: Stability of Motion. Springer, Berlin 1967

[HaLR89]   Hairer, E., Lubich, Ch., Roche, M.: The Numerical Solution of Differential-Algebraic Systems by Runge–Kutta Methods. Springer, Berlin 1989

[HaLW02]   Hairer, E., Lubich, C., Wanner, G.: Geometric Numerical Integration. Springer, Berlin 2002

[HaNW87]   Hairer, E., Nørsett, S.P., Wanner, G.: Solving Ordinary Differential Equations I. Nonstiff Problems. Springer, Berlin 1987

[HaW91]   Hairer, E., Wanner, G.: Solving Ordinary Differential Equations II. Stiff and Differential-Algebraic Problems. Springer, Berlin 1991

[Hak77]   Haken, H.: Synergetics. An Introduction. Springer, Berlin 1977

[Hale69]   Hale, J.K.: Ordinary Differential Equations. Wiley-Interscience, New York 1969

[HaK91]   Hale, J.K., Koçak, H.: Dynamics and Bifurcations. Springer, New York 1991

[HaPT00]   Hale, J.K., Peletier, L.A., Troy, W.C.: Exact homoclinic and heteroclinic solutions of the Gray-Scott model for autocatalysis. SIAM J. Appl. Math. **61** (2000) 102–130

[HaVL93]   Hale, J.K., Verduyn Lunel, S.M.: Introduction to Functional Differential Equations. Springer, New York 1993

[Hal86]   Halsey, T.C., Jensen, M.H., Kadanoff, L.P., Procaccia, I., Shraiman, B.I.: Fractal measures and their singularities: The characterization of strange sets. Phys. Rev. **A 33** (1986) 1141–1151

[Hao87]   Hao Bai-Lin (Ed.): Directions in Chaos. Vol 1. World Scientific, Singapore 1987

[Har61]   Harmon, L.D.: Studies with artificial neurons, I. Properties and functions of an artificial neuron. Kybernetik **1** (1961) 89–101

[HaFM89]   Harrison, R.G., Forysiak, W., Moloney, J.V.: Regular and chaotic dynamics of coherently pumped 3-level lasers. In: Engelbrecht, J. (Ed.): Nonlinear Waves in Active Media. Springer-Verlag Berlin, Heidelberg 1989

[Har64]   Hartman, P.: Ordinary Differential Equations. Wiley, New York 1964

[Has61]   Haselgrove, C.B.: The solution of nonlinear equations and of differential equations with two-point boundary conditions. Comput. J. **4** (1961) 255–259

[Has78]   Hassard, B.D.: Bifurcation of periodic solutions of the Hodgkin–Huxley model for the squid giant axon. J. Theor. Biol. **71** (1978) 401–420

[Has80]   Hassard, B.D.: BIFOR2, analysis of Hopf bifurcation in an ordinary differential system. Computer program. Dept. of Math., Buffalo 1980

[Has00]   Hassard, B.: An unusual Hopf bifurcation: the railway bogie. IJBC **10** (2000) 503–507

[HaKW81]   Hassard, B.D., Kazarinoff, N.D., Wan, Y.H.: Theory and Applications of Hopf Bifurcation. Cambridge University Press, Cambridge 1981

[HaS89]   Hassard, B.D., Shiau, L.J.: Isolated periodic solutions of the Hodgkin-Huxley equations. J. Theor. Biol. **136** (1989) 267–280

[HaZ94]   Hassard, B., Zhang, J.: Existence of a homoclinic orbit of the Lorenz system by precise shooting. SIAM J. Math. Anal. **25** (1994) 179–196

[HaK03]   Hasselblatt, B., Katok, A.: A First Course in Dynamics. Cambridge University Press, New York 2003

[Has75]   Hastings, S.P.: Some mathematical problems from neurobiology. Amer. Math. Monthly **82** (1975) 881–895

[Has82]   Hastings, S.P.: Single and multiple pulse waves for the FitzHugh-Nagumo equations. SIAM J. Appl. Math. **42** (1982) 247–260

[HaM75]   Hastings, S.P., Murray, J.D.: The existence of oscillatory solutions in the Field–Noyes model for the Belousov–Zhabotinskii reaction. SIAM J. Appl. Math. **28** (1975) 678–688

[Hea88]   Healey, T.J.: A group-theoretic approach to computational bifurcation problems with symmetry. Computer Methods in Applied Mechanics and Engineering **67** (1988) 257–295

[Hei06]   Heider, P.: Rating the performance of shooting methods for the computation of periodic orbits. IJBC **16** (2006) 199–206

[HeOR79]   Heinemann, R.F., Overholser, K.A., Reddien, G.W.: Multiplicity and stability of premixed laminar flames: An application of bifurcation theory. Chem. Eng. Sci. **34** (1979) 833–840

[HeP81]   Heinemann, R.F., Poore, A.B.: Multiplicity, stability, and oscillatory dynamics of the tubular reactor. Chem. Eng. Sci. **36** (1981) 1411–1419

[Hel80]   Helleman, R.H.G.: Self-generated chaotic behavior in nonlinear mechanics. In: Cohen, E.G.D. (Ed.): Fundamental Problems in Statistical Mechanics. Vol.5. North Holland, Amsterdam 1980

[Hen02]   Henderson, M.E.: Multiple parameter continuation: computing implicitly defined k-manifolds. IJBC **12** (2002) 451–476

[Hén76]   Hénon, M.: A two-dimensional mapping with a strange attractor. Commun. Math. Phys. **50** (1976) 69

[HeP83]   Hentschel, H.G.E., Procaccia, I.: The infinite number of dimensions of probabilistic fractals and strange attractors. Physica **8D** (1983) 435

[HeM85]   Hess, B., Markus, M.: The diversity of biochemical time patterns. Berichte Bunsen-Gesellschaft Physik. Chem. **89** (1985) 642–651

[Het89]   Hetzer, G.: A multiparameter sensitivity analysis of a 2D diffusive climate model. Impact of Computing in Science and Engineering **1** (1989) 327–393

[HeS91]   Hetzer, G., Schmidt, P.G.: Branches of stationary solutions for parameter-dependent reaction-diffusion systems from climate modeling. ISNM **97** (1991) 171–175

[HiMT03]   Higueras, I., März, R., Tischendorf, C.: Stability preserving integration of index-1 DAEs. Appl. Numer. Math. **45** (2003) 175-200

[HiG27]   Hildebrandt, T.H., Graves, L.M.: Implicit functions and their differentials in general analysis. A.M.S. Transactions **29** (1927) 127–153

[HiS74]   Hirsch, M.W., Smale, S.: Differential Equations, Dynamical Systems, and Linear Algebra. Academic Press, New York 1974

454   References

[HlKJ70]   Hlaváček, V., Kubíček, M., Jelinek, J.: Modeling of chemical reactors—
             XVIII. Stability and oscillatory behaviour of the CSTR. Chem. Eng.
             Sci. **25** (1970) 1441–1461
[HlMK68]   Hlaváček, V., Marek, M., Kubíček, M.: Modelling of chemical reactors
             X. Multiple solutions of enthalpy and mass balances for a catalytic
             reaction within a porous catalyst particle. Chem. Eng. Sci. **23** (1968)
             1083–1097
[HoH52]   Hodgkin, A.L., Huxley, A.F.: A quantitative description of membrane
             current and its application to conduction and excitation in nerve. J.
             Physiol. **117** (1952) 500–544
[HoS84]   Holder, E.J., Schaeffer, D.: Boundary conditions and mode jumping in
             the von Kármán equations. SIAM J. Math. Anal. **15** (1984) 446–458
[Hol77]   Holmes, P.J.: Bifurcations to divergence and flutter in flow-induced os-
             cillations: A finite dimensional analysis. J. Sound Vibration **53** (1977)
             471–503
[Hol79]   Holmes, P.J.: A nonlinear oscillator with a strange attractor. Phil.
             Trans. of the Royal Soc. London **A 292** (1979) 419–448
[HoK84a]   Holodniok, M., Kubíček, M.: New algorithms for evaluation of complex
             bifurcation points in ordinary differential equations. A comparative
             numerical study. Preprint 1981. Appl. Math. Comput. **15** (1984) 261–
             274
[HoK84b]   Holodniok, M., Kubíček, M.: DERPER—an algorithm for the con-
             tinuation of periodic solutions in ordinary differential equations. J.
             Comput. Phys. **55** (1984) 254–267
[HoMS85]   Honerkamp, J., Mutschler, G., Seitz, R.: Coupling of a slow and a
             fast oscillator can generate bursting. Bull. Mathem. Biology **47** (1985)
             1–21
[Hopf42]   Hopf, E.: Abzweigung einer periodischen Lösung von einer stationären
             Lösung eines Differentialsystems. Bericht der Math.-Phys. Klasse der
             Sächsischen Akademie der Wissenschaften zu Leipzig 94, 1942
[Hop75]   Hoppensteadt, F.: Mathematical Theories of Populations: Demogra-
             phics, Genetics and Epidemics. Society for Industrial Applied Mathe-
             matics, Philadelphia 1975
[Hoy06]   Hoyle, R.: Pattern Formation. An Introduction to Methods. Cam-
             bridge Univ. Press., Cambridge 2006
[Hsu87]   Hsu, C.S.: Cell-to-Cell Mapping: A Method of Global Analysis for
             Nonlinear Systems. Springer, New York 1987
[Hua68]   Huang, N.C., Nachbar, W.: Dynamic snap-through of imperfect visco-
             elastic shallow arches. J. Appl. Mech. **35** (1968) 289–296
[HuW92]   Hubbard, J.M., West, B.H.: MacMath, a Dynamical Systems Software
             Package for the Macintosh. Springer, 1992
[Hui91]   Huitfeldt, J.: Nonlinear Eigenvalue Problems—Prediction of Bifurca-
             tion Points and Branch Switching. Num. Anal. Group Report 17 (1991)
[HuiR90]   Huitfeldt, J., Ruhe, A.: A new algorithm for numerical path following
             applied to an example from hydrodynamical flow. SIAM J. Sci. Stat.
             Comput. **11** (1990) 1181–1192
[HuC03]   Hussein, A., Chen, K.: On efficient methods for detecting Hopf bifur-
             cation with applications to power system instability prediction. IJBC
             13 (2003) 1247–1262
[HuC04]   Hussein, A., Chen, K.: Fast computational methods for locating fold
             points for the power flow equations. J. Computational and Applied
             Mathematics **164** (2004) 419–430

[HuCW02]   Hussein, A., Chen, K., Wan, H.: On adapting test function methods for fast detection of fold bifurcations in power systems. IJBC **12** (2002) 179–185

[IoJ81]   Iooss, G., Joseph, D.D.: Elementary Stability and Bifurcation Theory. Springer, New York 1981

[IoL80]   Iooss, G., Langford, W.F.: Conjectures on the routes to turbulence via bifurcations. In: Helleman, H.G. (Ed.): Nonlinear Dynamics. New York Academy of Sciences, New York 1980

[Ise90]   Iserles, A.: Stability and dynamics of numerical methods for nonlinear ordinary differential equations. IMA J. Num. Anal. **10** (1990) 1–30

[IsBR88]   Isomäki, H.M., von Boehm, J., Räty, R.: Fractal boundaries of an impacting particle. Physics Letters **126** (1988) 484–490

[Izh01]   Izhikevich, E.M.: Synchronization of Ellliptic bursters. SIAM Review **43** (2001) 315-344

[JaHR83]   Janssen, R., Hlaváček, V., van Rompay, P.: Bifurcation pattern in reaction-diffusion dissipative systems. Z. Naturforschung **38a** (1983) 487–492

[JeR82]   Jensen, K.F., Ray, W.H.: The bifurcation behavior of tubular reactors. Chem. Eng. Sci. **37** (1982) 199–222

[Jep81]   Jepson, A.D.: Numerical Hopf bifurcation. Thesis, California Institute of Technology, Pasadena 1981

[JeSC91]   Jepson, A.D., Spence, A., Cliffe, K.A.: The numerical solution of nonlinear equations having several parameters. Part III : Equations with Z_2-symmetry. SIAM J. Numer. Anal. **28** (1991) 809–832

[JoT92]   Jones, D.A., Titi, E.S.: Determining finite volume elements for the 2D Navier-Stokes equations. Physica **60D** (1992) 165–174

[JoS77]   Jordan, D., Smith, P.: Nonlinear Ordinary Differential Equations. Oxford Univ. Press, Oxford 1977

[JoR83]   Jorgensen, D.V., Rutherford, A.: On the dynamics of a stirred tank with consecutive reactions. Chem. Eng. Sci. **38** (1983) 45–53

[Jos76]   Joseph, D.D.: Stability of Fluid Motions I and II. Springer, Berlin 1976

[JoS72]   Joseph, D.D., Sattinger, D.H.: Bifurcating time periodic solutions and their stability. ARMA **45** (1972) 79–109

[Kaas86]   Kaas-Petersen, Ch.: Chaos in a railway bogie. Acta Mechanica **61** (1986) 89-107

[KaMN89]   Kahaner, D., Moler, C., Nash, S.: Numerical Methods and Software. Prentice Hall, Englewood Cliffs 1989

[KaRV81]   Kahlert, C., Rössler, O.E., Varma, A.: Chaos in a CSTR with two consecutive first-order reactions, one exo, one endothermic. In: Proceedings Heidelberg workshop. Springer Series Chem. Physics 1981

[Kal35]   Kalecki, M.: A marcrodynamic theory of business cycles. Econometrica **3** (1935) 327–344

[KaRS92]   Kampowsky, W., Rentrop, P., Schmidt, W.: Classification and numerical simulation of electric circuits. Surv. Math. Ind. **2** (1992) 23–65

[KaY79a]   Kaplan, J.L., Yorke, J.A.: Preturbulence: A regime observed in a fluid flow model of Lorenz. Comm. Math. Phys. **67** (1979) 93–108

[KaY79b]   Kaplan, J.L., Yorke, J.: Chaotic behavior of multidimensional difference equations. In: Peitgen, H.-O., Walther, H.-O. (Eds.): Functional Differential Equations and Approximation of Fixed Points. Lecture Notes in Math. **730**. Springer, Berlin 1979

[Kap95]   Kapral, R.: Pattern formation in chemical systems. Physica **86D** (1995) 149–157

456   References

[Käs85]   Käser, G.: Direkte Verfahren zur Bestimmung von Umkehrpunkten nichtlinearer parameterabhängiger Gleichungssysteme. Diploma Thesis, Würzburg 1985
[KaH95]   Katok, A., Hasselblatt, B.: Introduction to the Modern Theory of Dynamical Systems. Cambridge University Press, New York 1995
[Kaz85]   Kazarinoff, N.D.: Pattern formation and morphogenetic fields. In: Antonelli, P.L. (Ed.): Mathematical Essays on Growth and the Emergence of Form. The University of Alberta Press, 1985
[KaD78]   Kazarinoff, N.D., van der Driessche, P.: A model predator–prey system with functional response. Math. Biosci. 39 (1978) 125–134
[KaS86]   Kazarinoff, N.D., Seydel, R.: Bifurcations and period doubling in E.N.Lorenz's symmetric fourth order system. Physical Review A 34 (1986) 3387–3392
[Kea83]   Kearfott, R.B.: Some general bifurcation techniques. SIAM J. Sci. Stat. Comput. 4 (1983) 52–68
[Kee79]   Keener, J.P.: Secondary bifurcation and multiple eigenvalues. SIAM J. Appl. Math. 37 (1979) 330–349
[Kee80]   Keener, J.P.: Waves in excitable media. SIAM J. App. Math. 39 (1980) 528–548
[Kel77]   Keller, H.B.: Numerical solution of bifurcation and nonlinear eigenvalue problems. In: [Rab77]
[Kel78]   Keller, H.B.: Global homotopies and Newton methods. In: deBoor, C., et al. (Eds.): Recent Advances in Numerical Analysis. Academic Press, New York 1978
[Kel81]   Keller, H.B.: Geometrically isolated nonisolated solutions and their approximation. SIAM J. Num. Anal. 18 (1981) 822–838
[Kel83]   Keller, H.B.: The bordering algorithm and pathfollowing near singular points of higher nullity. SIAM J. Sci. Stat. Comput. 4 (1983) 573–582
[KeA69]   Keller, J.B., Antman, S.: Bifurcation Theory and Nonlinear Eigenvalue Problems. Benjamin, New York 1969
[Kel67]   Kelley, A.: The stable, center-stable, center, center-unstable, and unstable manifolds. J. Diff. Eqs. 3 (1967) 546–570
[KeS83]   Kelley, C.T., Suresh, R.: A new acceleration method for Newton's method at singular points. SIAM J. Numer. Anal. 20 (1983) 1001–1009
[KeASP85] Kevrekidis, I.G., Aris, R., Schmidt, L.D., Pelikan, S.: Numerical computation of invariant circles of maps. Physica 16D (1985) 243–251
[KeJT91]   Kevrekidis, I.G., Jolly, M.S., Titi, E.S.: Preserving dissipation in approximate inertial forms for the Kuramoto–Shivashinsky equation. J. Dyn. Differ. Equations 3 (1991) 179–197
[KeNS90]   Kevrekidis, I.G., Nicolaenko, B., Scovel, J.C.: Back in the saddle again: A computer assisted study of the Kuramoto–Sivashinsky equation. SIAM J. Appl. Math. 50 (1990) 760–790
[Khi90]   Khibnik, A.: LINLBF: a program for continuation and bifurcation analysis of equilibria up to codimension three. in [RoDS90]
[KhBR92]   Khibnik, A., Borisyuk, R.M., Roose, D.: Numerical Bifurcation Analysis of a Model of Coupled Neural Oscillators. ISNM 104 (1992) 215–228. In: [AlBG92]
[KhKLN93] Khibnik, A., Kuznetsov, Y., Levitin, V., Nikolaev, E.: Continuation techniques and interactive software for bifurcation analysis of ODEs and iterated maps. Physica 62D (1993) 360–371
[Kir75]   Kirchgässner, K.: Bifurcation in nonlinear hydrodynamic stability. SIAM Review 17 (1975) 652–683

[KiS69]   Kirchgässner, K., Sorger, P.: Branching analysis for the Taylor pro-
             blem. Quart. J. Mech. Appl. Math. **22** (1969) 183–190
[KiS78]   Kirchgraber, U., Stiefel, E.: Methoden der analytischen Störungsrech-
             nung und ihre Anwendungen. Teubner, Stuttgart 1978
[Klo61]   Klopfenstein, R.W.: Zeros of nonlinear functions. J. ACM **8** (1961)
             366–373
[KnE96]   Knight, J.D., Ecker, H.: Simulation of nonlinear dynamics in magne-
             tic bearings with coordinate coupling. Summer Computer Simulation
             Conference, Portland 1996; Proceedings (Eds. Ingalls, V.W. et al.)
[KoLMR00]   Koon, W.S., Lo, M.W., Marsden, J.E., Ross, S.D.: Heteroclinic connec-
             tions between periodic orbits and resonance transitions in celestial me-
             chanics. Chaos **10** (2000) 427-469
[KoH73]   Kopell, N., Howard, L.N.: Plane wave solutions to reaction-diffusion
             equations. Studies Appl. Math. **52** (1973) 291–328
[Kow06]   Kowalczyk, P., di Bernardo, M., Champneys, A.R., Hogan, S.J., Ho-
             mer, M., Piiroinen, P.T., Kuznetsov, Y.A., Nordmark, A.: Two-para-
             meter discontinuity-induced bifurcations of limit cycles: classification
             and open problems. IJBC **16** (2006) 601–621
[Kra05]   Krauskopf, B., Osinga, H.M., Doedel, E.J., Henderson, M.E., Gucken-
             heimer, J., Vladimirsky, A., Dellnitz, M., Junge, O.: A survey of me-
             thods for computing (un)stable manifolds of vector fields. IJBC **15**
             (2005) 763–791
[KrC03]   Krise, S., Choudhury, S.R.: Bifurcations and chaos in a predator-prey
             model with delay and a laser-diode system with self-sustained pulsa-
             tions. CSF **16** (2003) 59-77
[KrK85]   Krug, H.-J., Kuhnert, L.: Ein oszillierendes Modellsystem mit autoka-
             talytischem Teilschritt. Z. Phys. Chemie **266** (1985) 65–73
[Kub75]   Kubíček, M.: Evaluation of branching points for nonlinear boundary-
             value problems based on GPM technique. Appl. Math. Comput. **1**
             (1975) 341–352
[Kub76]   Kubíček, M.: Algorithm 502. Dependence of solution on nonlinear sys-
             tems on a parameter. ACM Trans. of Math. Software **2** (1976) 98–107
[KuH84a]   Kubíček, M., Holodniok, M.: Evaluation of Hopf bifurcation points
             in parabolic equations describing heat and mass transfer in chemical
             reactors. Chem. Eng. Sci. **39** (1984) 593–599
[KuH84b]   Kubíček, M., Holodniok, M.: Numerical determination of bifurcation
             points in steady-state and periodic solutions—numerical algorithms
             and examples. ISNM **70** (1984) 247–270 In: [KüMW84]
[KuK83]   Kubíček, M., Klic, A.: Direction of branches bifurcating at a bifurca-
             tion point. Determination of starting points for a continuation algo-
             rithm. Appl. Math. Comput. **13** (1983) 125–142
[KuM79]   Kubíček, M., Marek, M.: Evaluation of limit and bifurcation points
             for algebraic equations and nonlinear boundary-value problems. Appl.
             Math. Comput. **5** (1979) 253–264
[KuRM78]   Kubíček, M., Ryzler, V., Marek, M.: Spatial structures in a reaction-
             diffusion system—detailed analysis of the "Brusselator." Biophys.
             Chem. **8** (1978) 235–246
[Kun88]   Kunkel, P.: Quadratically convergent methods for the computation of
             unfolded singularities. SIAM J. Numer. Anal. **25** (1988) 1392–1408
[Kun91]   Kunkel, P.: Augmented systems for generalized turning points. ISNM
             97 (1991) 231–236
[Kun00]   Kunze, M.: Non-Smooth Dynamical Systems. Springer Lecture Notes
             in Mathematics **1744**, Berlin 2000

[KuK97]   Kunze, M., Küpper, T.: Qualitative bifurcation analysis of a non-
            smooth friction-oscillator model. ZAMP **48** (1997) 87–101

[KüK84]   Küpper, T., Kuszta, B.: Feedback stimulated bifurcation. ISNM **70**
            (1984) 271–284. In: [KüMW84]

[KüMW84]   Küpper, T., Mittelmann, H.D., Weber H. (Eds.): Numerical Methods
            for Bifurcation Problems. Proceedings of a Conference in Dortmund,
            1983. ISNM **70**. Birkhäuser, Basel 1984

[KüST87]   Küpper, T., Seydel, R., Troger, H. (Eds.): Bifurcation: Analysis, Al-
            gorithms, Applications. Proceedings of a Conference in Dortmund.
            Birkhäuser, Basel 1987 ISNM **79**

[Kur00]   Kurths, J. (Ed.): Phase synchronization and its applications. IJBC
            10,10 (2000)

[Kut01]   Kutta, W.: Beitrag zur näherungsweisen Integration totaler Differen-
            tialgleichungen. Z. Math. Phys. **46** (1901) 435–453

[Kuz90]   Kuznetsov, Yu. A.: Computation of invariant manifold bifurcations.
            In: Roose, D., et al. (Eds.): Continuation and Bifurcations. Kluwer
            Academic Publishers (1990) 183–195

[Kuz98]   Kuznetsov, Yu.A.: Elements of Applied Bifurcation Theory. Second
            Edition. Springer, New York 1998

[KuRG03]   Kuznetsov, Yu.A., Rinaldi, S., Gragnani, A.: One-parameter bifurca-
            tions in planar Filippov systems. IJBC **13** (2003) 2157–2188

[Lam91]   Lambert, J.D.: Numerical Methods for Ordinary Differential Systems.
            Wiley, Chichester 1991

[LaL59]   Landau, L., Lifshitz, E.: Fluid Mechanics. Pergamon, Oxford 1959

[Lan77]   Langford, W.F.: Numerical solution of bifurcation problems for ordi-
            nary differential equations. Numer. Math. **28** (1977) 171–190

[Lan79]   Langford, W.F.: Periodic and steady-state mode interactions lead to
            tori. SIAM J. Appl. Math. **37** (1979) 22–48

[Lan81]   Langford, W.F.: A review of interactions of Hopf and steady-state
            bifurcations. In: Nonlinear Dynamics and Turbulence. Pitman, 1981

[Lan84]   Langford, W.F.: Numerical studies of torus bifurcations. ISNM 70
            (1984) 285–295. In: [KüMW84]

[LaH91]   Lauterborn, W., Holzfuss, J.: Acoustic chaos. IJBC **1** (1991) 13–26

[Lei00]   Leine, R.I.: Bifurcations in Discontinuous Mechanical Systems of
            Filippov-Type. PhD-dissertation, TU Eindhoven, 2000

[LeR05]   Lenci, S., Rega, G.: Heteroclinic bifurcations and optimal control in
            the nonlinear rocking dynamics of generic and slender rigid-blocks.
            IJBC **15** (2005) 1901–1918

[LeN91]   Leng, G., Namachchivaya, N. S.: Critical mode interaction in the pre-
            sence of external random excitation. AIAA J. of Guidance, Control
            and Dynamics **14** (1991) 770–777

[LeS85]   Levin, S.A., Segel, L.A.: Pattern generation in space and aspect. SIAM
            Review **27** (1985) 45–67

[Lia66]   Liapunov, A.M.: Stability of Motion. Academic Press, New York 1966

[LiM82]   Libchaber, A., Maurer, J.: A Rayleigh Bénard experiment: Helium in a
            small box. Proceedings of the NATO Advanced Studies Institute 1982

[LiL82]   Lichtenberg, A.J., Lieberman, M.A.: Regular and Stochastic Motion.
            Springer, Heidelberg-New York 1982

[List78]   List, S.E.: Generic bifurcation with application to the von Kármán
            equations. J. Diff. Eqs. **30** (1978) 89–118

[Lor63]   Lorenz, E.N.: Deterministic nonperiodic flow. J. Atmospheric Sci. **20**
            (1963) 130–141

[Lor84]   Lorenz, E.N.: The local structure of a chaotic attractor in four dimensions. Physica **13D** (1984) 90–104

[Lor92]   Lorenz, J.: Computation of invariant manifolds. In: Griffiths, D.F., Watson, G.A. (Eds.): Numerical Analysis 1991. Longman 1992

[Lory80]   Lory, P.: Enlarging the domain of convergence for multiple shooting by the homotopy method. Numer. Math. **35** (1980) 231–240

[Lust01]   Lust, K.: Improved numerical Floquet multipliers. IJBC **11** (2001) 2389–2410

[LuRSC98]   Lust, K., Roose, D., Spence, A., Champneys, A.R.: An adaptive Newton-Picard algorithm with subspace iteration for computing periodic solutions. SIAM J. Sci. Comput. **19** (1998) 1188–1209

[LuER01]   Luzyanina, T., Engelborghs, K., Roose, D.: Numerical bifurcation analysis of differential equations with state-dependent delay. IJBC **11** (2001) 737–753

[LyKB84]   Lyberatos, G., Kuszta, B., Bailey, J.E.: Discrimination and identification of dynamic catalytic reaction models via introduction of feedback. Chem. Eng. Sci. **39** (1984) 739–750

[Mac89]   Mackens, W.: Numerical differentation of implicitly defined space curves. Computing **41** (1989) 237–260

[MaG77]   Mackey, M.C., Glass, L.: Oscillation and chaos in physiological control systems. Science **197** (1977) 287–289

[Man83]   Mandelbrot, B.B.: The Fractal Geometry of Nature. Freeman, New York 1983

[MaCV03]   Manffra, E.F., Caldas, I.L., Viana, R.L.: Stabilizing periodic orbits in a chaotic semiconductor laser. CSF **15** (2003) 327–341

[MaS87]   Marek, M., Schreiber, I.: Formation of periodic and aperiodic waves in reaction-diffusion systems (1987). In: [KüST87]

[MaJ92]   Margolin, L.G., Jones, D.A.: An approximate inertial manifold for computing Burgers' equation. Physica **60D** (1992) 175–184

[Mar90]   Margolis, S.B.: Chaotic combustion of solids and high fluids near points of strong resonance. Sandia Report, Livermore 1990

[Mar91]   Margolis, S.B.: The transition to nonsteady deflagration in gasless combustion. Prog. Energy Combust. Sci. **17** (1991) 135–162

[MaM83]   Margolis, S.B., Matkowsky, B.J.: Nonlinear stability and bifurcation in the transition from laminar to turbulent flame propagation. Combustion Science and Technology **34** (1983) 45–77

[MaM85]   Margolis, S.B., Matkowsky, B.J.: Flame propagation in channels: Secondary bifurcation to quasi-periodic pulsations. SIAM J. Appl. Math. **45** (1985) 93–129

[MaH90]   Markus, M., Hess, B.: Isotropic cellular automaton for modelling excitable media. Nature **347** (1990) 56–58

[MaKH85]   Markus, M., Kuschmitz, D., Hess, B.: Properties of strange attractors in yeast glycolysis. Biophys. Chem. **22** (1985) 95–105

[MaS91]   Markus, M., Schäfer, C.: Spatially periodic forcing of spatially periodic oscillators. ISNM **97** (1991) 263–275

[MaM05]   Marquardt, W., Mönnigmann, M.: Constructive nonlinear dynamics in process systems engineering. Computers and Chemical Engng **29** (2005) 1265–1275

[MaM76]   Marsden, J.E., McCracken, M.: The Hopf Bifurcation and Its Applications. Springer, New York 1976

[MaM86]   Martin, S., Martienssen, W.: Circle maps and mode locking in the driven electrical conductivity of barium sodium niobate crystals. Phys. Rev. Lett. **56** (1986) 1522–1525

[MaS78]   Matkowsky, B.J., Sivashinsky, G.I.: Propagation of a pulsating reaction front in solid fuel combustion. SIAM J. Appl. Math. **35** (1978) 465–478

[MaN85]   Matthies, H.G., Nath, Chr.: Dynamic stability of periodic solutions of large scale nonlinear systems. Comput. Meth. Appl. Mech. Engng. **48** (1985) 191–202

[May76]   May, R.M.: Simple mathematical models with very complicated dynamics. Nature **261** (1976) 459–467

[McK70]   McKean, H.P.: Nagumo's equation. Advances in Math. **4** (1970) 209–223

[McW90]   McKenna, P.J., Walter, W.: Travelling waves in a suspension bridge. SIAM J. Appl. Math. **50** (1990) 703–715

[Mei00]   Mei, Zhen: Numerical Bifurcation Analysis for Reaction-Diffusion Equations. Springer, Berlin 2000

[Mei95]   Meinhardt, H.: The Algorithmic Beauty of Sea Shells. Springer, Berlin 1995

[MeR82]   Melhem, R.G., Rheinboldt, W.C.: A comparison of methods for determining turning points of nonlinear equations. Comput. **29** (1982) 201–226

[Men84]   Menzel, R.: Numerical determination of multiple bifurcation points. ISNM **70** (1984) 310–318. In: [KüMW84]

[MeS78]   Menzel, R., Schwetlick, H.: Zur Lösung parameterabhängiger nichtlinearer Gleichungen mit singulären Jacobi-Matrizen. Numer. Math. **30** (1978) 65–79

[MePL93]   Mettin, R., Parlitz, U., Lauterborn, W.: Bifurcation structure of the driven van der Pol oscillator. IJBC **3** (1993) 1529–1555

[Mey71]   Meyer, R.E.: Introduction to Mathematical Fluid Dynamics. Dover, New York 1971

[Mey75]   Meyer–Spasche, R.: Numerische Behandlung von elliptischen Randwertproblemen mit mehreren Lösungen und von MHD Gleichgewichtsproblemen. Report 6/141 of the Institut für Plasmaphysik, Garching 1975

[MeK80]   Meyer–Spasche, R., Keller, H.B.: Numerical study of Taylor-vortex flows between rotating cylinders. J. Comput. Phys. **35** (1980) 100–109

[Mil72]   Miller, W.: Symmetry Groups and Their Applications. Academic Press, New York 1972

[Min74]   Minorsky, N.: Nonlinear Oscillations. Krieger, Huntington 1974

[MiCR00]   Mithulananthan, N., Canizares, C.A, Reeve, J.: Indices to detect Hopf bifurcations in power systems. Proceedings of North American Power Symposium 2000

[MiR90]   Mittelmann, H.D., Roose, D. (Eds.): Continuation Techniques and Bifurcation Problems. Birkhäuser, Basel 1990, ISNM **92**

[MiW80]   Mittelmann, H.D., Weber, H. (Eds.): Bifurcation Problems and Their Numerical Solution. Proceedings of a Conference in Dortmund, 1980. Birkhäuser, Basel 1980, ISNM **54**

[Moi92]   Moiola, J.L.: On the computation of local bifurcation diagrams near degenerate Hopf bifurcations of certain types. IJBC **3** (1992) 1103–1122

[MoC96]   Moiola, J.L., Chen, G.: Hopf Bifurcation Analysis. A Frequency Domain Approach. World Scientific, Singapore 1996

[Moon80]   Moon, F.C.: Experiments on chaotic motions of a forced nonlinear oscillator: Strange attractors. J. Appl. Mech. **47** (1980) 638–644

[Moon87]   Moon, F.C.: Chaotic Vibrations. An Introduction for Applied Scientists and Engineers. John Wiley, New York 1987

[Moo80]   Moore, G.: The numerical treatment of non-trivial bifurcation points. Numer. Funct. Anal. Optim. **2** (1980) 441–472

[Moo95]   Moore, G.: Computation and parametrization of periodic and connecting orbits. IMA J. Numer. Anal. **15** (1995) 245–263

[MoS80]   Moore, G., Spence, A.: The calculation of turning points of nonlinear equations. SIAM J. Numer. Anal. **17** (1980) 567–576

[MoBC05]   Morel, C., Bourcerie, M., Chapeau-Blondeau, F.: Generating independent chaotic attractors by chaos anticontrol in nonlinear circuits. CSF **26** (2005) 541–549

[Mos66]   Moser, J.: On the theory of quasiperiodic motions. SIAM Review **8** (1966) 145–172

[Mur89]   Murray, J.D.: Mathematical Biology. Springer, Berlin 1989

[Nai59]   Naimark, J.: On some cases of periodic motions depending on parameters. Dokl. Akad. Nauk. SSSR **129** (1959) 736–739

[Nam90]   Namachchivaya, N.S.: Stochastic bifurcation. Appl. Math. and Comput. **38** (1990) 101–159

[Nam91]   Namachchivaya, N.S.: Co-dimension two bifurcations in the presence of noise. J. of Appl. Mech. **58** (1991) 259–265

[NaA87]   Namachchivaya, N.S., Ariaratnam, S.T.: Periodically perturbed Hopf bifurcation. SIAM J. Appl. Math. **47** (1987) 15–39

[NaT89]   Namachchivaya, N.S., Tien, W.M.: Bifurcation behavior of nonlinear pipes conveying pulsating flow. J. of Fluids and Structure **3** (1989) 609–629

[NaM79]   Nayfeh, A.H., Mook, D.T.: Nonlinear Oscillations. John Wiley, New York 1979

[Neu93a]   Neubert, R.: Über die Approximation von Lösungszweigen und die Entdeckung von Hopfschen Verzweigungspunkten in nichtlinearen Gleichungssystemen mit einem Parameter. PhD dissertation. Ulm 1993

[Neu93b]   Neubert, R.: Predictor-corrector techniques for detecting Hopf bifurcation points. IJBC **3** (1993) 1311–1318

[NeRT78]   Newhouse, S., Ruelle, D., Takens, F.: Occurrence of strange·axiom–A attractors near quasiperiodic flow on T^n, $n \leq 3$. Comm. Math. Phys. **64** (1978) 35

[NiMA06]   Nikolopoulos, S., Manis, G., Alexandridi, A.: Investigation of correlation dimension estimation in heartbeat time series. IJBC **16** (2006) 2481–2498

[NuY94]   Nusse, H.E., Yorke, J.A.: Dynamics: Numerical Explorations. Springer, New York 1994

[ObP81]   Oberle, H.J., Pesch, H.J.: Numerical treatment of delay differential equations by Hermite interpolation. Numer. Math. **37** (1981) 235–255

[Odeh67]   Odeh, F.: Existence and bifurcation theorems for the Ginzburg-Landau equations. J. Math. Phys. **8** (1967) 2351–2356

[OlCK03]   Oldeman, B.E., Champneys, A.R., Krauskopf, B.: Homoclinic branch switching: a numerical implementation of Lin's method. IJBC **13** (2003) 2977–2999

[Olm77]   Olmstead, W.E.: Extent of the left branching solution to certain bifurcation problems. SIAM J. Math. Anal. **8** (1977) 392–401

[OrR70]   Ortega, J.M., Rheinboldt, W.C.: Iterative Solution of Nonlinear Equations in Several Variables. Academic Press, New York 1970

[Ose68]   Oseledec, V.I.: A multiplicative ergodic theorem. Ljapunov characteristic numbers for dynamical systems. Trans. Moscow Math. Soc. **19** (1968) 197–231

[Osi03]   Osinga, H.M.: Nonorientable manifolds in three-dimensional vector fields. IJBC **13** (2003) 553–570

[OsL07]   Osterhage, H., Lehnertz, K.: Nonlinear time series analysis in epilepsy. IJBC **17** (2007) 3305–3323

[Ott81]   Ott, E.: Strange attractors and chaotic motion of dynamical systems. Rev. Modern Phys. **53** (1981) 655–671

[OttGY90]   Ott, E., Grebogi, C., Yorke, J.A.: Controlling chaos. Phys. Lett. Rev. **64** (1990) 1196–1199

[PaCFS80]   Packard, N.H., Crutchfield, J.P., Farmer, J.D., Shaw, R.S.: Geometry from a time series. Phys. Rev. Lett. **45** (1980) 712–716

[PaC89]   Parker, T.S., Chua, L.O.: Practical Numerical Algorithms for Chaotic Systems. Springer, New York 1989

[Par92]   Parlitz, U.: Identification of true and spurious Lyapunov exponents from time series. IJBC **2** (1992) 155–165

[PaC92]   Parlitz, U., Chua, L.O., Kocarev, Lj., Halle, K.S., Shang, A.: Transmission of digital signals by chaotic synchronization. IJBC **2** (1992) 973–977

[PaESL90]   Parlitz, U., Englisch, V., Scheffczyk, C., Lauterborn, W.: Bifurcation structure of bubble oscillators. J. Acoust. Soc. Am. **88** (1990) 1061–1077

[Pea93]   Pearson, J.E.: Complex patterns in a simple system. Science **261** (1993) 189–192

[Pea58]   Pearson, J.R.A.: On convection cells induced by surface tension. J. Fluid Mech. **4** (1958) 489–500

[PeC90]   Pecora, L.M., Carroll, T.L.: Synchronization in Chaotic Systems. Physical Rev. Letters **64** (1990) 821–824

[PePRS92]   Peinke, J., Parisi, J., Rössler, O.E., Stoop, R.: Encounter with Chaos. Springer, Berlin 1992

[PeP79]   Peitgen, H.-O., Prüfer, M.: The Leray-Schauder continuation method is a constructive element in the numerical study of nonlinear eigenvalue and bifurcation problems. In: Peitgen, H.-O., Walther, H.O. (Eds.): Functional Differential Equations and Approximation of Fixed Points. Lecture Notes in Math. **730**. Springer, Berlin 1979

[PeR86]   Peitgen, H.-O., Richter, P.H.: The Beauty of Fractals. Springer, Berlin 1986

[PeSS81]   Peitgen, H.-O., Saupe, D., Schmitt, K.: Nonlinear elliptic boundary value problems versus their finite difference approximations: Numerically irrelevant solutions. J. Reine Angew. Math. **322** (1981) 74–117

[PeOH81]   Peterson, J., Overholser, K.A., Heinemann, R.F.: Hopf bifurcation in a radiating laminar flame. Chem. Eng. Sci. **36** (1981) 628–631

[PeG84]   Petrillo, G.A., Glass, L.: A theory for phase locking of respiration in cats to a mechanical ventilator. Am. J. Physiol. **246** (1984) R311–R320

[Pfe92]   Pfeiffer, F.: On stick-slip vibrations in machine dynamics. Machine Vibration **1** (1992) 20–28

[PiRK01]   Pikovsky, A., Rosenblum, M., Kurths, J.: Synchronization. Cambridge University Press 2001

[Poi1885]   Poincaré, H.: Sur l'équilibre d'une masse fluide animée d'un mouvement de rotation. Acta Mathematica **7** (1885) 259–380

[PoM80]   Pomeau, Y., Manneville, P.: Intermittent transition to turbulence in dissipative dynamical systems. Commun. Math. Phys. **74** (1980) 189–197

[Pön85]   Pönisch, G.: Computing simple bifurcation points using a minimally extended system of nonlinear equations. Computing **35** (1985) 277–294

[PöS81]   Pönisch, G., Schwetlick, H.: Computing turning points of curves impli-
             citly defined by nonlinear equations depending on a parameter. Com-
             puting **26** (1981) 107–121
[PöS82]   Pönisch, G., Schwetlick, H.: Ein lokal überlinear konvergentes Verfah-
             ren zur Bestimmung von Rückkehrpunkten implizit definierter Raum-
             kurven. Report 07-30-77 TU Dresden, 1977. Numer. Math. **38** (1982)
             455–465
[PoHO95]   Popp, K., Hinrichs, N., Oestreich, M.: Dynamical behavior of a friction
             oscillator with simultaneous self and external excitation. Sādhanā **20**
             (1995) 627–654
[PoS78]   Poston, T., Stewart, I.: Catastrophe Theory and Its Applications. Pit-
             man, London 1978
[PrNR01]   Prasad, A., Negi, S.S., Ramaswamy, R.: Strange nonchaotic attractors.
             IJBC **11** (2001) 291–309
[PrSS83]   Preisser, F., Schwabe, D., Scharmann, A.: Steady and oscillatory ther-
             mocapillary convection in liquid columns with free cylindrical surface.
             J. Fluid Mech. **126** (1983) 545–567
[PrFTV89]   Press, W.H., Flannery, B.P., Teukolsky, S.A., Vetterling, W.T.: Nu-
             merical Recipes. Cambridge University Press 1989
[PuN04]   Putra, D., Nijmeier, H.: Limit cycling in an observer-based controlled
             system with friction: numerical analysis and experimental validation.
             IJBC **14** (2004) 3083–3093
[Pyr92]   Pyragas, K.: Continuous control of chaos by self-controlling feedback.
             Phys. Lett. **A170** (1992) 421–428
[QuSS00]   Quarteroni, A., Sacco, R., Saleri, F.: Numerical Mathematics. Sprin-
             ger, New York 2000
[Rab77]   Rabinowitz, P.H. (Ed.): Applications of Bifurcation Theory. Academic
             Press, New York 1977
[Ral65]   Ralston, A.: A First Course in Numerical Analysis. McGraw-Hill, New
             York 1965
[RaA87]   Rand, R.H., Armbruster, D.: Perturbation Methods, Bifurcation Theory
             and Computer Algebra. Springer, New York 1987
[RaS78]   Rauch, J., Smoller, J.: Qualitative theory of the FitzHugh–Nagumo
             equations. Advances in Math. **27** (1978) 12–44
[Ray77]   Ray, W.H.: Bifurcation phenomena in chemically reacting systems. In:
             [Rab77]
[Ray16]   Lord Rayleigh: On convection currents in a horizontal layer of fluid,
             when the higher temperature is on the under side. Philos. Mag. **32**
             (1916) 529–546
[Red78]   Reddien, G.W.: On Newton's method for singular problems. SIAM J.
             Numer. Anal. **15** (1978) 993–996
[Rega04]   Rega, G.: Nonlinear vibrations of suspended cables. Applied Mechanics
             Reviews **57** (2004) 443-478
[Rei91]   Reithmeier, E.: Periodic Solutions of Nonlinear Dynamical Systems.
             Springer, Berlin 1991
[Ren79]   Renardy, M.: Hopf-Bifurkation bei Lasern. Math. Meth. Appl. Sci. **1**
             (1979) 194–213
[Ren85]   Rentrop, P.: Partitioned Runge-Kutta methods with stiffness detection
             and stepsize control. Numer. Math. **47** (1985) 545–564
[ReRS89]   Rentrop, P., Roche, M., Steinebach, G.: The application of Rosenbrock-
             Wanner type methods with stepsize control in differential-algebraic
             equations. Numer. Math. **55** (1989) 545–563

[Rhe78]   Rheinboldt, W.C.: Numerical methods for a class of finite dimensional bifurcation problems. SIAM J. Numer. Anal. **15** (1978) 1–11

[Rhe80]   Rheinboldt, W.C.: Solution fields of nonlinear equations and continuation methods. SIAM J. Numer. Anal. **17** (1980) 221–237

[Rhe82]   Rheinboldt, W.C.: Computation of critical boundaries on equilibrium manifolds. SIAM J. Numer. Anal. **19** (1982) 653–669

[Rhe88]   Rheinboldt, W.C.: On the computation of multi-dimensional solution manifolds of parametrized equations. Numer. Math. **53** (1988) 165–181

[RhB83]   Rheinboldt, W.C., v. Burkardt, J.: A locally parametrized continuation process. ACM Trans. of Math. Software **9** (1983) 215–235

[RhRS90]   Rheinboldt, W.C., Roose, D., Seydel, R.: Aspects of continuation software. In: [RoDS90]

[RiT07]   Riaza, R., Tischendorf, C.: Qualitative features of matrix pencils and DAEs arising in circuit dynamics. Dynamical Systems **22** (2007) 107–131

[Rie96]   Riedel, K.: Stabilität periodischer Lösungen nicht-autonomer Systeme und Verzweigung bei Periodenverdopplung. Diploma Thesis, University Ulm (1996)

[Riks72]   Riks, E.: The application of Newton's method to the problem of elastic stability. Trans. ASME, J. Appl. Mech. **39** (1972) 1060–1065

[RiF84]   Rinzel, J., FitzHugh, R.: private communication, 1984

[RiK83]   Rinzel, J., Keener, J.P.: Hopf bifurcation to repetitive activity in nerve. SIAM J. Appl. Math. **43** (1983) 907–922

[RiM80]   Rinzel, J., Miller, R.N.: Numerical calculation of stable and unstable periodic solutions to the Hodgkin–Huxley equations. Math. Biosci. **49** (1980) 27–59

[RiT82a]   Rinzel, J., Terman, D.: Propagation phenomena in a bistable reaction-diffusion system. SIAM J. Appl. Math. **42** (1982) 1111–1137

[RiT82b]   Rinzel, J., Troy, W.C.: Bursting phenomena in a simplified Oregonator flow system model. J. Chem. Phys. **76** (1982) 1775–1789

[RoAM04]   Robbio, F.I., Alonso, D.M., Moiola, J.L.: On semi-analytic procedure for detecting limit cycle bifurcations. IJBC **14** (2004) 951-970

[Roo85]   Roose, D.: An algorithm for the computation of Hopf bifurcation points in comparison with other methods. J. Comput. Appl. Math. **12&13** (1985) 517–529

[RoDS90]   Roose, D., De Dier, B., Spence, A. (Eds.): Continuation and Bifurcation: Numerical Techniques and Applications. Kluwer Academic Publishers, Dordrecht 1990

[RoH83]   Roose, D., Hlaváček, V.: Numerical computation of Hopf bifurcation points for parabolic diffusion-reaction differential equations. SIAM J. Appl. Math. **43** (1983) 1075–1085

[RoH85]   Roose, D., Hlaváček, V.: A direct method for the computation of Hopf bifurcation points. SIAM J. Appl. Math. **45** (1985) 879–894

[RoP85]   Roose, D., Piessens, R.: Numerical computation of nonsimple turning points and cusps. Numer. Math. **46** (1985) 189–211

[RoPHR84]   Roose, D., Piessens, R., Hlaváček, V., van Rompay, P.: Direct evaluation of critical conditions for thermal explosion and catalytic reaction. Combustion and Flame **55** (1984) 323–329

[RoDH82]   Rosenblat, S., Davis, S.H., Homsy, G.M.: Nonlinear Marangoni convection in bounded layers. Part 1. Circular cylindric containers. J. Fluid Mech. **120** (1982) 91–122

[Rös76]   Rössler, O.E.: Different types of chaos in two simple differential equations. Z. Naturforschung **31a** (1976) 1664–1670

[Rue80]   Ruelle, D.: Strange attractors. Mathem. Intelligencer **2** (1980) 126–137
[Rue89]   Ruelle, D.: Elements of Differentiable Dynamics and Bifurcation Theory. Academic Press, San Diego 1989
[RuT71]   Ruelle, D., Takens, F.: On the nature of turbulence. Comm. Math. Phys. **20** (1971) 167–192
[Ruhe73]   Ruhe, A.: Algorithms for the nonlinear eigenvalue problem. SIAM J. Numer. Anal. **10** (1973) 674–689
[Run1895] Runge, C.: Über die numerische Auflösung von Differential Gleichungen. Math. Ann. **46** (1895) 167–178
[Sac65]   Sacker, R.S.: On invariant surfaces and bifurcations of periodic solutions of ordinary differential equations. Comm. Pure Appl. Math. **18** (1965) 717–732
[SaS85]   Sano, M., Saweda, Y.: Measurement of the Lyapunov spectrum from a chaotic time series. Phys. Rev. Lett. **55** (1985) 1082–1085
[Sat79]   Sattinger, D.H.: Group Theoretic Methods in Bifurcation Theory. Springer, Berlin 1979
[Sat80]   Sattinger, D.H.: Bifurcation and symmetry breaking in applied mathematics. Bull. Amer. Math. Soc. **3** (1980) 779–819
[SaY91]   Sauer, T., Yorke, J.A.: Rigorous verification of trajectories for the computer simulation of dynamical systems. Nonlinearity **4** (1991) 961–979
[ScK85]   Schaffer, W.M., Kot, M.: Do strange attractors govern ecological systems? BioSciences **35** (1985) 342–350
[ScTZ83]   Scheidl, R., Troger, H., Zeman, K.: Coupled flutter and divergence bifurcation of a double pendulum. Int. J. Non-linear Mech. **19** (1983) 163–176
[Sch77]   Scheurle, J.: Ein selektives Projektions-Iterationsverfahren und Anwendungen auf Verzweigungsprobleme. Numer. Math. **29** (1977) 11–35
[Sch86]   Scheurle, J.: Chaotic solutions of systems with almost periodic forcing. ZAMP **37** (1986) 12–26
[Sch87]   Scheurle, J.: Bifurcation of quasiperiodic solutions from equilibrium points of reversible systems. ARMA **97** (1987) 104–139
[ScM84]   Scheurle, J., Marsden, J.: Bifurcation to quasi-periodic tori in the interaction of steady state and Hopf bifurcation. SIAM J. Math. Anal. **15** (1984) 1055–1074
[ScS00]   Scheurle, J., Seydel, R.: A model of student migration. IJBC **10** (2000) 477–480
[Sch92]   Schmidt, P.G.: Hopf bifurcation in a class of ODE systems related to climate modeling. In: Wiener, J. et al. (Eds.): Ordinary and Delay Differential Equations. Longman, Harlow 1992
[Sch85]   Schneider, F.W.: Periodic perturbations of chemical oscillators: Experiments. Ann. Rev. Phys. Chem. **36** (1985) 347–378
[Schu84]   Schuster, H.G.: Deterministic Chaos. Physik-Verlag, Weinheim 1984
[Sch89]   Schwarz, H.R.: Numerical Analysis. Wiley, New York 1989
[Schw84]   Schwetlick, H.: Algorithms for finite-dimensional turning point problems from viewpoint to relationships with constrained optimization problems. ISNM **70** (1984) 459–479. In: [KüMW84]
[ScC87]   Schwetlick, H., Cleve, J.: Higher order predictors and adaptive stepsize control in path following algorithms. SIAM J. Numer. Anal. **24** (1987) 1382–1393
[SeN95]   Sekar, P., Narayanan, S.: Chaos in mechanical systems — a review. Sādhanā **20** (1995) 529-582

[Sey75]   Seydel, R.: Bounds for the lowest critical value of the nonlinear operator $-u'' + u^3$. ZAMP **26** (1975) 714–720

[Sey77]   Seydel, R.: Numerische Berechnung von Verzweigungen bei gewöhnlichen Differentialgleichungen. Ph.D. Thesis. 1977. Report TUM 7736, Mathem. Institut, Technische Univ. München, 1977

[Sey79a]   Seydel, R.: Numerical computation of branch points in ordinary differential equations. Numer. Math. **32** (1979) 51–68

[Sey79b]   Seydel, R.: Numerical computation of primary bifurcation points in ordinary differential equations. ISNM **48** (1979) 161–169. In: [AlCK79]

[Sey79c]   Seydel, R.: Numerical computation of branch points in nonlinear equations. Numer. Math. **33** (1979) 339–352

[Sey80]   Seydel, R.: Programme zur numerischen Behandlung von Verzweigungsproblemen bei nichtlinearen Gleichungen und Differentialgleichungen. ISNM **54** (1980) 163–175. In: [MiW80]

[Sey81a]   Seydel, R.: Numerical computation of periodic orbits that bifurcate from stationary .solutions of ordinary differential equations. Appl. Math. Comput. **9** (1981) 257–271

[Sey81b]   Seydel, R.: Neue Methoden zur numerischen Berechnung abzweigender Lösungen bei Randwertproblemen gewöhnlicher Differentialgleichungen. Habilitationsschrift, München 1981

[Sey83a]   Seydel, R.: Branch switching in bifurcation problems of ordinary differential equations. Numer. Math. **41** (1983) 93–116

[Sey83b]   Seydel, R.: Efficient branch switching in systems of nonlinear equations. ISNM **69** (1983) 181–191

[Sey83c]   Seydel, R.: BIFPACK—a program package for calculating bifurcations. Buffalo 1983. Current version 3.0, Ulm 1994

[Sey84]   Seydel, R.: A continuation algorithm with step control. ISNM **70** (1984) 480–494. In: [KüW84]

[Sey85a]   Seydel, R.: Calculating the loss of stability by transient methods, with application to parabolic partial differential equations. In: [AsR85]

[Sey85b]   Seydel, R.: Attractors of a Duffing equation—dependence on the exciting frequency. Physica **17D** (1985) 308–312

[Sey88]   Seydel, R.: From Equilibrium to Chaos. Practical Bifurcation and Stability Analysis. Elsevier, New York 1988

[Sey91]   Seydel, R.: On detecting stationary bifurcations. IJBC **1** (1991) 335–337

[Sey97]   Seydel, R.: Risk and bifurcation: towards a deterministic risk analysis. In: O. Renn (Ed.): Risk Analysis and Management in a Global Economy. Center of Technology Assessment, Stuttgart 1997

[Sey01]   Seydel, R.: Assessing voltage collapse. Latin Amer. Appl. Research **31** (2001) 171–176

[Sey04]   Seydel, R.: A new risk index. ZAMM **84** (2004) 850–855

[SeSKT91]   Seydel, R., Schneider, F.W., Küpper, T., Troger, H. (Eds.): Bifurcation and Chaos: Analysis, Algorithms, Applications. Birkhäuser, Basel 1991

[ShJ80]   Sharp, G.H., Joyner, R.W.: Simulated propagation of cardiac action potentials. Biophys. J. **31** (1980) 403–423

[Shaw81]   Shaw, R.: Strange attractors, chaotic behavior, and information flow. Zeitschrift Naturforschung **36a** (1981) 80–112

[ShiC05]   Shi, Y., Chen, G.: Chaotification of discrete dynamical systems governed by continuous maps. IJBC **15** (2005) 547–555

[ShN79]   Shimada, I., Nagashima, T.: A numerical approach to ergodic problem of dissipative dynamical systems. Progress of Theoretical Phys. **61** (1979) 1605–1616

[Shi93]   Shinbrot, T.: Chaos: Unpredictable yet controllable? Nonlinear Science Today **3** (1993) 1–8

[Sim88]   Simeon, B.: Homotopieverfahren zur Berechnung quasistationärer Lagen von Deskriptorformen in der Mechanik. Diploma Thesis TU München 1988

[SiV06]   Simon, P.L, Volford, A.: Detailed study of limit cycles and global bifurcations in a circadian rhythm model. IJBC **16** (2006) 349–367

[Sim75]   Simpson, R.B.: A method for the numerical determination of bifurcation states of nonlinear systems of equations. SIAM J. Numer. Anal. **12** (1975) 439–451

[Sma67]   Smale, S.: Differentiable dynamical systems. Bull. Amer. Math. Soc. **73** (1967) 747–817

[SmKLB83] Smith, C.B., Kuszta, B., Lyberatos, G., Bailey, J.E.: Period doubling and complex dynamics in an isothermal chemical reaction system. Chem. Eng. Sci. **38** (1983) 425–430

[SmD83]   Smith, M.K., Davis, S.H.: Instabilities of dynamic thermocapillary liquid layers. Part 1. Convective instabilities. J. Fluid Mech. **132** (1983) 119–144

[SmMK85]   Smooke, M.D., Miller, J.A., Kee, R.J.: Solution of premixed and counterflow diffusion flame problems by adaptive boundary value methods. In: [AsR85]

[SoG92]   Sommerer, J.C., Grebogi, C.: Determination of crisis parameter values by direct observation of manifold tangencies. IJBC **2** (1992) 383–396

[Sou00]   Souza, A.C.Z. de: Discussion on some voltage collapse indices. Electric Power Systems Research **53** (2000) 53–58

[Spa82]   Sparrow, C.: The Lorenz Equations: Bifurcations, Chaos and Strange Attractors. Springer, Berlin 1982

[SpJ84]   Spence, A., Jepson, A.D.: The numerical calculation of cusps, bifurcation points and isola formation points in two parameter problems. ISNM **70** (1984) 502–514. In: [KüW84]

[Spe93]   Spence, A., Wei, W., Roose, D., De Dier, B.: Bifurcation analysis of double Takens–Bogdanov points of nonlinear equations with a Z_2-symmetry. IJBC **3** (1993) 1141–1153

[SpW82]   Spence, A., Werner, B.: Non-simple turning points and cusps. IMA J. Numer. Anal. **2** (1982) 413–427

[Spi83]   Spirig, F.: Sequence of bifurcations in a three-dimensional system near a critical point. ZAMP **34** (1983) 259–276

[SpW91]   Spirig, F., Waldvogel, J.: Chaos in coorbital motion. In: Roy, A.E. (Ed.): Predictability, Stability and Chaos in n-Body Dynamical Systems. Plenum Publ., London 1991, 395–410

[Sta71]   Stakgold, I.: Branching of solutions of nonlinear equations. SIAM Review **13** (1971) 289–332

[Stä94]   Stämpfle, M.: Controlling chaos through iteration sequences and interpolation techniques. IJBC **4** (1994) 1697–1701

[Stä99]   Stämpfle, M.: Dynamical systems flow computation by adaptive triangulation methods. Comput. Visual. Sci. **2** (1999) 15–24

[Ste90]   Stelter, P.: Nichtlineare Schwingungen reibungserregter Strukturen. VDI-Verlag, Düsseldorf 1990

[StGSF08] Stiefs, D., Gross, T., Steuer, R., Feudel, U.: Computation and visualization of bifurcation surfaces. IJBC **18** (2008) 2191–2206

[StB80]   Stoer, J., Bulirsch, R.: Introduction to Numerical Analysis. Springer, New York 1980 (Third Edition 2002)

[Sto50]   Stoker, J.J.: Nonlinear Vibrations. Interscience, New York 1950

[StM88]   Stoop, R., Meier, P.F.: Evaluation of Lyapunov exponents and scaling functions from time series. J. Optical Soc. Amer. **B 5** (1988) 1037–1045

[Str88]   Strang, G.: Linear Algebra and Its Applications. Academic Press, New York 1988

[Str94]   Strogatz, S.H.: Nonlinear Dynamics and Chaos. Westview Press, Cambridge 1994

[StH98]   Stuart, A.M., Humphries, A.R.: Dynamical Systems and Numerical Analysis. Cambridge Univ. Press., Cambridge 1998

[Stu77]   Stuart, J.T.: Bifurcation theory in non-linear hydrodynamic stability. In: [Rab77]

[SuTW85]   Suchy, H., Troger, H., Weiss, R.: A numcrical study of mode jumping of rectangular plates. ZAMM **65** (1985) 71–78

[SwG78]   Swinney, H.L., Gollub, J.P.: The transition to turbulence. Phys. Today **31** (1978) 41–49

[SwG81]   Swinney, H.L., Gollub, J.P. (Eds.): Hydrodynamic Instabilities and the Transition to Turbulence. Springer, Berlin 1981

[Szy02]   Szydlowski, M.: Time to build in dynamics of economic models I: Kalecki's model. CSF **14** (2002) 697–703

[Szy03]   Szydlowski, M.: Time-to-build in dynamics of economic models II: models of economic growth. CSF **18** (2003) 355–364

[Tak74]   Takens, F.: Singularities of vector fields. Publ. Math. IHES **43** (1974) 47–100

[Tak81]   Takens, F.: Detecting strange attractors in turbulence. In: Rand, D.A., et al. (Eds.): Dynamical Systems and Turbulence. Lecture Notes in Math. **898**. Springer, New York 1981

[Tak93]   Takens, F.: Detecting nonlinearities in stationary time series. IJBC **3** (1993) 241–256

[Tass98]   Tass, P. et al.: Detection of $n{:}m$ phase locking from noisy data: application to magnetoencephalography. Physical Review Letters **81** (1998) 3291–3294

[Tay23]   Taylor, G.I.: Stability of a viscous liquid contained between two rotating cylinders. Philos. Trans. Roy. Soc. London **A 223** (1923) 289–343

[TaK91]   Taylor, M.A., Kevrekidis, I.G.: Some common dynamic features of coupled reacting systems. Physica **51D** (1991) 274–292

[Tém90]   Témam, R.: Inertial manifolds and multigrid methods. SIAM J. Math. Anal. **21** (1990) 154–178

[TePJ82]   Testa, J., Pérez, J., Jeffries, C.: Evidence for chaotic universal behavior of a driven nonlinear oscillator. Phys. Rev. Lett. **48** (1982) 714

[Thom72]   Thom, R.: Stabilité Structurelle et Morphogénèse. Benjamin, New York 1972

[ThH83]   Thompson, J.M.T., Hunt, G.W. (Eds.): Collapse: The buckling of structures in theory and practice. (IUTAM Symp., 1982). Cambridge University Press, Cambridge 1983

[ThS86]   Thompson, J.M.T., Stewart, H.B.: Nonlinear Dynamics and Chaos. Geometrical Methods for Engineers and Scientists. John Wiley, Chichester 1986

[TiKF02]   Titz, S., Kuhlbrodt, T., Feudel, U.: Homoclinic bifurcation in an ocean circulation box model. IJBC **12** (2002) 869–875

[Tro82]   Troger, H.: Über chaotisches Verhalten einfacher mechanischer Systeme. ZAMM **62** (1982) T18–T27

[Tro84]   Troger, H.: Application of bifurcation theory to the solution of nonlinear stability problems in mechanical engineering. ISNM **70** (1984) 525–546. In: [KüW84]

[TrKS85]   Troger, H., Kazani, V., Stribersky, A.: Zur Dimension Seltsamer At-
            traktoren. ZAMM **65** (1985) T109–T111
[TrS91]   Troger, H., Steindl, A.: Nonlinear Stability and Bifurcation Theory.
            Wien, Springer 1991
[TrZ84]   Troger, H., Zeman, K.: A nonlinear analysis of the generic types of loss
            of stability of the steady state motion of a tractor-semitrailer. Vehicle
            System Dynamics **13** (1984) 161–172
[Tur52]   Turing, A.M.: The chemical basis of morphogenesis. Phil. Trans. Roy.
            Soc. Lond. **B237** (1952) 37–72
[Ueda80]   Ueda, Y.: Steady motions exhibited by Duffing's equation—a picture
            book of regular and chaotic motions. Report, Institut of Plasma Phy-
            sics, Nagoya University, IPPJ-434, 1980
[UeA81]   Ueda, Y., Akamatsu, N.: Chaotically transitional phemomena in the
            forced negative-resistance oscillator. IEEE Trans. Circuits and Sys-
            tems CAS-28 (1981) 217–223
[UpRP74]   Uppal, A., Ray, W.H., Poore, A.B.: On the dynamic behavior of con-
            tinuous stirred tank reactors. Chem. Eng. Sci. **29** (1974) 967–985
[Ura67]   Urabe, M.: Nonlinear Autonomous Oscillations. Academic Press, New
            York 1967
[VaSA78]   Vaganov, D.A., Samoilenko, N.G., Abramov, V.G.: Periodic regimes of
            continuous stirred tank reactors. Chem. Eng. Sci. **33** (1978) 1133–1140
[Vai64]   Vainberg, M.M.: Variational Methods for the Study of Nonlinear Ope-
            rators. Holden-Day, San Francisco 1964
[VaT74]   Vainberg, M.M., Trenogin, V.A.: Theory of Branching of Solutions of
            Nonlinear Equations. Noordhoff, Leyden 1974
[Van27]   Van der Pol, B.: Forced oscillations in a circuit with nonlinear re-
            sistance (receptance with reactive triode). London, Edinburgh, and
            Dublin Phil. Mag. 3 (1927) 65–80
[Vel87]   van Veldhuizen, M.: A new algorithm for the numerical approximation
            of an invariant curve. SIAM J. Sci. Stat. Comput. **8** (1987) 951–962
[Vel64]   Velte, W.: Stabilitätsverhalten und Verzweigung stationärer Lösungen
            der Navier-Stokesschen Gleichungen. ARMA **16** (1964) 97–125
[Vel66]   Velte, W.: Stabilität und Verzweigung stationärer Lösungen der Navier-
            Stokesschen Gleichungen beim Taylor Problem. ARMA **22** (1966) 1–14
[Vol31]   Volterra, V.: Leçons sur la Théorie Mathématique de la Lutte pour la
            Vie. Gauthier-Villars, Paris 1931
[Wac78]   Wacker, H. (Ed.): Continuation Methods. Academic Press, New York
            1978
[Wac92]   Wacker, U.: Strukturuntersuchungen in der Wolkenphysik mit Nicht-
            linearen Parameterisierungsansätzen. Annalen der Meteorologie **27**
            (1992) 151–152
[WaS06]   Wagemakers, A., Sanjuán, M.A.F.: Building electronic bursters with
            the Morris–Lecar neuron model. IJBC **16** (2006) 3617–3630
[Wan78]   Wan, Y.-H.: Bifurcation into invariant tori at points of resonance.
            ARMA **68** (1978) 343–357
[WaAH92]   Wang, H., Abed, E.H., Hamdan, A.M.A: Is voltage collapse triggered
            by the boundary crisis of a strange attractor? Proc. Amer. Control
            Conference, Chicago 1992
[WaG96]   Wang, L., Girgis, A.A.: On-line detection of power system small dis-
            turbance voltage instability. IEEE Trans. Power Systems **11** (1996)
            1304-1313
[WaJ04]   Wang, R., Jing, Z.: Chaos control of chaotic pendulum system. CSF
            21 (2004) 201–207

[WaC08]   Wang, Z., Chau, K.T.: Anti-control of chaos of a permanent magnetic DC motor system for vibratory compactors. CSF **36** (2008) 694–708

[Wat79]   Watson, L.T.: An algorithm that is globally convergent with probability one for a class of nonlinear two-point boundary value problems. SIAM J. Numer. Anal. **16** (1979) 394–401

[WaL76]   Wazewska-Czyzewska, M., Lasota, A.: Mathematical problems of the dynamics of a system of red blood cells (in Polish). Ann. Polish Math. Soc. Ser. III, Appl. Math. **17** (1976) 23–40

[Web79]   Weber, H.: Numerische Behandlung von Verzweigungsproblemen bei gewöhnlichen Differentialgleichungen. Numer. Math. **32** (1979) 17–29

[Wei70]   Weinitschke, H.J.: Die Stabilität clliptischer Zylinderschalen bei reiner Biegung. ZAMM **50** (1970) 411–422

[Wei75]   Weiss, R.: Bifurcation in difference approximations to two-point boundary value problems. Math. Comput. **29** (1975) 746–760

[Wer84]   Werner, B.: Regular systems for bifurcation points with underlying symmetries. ISNM **70** (1984) 562–574. In: [KüMW84]

[Wer93]   Werner, B.: The numerical analysis of bifurcation problems with symmetries based on bordered Jacobians. In: Allgower, E.L., et al. (Eds.): Exploiting Symmetry in Applied and Numerical Analysis. American Math. Soc., Providence 1993

[Wer96]   Werner, B: Computation of Hopf bifurcation with bordered matrices. SIAM J. Numer. Anal. **33** (1996) 435–455

[WeJ91]   Werner, B., Janovsky, V.: Computation of Hopf branches bifurcating from Takens–Bogdanov points for problems with symmetries. ISNM **97** (1991) 377–388

[WeS84]   Werner, B., Spence, A.: The computation of symmetry-breaking bifurcation points. SIAM J. Numer. Anal. **21** (1984) 388–399

[Wig90]   Wiggins, S.: Introduction to Applied Nonlinear Dynamical Systems and Chaos. Springer, New York 1990

[WiR71]   Wilkinson, J.H., Reinsch, C.: Linear Algebra, Handbook for Automatic Computation. Springer, Berlin 1971

[Win75]   Winfree, A.T.: Unclockwise behaviour of biological clocks. Nature **253** (1975) 315–319

[Win90]   Winfree, A.T.: Stable particle-like solutions to the nonlinear wave equations of three-dimensional excitable media. SIAM Review **32** (1990) 1–53

[WoSSV85] Wolf, A., Swift, J.B., Swinney, H.L., Vastano, J.A.: Determining Lyapunov exponents from a time series. Physica **16D** (1985) 285–317

[Wol84]   Wolfram, S.: Universality and complexity in cellular automata. Physica **10D** (1984) 1–35

[XuST92] Xu, G., Steindl, A., Troger, H.: Nonlinear stability analysis of a bogie of a low-platform wagon. In: Sauvage, G. (Ed.): Dynamics of Vehicles on Road and Tracks. 653–665. Swetz and Zeitlinger, Amsterdam 1992

[Yam83]   Yamamoto, N.: Newton's method for singular problems and its application to boundary value problems. J. Math. Tokushima University **17** (1983) 27–88

[YeeS94] Yee, H.C., Sweby, P.K.: Global asymptotic behavior of iterative implicit schemes. IJBC **4** (1994) 1579–1611

[YoYM87] Yorke, J.A., Yorke, E.D., Mallet-Paret, J.: Lorenz-like chaos in a partial differential equation for a heated fluid loop. Physica **24D** (1987) 279–291

[Yu05]   Yu, P.: Closed-form conditions of bifurcation points for general differential equations. IJBC **15** (2005) 1467–1483

[Zee77]   Zeeman, E.C.: Catastrophe Theory. Selected Papers. Addison-Wesley, Reading 1977

[ZhW04]   Zhang, C., Wei, J.: Stability and bifurcation analysis in a kind of business cycle model with delay. CFS **22** (2004) 883–896

[ZhZZ06]   Zhang, C., Zu, Y., Zheng, B.: Stability and bifurcation of a discrete red blood cell survival model. CSF **28** (2006) 386–394

[ZhD90]   Zheng, Q., Dellnitz, M.: Schwingungen eines Ringoszillators — eine numerische Behandlung unter Berücksichtigung der Symmetrie. ZAMM **70** (1990) T135–T138

[ZiO82]   Zierep, J., Oertel, H. (Eds.): Convective Transport and Instability Phenomena. G. Braun, Karlsruhe 1982

[ZouKB06] Zou, Y., Küpper, T., Beyn, W.-J.: Generalized Hopf bifurcation for planar Filippov systems continuous at the origin. J. Nonlinear Sci. **16** (2006) 159–177

Index